Selected Titles in This Series

23 **Karen Hunger Parshall and Adrian C. Rice, Editors,** Mathematics unbound: The evolution of an international mathematical research community, 1800–1945, 2002

22 **Bruce C. Berndt and Robert A. Rankin, Editors,** Ramanujan: Essays and surveys, 2001

21 **Armand Borel,** Essays in the history of Lie groups and algebraic groups, 2001

20 Kolmogorov in perspective, 2000

19 **Hermann Grassmann,** Extension theory, 2000

18 **Joe Albree, David C. Arney, and V. Frederick Rickey,** A station favorable to the pursuits of science: Primary materials in the history of mathematics at the United States Military Academy, 2000

17 **Jacques Hadamard (Jeremy J. Gray and Abe Shenitzer, Editors),** Non-Euclidean geometry in the theory of automorphic functions, 1999

16 **P. G. L. Dirichlet (with Supplements by R. Dedekind),** Lectures on number theory, 1999

15 **Charles W. Curtis,** Pioneers of representation theory: Frobenius, Burnside, Schur, and Brauer, 1999

14 **Vladimir Maz′ya and Tatyana Shaposhnikova,** Jacques Hadamard, a universal mathematician, 1998

13 **Lars Gårding,** Mathematics and mathematicians: Mathematics in Sweden before 1950, 1998

12 **Walter Rudin,** The way I remember it, 1997

11 **June Barrow-Green,** Poincaré and the three body problem, 1997

10 **John Stillwell,** Sources of hyperbolic geometry, 1996

9 **Bruce C. Berndt and Robert A. Rankin,** Ramanujan: Letters and commentary, 1995

8 **Karen Hunger Parshall and David E. Rowe,** The emergence of the American mathematical research community, 1876–1900: J. J. Sylvester, Felix Klein, and E. H. Moore, 1994

7 **Henk J. M. Bos,** Lectures in the history of mathematics, 1993

6 **Smilka Zdravkovska and Peter L. Duren, Editors,** Golden years of Moscow mathematics, 1993

5 **George W. Mackey,** The scope and history of commutative and noncommutative harmonic analysis, 1992

4 **Charles W. McArthur,** Operations analysis in the U.S. Army Eighth Air Force in World War II, 1990

3 **Peter L. Duren et al., Editors,** A century of mathematics in America, part III, 1989

2 **Peter L. Duren et al., Editors,** A century of mathematics in America, part II, 1989

1 **Peter L. Duren et al., Editors,** A century of mathematics in America, part I, 1988

Mathematics Unbound:
The Evolution of an International Mathematical Research Community, 1800–1945

HISTORY OF MATHEMATICS
Volume 23

Mathematics Unbound:
The Evolution of an International Mathematical Research Community, 1800–1945

Karen Hunger Parshall
Adrian C. Rice
Editors

American Mathematical Society
London Mathematical Society

2000 *Mathematics Subject Classification.* Primary 01A55, 01A60, 01A70, 01A72, 01A73, 01A74, 01A80.

A complete listing of photograph and figure credits appears on page xxi.

Library of Congress Cataloging-in-Publication Data

Mathematics unbound : the evolution of an international mathematical research community, 1800–1945 / Karen Hunger Parshall, Adrian C. Rice, editors.
p. cm. — (History of mathematics, ISSN 0899-2428 ; v. 23)
Includes bibliographical references and index.
ISBN 0-8218-2124-5 (acid-free paper)
1. Mathematics—Research—History—19th century. 2. Mathematics—Research—History—20th century. 3. Mathematics—History—19th century. 4. Mathematics—History—20th century. I. Parshall, Karen Hunger, 1955– II. Rice, Adrian C. (Adrian Clifford), 1970– III. Series.

QA11.2 .M28 2002
510′.72—dc21 2002022744

∞ The paper used in this book is acid-free and falls within the guidelines
established to ensure permanence and durability.
The London Mathematical Society is incorporated under Royal Charter
and is registered with the Charity Commissioners.
Visit the AMS home page at URL: http://www.ams.org/

10 9 8 7 6 5 4 3 2 1 07 06 05 04 03 02

Contents

Acknowledgments

The present volume evolved from a three-day, international symposium, entitled "Mathematics Unbound: The Evolution of an International Mathematical Community, 1800–1945," and held at the University of Virginia 27-29 May, 1999 as part of the Mathematics Department's "Emphasis Year" in the history of mathematics. The conference, which brought together scholars from eleven different countries, received generous support from the National Science Foundation in the form of grant number SBR-9817933 as well as from the University of Virginia's Dean of the College of Arts and Sciences, Corcoran Department of History, and Department of Mathematics. This financial support notwithstanding, the conference would not have been a success without the tireless and ever-cheerful practical assistance of the Mathematics Department's administrative staff, Connie Abell, Mary Cline, Julie Riddleberger, and Joyce Stevens.

The last twenty-seven months have found us all—authors, referees, and editors—as part of a truly international, collaborative effort that has united scholars at all levels of professional development. The authors have worked faithfully and conscientiously to produce chapters that responded meaningfully to the critical comments and suggestions of both the referees and the editors. For their part, the referees provided thoughtful, detailed, and, above all, prompt reports with an eye toward thematic tightness. The ensuing dialogue has resulted in a series of chapters that have been almost surprising in their interconnection and mutual complementarity.

As editors, we thank not only the authors and referees but also the numerous people who have helped in the production of the finished volume. First and foremost, Julie Riddleberger labored under the unforgiving yoke of *LaTeX*'s indexing program, keying the entire index with remarkably good cheer in the face of technical adversity! At the American Mathematical Society, Chris Thivierge of the HMATH series, Vickie Ancona in the Production Department, and Barbara Beeton in Author Support have been constant sources of information, help, and guidance. Our thanks also go to Allen Debus, Ivor Grattan-Guinness, and Brian Parshall for their constructive remarks and observations on various parts of the manuscript. They seem never to tire of requests from their friends (or their spouse as the case may be) for their expert advice.

Our work on this project has been stimulating, both professionally and intellectually. We hope that the end-product will not only prove illuminating to our readers but also provide an impetus for future investigations into the fascinating historical question of the internationalization of research-level mathematics.

Charlottesville, Virginia Karen Hunger Parshall
10 August, 2001 Adrian C. Rice

List of Contributors

Thomas Archibald is Professor in the Department of Mathematics and Statistics at Acadia University in Wolfville, Canada. He studied at the University of Waterloo, York University, and the University of Toronto, obtaining a Ph.D. from the last of these in 1987. His current research concerns mathematics in France in the late nineteenth century. Other interests include the history of the relationship between mathematics and physics, particularly in the nineteenth century, and the history of mathematics in Canada.
Address: Department of Mathematics and Statistics, Acadia University, Wolfville, NS, B0P 1X0, Canada.
E-mail: tom.archibald@acadiau.ca

Elena Ausejo earned her Ph.D. in mathematics at the University of Zaragoza in 1991 and currently serves as Associate Professor of the History of Science at her *alma mater*. The Secretary General of the Organizing Committee of the XIX International Congress of the History of Science (Zaragoza, 1993) and former Secretary General of the Spanish Society for the History of Science and Technology (1986–1993), she is Vice President of the Spanish Society for the History of Science and Technology. Her research focuses on the history of mathematics and science (primarily in Spain) in the nineteenth and twentieth centuries and on scientific institutionalization. She is the author of *Por la Ciencia y por la Patria: La institucionalizacion cientifica en España en el primer tercio del siglo XX* (Siglo XXI de España Editores S.A., 1993) and has edited (with Mariano Hormigón) the books, *Messengers of Mathematics: European Mathematical Journals 1800-1946* (Siglo XXI, 1993) and *Paradigms and Mathematics* (Siglo XXI, 1996), among others.
Address: Facultad de Ciencias (Matematicas), Ciudad Universitaria, E-50009 Zaragoza, Spain.
E-mail: ichs@posta.unizar.es

June E. Barrow-Green is a Research Fellow in the History of Mathematics at the Open University. Her research interests are the history of mathematics of the nineteenth and twentieth centuries, in particular, dynamical systems theory. She is the author of *Poincaré and the Three Body Problem* (AMS/LMS HMATH, 1997) and is currently studying the work of George Birkhoff. She is also concerned with the use of databases in the history of mathematics and with the role of the Web in history of mathematics education.
Address: Faculty of Mathematics, The Open University, Milton Keynes MK7 6AA, United Kingdom.
E-mail: j.e.barrow-green@open.ac.uk

Aldo Brigaglia earned his degree in physics at the University of Palermo, where he is now Professor of "Matematiche Complementari." Since the early 1980s, his main area of research has been the history of science, particularly the history of mathematics in Sicily and the history of geometry and its interrelations with algebra. He has published, with Guido Masotto, *Il Circolo matematico di Palermo* (Dedalo, 1982) and, with Ciro Ciliberto, *Italian Algebraic Geometry Between the Two World Wars* (Circolo matematico di Palermo, 1995), as well as articles on the history of mathematics.
Address: Dipartimento di Matematica, Università di Palermo, Via Archirafi 34, I-90123 Palermo, Italy.
E-mail: brig@dipmat.math.unipa.it

Joseph W. Dauben is Professor of History and History of Science at Lehman College, City University of New York (CUNY) and is a member of the Ph.D. Program in History at the Graduate Center of CUNY. Formerly editor of *Historia Mathematica* and a previous Chair of the International Commission on History of Mathematics, he is also the author of biographies of Georg Cantor (Harvard, 1979) and Abraham Robinson (Princeton, 1995). He made his first trip to China as a visiting lecturer for four months under the joint auspices of the U.S. National Academy of Sciences and the Chinese Academia Sinica in 1988 and is currently involved in a joint venture with colleagues in China, Taiwan, and Singapore to produce a critical English edition of the *Ten Classics of Ancient Chinese Mathematics.* He is also interested in questions of the transmission of mathematics between China and other countries, both in antiquity, and more recently, as modern mathematics began to develop in China in the late Qing Dynasty, and in the subsequent course of the twentieth century.
Address: Ph.D. Program in History, The Graduate Center, The City University of New York, 365 Fifth Avenue, New York, New York 10016, United States.
E-mail: jdauben@worldnet.att.net

Sloan Evans Despeaux earned a B.A. in mathematics at Francis Marion Univerisity and an M.S. in mathematics at Florida State University. She is presently completing her Ph.D. in mathematics at the University of Virginia. Her forthcoming thesis is entitled "The Development of a Publication Community: Nineteenth-century Mathematics in British Scientific Journals."
Address: Department of Mathematics, University of Virginia, Kerchof Hall, P.O. Box 400137, Charlottesville, Virginia 22904-4137, United States.
E-mail: sed3v@virginia.edu

Della Dumbaugh Fenster is Assistant Professor of Mathematics at the University of Richmond. Her research focuses on the history of American mathematics, particularly algebra and number theory, in the late nineteenth and early twentieth centuries. Her current work considers the role of the Carnegie Institution in the development of American mathematics.
Address: Department of Mathematics and Computer Science, University of Richmond, Richmond, Virginia 23173, United States.
E-mail: dfenster@richmond.edu

Hélène Gispert teaches the history of science at the *Institut universitaire de formation des maîtres de Versailles* and at the Université d'Orsay. A member of

the Group on the History and Diffusion of the Sciences (GHDSO), she works on the history of mathematics in France at the end of the nineteenth century as well as on the history of education and on the diffusion of the sciences in France during the Third Republic.
Address: Institut universitaire de formation des maîtres de Versailles, Groupe d'histoire et de diffusion des sciences d'Orsay, Bâtiment 407, Centre universitaire, F-91405 Orsay Cedex, France.
E-mail: helene.gispert@ghdso.u-psud.fr

Ivor Grattan-Guinness is Professor of the History of Mathematics and Logic at Middlesex University, England. He was editor of the history of science journal, *Annals of Science*, from 1974 to 1981. In 1979, he founded the journal, *History and Philosophy of Logic*, and edited it until 1992. He edited the *Companion Encyclopedia of the History and Philosophy of the Mathematical Sciences* (2 vols., Routledge, 1994) and published *The Fontana History of the Mathematical Sciences: The Rainbow of Mathematics* (Fontana, 1997) and *The Search for Mathematical Roots, 1870–1940: Logics, Set Theories, and the Foundations of Mathematics from Cantor through Russell to Gödel* (Princeton, 2000). He is the Associate Editor for mathematicians and statisticians for the *New Dictionary of National Biography*, to be published in 2004. He is editing for Elsevier a collection of essays on *Landmark Writings in Western Mathematics, 1640-1940*, also to appear in 2004.
Address: 43, St. Leonard's Road, Bengeo, Herts SG14 3JW, United Kingdom.
E-mail: ivor2@mdx.ac.uk

Jeremy J. Gray is Senior Lecturer in Mathematics at the Open University, where he is the Director of the Centre for the History of the Mathematical Sciences. He is also an Affiliated Research Scholar in the Department of History and Philosophy of Science at the University of Cambridge, England. He works on the history of mathematics in the nineteenth and twentieth centuries, with a particular interest in complex function theory and geometry, and also on issues in the philosophy and social significance of mathematics. His most recent publication is *The Hilbert Challenge* (Oxford, 2000).
Address: Faculty of Mathematics, The Open University, Milton Keynes MK7 6AA, United Kingdom.
E-mail: j.j.gray@open.ac.uk

Mariano Hormigón earned degrees in mathematics before taking his Ph.D. in Philosophy at the Universidad Autónoma de Madrid in 1982. He has been Professor of the History of Science in the Science Faculty of the University of Zaragoza since 1986. President of the Organizing Committee of the XIX International Congress of the History of Science (Zaragoza, 1993) and Past President of the Spanish Society for the History of Science and Technology (1984–1993), he is a member of the Académie internationale d'histoire des sciences (since 1993) and of the International Commission on History of Mathematics (since 1997). He has served as the editor of *LLULL*, the journal of the Spanish Society for the History of Science and Technology, since 1981. He has edited (with Elena Ausejo), *Messengers of Mathematics: European Mathematical Journals 1800-1946* (Siglo XXI de España

Editores S.A., 1993) and *Paradigms and Mathematics* (Siglo XXI, 1996). His research centers on the history of mathematics and science (primarily in Spain) in the nineteenth and twentieth centuries and on science and ideology.
Address: Faculdad de Ciencias (Matematicas), Ciudad Universitaria, E-50009 Zaragoza, Spain.
E-mail: hormigon@posta.unizar.es

Olli Lehto received his Ph.D. in mathematics in 1949 at the University of Helsinki, where he served as Associate Professor (1956–1961), Professor (1961–1988), Rector (1983–1988), and Chancellor (1988–1993). He was President of the International Congress of Mathematicians in 1978, a Member of the Executive Committee of the International Mathematical Union (1975–1990), the Union's Secretary (1983–1990), a Member of the Administrative Board of the International Association of Universities (1985–1995), and the Association's Vice-President (1990–1995). He is both a member (1962) and an honorary member (2001) of the Finnish Academy of Science and Letters. He acted as President of the Delegation of the Finnish Academies (1979–1999), was Chair of the Board of the Finnish Cultural Foundation (1991–1994), and the latter's Honorary President in 1998. In mathematics, his special field is complex analysis and, after retirement, also the history of mathematics. He is the author of *Mathematics Without Borders: A History of the International Mathematical Union* (Springer-Verlag, 1998).
Address: Department of Mathematics, University of Helsinki, P.O. Box 4 (Yliopistonkatu 5), 00014 Helsinki, Finland.
E-mail: oelehto@cc.helsinki.fi

Jesper Lützen obtained a Master's Degree (1976) and a Ph.D. (1980) from the Department of History of Science at Aarhus University. A Habilitation followed in 1990 from Copenhagen University. He is currently Professor of History of the Exact Sciences in the Department of Mathematics, Copenhagen University. The author of *Joseph Liouville: Master of Pure and Applied Mathematics* (Springer-Verlag, 1990) and *The Prehistory of the Theory of Distributions* (Springer-Verlag, 1980), he is currently working on a book on Heinrich Hertz's *Principles of Mechanics*.
Address: Mathematics Institute, Copenhagen University, Universitetsparken 5, 2100 Copenhagen, Denmark.
E-mail: lutzen@math.ku.dk

Laura Martini earned her *laurea* from the University of Siena in 1997 with a thesis on "Cesare Arzelà's Lectures on Galois Theory, Bologna 1886–1887" directed by Laura Toti Rigatelli. She is presently at work on a Ph.D. in mathematics at the University of Virginia. Her main research interests lie in the history of algebra in the nineteenth century and in the social development of mathematics in nineteenth-century Italy.
Address: Department of Mathematics, University of Virginia, Kerchof Hall, P.O. Box 400137, Charlottesville, Virginia 22904-4137, United States.
E-mail: lm4x@virginia.edu

Karen Hunger Parshall is Professor of History and Mathematics at the University of Virginia, a former editor of *Historia Mathematica*, and the current Chair of the International Commission on History of Mathematics. Her research focuses on the history of nineteenth- and twentieth-century mathematics with a special focus

on developments in algebra. In 1994, her book with David Rowe, *The Emergence of the American Mathematical Research Community, 1876–1900: J. J. Sylvester, Felix Klein, and E. H. Moore*, appeared in the AMS/LMS HMATH series. Since then, she has co-edited with Paul Theerman, *Experiencing Nature: Proceedings of a Conference in Honor of Allen G. Debus* (Kluwer, 1997), and has published the book, *James Joseph Sylvester: Life and Work in Letters* (Oxford, 1998).
Address: Departments of History and Mathematics, University of Virginia, Kerchof Hall, P.O. Box 400137, Charlottesville, Virginia 22904-4137, United States.
E-mail: khp3k@virginia.edu

Adrian C. Rice received a B.Sc. in mathematics from University College London in 1992 and a Ph.D. in the history of mathematics from Middlesex University in 1997 for a thesis on Augustus De Morgan and the development of university-level mathematics in nineteenth-century London. He is Assistant Professor of Mathematics at Randolph-Macon College in Ashland, Virginia, where his research focuses on the nineteenth- and early twentieth-century British mathematical research community with especial reference to the role of the London Mathematical Society.
Address: Department of Mathematics, Randolph-Macon College, Ashland, Virginia 23005-5505, United States.
E-mail: arice4@rmc.edu

Chikara Sasaki studied mathematics at Tohoku University and the history of science at Princeton University. He received his Ph.D. from Princeton for a thesis on "Descartes's Mathematical Thought." He is now Professor of History and Philosophy of Science at the University of Tokyo and a member of the Executive Committee of the International Commission on the History of Mathematics. He was awarded the Suntory Prize for Social Sciences and Humanities for the year 1993.
Address: Department of History and Philosophy of Science, Graduate School of Arts and Sciences, University of Tokyo, 3-8-1, Komaba, Meguro-ku, Tokyo 153-0041, Japan.
E-mail: ch-sasaki@msi.biglobe.ne.jp

Sanford L. Segal is Professor of Mathematics at the University of Rochester and was Chair of the department there from 1979 to 1987. He obtained his Ph.D. in Mathematics in 1963 from the University of Colorado (supervised by Sarvadaman Chowla) and since then has pursued research in mathematics, the history of mathematics, and mathematics education. He has been a visiting lecturer at the University of Nottingham (1972–1973) and a research fellow at the Federal University of Rio de Janeiro (summer 1982). He received Fulbright grants in 1958–1959 to study in Mainz, Germany and in 1965–1966 for research at the University of Vienna. More recently, he received an Alexander von Humboldt fellowship to Germany in 1988. Work during this last and succeeding trips to Germany were in preparation for a book, *Mathematicians under the Nazis*, to be published shortly. He is also the author of the book, *Nine Introductions in Complex Analysis* (Elsevier, 1981).
Address: Department of Mathematics, University of Rochester, Rochester, New York 14627, United States.
E-mail: ssgl@math.rochester.edu

Reinhard Siegmund-Schultze studied mathematics at the University of Halle. Since 1975, his research has focused on the history of mathematics, primarily on the history of functional analysis and the social history of mathematics in the Third Reich. He is author of *Mathematische Berichterstattung in Hitlerdeutschland: Der Niedergang des Jahrbuchs über die Fortschritte der Mathematik (1869–1945)* (Vandenhoeck & Ruprecht, 1993), *Mathematiker auf der Flucht vor Hitler: Quellen und Studien zur Emigration einer Wissenschaft* (Vieweg, 1998), and *Rockefeller and the Internationalization of Mathematics Between the World Wars* (Birkhäuser Verlag, 2001). He is currently Professor for History of Mathematics at Agder University College in Kristiansand, Norway.
Address: Department of Mathematics, Agder University College, Serviceboks 422, N-4604 Kristiansand, Norway.
E-mail: Reinhard.Siegmund-Schultze@hia.no

Yibao Xu studied the history of Chinese mathematics at the Institute for the History of Science, Inner Mongolia Normal University, Huhhot, China, where he received his M.Sc. in July, 1991. Subsequently, he worked as an editor at the Office of Jiangxi Local History Editing Committee, Nanchang, China. In August, 1995, he entered the Graduate School and University Center of the City University of New York (CUNY) to study the history of science and is now a doctoral candidate in the Ph.D. Program in History, CUNY.
Address: Ph.D. Program in History, The Graduate Center, The City University of New York, 365 Fifth Avenue, New York, New York 10016, United States.
E-mail: xuyibao@hotmail.com

Photograph and Figure Credits

We gratefully acknowledge the kindness of the following for granting these permissions:

Cambridge University Library

Photograph of the Rue des Nations at the Paris Exhibition of 1900; cover and p. 206; From *The Paris Exhibition, 1900.* Ed. D. Croal Thomson. Assisted by Herbert E. Butler and E. G. Halton. London: H. Virtue and Co., 1901. Facing page 111; By permission of the Syndics of Cambridge University Library.

Institut Mittag-Leffler

Photograph of Gösta Mittag-Leffler (1846–1927); p. 138; By permission of the Institut Mittag-Leffler.

Professor Ren Nanheng

Photograph of the Annual Meeting of the Chinese Mathematical Society on the Occasion of Its Fiftieth Anniversary, Fudan University, Shanghai, December 1985; p. 278; Reproduced by permission of Professor Ren Nanheng of the Chinese Mathematical Society.

Springer-Verlag

Sketch of L. E. Dickson, made during his lecture at the International Congress of Mathematicians, Strasbourg, p. 312; From *International Mathematical Congresses: An Illustrated History.* By Donald J. Albers, Gerald L. Alexanderson, and Constance Reid. New York: Springer-Verlag, 1987.

The following images are in the public domain:

Cartoon "Demonstriere zu Hause!" by D. Gulbransson; p. 368; from *Simplicissimus* (1919). This cartoon also appeared in Paul Forman. "Scientific Internationalism and the Weimar Physicists: The Ideology and Its Manipulation in Germany after World War I." *Isis* **64** (1973):151-180 on p. 150.

List of Editorial Board members of *Acta Mathematica* in 1882; p. 141; from *Acta Mathematica* **1** (1882).

Paragraph from Section 150 of Cesare Arzelà's lectures on Galois theory; p. 172; from the Bortolotti Collection, Mathematics Library, University of Bologna.

Photograph of Charles Hermite; p. 122; from *Œuvres de Charles Hermite.* Ed. Émile Picard. 4 Vols. Paris: Gauthier-Villars, 1905–1917, 3: frontispiece.

Table of Contents of the first volume of *Acta Mathematica*; p. 145; from *Acta Mathematica* **1** (1882).

Title page of Yonagawa Shunsan's *Yôsan Yôhô* from 1857; back cover and p. 233.

Yonagawa's explanation of four fundamental arithmetical operations and of proportion in *Yôsan Yôhô*; p. 234; appears on folio 7r.

CHAPTER 1

The Evolution of an International Mathematical Research Community, 1800–1945: An Overview and an Agenda

Karen Hunger Parshall
University of Virginia (United States)
and
Adrian C. Rice
Randolph-Macon College (United States)

Introduction

Few would disagree with the characterization of today's intellectual community, indeed of society itself, as "international."[1] In these days of instant global communication via satellites and the World Wide Web, few boundaries exist to the dissemination of ideas. Scholars may receive the entirety of their formal training, undertake doctoral and post-doctoral research, and receive academic employment—in fact, pursue their entire careers—all outside their country of origin. Academic research (sometimes carried out in collaboration by researchers on different continents) is swiftly distributed via journals, both electronic and "real" to a potentially worldwide audience. One might say—updating Thomas Jefferson's concept to the twenty-first century—that the world has become a vast *global* "academical village."

As a constituent of this broad scholarly body, the community of mathematicians shares these characteristics and, like most of its academic peers, could be forgiven for taking the international nature of its field for granted. Its members attend meetings and conferences the world over, participate in exchange programs with foreign institutions, publish papers in globally circulated journals editorially based in a host of countries, and draw from the work of mathematicians worldwide, all with the principal objective of advancing their discipline. It does not take extensive knowledge of the history of mathematics—or of humankind for that matter—to know that this is a recent phenomenon. More particularly, it seems to have been a characteristic of the late twentieth century.

Or was it? After all, it is easy to argue that mathematicians with an international bent are not a new phenomenon. Gerbert d'Aurillac (later Pope Sylvester II) traveled widely around the end of the tenth century in search of competent teachers of the then new Islamic mathematics and astronomy, while Leonardo Fibonacci discovered the ease of Hindu-Arabic numerals and the products of medieval, Islamic mathematical scholarship on his journeys around the Mediterranean in the late

[1] While the word "international" would be most commonly used in everyday parlance, scholars have coined a plethora of neologisms in their efforts to analyze this concept. See below on some of this terminology.

twelfth and early thirteenth centuries. Somewhat later in the thirteenth century, the Oxford natural philosopher, Roger Bacon, studied for a time in Paris. More recent examples of mathematicians who spent extended periods away from their homeland also come quickly to mind: in the seventeenth century, Johannes Kepler in Prague and Christiaan Huygens in Paris; in the eighteenth century, Leonhard Euler, first in St. Petersburg, then in Berlin, then back in St. Petersburg; and Turin native, Joseph Louis Lagrange, in Berlin between 1766 and 1787 and then in Paris from 1787 until his death in 1813. Publication crossed national boundaries as well; the famous (or infamous) case of the Dutch publication of Galileo's *Discourses on Two New Sciences* in 1638 is but one example. Finally, as early as 1798–1799, there was even an arguably international scientific congress, held in Paris, on the question of standardizing weights and measures.[2]

In what ways, then, does today's internationalized climate differ from that suggested by these earlier precedents? What historical factors shaped the modern, international world of mathematics, and when? How does—indeed *does*—the case of mathematics and its internationalization differ from those of the other sciences? This chapter lays out a framework for answering these questions as it serves to introduce the analysis of the evolution of the international mathematical community in the chapters to follow.

Internationalization, Internationalism, Transnationalism, Supranationalism, Multinationalism, Denationalization, . . . : What's in a Word?

The first issue that must be addressed is one of terminology, and the first distinction to make is between "internationalization" and "internationalism." The suffix "ization" connotes a process, while "ism" refers to a doctrine. Much of the work that has been done thus far in the history and sociology of science on the national/international issue has, in fact, focused more on the doctrine or ideology of internationalism—that is, the belief that governments and individuals could act in concert to end such evils as war and famine—than on the actual process of internationalization—namely, the development of a globalized community of individuals sharing many of the same values and goals. These are, of course, by no means mutually exclusive from an analytical point of view; internationalism in the period between the two World Wars, for example, both aided and abetted the process of internationalization, whereas it hardly existed as a doctrine in the same politically charged sense before 1870. In the context of the present volume, however, it is internationalization and not internationalism that provides the more meaningful conceptualization.

The broader literature also contains a host of terms that have been coined in an effort to analyze what may be called the fine structure of the historical and sociological processes of internationalization over time. The term "international" literally means "between or among nations," but scholars have juxtaposed it with terms like "transnational" ("across nations"), "supranational" ("above or transcending nations"), "multinational" ("more than one nation"),[3] and even "denational," in the sense of moving away from nations altogether as an organizing principle in

---

[2]Maurice Crosland, "The Congress on Definitive Metric Standards, 1798–1799: The First International Scientific Conference?" *Isis* **60** (1969):226-231.

[3]Several authors of the chapters below utilize the term "multinational," especially in analyzing the first half of the nineteenth century. It, like "transnational" (see below), suggests an absence

the sciences. In her work on international scientific relations from the Franco-Prussian War in 1870–1871 to the outbreak of World War I in 1914, for example, Brigitte Schröder (later Schröder-Gudehus) utilized the terminology "international" but sought to maintain a distinction between it and "transnational."[4] She grouped what she termed the "international" scientific activities of that period into four categories, dependent both on their "degree of internationality" and on "the relative distance that separate[d] them from political power": national scientific research institutions founded abroad; institutions, associations, or scientific enterprises organized on an inter-state ("*interétatique*") basis; scientific relations and cultural diplomacy; and transnational relations such as associations, congresses, or other scientific enterprises.[5] Since, in her analysis, the term "international" was equivalent to "inter-state," the distinction between these synonyms and "transnational" allowed her to explore the extent to which "international" activites were, in fact, cooperative, that is, between as opposed to across nations.

Paul Forman maintained and extended this distinction in his seminal article on "Scientific Internationalism and the Weimar Physicists." There, he examined the ideology of scientific internationalism by means of the particular case of physicists in Weimar Germany. For Forman, scientific internationalism hinged, first and foremost, on the belief among scientists in the universality of scientific knowledge, yet, however paradoxically, this internationalism was essentially nationalistic both in its foundations and in its function.[6] As he put it, "[t]he propositions and rhetoric asserting the reality and necessity of supranational agreement on scientific doctrine, of transnational social intercourse among scientists, and of international collaboration in scientific work are thus regarded here as tenets of scientific ideology; we may call these tenets the ideology of scientific internationalism."[7] He thus (consciously) expanded Schröder-Gudehus's terminology by adding the notion of "supranational" and thereby isolated a dimension of scientific internationalism (and, it can be argued, scientific internationalization) that transcends national boundaries altogether.[8]

Taken together, the studies by Schröder-Gudehus and Forman cover the period from 1870 to World War II.[9] Efforts to analyze science in the postwar era have led to further terminological refinements and thus additional analytical concepts. In the opening chapter of their edition, *Denationalizing Science: The Contexts*

---

of explicit collaboration or cooperation. See, for example, Ivor Grattan-Guinness's distinction in the next chapter.

[4]Brigitte Schröder, "Caractéristiques des relations scientifiques internationales, 1870–1914," *Cahiers d'histoire mondiale/Journal of World History* **10** (1966):161-177.

[5]"le degré de leur 'internationalité' " and "la distance relative qui les sépare du pouvoir politique." *Ibid.*, p. 162.

[6]Paul Forman, "Scientific Internationalism and the Weimar Physicists: The Ideology and Its Manipulation in Germany after World War I," *Isis* **64** (1973):151-180 on p. 152.

[7]*Ibid.*, p. 155. Note that this differs from the definition of internationalism given above.

[8]For examples of supranational factors affecting the internationalization of mathematics, see below.

[9]Much work has been done on this period, although mathematics scarcely figures into the accounts. See, for example, Brigitte Schröder-Gudehus, *Deutsche Wissenschaft und internationale Zusammenarbeit* (Geneva: Dumaret & Golay, 1966); Brigitte Schröder-Gudehus, *Les scientifiques et la paix: La communauté scientifique internationale au cours des années 20* (Montreal: Presses de l'Université de Montréal, 1978); and Elisabeth Crawford, *Nationalism and Internationalism in Science, 1880–1939: Four Studies of the Nobel Population* (Cambridge: University Press, 1992), among other works.

of International Scientific Practice, Elisabeth Crawford, Terry Shinn, and Sverker Sörlin periodize the history of what they call "cross-boundary science" in this way: the Middle Ages to the French Revolution, that is, the period before the rise of the nation-state; the mid-nineteenth century to the outbreak of World War I or the age of nationalism; and the period from 1914 to the present or the period characterized by increasing "denationalization" especially from the 1980s on.[10] In their view, "transnational science," that is, "activities involving persons, equipment or funds from more than one country," has come to predominate over national science.[11] They thus see a progressive "denationalization" of science under way in that nations are increasingly divesting their scientific capital, industry is picking up a significant proportion of the slack, and regional cooperative organizations (such as those supported by the European Union) are playing a greater and greater role.[12] As an analytic concept, then, denationalization derives principally from economic concerns.

This discussion of "isms," "izations," and their corresponding adjectives underscores the complexity of attempting to come to terms with the diverse components of the historical process that has resulted in today's international mathematical research community. Here, we consciously use the word "international" because the chapters to follow explore what we view as the process of the internationalization from 1800 to 1945 not only of the body of knowledge known as mathematics but also of the community of mathematicians and its practices. At times, and especially before 1870, that process has had a trans- or what we will term a multinational character (see especially Ivor Grattan-Guinness's chapter below); at times, it has reflected an ideological component of internationalism (see the chapters by Reinhard Siegmund-Schultze, Sanford Segal, and Olli Lehto); at times, it has been driven by strong nationalistic impulses in Forman's sense (see especially the chapters by Jesper Lützen, Hélène Gispert, and Della Fenster); at times, other factors have been at play.

Internationalization, then, is itself a term with a history, with a meaning that has changed over time. In the context of mathematics, whether considering the French dominance of the first half of the nineteenth century, or the Prussianization of mathematics especially from the 1870s through World War I (compare Chikara Sasaki's chapter on Japan below), or the Americanization that began in China as early as the 1910s and that has arguably followed more globally after 1945 (see the chapters by Joseph Dauben and Xu Yibao for more on China), the end result has been a language of mathematics used for communication by mathematicians worldwide, a range of research agendas recognized and engaged in by mathematicians worldwide, a subject matter and practice shared *among* mathematicians worldwide, hence an *inter*nationalization in the etymological sense laid out above.

---

[10] Elisabeth Crawford, Terry Shinn, and Sverker Sörlin, "The Nationalization and Denationalization of the Sciences: An Introductory Essay," in *Denationalizing Science: The Contexts of International Scientific Practice*, ed. Elisabeth Crawford, Terry Shinn, and Sverker Sörlin (Dordrecht: Kluwer Academic Publishers, 1993), pp. 1-42 on p. 6.

[11] *Ibid.*, p. 1.

[12] *Ibid.*, pp. 1-6. The science that Crawford, Shinn, and Sörlin have in mind here, however, is big science, expensive science most effectively financed by pooled resources. Mathematics clearly does not fit this model, so it will remain for historians of late twentieth-century science to decide whether or to what extent the concept of "denationalization" applies to mathematics. The case of late twentieth-century mathematics may well provide an instructive counterexample to generalizations about science as a whole such as those Crawford, Shinn, and Sörlin present.

The Timeframe: 1800–1945

One term left conspicuously undefined at this point is "nationalization." It hardly makes sense to talk about "internationalization" without the modern notion of the nation-state, and that concept arose, for the most part, in the nineteenth century. While it is certainly true that countries like France, Spain, Sweden, and Denmark were governmentally unified and centralized before 1800, the Italian states did not unify until 1861 with the German states following after Austria's loss in 1866 at Königgrätz and France's crushing defeat in the Franco-Prussian War five years later. The rise of the nation-state brought with it a sense of national culture that encompassed—in addition to the arts and literature—science. A proof of the strength of one's nation became the strength of one's national science, with some countries, like France in the first half of the nineteenth century or Germany between 1870 and World War I, viewed as the standard-bearers or what has been termed the "center" and others deemed on the "periphery."[13] National science and professionalized scientific communities of varying strengths resulted from this competition and strong sense of national self-interest.

As this discussion makes clear, there is yet another dimension to our operative definition of internationalization; it must be viewed against a backdrop of nationalization. In that light, the examples given earlier of what might be seen as internationalization in mathematics (and the other sciences) do not fit our definition. Gerbert d'Aurillac, Leonardo Fibonacci, and Roger Bacon were all scholars active in the Middle Ages, a period in which the nation-state had not yet come into existence. Theirs was an epoch characterized in Europe by the dominance of Catholic theology, not by national, governmental interests. Moreover, the universities shared a common curriculum that made them in some sense interchangeable. It was not unusual for scholars to travel from Oxford to Paris to Padua to Toledo to teach or to pursue their studies. Rather than reflecting internationalization in our sense, then, the Middle Ages may be seen as a period of what has been termed "transnational universalism."[14]

This gradually began to change during the period of the Scientific Revolution, the years roughly from 1450 to 1700, as local rulers established universities as well as academies and court centers for research with local, regional, and increasingly national interests in mind. The German Kepler went to Prague in 1600 to work with Tycho Brahe, who had left his native Denmark and the patronage of the Danish monarch to reestablish his research facility at the court of Rudolph II; when Tycho died in 1601, Kepler soon assumed his position. Later in the seventeenth

---

[13]On the analytical device of center/periphery, see, for example, Rainald von Gizycki, "Center and Periphery in the International Scientific Community: Germany, France and Great Britain in the 19th Century," *Minerva* **11** (1973):474-494. In some historiographical circles, however, the notion of center/periphery is seen as privileging the center and devaluing the periphery. Numerous recent studies on the history of empire reflect this view. For one example in the history of science, see Richard Drayton, *Nature's Government: Science, Imperial Britain, and the 'Improvement' of the World* (New Haven: Yale University Press, 2000).

In the case of research mathematics, where mathematicians in given countries recognized *themselves* as on the periphery or in the center, center/periphery analyses retain their value. Elena Ausejo and Mariano Hormigón use this tool to good effect in the case of Spain in their chapter below as does Hélène Gispert in her analysis of France between the Franco-Prussian War and World War I.

[14]Crawford, Shinn, and Sörlin, "Nationalization and Denationalization," p. 7. The discussion in this and the next paragraph follows the framework they set out on pp. 7-11.

century, Louis XIV founded the *Académie des sciences* in Paris in 1666. In light of his intention for the new institution to reflect the glory of the "Sun King," Louis sought the best scientists in Europe to staff his academy. His first hire and the man he intended to head his new venture was the Dutch natural philosopher, Christiaan Huygens. This same kind of royal patronage attracted Leonhard Euler successively to the courts of Catherine I in St. Petersburg, Frederick the Great in Berlin, and Catherine the Great back in St. Petersburg in the eighteenth century, as it drew Joseph Louis Lagrange from his native Turin first to Berlin and then to Paris. As especially the case of France makes clear, by 1800, the nation-state increasingly defined the primary venue in which scientific activity took place.

Thus, 1800 marks a natural starting point for the present study. Nation-states grew and expanded throughout the course of the nineteenth century; the empire-building particularly of the British, French, and Germans on other continents worldwide reinforced the model; science and other intellectual and cultural activities were nurtured and professionalized in national contexts in western Europe, Russia, the United States, Japan, China, and elsewhere. Yet, the nineteenth century and the years through the interwar period also witnessed the rise of the doctrine of internationalism as well as the process of internationalization in many fields of endeavor. Virtually no area was immune to internationalist impulses. Communications, transportation, agriculture, trade and industry, labor, language, law, science, health care, religion, and humanitarian and peace-oriented efforts all moved beyond national boundaries to define a common international arena characterized by international organizations and international congresses and bound by an evolving ideology of internationalism.[15] And while the atrocities of World War I may have dampened if not destroyed the idealism of that ideology, they did not halt internationalization, which was already well under way within and beyond the borders of western Europe. By 1945 and the end of World War II, the physical displacement of European mathematicians to Britain, to the United States, and elsewhere marked at least a symbolic culmination of that on-going process.[16]

The Internationalization of a Mathematical Research Community, 1800–1945: A First Vintage

To date, much historical scholarship has focused on the development of national mathematical communities and on national mathematical developments.[17] Owing

[15]Francis S. L. Lyons documented the spread of internationalism in all of these areas in his book, *Internationalism in Europe, 1815–1914* (Leyden: A. W. Sythoff, 1963). According to Lyons, some 1,978 international organizations were founded from 1815 to 1914 with almost 3,000 international meetings or congresses held between 1840 and the outbreak of World War I. See *ibid.*, p. 12.

[16]As noted above, Crawford, Shinn, and Sörlin, with big science in their sights, periodize the development of international scientific practice with a breaking point in 1914. They justify this on the ground that after World War I science became part of national foreign policies, owing, first, to the development of what they call "science-intensive weaponry" and, second, to the rise of totalitarian regimes in which scientists were regarded as national resources. See Crawford, Shinn, and Sörlin, "Nationalization and Denationalization," pp. 21-22.

While we do not dispute this for some sciences, the case of mathematics does not seem to fit well within this largely economically driven model. For mathematics, especially in light of the mass emigration of mathematicians prior to World War II, the year 1945 makes for a more natural periodization.

[17]For a sense of the wide range of national studies, see the references to the individual chapters below.

perhaps to the linguistic difficulties associated with a comparative, international research focus as well as to the difficulties in assembling source materials from a wide range of countries, relatively little work has been directed toward coming to terms with the internationalization of the field.[18] To begin to redress this and in an effort to set an agenda for further work on the internationalization of mathematics, the present volume surveys—from numerous perspectives—the evolution of an international mathematical research community from 1800 to 1945.[19] It offers what Francis Bacon might have called a "first vintage" of the process of the internationalization of both mathematics and such a community.

In his book, the *Novum Organon* of 1620, Bacon put forth his conception of how scientific inquiry should proceed. His was an inductive method based on careful observation of individual instances, comparison and contrast between those instances, and a first, interpretive step or "first vintage" to begin to reach an understanding of nature. The example he considered in the *Novum Organon* was heat; his question was "what is its nature?"; and his first vintage was that "[h]eat is a motion, expansive, restrained, and acting in its strife upon the smaller particles of bodies. But the expansion is thus modified: while it expands all ways, it has at the same time an inclination upward. And the struggle in the particles is modified also; it is not sluggish, but hurried and with violence."[20] The close examination and comparison of particular cases thus yielded a nuanced description of a physical phenomenon. It is in this Baconian spirit that the case studies presented below seek to induce a

[18]Some efforts have been made in this direction, however. See, for example, Karen Hunger Parshall, "Mathematics in National Contexts (1875–1900): An International Overview," in *Proceedings of the International Congress of Mathematicians: Zürich*, 2 vols. (Basel: Birkhäuser Verlag, 1995), 2:1581-1591, and Catherine Goldstein, Jeremy J. Gray, and Jim Ritter, ed., *L'Europe mathématique—Mathematical Europe* (Paris: Édition de la Maison des sciences de l'homme, 1996) for two studies very different in scope that nevertheless both analyze the situations in a wide range of countries.

Other studies have approached internationalization from a particular optic, such as Karen Hunger Parshall and Eugene Seneta from the point of view of reputation in "Building an International Reputation: The Case of J. J. Sylvester," *American Mathematical Monthly* **104** (March 1997):210-222, and Reinhard Siegmund-Schultze from the point of view of the role of external funding in *Rockefeller and the Internationalization of Mathematics Between the World Wars* (Basel: Birkhäuser Verlag, 2001).

Finally, some work has concentrated on the interactions and interrelations between two countries, for example, Eduardo Ortiz, "El rol de las revistas matemáticas intermedias en el establecimiento de contactos entre las comunidades de Francia y España hacia fines del siglo XIX," in *Contra los titanes de la rutina/Contre les titans de la routine*, ed. Santiago Garma, Dominique Flament, and Victor Navarro (Madrid: Comunidad de Madrid/C.S.I.C., 1994), pp. 367-382; Reinhard Siegmund-Schultze, "The Emancipation of Mathematical Research Publishing in the United States from German Dominance (1878–1945)," *Historia Mathematica* **24** (1997):135-166, and " 'Scientific Control' in Mathematical Reviewing and German-U.S.-American Relations Between the Two World Wars," *Historia Mathematica* **21** (1994):306-329; Gert Schubring, "Differences in the Involvement of Mathematicians in the Political Life in France and in Germany," *Bollettino di storia delle scienze matematiche* **15** (1995):61-83; Dianzhou Zhang and Joseph W. Dauben, "Mathematical Exchanges Between the United States and China: A Concise Overview," in *The History of Modern Mathematics*, vol. 3, ed. Eberhard Knobloch and David E. Rowe (Boston: Academic Press, Inc., 1994), pp. 263-297; and Li Wenlin and Jean-Claude Martzloff, "Aperçu sur les échanges mathématiques entre la Chine et la France (1880–1949)," *Archive for History of Exact Sciences* **53** (1998):181-200.

[19]The notion of evolution, however, in no way implies a whiggish idea of progress, nor does it suggest that the process has been smooth or without setback.

[20]Francis Bacon, *The New Organon and Related Writings*, ed. Fulton H. Anderson (Indianapolis: Bobbs-Merrill Company, Inc., 1960), p. 162.

sense of the complex of factors that shaped the international research community of mathematicians.[21]

As noted above, nationalization contextualizes internationalization, and the case of mathematics is no exception. While national communities of mathematicians emerged throughout the period from 1800 to 1945, certain key countries affected mathematical developments elsewhere, and these standard-bearers changed over time. As the two opening chapters attest, France was the dominant mathematical influence for the first half of the nineteenth century. Ivor Grattan-Guinness explains, however, that the supremacy the French held so firmly in 1800 soon disappeared; Germany had supplanted France by mid-century. Germany's influence in a wide variety of national venues—late nineteenth-century Spain and France, Meiji Japan, and China and the United States in the early twentieth century—comes through clearly in the chapters by Elena Ausejo and Mariano Hormigón, Thomas Archibald, Chikara Sasaki, Joseph Dauben, and Della Fenster, respectively. In the twentieth century, other mathematical nations exerted their influence internationally. For example, as both Dauben and Xu Yibao show, the mathematical communities of Great Britain and the United States played a definite role in shaping mathematical institutions and tastes in China.

While these case studies demonstrate the effects of nationalization on internationalization, that is, the effects of dominant, national mathematical communities on the development of other national communities, there was also a reciprocal effect. The increased internationalization of mathematics in the nineteenth century proved, in some cases, of crucial importance in the emergence of national communities. As is clearly seen in the chapters by Ausejo and Hormigón for Spain and by Dauben, Xu, and Sasaki for China and Japan, the building of individual, national mathematical communities was greatly facilitated by the importation of foreign mathematicians, methods, and ideas. Moreover, this was true not just for "peripheral" nations. As Hélène Gispert argues, it was precisely *because* of developments abroad that the French realized the urgent need to form their own national mathematical society in the 1870s.

Gispert also highlights another feature crucial to the period from 1800 to 1945: the growth of mathematics as a profession. The process of professionalization, like that of internationalization, is tangled. Relative to mathematics (and other fields), it took place initially in individual national venues and involved, first and foremost, changes in higher education. Prussia was the model here. Educational reforms spearheaded by Wilhelm von Humboldt in the opening decades of the nineteenth century emphasized research as well as academic freedom at the university level. By mid-century, this had evolved into a system geared not only toward the production of new knowledge but also toward the training of future researchers.

---

[21]This concept, "research community of mathematicians," requires some definition. While the notion clearly implies a plurality of individuals identifiable as a group, this group does not necessarily act in concert. Moreover, some belong to the community; some do not. Here, the trait that distinguishes membership is "being a research mathematician." This, however, is an historical concept, the meaning of which has changed over time. Early in the nineteenth century, to be considered a research mathematician, it was enough that one had sufficient interest in the subject to write (but not necessarily to publish) books, papers, or articles on the subject. By the twentieth century, however, publication had become a *sine qua non.*

Although the present study consciously focuses on the research community, a parallel study on mathematical pedagogy and the community of mathematics educators would also be highly illuminating.

Research and publication became key criteria for the establishment of reputation in the field at the same time that specialization not only sharpened the different mathematical areas but also gave rise to new positions within higher education. As mathematicians gained a sense of identity, moreover, they joined together in national societies and, by the end of the century, in international congresses to share and communicate their work. Their sense of national identity, moreover, often contributed to competition as "peripheral" nations sought either to join or to displace those at the center (see, for example, Della Fenster's analysis of the motivations behind some of Leonard Dickson's international efforts). All of these factors—shared educational ideals, research, publication outlets, societies and congresses, the existence of jobs, competition—contributed to the professionalization of mathematics both nationally and internationally during our period.[22]

One sure measure of the growth of the mathematical profession was the increase in the number of outlets for the dissemination of mathematical work. These typically took the form of learned societies and journals, initially aimed at a local, or at most at a national, clientele. However, as the chapters by Jesper Lützen and Sloan Despeaux reveal, even relatively early in our period, contributions from overseas were not only quickly forthcoming but were, in many cases, actively solicited. By the end of the century with the foundation of intentionally international research journals—chiefly *Acta Mathematica* and the *Rendiconti del Circolo matematico di Palermo*—the international dissemination of mathematical research through journal publication was well on its way to becoming the norm.

The growth of national communities, then, with its concomitant professionalization played a fundamental role in the internationalization of the research community of mathematicians. These national communities, however, operated within and were affected by the broader spheres of both national and international politics. Not surprisingly, politics had a tremendous—and not always positive—impact on the development of an international mathematical community. In his chapter on the *Circolo matematico di Palermo*, for example, Aldo Brigaglia traces the history of an organization that could lay claim to being the first truly international mathematical society, but whose influence was prematurely cut short. An innovative and progressive body, the *Circolo* was founded in 1884 and served as a model of international cooperation and collaboration during the early years of the twentieth century, before being virtually destroyed by the policies of Italy's fascist regime in the 1930s. Another twentieth-century example is Nazi Germany; the effects of its policies dominate the chapters by Reinhard Siegmund-Schultze and Sanford Segal, and they helped shape the tortured, early history that Olli Lehto presents of the International Mathematical Union (IMU).

The IMU and the convoluted history of its inception and eventual demise highlight another political aspect of the internationalization of mathematics, namely, diplomacy in international mathematical relations. This comes through not only in Lehto's description of the negotiations and wranglings that led to the first International Congresses of Mathematicians and eventually to the IMU itself in 1920 but also in June Barrow-Green's chapter on the Swedish mathematical ambassador, Gösta Mittag-Leffler. In his efforts to set up and nurture *Acta Mathematica*, Mittag-Leffler provides an object lesson that tact and discretion when dealing with

[22]Compare Parshall, "Mathematics in National Contexts: An International Overview," pp. 1582-1583 and 1589.

other mathematicians on the international stage has proved every bit as vital as it has in the broader world of international politics.

Mittag-Leffler, moreover, is but one of several examples in this book of prominent individuals who have played decisive and influential roles in the formation of an international community of mathematicians. He animated *Acta*; the Italian Giovan Battista Guccia played a similar role in the first thirty years of the *Circolo matematico di Palermo*. Others, such as Charles Hermite in France, worked to transmit foreign (particularly German) mathematics into their home countries, while still others, like Leonard Dickson in the United States, were committed to the notion that the transmission should be at least two-way. Thus, distinct individuals proved highly significant in the development of an international community.

Up to this point, the analysis of the internationalization of the mathematical research community has turned on the political notion of the nation-state and on its consequences, but there is a distinct supranational, apolitical, intellectual component to this internationalization process, namely, the content of mathematics itself. Mathematicians in national contexts share educational experiences and, hence, research goals and agendas.[23] As mathematics moves beyond national boundaries, these goals and agendas become more universally held. The subject matter—the language of mathematics—comes to unite mathematicians regardless of their national loyalties; the subject matter becomes supranational; it transcends national boundaries altogether.

The free exchange of mathematical ideas across national borders has contributed to this supranationalization of mathematics and so to the internationalization of the mathematical research community. This is evident in the general dissemination of ideas and in the resulting extension of the research frontiers. For example, Jesper Lützen documents how the French mathematician, Joseph Liouville, first absorbed and then went significantly beyond the foreign—especially German—mathematical literature he read in the 1830s and 1840s; the new mathematical results he produced were neither French nor German; they belonged to a supranational realm of mathematics. Laura Martini demonstrates how Cesare Arzelà combined the best available sources in the 1880s to construct the first lecture course in Galois theory to be delivered in Italy; the course was given in Italian, but the mathematical presentation, drawn from a variety of French and German sources, transcended national boundaries. And Chikara Sasaki chronicles the career of the twentieth-century Japanese mathematician, Takagi Teiji, who made fundamental contributions to the class field theory he had been exposed to in Germany; the initial development of the theory had taken place primarily in Germany, but Takagi's brilliant extension of it in virtual isolation from printed sources demonstrated that national boundaries were irrelevant for its further growth. These examples aside, the internationalization of mathematical research does not always imply its supranationalization; Jeremy Gray, for instance, details the intense *inter*national debate *among* mathematicians from a number of nations in the opening decades of the

[23]For just two examples of studies on national differences in nineteenth-century mathematics before the internationalization of the field, see Karen Hunger Parshall, "Toward a History of Nineteenth-century Invariant Theory," in *The History of Modern Mathematics*, ed. David E. Rowe and John McCleary, 2 vols. (Boston: Academic Press, Inc., 1989), 1:157-206, and Eugene Seneta, Karen Hunger Parshall, and François Jongmans, "Nineteenth-century Developments in Geometric Probability: J. J. Sylvester, M. W. Crofton, J. É. Barbier, and J. Bertrand," *Archive for History of Exact Sciences* **55** (2001):501-524.

twentieth century that ultimately failed to settle the issue of the most appropriate foundation for mathematics.

In this light, then, what is a first vintage of the process of the internationalization of the mathematical research community? To mimic Bacon, the internationalization of the mathematical research community involves the nationalization of mathematical communities and the emergence of dominant communities that are then emulated as well as the supranational quality of mathematics itself. But the nationalization is thus modified; first, it involves professionalization which itself encompasses, among other things, a shared sense of purpose, the existence of jobs, and the production of journals and the creation of societies for the dissemination of ideas; second, it involves national and international politics and diplomacy in international mathematical relations; and third, it involves the individual and individual initiatives. But the supranational quality of mathematics is also modified; while mathematical traditions may grow in national contexts, national boundaries need not confine mathematical ideas and agendas. Mathematics as a body of knowledge has been key in the internationalization of the mathematical research community; it cannot be factored out of the equation, if we are to understand this process fully.

A Second Vintage and Beyond

For Bacon, the first vintage was merely a "commencement of interpretation."[24] He thus called for going beyond the first vintage with additional research aimed at further refining and sharpening that interpretation. The first vintage we present here should follow his precept as well, and there are at least two directions in which new research on the internationalization of the mathematical research community could and should go.

First, the formation of national communities in regions not covered by the present study and the process by which these national communities did (or possibly did not) internationalize may add significant, new dimensions—additional modifications of our first vintage—to our understanding of the internationalization process. Although June Barrow-Green touches on some of the issues affecting Russia's participation in an emergent, international mathematical community at the end of the nineteenth century in her discussion of Sonya Kovalevskaya's participation in *Acta Mathematica*, the case of Russia, and later the Soviet Union, would undoubtedly uncover deeper or different political factors and motiviations involved in internationalization than those uncovered here. Likewise, an analysis of the extent to which dominant religious cultures—such as Islam in countries of the Middle East—have affected the formation of national mathematical communities and their participation internationally might point to religion as a previously unconsidered factor in the internationalization process. Finally, a focus on national communities that eventually emerged from colonial ones—as in Central and South America, India, Africa, and Australasia—would provide interesting insights into how fully the colonies (later nations) shared in the mathematical research agendas and internationalizing impulses of the imperial, ruling nation.

Second, research on the internationalization of the mathematical research community should engage and challenge the existing and emerging literature on the internationalization of science, even if mathematics and its practitioners are rarely, if ever, considered in that literature. To date most research on internationalization

[24] Bacon, p. 156.

has been based on the physical and biological sciences—the sciences that went "big" in the latter half of the twentieth century and that have closer ties to governmental, military, or medical concerns. Yet, as Bacon has taught us, unless generalizations are drawn from as wide a range of instances as possible, they risk being misleading at best, faulty at worst. A consideration of the case of mathematics and an incorporation of the lessons it teaches may thus help broaden and sharpen our understanding of the internationalization process.

Consider, as but one example, the analysis that Elisabeth Crawford gives in her article, "The Universe of International Science: 1880–1939," especially of international science during the interwar years.[25] The article opens with an interesting and insightful "conceptual map of 'international science' " that she arranges along two perpendicular axes.[26] The horizontal axis delineates between "non-science-specific" and "science-specific" factors in international science, while the vertical axis orders a continuum of factors from "ideological-cultural" to "social-organizational." Science-specific factors roughly correspond to non-science-specific ones up and down the vertical axis. Crawford thus argues that the non-science-specific notion of universalism shaped the science-specific "universalist ethos in science," which "holds that the acceptance or rejection of knowledge claims is totally independent of the personal attributes of those who make them, their nationality, race, religion, or social class."[27] Similarly, internationalism informed scientific internationalism, an ideal that "went further than the belief that knowledge could transcend national boundaries" to "an active internationalism that put scientists in the vanguard of those who worked for the betterment of the human condition, whether it be material or intellectual."[28] Other non-science-specific factors that Crawford finds in the emergence of international science are war and conquest, trade and commerce, and international organizations that have their science-specific manifestations in global science vs. "world science" (where "world science" describes "the domination that a world power (or one aspiring as such) will exert over cultural-scientific activities in the territories it occupies or in its client states"), colonial science, international scientific organizations, and international research facilities.[29] As a general framework, this seems sufficiently broad to encompass all of the sciences, including mathematics, even though some of the non-science-specific factors may have had more influence on the evolution of an international mathematical community than others.

After a brief section on the growth of international science before World War I, Crawford moves on to characterize the interwar period from 1918 to 1939 as a period of "international science without internationalism."[30] As we noted above, for the

---

[25]Elisabeth Crawford, "The Universe of International Science: 1880–1939," in *Solomon's House Revisited: The Organization and Institutionalization of Science*, ed. Tore Frängsmyr (Canton, MA: Science History Publications and The Nobel Foundation, 1990), pp. 251-269, especially on pp. 261-265.

[26]*Ibid.*, p. 253.

[27]*Ibid.* This is consonant with the supranational aspect of the internationalization of the mathematical research community that we isolated above.

[28]*Ibid.*, p. 254.

[29]*Ibid.*, p. 255.

[30]*Ibid.*, p. 261.

case of mathematics, while the doctrine of internationalism had its adherents in figures such as Giovan Battista Guccia and Gösta Mittag-Leffler,[31] it was the process of internationalization—with the component, supranational quality of mathematics as a body of knowledge—and not internationalism that most drove mathematical developments in the interwar period as well as before and after. Crawford's insistence on the ideology of internationalism as an analytical tool forces her to ask a question that is largely irrelevant for the case of mathematics: "What then replaced such an ideology?"[32] She answers that it was the internationalization of scientific disciplines arising from the increasing specialization of science.

So what about mathematics in all of this? The body of literature on the history of mathematics in the nineteenth century demonstrates categorically that mathematics was highly specialized—for better or for worse—well before World War I, in fact well before the end of the nineteenth century. Moreover, mathematicians, perhaps more than other scientists, developed a common language over the course of the nineteenth century that allowed them to participate in shared research agendas.[33] They also shared publication outlets like the internationally oriented *Acta Mathematica* and more properly national journals like Crelle's *Journal für die reine und angewandte Mathematik*. In this light, Crawford's suggestion that "successfully 'going international' " in the interwar period "would require that there was a tradition of international collaboration in the discipline, an 'agenda' for research held in common by scientists of different countries, a common international 'language' that kept national 'styles' and predilections within bounds, and some preexisting international structures for communication and validation of new knowledge"[34] actually applies to mathematics already by 1890, if, as seems reasonable, the shared research agendas in mathematics may be seen as constituting "international collaboration" in Crawford's sense. The implications that then follow are that the other sciences were decades behind mathematics and that many of the political considerations that dominated the opening decades of the twentieth century were all but irrelevant to the internationalization process. The present volume shows, however, that these conclusions are false; in so doing, it suggests that Crawford's characterization of "going international" in the interwar period is inadequate.

What seems largely missing from earlier studies on the internationalization of science, and what Crawford seems to want to stress for the interwar period but not for the earlier periods, is a meaningful and concerted analysis of the role of the science itself in the process. In fact, she asserts that "[t]heoretical atomic physics, especially quantum mechanics, holds the most promise as an area of study of *all* the elements that went into the internationalization of a discipline," namely, the existence of elder statesmen, active younger leaders and centers of teaching and research, arrangements and funding of exchanges of post-doctoral fellows and full-fledged researchers, journals, regular meeting places, "and last but not least, the ability to endow the process with a higher purpose that was no longer the

---

[31] G. H. Hardy could also be included in this list. Compare the chapters below by Sanford Segal and Olli Lehto.

[32] *Ibid.*, p. 263.

[33] On the importance of mathematics as a shared language, see, for example, Elizabeth Garber's book, *The Language of Physics: The Calculus and the Development of Theoretical Physics in Europe, 1750–1914* (Boston: Birkhäuser Verlag, 1999).

[34] Crawford, "The Universe of International Science," p. 264.

internationatist ideology of the *fin-de-siècle*, but was now the quest for pure knowledge."[35] By these criteria, the case of mathematics would seem to hold equal or greater promise. Pure knowledge has been its quest since at least the nineteenth century. The case of mathematics would seem to show that—regardless of the historical period under consideration—to discount or to overlook the science and the production of scientific knowledge as a key motivating factor in the life of the scientist can result in over-hasty generalization or misdirected efforts.

The chapters that follow highlight mathematics as a case study in the internationalization of a scientific field. It is hoped that they shed light on that process as it applies not only to mathematics but to the other sciences as well.

References

Bacon, Francis. *The New Organon and Related Writings.* Ed. Fulton H. Anderson. Indianapolis: Bobbs-Merrill Company, Inc., 1960.

Crawford, Elisabeth. *Nationalism and Internationalism in Science, 1880–1939: Four Studies of the Nobel Population.* Cambridge: University Press, 1992.

———. "The Universe of International Science: 1880–1939." In *Solomon's House Revisited: The Organization and Institutionalization of Science.* Ed. Tore Frängsmyr. Canton, MA: Science History Publications and The Nobel Foundation, 1990, pp. 251-269.

Crawford, Elisabeth; Shinn, Terry; and Sörlin, Sverker, Ed. *Denationalizing Science: The Contexts of International Scientific Practice.* Dordrecht: Kluwer Academic Publishers, 1993.

———. "The Nationalization and Denationalization of the Sciences: An Introductory Essay." In *Denationalizing Science: The Contexts of International Scientific Practice.* Ed. Elisabeth Crawford, Terry Shinn, and Sverker Sörlin. Dordrecht: Kluwer Academic Publishers, 1993, pp. 1-42.

Crosland, Maurice. "The Congress on Definitive Metric Standards, 1798–1799: The First International Scientific Conference?" *Isis* **60** (1969):226-231.

Drayton, Richard. *Nature's Government: Science, Imperial Britain, and the 'Improvement' of the World.* New Haven: Yale University Press, 2000.

Forman, Paul. "Scientific Internationalism and the Weimar Physicists: The Ideology and Its Manipulation in Germany after World War I." *Isis* **64** (1973):151-180.

Frängsmyr, Tore, Ed. *Solomon's House Revisited: The Organization and Institutionalization of Science.* Canton, MA: Science History Publications and The Nobel Foundation, 1990.

Garber, Elizabeth. *The Language of Physics: The Calculus and the Development of Theoretical Physics in Europe, 1750–1914.* Boston: Birkhäuser Verlag, 1999.

Gizycki, Rainald von. "Center and Periphery in the International Scientific Community: Germany, France and Great Britain in the 19th Century." *Minerva* **11** (1973):474-494.

Goldstein, Catherine; Gray, Jeremy J.; and Ritter, Jim, Ed. *L'Europe mathématique—Mathematical Europe.* Paris: Édition de la Maison des sciences de l'homme, 1996.

[35] *Ibid.*, pp. 264-265; her emphasis.

Li Wenlin and Martzloff, Jean-Claude. "Aperçu sur les échanges mathématiques entre la Chine et la France (1880–1949)." *Archive for History of Exact Sciences* **53** (1998):181-200.

Lyons, Francis S. L. *Internationalism in Europe, 1815–1914*. Leyden: A. W. Sythoff, 1963.

Ortiz, Eduardo. "El rol de las revistas matemáticas intermedias en el establecimiento de contactos entre las comunidades de Francia y España hacia fines del siglo XIX." In *Contra los titanes de la rutina/Contre les titans de la routine*. Ed. Santiago Garma, Dominique Flament, and Victor Navarro. Madrid: Comunidad de Madrid/C.S.I.C., 1994, pp. 367-382.

Parshall, Karen Hunger. "Mathematics in National Contexts (1875-1900): An International Overview." In *Proceedings of the International Congress of Mathematicians: Zürich*. 2 Vols. Basel: Birkhäuser Verlag, 1995, 2:1581-1591.

———. "Toward a History of Nineteenth-century Invariant Theory." In *The History of Modern Mathematics*. Ed. David E. Rowe and John McCleary. 2 Vols. Boston: Academic Press, Inc., 1989, 1:157-206.

Parshall, Karen Hunger and Seneta, Eugene. "Building an International Reputation: The Case of J. J. Sylvester (1814–1897)." *American Mathematical Monthly* **104** (March 1997):210-222.

Schröder(-Gudehus), Brigitte. "Caractéristiques des relations scientifiques internationales, 1870–1914." *Cahiers d'histoire mondiale/Journal of World History* **10** (1966):161-177.

———. *Deutsche Wissenschaft und internationale Zussamenarbeit*. Geneva: Dumaret & Golay, 1966.

———. *Les scientifiques et la paix: La communauté scientifique internationale au cours des années 20*. Montreal: Presses de l'Université de Montréal, 1978.

Schubring, Gert. "Differences in the Involvement of Mathematicians in the Political Life in France and in Germany." *Bollettino di storia delle scienze matematiche* **15** (1995):61-83.

Seneta, Eugene; Parshall, Karen Hunger; and Jongmans, François. "Nineteenth-century Developments in Geometric Probability: J. J. Sylvester, M. W. Crofton, J. É. Barbier, and J. Bertrand." *Archive for History of Exact Sciences* **55** (2001):501-524.

Siegmund-Schultze, Reinhard. "The Emancipation of Mathematical Research Publishing in the United States from German Dominance (1878–1945)." *Historia Mathematica* **24** (1997):135-166.

———. *Rockefeller and the Internationalization of Mathematics Between the World Wars*. Basel: Birkhäuser Verlag, 2001.

———. " 'Scientific Control' in Mathematical Reviewing and German-U.S.-American Relations Between the Two World Wars." *Historia Mathematica* **21** (1994):306-329.

Zhang Dianzhou and Dauben, Joseph W. "Mathematical Exchanges Between the United States and China: A Concise Overview." In *The History of Modern Mathematics*. Vol. 3. Ed. Eberhard Knobloch and David E. Rowe. Boston: Academic Press, Inc., 1994, pp. 263-297.

CHAPTER 2

The End of Dominance: The Diffusion of French Mathematics Elsewhere, 1820–1870

Ivor Grattan-Guinness
Middlesex University (United Kingdom)

Multinationalism vs. Internationalism

The principal vehicles for international collaboration in mathematics during the nineteenth century date from its last decades, and other chapters in this volume chart many of the details, such as the launch of mathematical journals with wide authorship and the International Congresses of Mathematicians. Earlier on, however, little activity of this kind was evident. Nevertheless a substantial amount of *multinational* reaction and interaction can be found, which, with special attention to one aspect, will be surveyed here. In this kind of process, a source from one country (or state or kingdom) is used in another, but without any launch or continuation of collaboration on a larger scale or longer-running cooperation of the kind that distinguishes the later *international* activities. To give an important example used below, the translation of a book from one language to another is a multinational process, but its quality may cause it to gain international importance. To use an analogy from set theory, multinationality is like the extensional definition of a set by its members, while internationality resembles an intensional definition in terms of some property.[1]

The aspect given special attention here is the changing place of France, which enjoyed a remarkable dominance in mathematics from the 1780s until the 1820s, with Paris by far the leading center for the subject in the world. However, partly in reaction to French achievements, other countries began to produce significant mathematicians and/or to revise their institutions and curricula, especially from the mid-1810s. Mathematicians visited other countries (especially after the fall of Napoleon in 1816) and often developed warm personal friendships, publishing in each others' journals on occasion, preparing translations, etc. Indeed, from the early 1800s, mathematics has been a much more international activity. Thus, in the context of the present volume, this chapter aims to set the scene for subsequent ones by discussing various manifestations of the change in the mathematical landscape that occurred in the period from 1820 to 1870, a period in which the foundations were laid for what could eventually be called an international mathematical community.

The chapter opens by reviewing in various ways the loss of French dominance in mathematics, especially via the diffusion and translation of French books into other languages. It also assesses the place of mathematics in the apparent decline of

[1] On the diverse scholarly terminology associated with internationalization, recall the discussion in the chapter by Karen Parshall and Adrian Rice above.

French science after 1830, emphasizing the *incompleteness of our historical knowledge* of these developments and providing historical existence theorems rather than proofs of a thesis. In some compensation, the chapter proceeds to a quintet of case studies in which both French and non-French (especially German) mathematics played significant roles.

The first two case studies come from pure mathematics, the next two deal with applications, and the last considers the reception of the work of Gauss. They have been chosen not only for their individual merits but also for the *differences* between them, both in subject matter and in manners of the analysis they require. None of them has received the historical attention it deserves, and each would be a candidate topic for at least one doctorate. As a result, some of the questions these areas raise can only be partially answered here. Incomplete though our historical knowledge of these topics may be, it nonetheless provides us with useful illustrations of the multinational diffusion of mathematical ideas during the decades from 1820 to 1870.

French Dominance

The deaths of Jean d'Alembert, Daniel Bernoulli and Leonhard Euler in the early 1780s led naturally to the primacy of Paris in the mathematical world, with residents such as Gaspard Monge, Pierre Simon Laplace, Adrien-Marie Legendre, and, after 1787, Joseph Louis Lagrange.[2] The Revolution two years later led to major changes in the institutional provision of mathematics for both teaching and research during the next decade: the new *École polytechnique* from 1794,[3] and the specialist engineering schools for further training. Some of the latter were civil, especially the *École des ponts et chaussées*[4] and the *École des mines* in Paris. The others were military and mostly based in the provinces. Of these, the most significant was the *École d'artillerie et génie*, founded in Metz in 1802 as a fusion of schools previously in Mézières and Châlons-sur-Marne.[5] An important but unintended consequence of the founding of the *École polytechnique* resulted from the policy of recruiting students from across the country independent of personal resources. From the start, there emerged a remarkable string of graduates who became major mathematicians in their own right; many returned as teachers at or examiners for the institutions from which they had graduated.

In addition, various organizations were in place to promote research. Led by the *Académie des sciences* (which took a different name during the revolutionary period), they included some governmental bodies such as the *Bureau des longitudes*, with several being attached to the Army and Navy. Also important was the private *Société philomatique de Paris*, where much original work first saw print.

---

[2]Most of the material here and in the next two sections is based upon my detailed study of the period, where many references to historical literature are given: Ivor Grattan-Guinness, *Convolutions in French Mathematics, 1800–1840*, 3 vols. (Basel: Birkhäuser Verlag and Berlin: Deutscher Verlag der Wissenschaften, 1990), especially chapters 2-6 for the situation around 1800. The book is cited below as *Convolutions*.

[3]Still a fine history of the early decades of the school is Ambroise Fourcy, *Histoire de l'École polytechnique* (Paris: École polytechnique, 1828; reprint ed., Paris: Belin, 1987).

[4]Antoine Picon, *L'invention de l'ingénieur moderne: L'École des ponts et chaussées 1747–1851* (Paris: Presses de l'École nationale des ponts et chaussées, 1992).

[5]Bruno Belhoste and Antoine Picon, ed., *L'École d'application de l'artillerie et du génie de Metz (1802–1870): Enseignement et recherches: Actes de la journée d'étude de 2 novembre 1995* (Paris: Ministère de la culture, Direction du patrimoine, Musée des plans-reliefs, 1996).

Table 1 lists the main mathematicians who had gained prominence by the 1840s. By far the dominant branches of mathematics were the calculus, with its extension into mathematical analysis, and mechanics, with its extension into mathematical physics.[6] The cohort is divided by academic interests into two groups. The first, the *ingénieurs savants*, concentrated on contexts inspired by engineering and technology, often serving as professional civil or military engineers—instead of or as well as academics—and publishing in a remarkably wide range of journals.[7] The second specialized in so-called pure mathematics and the more general aspects of applications (including the extension of mechanics into mathematical physics).

TABLE 1: PRINCIPAL FRENCH MATHEMATICIANS 1780s–1840s

Calculus/Mechanics/Engineering	Mathematical Analysis/Physics
C. S. J. Bossut (1730–1814)	A. M. Ampère (1775–1836)
L. Carnot (1753–1823)	J. P. M. Binet (1786–1856)
B. P. E. Clapeyron (1797–1864)	J. B. Biot (1774–1862)
C. P. M. Combes (1801–1872)	A. L. Cauchy (1789–1857)
G. G. Coriolis (1792–1843)	J. M. C. Duhamel (1797–1872)
J. B. J. Delambre (1749–1822)	J. B. J. Fourier (1768–1830)
F. P. C. Dupin (1784–1873)	A. J. Fresnel (1788–1827)
L. B. Francoeur (1773–1849)	S. Germain (1776–1831)
P. S. Girard (1765–1836)	J. L. Lagrange (1736–1813)
J. N. P. Hachette (1769–1834)	G. Lamé (1795–1870)
G. Monge (1746–1818)	P. S. Laplace (1749–1827)
A. J. Morin (1795–1880)	A. M. Legendre (1752–1833)
C. L. M. H. Navier (1785–1836)	E. L. Malus (1775–1812)
T. Olivier (1793–1853)	L. Poinsot (1777–1859)
J. V. Poncelet (1788–1867)	S. D. Poisson (1781–1840)
G. Riche de Prony (1755–1839)	P. G. le D. de Pontécoulant (1795–1874)
L. Puissant (1769–1843)	C. Sturm (1803–1855)

The strength of the links between engineering and science, especially the high status given to engineering institutions, was (and remains) a peculiarly French feature and represents an example of a key difference between the national setting of France and those of other nations. Another important difference concerns the status of universities. Elsewhere they were the dominant places for higher education and important for stimulating research, but in France the *Université* system, when finally refounded in 1808, was centered on the structure of schools (in the usual

[6]For a detailed, overall account of these extensions, see *Convolutions*, chapters 8-17.

[7]Ivor Grattan-Guinness, "The *ingénieur savant*, 1800–1830: A Neglected Figure in the History of French Mathematics and Science," *Science in Context* **6** (1993):405-433.

sense of the term), and the faculty teaching at its top level was distinctly secondary to that of the engineering schools.[8]

The main mathematical topics not covered in Table 1 are projective and related geometry, where the chief name to add is Michel Chasles (since descriptive geometry was tied to mechanics); number theory, then still a specialist topic; and probability theory, also rather specialized and not yet of high status. Another list could be composed of more minor figures who, nevertheless, made notable contributions and would have stood out in any other country.

Already at the start of the century there were notable mathematicians elsewhere in Europe, although the list is strikingly short. First and foremost, there was Carl Friedrich Gauss, yet even his work poses an historical problem that will be tackled in the fifth case study below. Giovanni Malfatti, Paolo Ruffini, and Robert Woodhouse probably come next, joined later by Friedrich Bessel. The cohort begins to emerge in some quantity only from the late-1810s onwards. England produced Charles Babbage, John Herschel, and George Peacock, followed by George Biddell Airy, William Whewell, and Augustus De Morgan. In Ireland, William Rowan Hamilton was outstanding but not alone. The new generation of Germans included Carl Jacobi, Peter Gustav Lejeune Dirichlet, August Ferdinand Möbius, Karl von Staudt, Jakob Steiner, and Julius Plücker. Among other nationalities, there was Niels Henrik Abel from Norway, Mikhail Vasilevich Ostrogradsky and Viktor Yakovlevich Bunyakoffsky from Russia, and Giovanni Plana (a *polytechnicien*) and Gabrio Piola from Italy. New work was often much inspired by the French literature. Indeed, in the 1820s, Dirichlet, Ostrogradsky, and Bunyakoffsky passed several highly formative years in Paris (Abel's stay was much shorter and less useful), while Babbage and Herschel were among those who visited several times. Further figures emerged from the 1830s onwards, in Britain and in the German and Italian states. A few are mentioned below, and several appear elsewhere in this volume.

Translating the French

One of the main sources of French influence elsewhere was the distribution of French books and journals.[9] Many textbooks and monographs were also written, especially by the mathematicians just named but also by several others, and the

[8] Ivor Grattan-Guinness, "*Grandes écoles, petite université*: Some Puzzled Remarks on Higher Education in Mathematics in France, 1795–1840," *History of Universities* **7** (1988):197-225, and Terry Shinn, "The French Science Faculty System, 1808–1914: Institutional Changes and Research Potential in Mathematics and the Physical Sciences," *Historical Studies in the Physical Sciences* **10** (1979):271-332.

[9] Information on print-runs is, however, always very difficult to find. While researching the French literature, *no* archives were found for any of the publishers involved (the main ones being Bachelier, Courcier, Duprat, Firmin Didot, and Klostermann). However, a very little-known—but large and precious—collection of declarations by printers required by the Bourbon government after 1816, survives in the *Archives nationales.* The files are held at F^{18}, 43-119, with registers at F^{18*}, II 1-182. There, both print-run and client information suggest that monographs might gain around 1,000 copies, textbooks on widely studied topics somewhat more, (for example, the first edition of Poisson's *Traité de mécanique* appeared in 1811 with a second edition in 1833 of 5,000 copies each time), and elementary textbooks could appear in runs of 3,000 every few years. For earlier French items and other countries, no comparable source seems to exist, and data are hard to find.

The history of scientific publishers for the period 1780–1870 is also very fragmentary (in contrast to our knowledge of Enlightenment France). For a valuable recent study of the Cambridge

French influence was often considerable. In addition, many translations of French originals were produced. (See the Appendix for a list of translations.[10])

Among the types of books translated, two major categories merit attention here: research monographs, usually devoted to one group of topics, and textbooks for use at the university or equivalent level, often in later years of study. Apart from a few "complete courses" in the first category, the more elementary textbooks for use at schools up to the stage of candidacy for higher education (comprising arithmetic, algebra, elementary geometry, trigonometry, basic calculus, and maybe probability theory) do not enter here.[11] At this level, Étienne Bézout continued to be popular, but he was supplanted by Sylvestre Lacroix,[12] whose more elementary books received a wide range of translations (for example, into Greek and Polish).

Some translations gained importance in their own right. Among those in English two stand out: Nathaniel Bowditch's translation of Laplace's *Traité de mécanique céleste*, which became internationally recognized;[13] and that of the Analytical Society at Cambridge of Lacroix's *Traité élémentaire* on the calculus (1816), a book set at a far higher level than the others in his course and so included in the Appendix.[14]

A few books were translated more than once into a language, sometimes from different editions. For example, two German translations of the large study of forts by Carnot and of the wonderful textbook on statics by Poinsot appeared within a few years. Since Germany did not exist as a unified country at the time of these translations—although Prussia, Bavaria, and Saxony were important states (among many which were not)—the duplication may reflect an absence of trading

house of Deighton, see Jonathan Topham, "Two Centuries of Cambridge Publishing and Bookselling: A Brief History of Deighton, Bell and Co., 1778–1998, with a Checklist of the Archive," *Transactions of the Cambridge Bibliographical Society* **11** (1998):349-403. Among a sparse literature, for Britain see also William H. Brock and A. Jack Meadows, *The Lamp of Learning: Taylor and Francis and the Development of Science Publishing*, 2nd ed. (London: Taylor and Francis, 1998), although mathematics was not a major concern for them; and for Spanish-English links, see Elena Ausejo and Mariano Hormigón, "Mathematics for Independence: From Spanish Liberal Exile to the Young American Republics," *Historia Mathematica* **26** (1999):314-326.

Since the scientific publication industry was large, especially in France, and already fairly multinational even before 1800, much remains to be discovered concerning not only print-runs and means of selling these books and journals but also the process of acceptance or rejection of manuscripts and the use of the books as required reading on courses.

[10]The Appendix lists the main French books together with translations. Some of the authors are more minor figures not included in Table 1, and the full range of topics is covered. When known, it is indicated if a translation is incomplete or free.

Sometimes license was taken. For example, the translator into Dutch of Lacroix's elementary book on algebra replaced without warning the original text on complex numbers with the current Netherlands doctrine! I thank Danny Beckers of Nijmegen University for this information.

[11]A full list of such books would provide information about the multi- and international development of mathematics education: yet another research topic awaiting a dissertation.

The list presented in the Appendix is no doubt incomplete, but hopefully no major book or translation has escaped. It is hoped that the material contained there provides a substantial starting point for bibliographical data.

[12]Gert Schubring, "On the Methodology of Analysing Historical Textbooks: Lacroix As Textbook Author," *For the Learning of Mathematics* **7** (3) (1987):41-51.

[13]Pierre Simon Laplace, *Celestial Mechanics*, 4 vols., trans. Nathaniel Bowditch (with notes and appendices) (Boston: Hillard, Gray, Little, and Wilkins, 1829–1839; reprint ed., New York: Chelsea Publishing Co., 1969).

[14]Sylvestre François Lacroix, *An Elementary Treatise on the Differential and Integral Calculus* (Cambridge: Deighton, 1816).

arrangements between the states involved, an ignorance of recent publication, or a determination to complete a translation already well advanced despite the competition. The same possibilities may cover the passion for translating Legendre's textbook on geometry in the Italian states.

The cases of multiple translation exemplify the towering place of German translations. Many of these were effected in Prussia, in the context of university reforms from 1810 onwards.[15] The steady rise in the state of German mathematics—and, indeed, of science in general—from the 1820s is a well-known theme. Two German translators particularly stand out: August Leopold Crelle and Christian Heinrich Schnuse.

Crelle is remembered almost exclusively for the research-level mathematics journal he launched in Berlin in 1826, the *Journal für die reine und angewandte Mathematik*.[16] Although an engineer by profession (he also edited a *Journal für Baukunst*), he aspired to the heights of pure mathematics and translated the major treatises of Lagrange as well as Legendre's textbook on geometry. In addition, he encouraged the training of younger compatriot mathematicians such as Dirichlet by helping them gain support to study in Paris.[17]

More striking is Christian Heinrich Schnuse. A graduate in the mid-1830s from Göttingen University (where he had taken a course in geodesy from Gauss), he was soon appointed to a school in Braunschweig. A decade later, he moved to Heidelberg, and after another decade to Munich. For twenty years from the mid-1830s, he not only wrote several textbooks of his own but also translated twenty-one French works into German.[18] Most of these were textbooks across a wide range of topics. His authors included Arago (astronomy), Biot (statics), Cauchy (differential calculus, differential geometry), Cournot (mathematical analysis, probability theory), Fourier (equations), Chasles (geometry), Lamé (physics), Moigno (integral calculus), Pambour (two books on steam engines: Crelle translated another one), Olivier (gears), Poisson (probability theory), and Poncelet (engineering mechanics). In addition, mostly in the 1850s, he turned to translating English books, including Herschel (exercises on the calculus) and George Boole (difference equations).

[15]Still, the profusion of these translations for mathematics does not seem to have been much considered. An astonishing (though only partly mathematical) case is the sixteen-volume edition of the collected works of Arago, contemporary with the French original. The mathematical side of these developments is well outlined in Gert Schubring, *Die Entstehung des Mathematiklehrerberufs im 19. Jahrhundert: Studien und Materialien zum Prozeß der Professionalisierung in Preußen (1810–1870)*, 2nd ed. (Weinheim: Deutscher Studien Verlag, 1991). The fact that very few translations appeared in the Austro-Hungarian Empire also deserves study, since no Czech translations seem to have been made there either.

A large list of textbooks and monographs in pure mathematics for the entire century, mostly German but also some others, is provided in Ernst Wölffing, "Mathematischer Bücherschatz: 1. Teil [and only] reine Mathematik," *Abhandlungen zur Geschichte der mathematischen Wissenchaften* **20** (1903): xxxvi and 416 pp.

[16]Wolfgang Eccarius, "August Leopold Crelle als Herausgeber wissenschaftlicher Fachzeitschriften," *Annals of Science* **33** (1976):229-261; similar version in *Journal für die reine und angewandte Mathematik* **286-287** (1976):5-25.

[17]Wolfgang Eccarius, "August Leopold Crelle als Förderer bedeutender Mathematiker," *Jahresbericht der Deutschen Mathematiker-Vereinigung* **79** (1977):137-174.

[18]Ivo Schneider, "Christian Heinrich Schnuse als Übersetzer mathematischer, naturwissenschaftlicher und technischer Literatur" and "Bibliographie," *Aus dem Antiquariat* (1982): A205-A221 and A256-A261.

Despite the fact that several translators of French works were American (Bowditch was singled out above), research-level mathematics remained modest in the United States until the 1880s.[19] Elsewhere, modest starts in translation were made in Spain[20] and in Greece,[21] although most of the books chosen were at the elementary level. Similar stories could presumably be told for other countries.[22]

A point of significance is the decreasing number of translations after around 1850, doubtless a sign of the growing number of authors in other countries competent to write their own books. The (apparent) absence of translations is sometimes striking, however. For instance, translations of the treatises on topography and geodesy by Louis Puissant are noticeably missing.[23] Largely forgotten now, Puissant was then renowned as the leading practitioner of these topics. Perhaps other countries already had local personnel (probably military) capable of producing their own books, in which his were used and acknowledged. The silence over Bélidor's engineering treatises contrasts with his popularity in German in the eighteenth century; presumably even the updated editions prepared by Navier were seen as *passé*.

Papers were also translated, but no compact means of reference is possible; most of the information can be retrieved from the *Royal Society Catalogue of Scientific Papers.* (Translations of French papers in book form are included in the Appendix.) Moreover, the translation of mathematical papers was a lesser activity than practiced for sciences such as physics and chemistry. Thus, those papers translated often concerned mechanics or mathematical physics, such as those by Ampère, Fresnel, and Poisson in the *Annales de chimie et de physique* that appeared in English in the *Quarterly Journal of Science* and in German in the *Annalen der Physik.* From 1837 until 1853, *Taylor's Scientific Memoirs* was a regular venue for English translations, although specifically mathematical papers were not included.[24]

The most intriguing translations of papers issued from Johann Philip Gruson in the late 1810s. Already the translator of books on the calculus by Lacroix and Lagrange, he rendered into German two papers of 1806 by Marc-Antoine Parseval, one of which contained a version of "Parseval's formula." When Gruson reached the French passage "M. A. Parseval," he rendered it each time as "J. P. Gruson" and

---

[19] Karen Hunger Parshall and David E. Rowe, *The Emergence of the American Mathematical Research Community, 1876–1900: J. J. Sylvester, Felix Klein, and E. H. Moore*, HMATH, vol. 8 (Providence: American Mathematical Society and London: London Mathematical Society, 1994).

[20] Fernando Vea, "On the Influence of French Mathematics Textbooks on the Establishment of the Liberal Education System in Spain (1845-1868)," in *Paradigms in Mathematics*, ed. Elena Ausejo and Mariano Hormigón, (Zaragoza: Siglo XXI de España Editores, 1996), pp. 365-390.

[21] The Greek initiative was led by Ioannis Carandinos (1784–1834), who was encouraged in 1812 by Dupin when spending some time in Athens. In 1820, he was an external auditor of courses at the *École polytechnique.* See Christine Phili, "La reconstruction des mathématiques en Grèce: L'apport de Ioannis Carandinos (1784–1834)," in *L'Europe mathématique–Mathematical Europe*, ed. Catherine Goldstein, Jeremy J. Gray, and Jim Ritter (Paris: Éditions de la Maison des sciences de l'homme, 1996), pp. 305-319. Some information on other countries is provided in this book.

[22] While various national histories have been written, the place of French mathematics and translations has not been addressed in sufficient detail to be summarized here.

[23] The astronomer, Heinrich Schumacher, was translating Puissant's *Géodésie* of 1805 by 1810, although for seemingly financial reasons he could not find a publisher. See Karin Reich, *Im Umfeld der "Theoria motus"* (Göttingen: Vandenhoeck & Ruprecht, 2001), p. 43.

[24] Other journals also carried translations. For example, the *Philosophical Magazine* published Thomas Archer Hirst's translation of Poinsot's work on the percussion of bodies. Compare Sloan Despeaux's chapter below.

published the fruits of his labors in the hardly minor venue of the Berlin Academy.[25] In a rather nice touch, he helped the reader by including French versions of certain technical terms.

Another source of multinational contact was book reviews, of which there were quite a number, in several countries. In addition, from 1823 to 1832 in Paris, there existed an amazingly comprehensive journal reviewing books and papers in all sciences and technology worldwide. Eight series ran, at around 500 or 750 copies each.[26] Known as the "Bulletin de Ferrusac" after its general editor, the naturalist and geographer André-Étienne-Just-Pascal-Joseph-François d'Audebart, Baron de Ferrusac, it ceased because the government that succeeded the Bourbons after the 1830 Revolution withdrew its subscriptions, and thus destroyed its economic basis. Nothing comparable has ever run since, although for mathematics and physics reviewing, *Jahrbücher* started in Germany in the late 1860s—just in time for the Franco-Prussian War of 1870–1871 to handicap the availability of French material.[27]

Déclin?

Given the pervasiveness of translations of French mathematical texts between 1820 and 1870, it is natural to ask to what extent did French mathematics decline during this period? The question, well-known to all familiar with nineteenth-century French cultural history, was asked by the French themselves at the time. Indeed, the possible causes have defined a branch of French historiography since at least the 1870s. The remarks here focus on the mathematics, which, as usual, is ignored by historians of science.[28]

The claims of decline contain important elements of truth. For example, the French were often slow to respond to foreign mathematics (and other sciences); French translations from other languages make a very short list. This, however, reflects more the insularity of *Paris* than of France. Claims of decline were made within the *Facultés*, partly to encourage more governmental spending on higher

[25]Johann Philip Gruson, "Über Reihen und vollständigen Integration ... " and "Allgemeine Methode mittelst bestimmter Integralien die durch den Lagrange'schen Lehrsatz gegebene Reihe zu summiren," *Abhandlungen der Akademie der Wissenchaften zu Berlin, Mathematische Klasse* (1812–1813):23-30 and 31-44. Compare, respectively, Marc-Antoine Parseval, "Mémoire sur les séries et sur l'intégration complète d'une équation aux différences partielles du second ordre, à coefficiens constans" and "Méthode générale pour sommer ... la suite donnée par le théorème de M. Lagrange ... ," *Mémoires présentés à l'Institut ... par divers savans, classe des sciences mathématiques et physiques*, 1st ser. **1** (1806):638-648 and 567-586.

[26]On the mathematics series, see René Taton, "Les mathématiques dans le 'Bulletin de Ferrusac'," *Archives internationales d'histoire des sciences* **26** (1947):100-125.

[27]See Hermann Felix Müller, "Das Jahrbuch über die Fortschritte der Mathematik 1869–1904," *Bibliotheca mathematica* (3) **5** (1905): 292-297. On the French situation, see Héléne Gispert, "Sur la production mathématique française en 1870," *Archives internationales d'histoire des sciences* **35** (1985):380-399.

[28]A modern basic source for this thesis is Joseph Ben-David, "The Rise and Decline of France as a Scientific Centre," *Minerva* **8** (1970):160-179. Mathematics is completely omitted there, as also in responses such as Dorinda Outram, "Politics and Vocation: French Science 1793–1830," *British Journal for the History of Science* **13** (1980):27-43. Among the more valuably influenced, Nicole Hulin-Jung gives a good survey of educational policy during the Second Empire (1850s–1871) in *L'organisation de l'enseignement des sciences: La voie ouverte par le Second Empire* (Paris: Comité des travaux historiques et scientifiques, 1989). Maurice P. Crosland treats the institutional place of the *Académie des sciences* during the whole century in *Science Under Control: The French Academy of Sciences* (Cambridge: University Press, 1992).

education; as noted above, the *Facultés* were secondary to the engineering schools. When the latter establishments are admitted into the context, and the place of engineering within science is recognized, a more impressive picture emerges. Nevertheless, after the 1850s the caliber of new *ingénieurs savants* seemed to decrease somewhat, so that the link lost some of its strength. The defeats of the army in the Franco-Prussian War in 1871 further excited questions about the quality of officer training.[29]

On the other hand, in various respects the case for decline seems to be much exaggerated. For example, the contributions of men like Fourier, Ampère, Cauchy, Navier, and Fresnel in the 1820s had been spectacularly *pioneering*. Later work filled in details and extended theories that they had established, but the tasks were not necessarily "easier." Again, having risen so high then (and not only in mathematics), the only way forward for France was down. Nevertheless, between 1850 and 1870, the cohort of successors included Charles Hermite, Victor Puiseux, Émile Mathieu, Camille Jordan, and Gaston Darboux, certainly an impressive list. They were followed, in turn, by figures such as Pierre Duhem and Henri Poincaré. Moreover, the creation of the *Société mathématique de France* just after the Franco-Prussian War also increased the professional base. In other words, the decline in mathematics resembles an optical illusion caused by the rise of other countries. Throughout this period, levels of work were still achieved in France that exceeded those of most other countries.

France, then, was still a mathematical power to be reckoned with throughout the first three-quarters of the nineteenth century, but other nations—notably Germany and, to a lesser extent, Italy and Great Britain—also entered onto the international mathematical scene. Mathematics was no longer dominated solely by one nation; it increasingly reflected the achievements of a multinational effort. A look at some of the mathematical areas of interest during this period reveals the complexity of the interplay among the nations contributing to the research.

Case Study 1: Real-Variable Analysis

One of the chief French mathematical contributions at this time was the creation of the theory of real-variable functions. Taking a general theory of limits as his basis, Cauchy defined the derivative of a function $f(x)$ as the limiting value of any sequence of pertaining difference quotients, and the integral as that of any sequence of rectilinear approximating areas between $f(x)$ and the x-axis: in neither case did he assume that the limiting value existed. He also used limit theory to define the continuity of $f(x)$ and the convergence of infinite series, and thereby launched mathematical analysis in the modern sense of the term (including its unfortunate use of the word "analysis;" in terms of the traditional Greek distinction between analysis and synthesis, Cauchy's proof methods were synthetic). He raised standards of rigor in the subject so that, for example, the so-called "fundamental theorem of calculus" (not his name) was, for the first time, actually a theorem, adorned with sufficient conditions for $f(x)$ to satisfy. Indeed, Cauchy systematized the inclusion of necessary and/or sufficient conditions in theorems in general. (He also brought in the systematized numbering of equations in a paper or book.) As

[29]On the impact of the Franco-Prussian War on French mathematics, see the chapters by Thomas Archibald and Hélène Gispert in this volume.

mathematics it is impressive and beat the competition for foundations. However, it was by no means certain that it would be accepted in France, let alone elsewhere.

Cauchy presented his approach in detail in courses at the *École polytechnique* from the mid-1810s, but both staff and students hated it. In April 1821, the students actually walked out of Cauchy's class.[30] Changes were requested of him, but to no avail. The situation was perhaps complicated by political circumstances, which had become very tense from 1820 when the heir to the Bourbon throne was assassinated. One of Cauchy's passions, after all, was his adherence to the monarchy and to its attendant Catholicism.[31] The predicament at the school was solved by the Revolution of July 1830, when the Bourbons were expelled, and Cauchy followed the royal family into exile as mathematics tutor to the pretender to the throne (the son of the assassinated heir). Cauchy had written four textbooks, but none was adopted by the school or the *Facultés*. After his departure, rather watered-down versions of his approach were taught by successors: Coriolis briefly, then Navier to 1836, Duhamel to 1839, and Sturm to his death in 1855. Sturm followed Cauchy in the greatest detail (and elaboration) and was partly responsible for the wider adoption of the approach. Even then, his textbook was only published posthumously. In others' texts, such as the continuously published Lacroix, simpler use of limits was used, and the earlier approaches of Euler using differentials and differential coefficients and of Lagrange with power series stayed buoyant, not least for their superior utility in applications.

The principal vindication of Cauchy came not from France but with the lecture courses that Weierstraß launched at Berlin after his rapid rise to fame in the mid-1850s. The refinements introduced by Weierstraß and his followers are various. For example, single limit theory became multiple limit theory, the least upper bound was distinguished from the upper limit, definitions were offered for irrational numbers, and convergence of infinite series was split into modes of (non-)uniformity.[32] The origins in Weierstraß himself are also pretty clear,[33] but what was the *general* pre-history and background? The question is especially interesting because it focuses upon Germany, where so many French books were translated. The role of the educator Karl Schellbach in Berlin is worth noting, as several Weierstrassians also studied with him.[34]

Cauchy's theory gradually gained support throughout Europe and was developed mainly by Weierstraß and his followers into the greatly refined form summarized above. But what history lies behind the word "gradually"? The answer

---

[30] For the content and reception of Cauchy's teaching, see *Convolutions*, chapters 10-11 (the walkout is described on pp. 710-712) and the documents on pp. 1337-1340. See also Ivor Grattan-Guinness, "Recent Researches in French Mathematical Physics of the Early 19th Century," *Annals of Science* **37** (1981):663-690 on pp. 680-682. On the later history, see especially Umberto Bottazzini, *The Higher Calculus: A History of Real and Complex Analysis from Euler to Weierstraß* (New York: Springer-Verlag, 1986), chapters 3 and 5.

[31] See Bruno Belhoste, *Augustin-Louis Cauchy: A Biography* (New York: Springer-Verlag, 1991), especially chapters 5 and 8-11.

[32] Ivor Grattan-Guinness, *The Development of the Foundations of Mathematical Analysis from Euler to Riemann* (Cambridge, MA: The MIT Press, 1970), chapter 6, and Joseph W. Dauben, *Georg Cantor: His Mathematics and Philosophy of the Infinite* (Cambridge, MA: Harvard University Press, 1979; reprint ed., Princeton: University Press, 1990), chapter 1.

[33] Pierre Dugac, "Éléments d'analyse de Karl Weierstraß," *Archive for History of Exact Sciences* **10** (1973):41-176.

[34] Hermann Felix Müller, "Karl Schellbach: Rückblick auf sein wissenschaftliches Leben," *Abhandlungen zur Geschichte der mathematischen Wissenchaften* **20** (1905):1-86.

may well be quite messy, with an author or teacher maybe adopting a mixture of the three approaches to his own satisfaction. This seems to have been the case, for example, in England, not a major country for mathematical analysis but with an enthusiasm for algebras that granted a good place to differential operators, a theory that took its parentage from Lagrange's approach.[35] An early and significant book was De Morgan's *The Differential and Integral Calculus* (1836–1842). In Cauchyesque style, he began with an outline of the theory of limits and gave the basic definitions, but he made no mention of Cauchy in these places, and he even used Euler's name "differential coefficient" for the derivative. He also treated algebraic topics such as "divergent developments" of infinite series, which Cauchy had declared to be illegitimate.[36] He did not treat limit theory itself, since he regarded it as part of algebra and so had already presented it in a textbook on the subject.[37] An incomplete reading of contemporary German textbooks suggests that similar partial and varying selections from the new doctrine were also adopted there. A full multinational study of the reception of Cauchy's mathematical analysis before the epoch of Weierstraß would form a fine example of the diffusion abroad of an important French theory offered in competition with theories already well established.

Case Study 2: Complex-Variable Analysis

This case study differs from its predecessor in an interesting way. Instead of a new French theory competing with well-established alternatives, new German theories were proposed to compete with a moderately well-established French original (and with each other). In a seminal paper of 1814, Cauchy had aimed to regularize the uncritical use of $\sqrt{-1}$ in evaluating definite integrals by Euler, Laplace, Legendre, and others by offering criteria for the legitimacy or otherwise of the procedure. At this stage, the variable was complex, but the limits on the integral were real, and Cauchy developed his theory in close imitation of integral-variable calculus as much as possible: limits, continuity of functions, and the integral itself. But the analogy was put under severe test from the mid-1820s when he also allowed the limits of the integral to be complex, so that the variable could traverse various paths from one complex value A to another one B. We avoid mention of the complex plane, as Cauchy did: his desire for rigor demoted geometry from mathematical analysis, both complex and real. However, he talked about "neighboring curves" between the points corresponding to A and B, in a naïve and unclear invocation of topology. By 1830, he had rethought the notion of two different paths between A and B into a closed contour in effect from A to A via B, and the residue calculus was more prominent. But then the disruption of 1830 greatly delayed publication and maybe even the development of his theory. Only in the 1840s did he admit the complex plane and truly think of such curves as contours (a move that Sturm and Liouville had already taken independently).

The analogy between Cauchy's real- and complex-variable calculi was thus reduced, and their histories are very different. Cauchy's paper of 1814 had been published (with addenda) only in 1827, and the later stages appeared between the early

[35] Maria Panteki, "Relationships Between Algebra, Differential Equations and Logic in England: 1800–1860" (London: C.N.A.A. doctoral dissertation, 1992), chapters 2-5.

[36] Augustus De Morgan, *The Differential and Integral Calculus* (London: Taylor and Walton, 1836–1842).

[37] Augustus De Morgan, *Elements of Algebra*, 1st ed. (London: Taylor and Walton, 1835).

1820s and the mid-1840s in one short book, various papers, and some lithographs. He neither taught the theory at the *École polytechnique* (thank heavens), nor wrote a textbook on the subject.[38]

So far so French, with Cauchy's as the only *theory* of complex variables. Beginning in the 1850s, however, two alternative approaches emerged: a weird and wonderful vision from Bernhard Riemann of surfaces with cuts in them and with Cauchy's criterion of 1814 for definite integrals elevated into a main principle as the "Cauchy-Riemann equations," and a severely analytical approach due to Weierstraß that was based upon power series and analytic continuation. Riemann had also envisaged the latter notion, but otherwise his theory was fundamentally different. Weierstraß called his play with surfaces a "geometric fantasy," but had he lived Riemann might have retorted that Weierstraß worked in an analytical strait jacket, especially given the integral's minimal role.

Now three approaches were available, one French and two German, a trio as in real-variable theory and, again, explicitly in competition. It would appear that the German invasion was successful—even in France, to some extent, and certainly elsewhere—and that Cauchy's approach became somewhat subordinated to a source of (important) theorems principally about residues.[39] As for competition over foundations, some authors used the three approaches quite pragmatically. One example (albeit not a very impressive one mathematically) was a large treatise by the Englishman Andrew Russell Forsyth.[40] But the main activity had been taking place on the Continent, the story of which has yet to be told. The case of Italy is particularly intriguing, not only for the quality of its analysts but also because of the role of their personal contact with the convalescing Riemann in the 1860s. A complication is that before the proposals of Riemann and Weierstraß, complex variables were being used in the theory of elliptic functions, which was very much a German specialty from the 1830s onwards.[41]

Case Study 3: From Energy Mechanics to Energetics

In relying solely upon power series, Lagrange's approach to the (real-variable) calculus was an example of his desire to reduce mathematical theories of all kinds to algebraic form. He applied the same methodology to mechanics, gradually developing a version based upon the principles of virtual "velocities" and least action that he presented in detail in his *Méchanique analitique* (1788). There were no diagrams (a property of almost all his mathematics), as he famously announced in

[38]For the content and development of Cauchy's approach, see *Convolutions*, chapters 10-11 *passim*, and especially Frank Smithies, *Cauchy and the Creation of Complex Function Theory* (Cambridge: University Press, 1997). Cauchy's own literature is too complex (as it were) to list here.

[39]On aspects of the later history, see Ivan Yurevich Timchenko, "Osnovaniya teorii analyticheskikh' funkysii," *Zapiski matamaticheskago otdeleniya novorossiiskugo obshchestva estestvoistitatelei* **12** (1892):1-256; **16** (1899):1-216; and **19** (1899):i-xv, 1-183 (reprint ed., Odessa: n.p., 1899); Erwin Neuenschwander, "Studies in the History of Complex Function Theory II: Interactions Among the French School, Riemann, and Weierstraß," *Bulletin of the American Mathematical Society* n.s. **2** (1981):85-105; Jeanne Peiffer, "Joseph Liouville (1809–1882): Ses contributions à la théorie des fonctions d'une variable complexe," *Revue d'histoire des sciences* **36** (1983):209-248; and Bottazzini, *Higher Calculus*, chapters 4 and 6.

[40]Andrew Russell Forsyth, *Theory of Functions of a Complex Variable* (Cambridge: University Press, 1893).

[41]Christian Houzel, "Fonctions elliptiques et intégrales abéliennes," in *Abrégé d'histoire des mathématiques 1700–1900*, ed. Jean A. Dieudonné, 2 vols. (Paris: Hermann, 1978), 2:1-113.

the preface: diagrams and geometry could not avoid features of the *special* cases that were unavoidably pictured, while, by contrast, algebra guaranteed both rigor and generality. The main areas tackled included the dynamics of systems of point-masses under one or more centers of attraction, the rotation of continuous solid bodies, small oscillations, hydrodynamics, and some elasticity theory. The emphasis lay upon dynamics, although d'Alembert's principle was adopted in order to claim that any dynamical situation could be reduced to a statical one.[42]

So far so impressive, but there were limitations. One lay in the restricted capacity to generate *new* theories, although a brilliant counterexample was provided by Lagrange's collaboration in the late 1800s with a junior follower to produce the "Lagrange" and "Poisson" brackets theory of solutions to the Newtonian equations of motion.[43] Another concerned the restriction to equilibrate situations by the assumption that every work expression assumed a potential (to use later language). Even before the treatise appeared, it was challenged by Lazare Carnot. Drawing upon his engineering experience, Carnot realized that mechanics had to handle cases of impact and disequilibrium such as water hitting the blade of a waterwheel. He thus took up the energy approach and also adopted a markedly geometric style. In his *Essai sur les machines en général* (1783), he proposed, in quite a general way, that the loss of "live forces" on impact went into the product of force times distance.[44] This notion was called "work" in the 1820s by Coriolis when he, and also Navier and Poncelet, extended his insights in their respective teaching at the *École polytechnique*, *École des ponts et chaussées*, and Metz. Much research was done on water flow, the efficiency of water-wheels and turbines, modes of action of steam and hot air engines,[45] the operation of drawbridges, the design of iron bridges, properties of friction in large apparatus, and the work rate of men and animals. In all cases, notions corresponding to kinetic energy and/or to work were formulated and related.[46] A great French achievement, it was furthered by engineers in other countries.[47] At the same time, however, Lagrange's approach to mechanics also gained favor, and not only with Poisson, as is hinted by the translations of his treatise and the second edition of the 1810s.

The clash between the interests shown in the two columns of Table 1 is very stark here. The competing approaches to mechanics were incompatible, with one advocating equilibrium and conservation, the other impact and energy loss. The

---

[42]Joseph Louis Lagrange, *Méchanique analitique*, 1st ed. (Paris: Desaint, 1788).

[43]The details are described in *Convolutions*, pp. 371-386. On Lagrange's approach in general, see *op. cit.*, chapter 5, and on its gradual development in Lagrange, see Craig Fraser, "J. L. Lagrange's Early Contributions to the Principles and Methods of Mechanics," *Archive for History of Exact Sciences* **28** (1983):197-241.

[44]Wilson L. Scott, *The Conflict Between Atomism and Conservation Theory 1644 to 1860* (London: MacDonald and New York: Elsevier, 1970), book 2.

[45]The main thrust here is the eventual recognition of the proposals Lazare Carnot's son, Sadi, made in 1826. See Pietro Redondi, *L'accueil des idées de Sadi Carnot ...* (Paris: J. Vrin, 1980). The mathematical treatment by Clapeyron, "Mémoire sur la puissance motrice de la chaleur," *Journal de l'École polytechnique*, ser. 1, **14**, cah. 23 (1834):153-190 is one of the few papers with such content that was translated in *Taylor's Scientific Memoirs* (**1** (1837):347-376).

[46]*Convolutions*, chapter 16.

[47]The best approximation to a history of this story is in Moritz Rühlmann, *Vorträge zur Geschichte der theoretischen Maschinenlehre und der damit in Zusammenhang stehenden mathematischen Wissenschaften*, part 2 (Braunschweig: Schwetschke, 1885). An important aspect of the history of French engineering mechanics is treated in Bruno Belhoste *et al.*, *Le moteur hydraulique en France au XIXe siècle* (Paris: Belin, 1990).

reconciliation came later, mainly in the hands of physicists and drawing upon the newly developing mathematical physics. Work expressions *could* admit a potential, but now they encompassed terms referring to *all* kinds of physical effects and not just to mechanical ones. This was the science of "energetics," as it came to be known from mid-century, especially with William Thomson and Peter Guthrie Tait in Scotland and Hermann von Helmholtz in Germany. The local story has been told often,[48] but the mathematical background is much less well studied. Once again, the special feature here is the transmission of French initiatives to other countries. In particular, how did the Lagrangian and Carnotian approaches develop abroad after the 1820s? Only Hamilton's extension of Lagrange's basic equations has been examined in detail; the engineering side is much neglected. One factor is the role of potential theory, then a newly growing multinational subject, especially after Thomson publicized the remarkable contributions of George Green in the 1840s.[49] Here, the French contributions were *not* outstanding, although Mathieu ultimately became a distinguished figure in potential theory—and remained isolated in the provinces of his own country.

Case Study 4: Celestial Mechanics, Especially Perturbations

While Lagrange sought to further his own approach to mechanics, he used the other ones when appropriate. An important case concerned determining the orbits of planets, where he followed Newton's laws (which were theorems in his mechanics). A planet P was "perturbed" from its elliptical orbit around the Sun by the effect of inverse-square central forces acting from other planets and from satellites. These perturbations could, in principle, be large enough to send P out of the ecliptic or to move it in a quite different inclination. The passage of a comet nearby might also produce such disturbances. Newton was not worried, seeing an opportunity for God to restore equilibrium upon P. Protestant Euler held the same view; however, he also provided the mathematical means to develop an alternative position.

Perturbation theory was technically very cumbersome because inverse powers of distances in various directions and angular orientations were handled together. During the 1740s, Euler made a brilliant simplification in expressing these distances in polar coordinates, so that the sum of their appropriate powers could be rendered as power series of an angle pertaining to P and thence as trigonometric series in multiples of it.[50] In the 1770s, Lagrange introduced new independent variables that cast the equations of motion in an anti-symmetric form, thus greatly facilitating solution. In addition, by assuming that the planets moved in the same direction around the Sun, he reduced the stability problem for P to a form, which, in terms of matrix theory (the development of which was, indeed, partly inspired by this problem), required proof that all the latent roots of the matrix associated with the coefficients of the expansion for P were real as were their associated latent

---

[48]The better historical treatments are the older ones: Max Planck, *Das Princip der Erhaltung der Energie* (Leipzig: B. G. Teubner Verlag, 1887), pp. 1-91, and Artur Erich Haas, *Die Entwicklungsgeschichte des Satzes von der Erhaltung der Kraft* (Vienna: Hölder, 1909; Italian trans., Pavia: La Goliardica, 1990).

[49]Max Bacharach, *Abriß zur Geschichte der Potentialtheorie* (Würzburg: Thein, 1883).

[50]Leonhard Euler, "Recherches sur le mouvement des corps célestes en général," *Mémoires de l' Académie royale des sciences de Berlin* **3** (1747):93-143; reprinted in *Opera omnia*, ser. 1 (Leipzig and Berlin: B. G. Teubner Verlag, 1911–), 25:1-44. For commentary, see Curtis Wilson, "Perturbation and Solar Tables from Lacaille to Delambre," *Archive for History of Exact Sciences* **22** (1980):53-188 and 189-304.

vectors.[51] The theory was developed to the first degree in masses of the planets; the extension to the second degree led to the brackets theory mentioned earlier.

Laplace took up this approach, adapting it specifically for our planetary system.[52] Thereafter, the use of these trigonometric series expansions (not to be confused with Fourier series) became an obsession in French celestial mechanics, not only for the stability problem[53] but also for perturbation theory in general. Successors such as Poisson, the Baron de Damoiseau, Pontécoulant, and Urbain Leverrier used them routinely. The apotheosis is the theory of the Moon offered by Charles Delaunay in the 1860s, where in two volumes of the *Mémoires* of the *Académie des sciences*, he filled nearly 1,900 pages with expansions in hundreds of trigonometric terms of the lunar variables, followed by the examination of the order of magnitude of each.[54]

Magnifique, mais ce n'est pas le ciel. Already before 1800, German astronomers had been following a different methodology, preferring approximate but compact and feasible means of tackling problems in celestial mechanics: Wilhelm Olbers on cometary paths (1797), phenomena which were never easily conducive to analysis by series expansion;[55] Gauss with his book *Theoria motus corporum coelestium* (1809);[56] Bessel on secular variations (1810);[57] and Johann Franz Encke on the use of the least squares method (1822), for which he became a paladin.[58] This last case is particularly interesting for our theme. Priority over the method led in the 1810s to unpleasant conversations between Legendre, Gauss, and Laplace, the latter two bringing in probabilistic considerations.[59]

---

[51] Joseph Louis Lagrange, "Recherches sur les équations séculaires des mouvements des noeuds et des inclinaisons des orbites des planètes," *Mémoires de l' Académie des sciences* (1774–1778):97-174; reprinted in *Oeuvres*, ed. Joseph Serret, 14 vols. (Paris: Gauthier-Villars, 1867–1892), 6:635-709.

[52] See especially Pierre Simon Laplace, *Traité de mécanique céleste*, vol. 1 (Paris: Duprat, 1799), book 2, chapter 7.

[53] Thomas Hawkins, "Cauchy and the Spectral Theory of Matrices," *Historia Mathematica* **2** (1975):1-29.

[54] Charles Delaunay, "Théorie du mouvement de la Lune," *Mémoires de l'Académie des sciences* **28** (1860):xxviii + 883 pp., and **29** (1867):xi + 931 pp. There is no connected history of these developments, but much of the physical astronomy is nicely captured in Robert Grant, *History of Physical Astronomy, from the Earliest Ages to the Middle of the Nineteenth Century* (London: Boh, 1852; reprint ed., New York: Johnson Reprint Co., 1966), especially chapters 8-11.

[55] Wilhelm Olbers, *Abhandlung über die ... Methode die Bahn eines Kometen aus einigen Beobachtungen zu berechnen* (Weimar: Industrie Comptoire, 1797; reprint ed., 1847).

[56] Carl Friedrich Gauss, *Theoria motus corporum coelestium in sectionibus conicis solem ambientium* (Hamburg: Perthus and Besser, 1809); reprinted in *Werke*, ed. Königliche Gesellschaft der Wissenschaften zu Göttingen, 12 vols. (Göttingen: Dieterischen Universitätsdruckerei, 1863–1929), 7:1-288 (hereinafter cited as *Werke*). On the context of publication of this book, see Reich.

[57] Friedrich Wilhelm Bessel, "Entwicklung einer allgemeinen Methode, die Störungen der Kometen zu berechnen," in *Untersuchungen über die scheinbare und wahre Bahn des im Jahre 1807 erscheinenen grossen Cometen* (Königsberg: n.p., 1810), pp. 43-65 and 77-78; reprinted in *Abhandlungen von Friedrich Wilhelm Bessel*, ed. Rudolf Engelmann, 3 vols. (Leipzig: W. Englemann, 1875–1876), 1:20-27.

[58] Johann Franz Encke, "Über einen merkwürdigen Cometen ... ," *Astronomisches Jahrbuch* (1822):180-202; also in *Astronomische Nachrichten* **1** (1823), columns 371-376, 411-420, and 473-480. For commentary, see Eberhard Knobloch, "Historical Aspects of the Foundations of Error Theory," in *The Space of Mathematics*, ed. Javier Echeverria *et al.* (Berlin and New York: de Gruyter, 1992), pp. 253-279.

[59] On Gauss's contributions, see Oscar Sheynin, "The Discovery of the Principle of Least Squares," *Historia scientarum* **8** (1999):249-264.

During the succeeding decades, the French approach continued to be followed for various important topics such as perturbations and stability, but German methods seem to have grown in popularity not only with compatriots such as Peter Andreas Hansen but also, among others, Airy in England. A complicated story awaits the unraveling, but would be an especially fine example of clashes between "exact" and numerical applied mathematics, the latter spiced also with some statistics.[60] The issue of competition between French and non-French mathematics is sharpest here among our case studies, even more than with complex-variable analysis, since the philosophical issue of exactitude versus pragmatic approximation is also prominent. Both stories have a common factor, however. The French "lost."

Case Study 5: The Influence of Gauss

The controversy over the least squares method is an example where Gauss's work gained wide attention. Interestingly, this was not normally the case, and his major mathematical publications seem to have met with limited response until about 1840. He published regularly, including many shortish astronomical papers in the *Monatliche Correspondenz*. His often repeated phrase "*pauca sed matura*"[61] about his output is a conceit, although he did keep back much of his most innovative work.

Gauss's first substantial piece was the book, *Disquisitiones arithmeticæ*, which raised new standards in number theory.[62] It was soon even translated into French (itself a rare event).[63] The response, however, was slight, although Cauchy developed certain features (using either the original or the translation). An explanation for this neglect was alluded to above: number theory was then still recondite, attracting only very few mathematicians, although often the best. Similarly, Gauss's first three attempted proofs of the fundamental theorem of algebra (1797, 1816, 1816) also did not encourage many others, perhaps because of the difficulties he exposed.

The lack of reaction to Gauss's other publications is harder to understand. His book on orbits was soon warmly reviewed at length by Delambre,[64] but the French neither much used his methods nor limited their passion for trigonometric series. Again, a paper on the hypergeometric series, published by the Göttingen Academy, prefigured some features of Cauchy's real-variable analysis in its handling of the

---

[60]The literature on this competition seems to be remarkably sparse in modern writings. The best single source is various articles in volume 6, part 2 of the *Encyklopädie der mathematischen Wissenschaften* (1905–1934). A fine account of physical astronomy up to the time of writing is given in Grant, *History of Physical Astronomy*; the mathematics is at best paraphrased, but much can be learned there. The French tradition up to the 1840s is treated in *Convolutions*, chapters 17-18 *passim*. Among later treatises that sum up further developments, Felix Tisserand, *Traité de mécanique céleste*, 4 vols. (Paris: Gauthier-Villars, 1889–1896) is authoritative, although naturally somewhat guided by its French heritage.

[61]"Few, but ripe."

[62]Carl Friedrich Gauss, *Disquisitiones arithmeticæ* (Leipzig: Fleischer, 1801); reprinted as *Werke*, vol.1.

[63]Carl Friedrich Gauss, *Recherches arithmétiques* (Paris: Courcier, 1807). The translator, A. C. M. Poullet de Lisle (1770–1849), is unusual as a graduate of the *École polytechnique* who made his career in the *Université* system. See Baldassare Boncompagni, "Intorno alla vita ed ai lavori di Antonio Carlo Marcellino Poullet-Delisle," *Bullettino di bibliografia e di storia delle scienze matematiche e fisiche* **6** (1883):670-678.

[64]Jean-Baptiste-Joseph Delambre, Review of Gauss's *Theoria motus*, *Connaissance des temps* (1812):344-394.

continuity of functions, the convergence of series, and continued fractions.[65] Yet Cauchy seems not to have known it; the first significant response came in Crelle's *Journal* from the young Ernst Kummer, who, in 1836, examined properties of the various summation functions that Gauss had found.[66] Similarly, the full significance of a major paper of 1828 by Gauss on differential geometry[67] was not grasped until Riemann generalized the theory in 1854 to study the intrinsicality of manifolds,[68] although Liouville had published a French translation of the paper in 1850.[69]

There was then no doubt about Gauss's eminence as a mathematician, but he had kept much to himself and generated neither a group of students nor important lecture courses. The response to his work seems to have been limited until the 1840s, when some items were translated into French (usually from Latin) and when the international *Magnetisches Verein* that he led with Wilhelm Weber gave him a broader multinational audience.[70]

The conundrum of the reception of one of the greatest mathematicians of the nineteenth century differs from the previous four case studies in dealing with influence *upon* the French instead of the normal *vice versa* of the period. Since it also exemplifies the slowness with which important work sometimes diffused multinationally, and even in its home territory, this case study perhaps makes a fitting end to this survey of historical tasks still at hand.

Concluding Remarks

The quintet of case studies shows various ways in which multinational diffusion and interaction did (not) occur in mathematics in the period 1820–1870, when the French steadily declined from their dominating position: the adoption of French theories, but maybe only in parts; the proposal of alternative theories, set in competition with French forebears; and the fusion of French and non-French theories into "super-theories" such as energetics with potential theory. Stories similar to these exist in other contemporary developments. Doubtless others await detection, as do cases where the ways described here were *not* followed. In one such type, a mathematical theory was (partly) introduced in more minor countries and then diffused in the major ones, among which France need *not* be given any special attention. The tardy adoption of non-Euclidean geometries is a well-known example

---

[65] Carl Friedrich Gauss, "Disquisitiones generales circa seriam infinitam . . . ," *Commentarii Societatis Regiae Scientiae Göttingen*, Mathematische Klasse, **2** (1811-1813), 46 pp.; reprinted in *Werke*, 3:123-162.

[66] Ernst Kummer, "Über die hypergeometrische Reihe . . . ," *Journal für die reine und angewandte Mathematik* **15** (1836):39-83 and 127-172; reprinted in *Collected Works*, ed. André Weil, 2 vols. (Berlin: Springer-Verlag, 1975), 2:75-166.

[67] Carl Friedrich Gauss, "Disquisitiones generales circa superficies curvas," *Commentarii Societatis Regiae Scientiae Göttingen* **6** (1828):99-146; reprinted in *Werke*, 4:217-258.

[68] Bernhard Riemann, "Über die Hypothesen, welche der Geometrie zu Grunde liegen" (1854), in *Gesammelte mathematische Werke und wissenschaftliche Nachlaß*, ed. Richard Dedekind and Heinrich Weber, 2nd ed. (Leipzig: B. G. Teubner Verlag, 1892), pp. 272-287.

[69] See Gaspard Monge, *Applications de l'analyse à la géométrie*, 5th ed., ed. Joseph Liouville (Paris: Bachelier, 1850), pp. 505-546.

[70] While there are many useful articles on Gauss's work and several biographies, none handles the chronology of his influence in detail. The best available single source is the "Materialien" published in the early decades of the twentieth century as a supplement to volume 10 of his *Werke*. Also useful is Klein's survey of mathematics in the nineteenth century, *Vorlesungen über die Entwicklung der Mathematik im 19. Jahrhundert*, pt. 1 (Berlin: Springer-Verlag, 1926; reprint ed., New York: Chelsea Publishing Co., n.d.).

where the French played second to the Germans in role and to the British and Italians in attention.[71]

These are concluding remarks without a conclusion. The main purpose of this chapter has been to propose questions in an area of nineteenth-century mathematics which, despite much fruitful work in recent decades, often remains obscure and rarely studied: the diffusion elsewhere of French works, as summarized in the chapter's first half and well exemplified in the appendix of translations, most of which are forgotten. The thrust has been in information, with the only conclusion being that a large body of material has hitherto received but a small amount of analysis and that our understanding of the mathematics in the nineteenth century would be much enhanced by a detailed examination of the transition from the situation where one dominating nation gave way to an ensemble—in other words, the genesis of the evolution of an international mathematical community.

References

Ausejo, Elena and Hormigón, Mariano. "Mathematics for Independence: From Spanish Liberal Exile to the Young American Republics." *Historia Mathematica* **26** (1999):314-326.

Bacharach, Max. *Abriß zur Geschichte der Potentialtheorie.* Würzburg: Thein, 1883.

Belhoste, Bruno. *Augustin-Louis Cauchy: A Biography.* New York: Springer-Verlag, 1991.

Belhoste, Bruno, *et al.* *Le moteur hydraulique en France au XIXe siècle.* Paris: Belin, 1990.

Belhoste, Bruno and Picon, Antoine, Ed. *L'École d'application de l'artillerie et du génie de Metz (1802–1870): Enseignement et recherches. Actes de la journée d'étude de 2 novembre 1995.* Paris: Ministère de la culture, Direction du patrimoine, Musée des plans-reliefs, 1996.

Ben-David, Joseph. "The Rise and Decline of France as a Scientific Centre." *Minerva* **8** (1970):160-179.

Bessel, Friedrich Wilhelm. "Entwicklung einer allgemeinen Methode, die Störungen der Kometen zu berechnen." In *Untersuchungen über die scheinbare und wahre Bahn des im Jahre 1807 erscheinenen grossen Cometen.* Königsberg: N.p., 1810, pp. 43-65 and 77-78. Reprinted in *Abhandlungen von Friedrich Wilhelm Bessel.* Ed. Rudolf Engelmann. 3 Vols. Leipzig: W. Englemann, 1875–1876, 1:20-27.

Bonola, Roberto. "Index operum ad geometriam absolutam spectantium." In *Ioannis Bolyai in memoriam.* Budapest: University, 1902, pp. 83-154.

Boncompagni, Baldassare. "Intorno alla vita ed ai lavori di Antonio Carlo Marcellino Poullet-Delisle." *Bullettino di bibliografia e di storia delle scienze matematiche e fisiche* **6** (1883):670-678.

Bottazzini, Umberto. *The Higher Calculus: A History of Real and Complex Analysis from Euler to Weierstraß.* New York: Springer-Verlag, 1986.

Brock, William H. and Meadows, A. Jack. *The Lamp of Learning: Taylor and Francis and the Development of Science Publishing.* 2nd Ed. London: Taylor and Francis, 1998.

[71]See, for example, Roberto Bonola, "Index operum ad geometriam absolutam spectantium," in *Ioannis Bolyai in memoriam* (Budapest: University, 1902), pp. 83-154.

Clapeyron, Benoît. "Mémoire sur la puissance motrice de la chaleur." *Journal de l'École polytechnique*, Ser. 1. **14** Cah. 23 (1834):153-190. English trans. *Taylor's Scientific Memoirs* **1** (1837):347-376.

Crosland, Maurice P. *Science Under Control: The French Academy of Sciences.* Cambridge: University Press, 1992.

Dauben, Joseph W. *Georg Cantor: His Mathematics and Philosophy of the Infinite.* Cambridge, MA: Harvard University Press, 1979. Reprint Ed. Princeton: University Press, 1990.

Delambre, Jean-Baptiste-Joseph. Review of Gauss's *Theoria motus. Connaissance des temps* (1812):344-394.

Delaunay, Charles. "Théorie du mouvement de la Lune." *Mémoires de l'Académie des sciences* **28** (1860):xxviii + 883 pp., and **29** (1867):xi + 931 pp.

De Morgan, Augustus. *The Elements of Algebra.* London: Taylor and Walton, 1835.

———. *The Differential and Integral Calculus.* London: Taylor and Walton, 1836–1842.

Dugac, Pierre. "Éléments d'analyse de Karl Weierstraß." *Archive for History of Exact Sciences* **10** (1973):41-176.

Eccarius, Wolfgang. "August Leopold Crelle als Förderer bedeutender Mathematiker." *Jahresbericht der Deutschen Mathematiker-Vereinigung* **79** (1977): 137-174.

———. "August Leopold Crelle als Herausgeber wissenschaftlicher Fachzeitschriften." *Annals of Science* **33** (1976):229-261. Similar version in *Journal für die reine und angewandte Mathematik* **286-287** (1976):5-25.

Encke, Johann Franz. "Über einen merkwürdigen Cometen ... ." *Astronomisches Jahrbuch* (1822):180-202. Also in *Astronomische Nachrichten* **1** (1823), cols. 371-376, 411-420, and 473-480.

Euler, Leonhard. "Recherches sur le mouvement des corps célestes en général." *Mémoires de l'Académie royale des sciences de Berlin* **3** (1747):93-143. Reprinted in *Opera omnia.* Ser. 1. Leipzig and Berlin: B. G. Teubner Verlag, 1911–, 25:1-44.

Forsyth, Andrew Russell. *Theory of Functions of a Complex Variable.* Cambridge: University Press, 1893.

Fourcy, Ambroise. *Histoire de l'École polytechnique*, Paris: École polytechnique, 1828. Reprint Ed. Paris: Belin, 1987.

Fraser, Craig. "J. L. Lagrange's Early Contributions to the Principles and Methods of Mechanics." *Archive for History of Exact Sciences* **28** (1983):197-241.

Gauss, Carl Friedrich. *Disquisitiones arithmeticæ.* Leipzig: Fleischer, 1801. Reprinted in *Werke.* Ed. Königliche Gesellschaft der Wissenscaften zu Göttingen. 12 Vols. Göttingen: Dieterischen Universitätsdruckerei, 1863–1929. Vol. 1. *Recherches arithmétiques.* Trans. A. C. M. Poullet de Lisle. Paris: Courcier, 1807.

———. "Disquisitiones generales circa seriam infinitam ... ." *Commentarii Societatis Regiae Scientiae Göttingen.* Mathematische Klasse. **2** (1811–1813), 46 pp. Reprinted in *Werke*, 3:123-162.

———. "Disquisitiones generales circa superficies curvas." *Commentarii Societatis Regiae Scientiae Göttingen.* Mathematische Klasse. **6** (1828):99-146.

Reprinted in *Werke*, 4:217-258. French Trans. by Joseph Liouville in Gaspard Monge. *Applications de l'analyse à la géométrie*. 5th Ed. Paris: Bachelier, 1850, pp. 505-546.

———. *Theoria motus corporum coelestium in sectionibus conicis solem ambientium*. Hamburg: Perthus and Besser, 1809. Reprinted in *Werke*, 7:1-288.

Gispert, Hélène. "Sur la production mathématique française en 1870." *Archives internationales d'histoire des sciences* **35** (1985):380-399.

Grant, Robert. *History of Physical Astronomy, from the Earliest Ages to the Middle of the Nineteenth Century*. London: Boh, 1852. Reprint Ed. New York: Johnson Reprint Co., 1966.

Grattan-Guinness, Ivor. *Convolutions in French Mathematics, 1800–1840*. 3 Vols. Basel: Birkhäuser Verlag and Berlin: Deutscher Verlag der Wissenschaften, 1990.

———. *The Development of the Foundations of Mathematical Analysis from Euler to Riemann*. Cambridge, MA: The MIT Press, 1970.

———. "*Grandes écoles, petite université*: Some Puzzled Remarks on Higher Education in Mathematics in France, 1795–1840." *History of Universities* **7** (1988):197-225.

———. "The *ingénieur savant*, 1800–1830: A Neglected Figure in the History of French Mathematics and Science." *Science in Context* **6** (1993):405-433.

———. "Recent Researches in French Mathematical Physics of the Early 19th Century." *Annals of Science* **37** (1981):663-690.

Gruson, Johann Philip. "Allgemeine Methode mittelst bestimmter Integralien die durch den Lagrange'schen Lehrsatz gegebene Reihe zu summiren." *Abhandlungen der Akademie der Wissenchaften zu Berlin, Mathematische Klasse* (1812–1813):31-44.

———. "Über Reihen und vollständigen Integration ... ." *Abhandlungen der Akademie der Wissenchaften zu Berlin, Mathematische Klasse* (1812–1813): 23-30.

Haas, Artur Erich. *Die Entwicklungsgeschichte des Satzes von der Erhaltung der Kraft*. Vienna: Hölder, 1909. Italian Trans. Pavia: La Goliardica, 1990.

Hawkins, Thomas. "Cauchy and the Spectral Theory of Matrices." *Historia Mathematica* **2** (1975):1-29.

Houzel, Christian. "Fonctions elliptiques et intégrales abéliennes." In *Abrégé d'histoire des mathématiques 1700–1900*. Ed. Jean A. Dieudonné. 2 Vols. Paris: Hermann, 1978, 2:1-113.

Hulin-Jung, Nicole. *L'organisation de l'enseignement des sciences: La voie ouverte par le second Empire*. Paris: Comité des travaux historiques et scientifiques, 1989.

Klein, Felix. *Vorlesungen über die Entwicklung der Mathematik im 19. Jahrhundert*. Pt. 1. Berlin: Springer-Verlag, 1926. Reprint Ed. New York: Chelsea Publishing Co., n.d.

Knobloch, Eberhard. "Historical Aspects of the Foundations of Error Theory." In *The Space of Mathematics*. Ed. Javier Echeverria *et al.* Berlin and New York: de Gruyter, 1992, pp. 253-279.

Kummer, Ernst. "Über die hypergeometrische Reihe ... ." *Journal für die reine und angewandte Mathematik* **15** (1836):39-83 and 127-172. Reprinted in *Collected Works*. Ed. André Weil. 2 Vols. Berlin: Spinger-Verlag, 1975, 2:75-166.

Lacroix, Sylvestre François. *An Elementary Treatise on the Differential and Integral Calculus.* Cambridge: Deighton, 1816.

Lagrange, Joseph Louis. *Celestial Mechanics.* Trans. Nathaniel Bowditch (with Notes and Appendices). 4 Vols. Boston: Hillard, Gray, Little, and Wilkins, 1829–1839. Reprint Ed. New York: Chelsea Publishing Co., 1969.

———. *Méchanique analitique.* 1st Ed. Paris: Desaint, 1788.

———. "Recherches sur les équations séculaires des mouvements des noeuds et des inclinaisons des orbites des planètes." *Mémoires de l'Académie des sciences* (1774–1778):97-174. Reprinted in *Oeuvres.* Ed. Joseph Serret. 14 Vols. Paris: Gauthier-Villars, 1867–1892, 6:635-709.

———. *Traité de mécanique céleste.* Vol. 1. Paris: Duprat, 1799.

Monge, Gaspard. *Applications de l'analyse à la géométrie.* 5th Ed. Ed. Joseph Liouville. Paris: Bachelier, 1850.

Müller, Hermann Felix. "Das Jahrbuch über die Fortschritte der Mathematik 1869–1904." *Bibliotheca mathematica* (3) **5** (1905):292-297.

———. "Karl Schellbach: Rückblick auf sein wissenschaftliches Leben." *Abhandlungen zur Geschichte der mathematischen Wissenchaften* **20** (1905):1-86.

Neuenschwander, Erwin. "Studies in the History of Complex Function Theory II: Interactions Among the French School, Riemann, and Weierstraß." *Bulletin of the American Mathematical Society.* New Ser. **2** (1981):85-105.

Olbers, Wilhelm. *Abhandlung über die ... Methode die Bahn eines Kometen aus einigen Beobachtungen zu berechnen.* Weimar: Industrie Comptoire, 1797.

Outram, Dorinda. "Politics and Vocation: French Science 1793–1830." *British Journal for the History of Science* **13** (1980):27-43.

Panteki, Maria. "Relationships Between Algebra, Differential Equations and Logic in England: 1800-1860." London: C.N.A.A. Doctoral Dissertation, 1992.

Parseval, Marc-Antoine. "Mémoire sur les séries et sur l'intégration complète d'une équation aux différences partielles du second ordre, à coefficiens constans." *Mémoires présentés à l'Institut ... par divers savans, classe des sciences mathématiques et physiques* (1) **1** (1806):638-648.

———. "Méthode générale pour sommer ... la suite donnée par le théorème de M. Lagrange ... ." *Mémoires présentés à l'Institut ... par divers savans, classe des sciences mathématiques et physiques* (1) **1** (1806):567-586.

Parshall, Karen Hunger and Rowe, David E. *The Emergence of the American Mathematical Research Community, 1876–1900: J. J. Sylvester, Felix Klein, and E. H. Moore.* HMATH. Vol. 8. Providence: American Mathematical Society and London: London Mathematical Society, 1994.

Peiffer, Jeanne. "Joseph Liouville (1809–1882): Ses contributions à la théorie des fonctions d'une variable complexe." *Revue d'histoire des sciences* **36** (1983):209-248.

Phili, Christine. "La reconstruction des mathématiques en Grèce: L'apport de Ioannis Carandinos (1784–1834)." In *L'Europe mathématique–Mathematical Europe.* Ed. Catherine Goldstein, Jeremy J. Gray, and Jim Ritter. Paris: Éditions de la Maison des sciences de l'homme, 1996, pp. 305-319.

Picon, Antoine. *L'invention de l'ingénieur moderne: L'École des ponts et chaussées 1747–1851.* Paris: Presses de l'École nationale des ponts et chausses, 1992.

Planck, Max. *Das Princip der Erhaltung der Energie.* Leipzig: B. G. Teubner Verlag, 1887.

Redondi, Pietro. *L'accueil des idées de Sadi Carnot ...* . Paris: J. Vrin, 1980.

Reich, Karin. *Im Umfeld der "Theoria motus"*. Göttingen: Vandenhoeck & Ruprecht, 2001.

Riemann, Bernhard. "Über die Hypothesen, welche der Geometrie zu Grunde liegen" (1854). In *Gesammelte mathematische Werke und wissenschaftliche Nachlaß*. Ed. Richard Dedekind and Heinrich Weber. 2nd Ed. Leipzig: B. G. Teubner Verlag, 1892, pp. 272-287.

Rühlmann, Moritz. *Vorträge zur Geschichte der theoretischen Maschinenlehre und der damit in Zusammenhang stehenden mathematischen Wissenschaften*. Part 2. Braunschweig: Schwetschke, 1885.

Schneider, Ivo. "Christian Heinrich Schnuse als Übersetzer mathematischer, naturwissenschaftlicher und technischer Literatur" and "Bibliographie." *Aus dem Antiquariat* (1982):A205-A221 and A256-A261.

Schubring, Gert. *Die Enstehung des Mathematiklehrerberufs im 19. Jahrhundert: Studien und Materialien zum Prozeß der Professionalisierung in Preußen (1810–1870)*. 2nd Ed. Weinheim: Deutscher Studien Verlag, 1991.

______. "On the Methodology of Analysing Historical Textbooks: Lacroix As Textbook Author." *For the Learning of Mathematics* **7** (3) (1987):41-51.

Scott, Wilson L. *The Conflict Between Atomism and Conservation Theory 1644 to 1860*. London: MacDonald and New York: Elsevier, 1970.

Sheynin, Oscar. "The Discovery of the Principle of Least Squares." *Historia Scientarum* **8** (1999):249-264.

Shinn, Terry. "The French Science Faculty System, 1808–1914: Institutional Changes and Research Potential in Mathematics and the Physical Sciences." *Historical Studies in the Physical Sciences* **10** (1979):271-332.

Smithies, Frank. *Cauchy and the Creation of Complex Function Theory*. Cambridge: University Press, 1997.

Taton, René. "Les mathématiques dans le 'Bulletin de Ferrusac'." *Archives internationales d'histoire des sciences* **26** (1947):100-125.

Timchenko, Ivan Yurevich. "Osnovaniya teorii analyticheskikh' funkysii." *Zapiski matamaticheskago otdeleniya novorossiiskugo obshchestva estestvoistitatelei* **12** (1892):1-256; **16** (1899):1-216; and **19** (1899):i-xv and 1-183. Reprint Ed. Odessa: N.p. 1899.

Tisserand, Felix. *Traité de mécanique céleste*. 4 Vols. Paris: Gauthier-Villars, 1889–1896.

Topham, Jonathan. "Two Centuries of Cambridge Publishing and Bookselling: A Brief History of Deighton, Bell and Co., 1778-1998, with a Checklist of the Archive." *Transactions of the Cambridge Bibliographical Society* **11** (1998):349-403.

Vea, Fernando. "On the Influence of French Mathematics Textbooks on the Establishment of the Liberal Education System in Spain (1845-1868)." In *Paradigms in Mathematics*. Ed. Elena Ausejo and Mariano Hormigón. Zaragoza: Siglo XXI de España Editores, 1996, pp. 365-390.

Wilson, Curtis. "Perturbation and Solar Tables from Lacaille to Delambre." *Archive for History of Exact Sciences* **22** (1980):53-188 and 189-304.

Wölffing, Ernst. "Mathematischer Bücherschatz: 1. Teil [and only] reine Mathematik." *Abhandlungen zur Geschichte der mathematischen Wissenchaften* **20** (1903):xxxvi and 416 pp.

Appendix

Major French Books and Their Translations, 1780s–1900[72]

General (Collections of) Textbooks

Bossut, *Cours de mathématiques.* 1st ed. 1782—5th ed. 1800. Italian (part): 1802–1803.

Francoeur, *Cours complet.* 1st ed. 1809—4th ed. 1837. German: 1815 (Denmark), 1838 (Switzerland). English: 1829–1830. Italian: 1827, 1840–1841, 1852. Spanish: 1853–1855. Russian: ⊙.

Lacroix, *Cours complet.* 1802+. [Aggregation of books, much translated individually.] German: 1814. Spanish (part): 1826.

Bézout (ed. Peyrard), *Cours ... marine et de l'artillerie.* 1st ed. 1798—4th ed. 1808. German (part): 1820.

Puissant *et al.*, *Cours ... écoles ... militaires.* 1st ed. 1809—3rd ed. 1853.

Algebra

Fourier, *Analyse des équations.* 1831†. German: 1846 (Schnuse).

Lacroix, *Complément des élémens d'algèbre.* 1st ed. 1799—5th ed. 1815. German: 1804, 1811.

Lagrange, *Résolution des équations.* 1st ed. 1798—3rd ed. 1826. German: 1824 (Crelle).

Legendre, *Essai ... nombres.* 1st ed. 1798—3rd ed. 1830. German (part): 1829, 1886.

Sturm, "Résolution des équations" [paper]. 1835. English: 1835*.

Probability

Bicquilley, *Calcul des probabilités.* 1783. German: 1788.

Condorcet, *Calcul des probabilités.* 1805†.

Cournot, *Des chances et des probabilités.* 1843. German: 1849 (Schnuse).

Lacroix, *Traité élémentaire des probabilités.* 1st ed. 1816—4th ed. 1833. German: 1818.

Laplace, *Essai ... probabilités.* 1st ed. 1816—4th ed. 1819—6th ed. 1840†. German: 1819, 1886. Italian: 1820.

Laplace, *Théorie analytique des probabilités.* 1st ed. 1814–3rd ed. 1820.

Poisson, *Recherches ... probabilité ...* 1837. German: 1841 (Schnuse).

---

[72]In this appendix, the short title of the books listed has been used, with changes of title over editions usually ignored. Reprints are not recorded, although some "editions" are of this kind. Most lithograph editions and pirated Belgian printings have been excluded.

The edition on which translations were based is not recorded. It should not be assumed that a particular translation was made from the latest edition then available. Also, the dates given are not always unique.

The following abbreviations have been used: (supp) = supplement, † = posthumous, * = influential translation, ⊙ = edition seen mentioned, but details not retrieved.

Sources: Standard major catalogs (*Bibliothèque nationale*, British Library, Library Union Catalog, etc.); catalogs of various institutions and of booksellers; obituaries and biographies; J. C. Poggendorff, *Biographisch-literarisches Handwörterbuch für Mathematik, Astronomie, Physik mit Geophysik, Chemie, Kristallographie und verwandte Wissensgebiete* (Leipzig: J. A. Barth, 1863–1904); German data from *Gesamtverzeichnis der deutschsprachigen Schrifttums 1700–1910*, 161 volumes (Munich: Saur, 1979–1987).

Geometry

Adhémar, *Traité de la coupe des pierres.* 1st ed. 1837—5th ed. 1883. German: 1840.

Adhémar, *Traité de géométrie descriptive.* 1st ed. 1834—2nd ed. 1847. German: 1845, 1886.

Biot, *Géométrie analytique.* 1st ed. 1802—8th ed. 1834. German: 1817. American: 1840.

Carnot, *Géométrie de position.* 1803. German: 1808–1810.

Chasles, *Historique ... méthodes en géométrie.* 1837. German: 1839.

Chasles, *Traité de géométrie.* 1st ed. 1852—2nd ed. 1880. German [free]: 1856 (Schnuse).

Comte, *Géométrie analytique.* 1843.

Dupin [See under "Mechanics"].

Lacroix, *Élémens de géométrie.* 1st ed. 1799—9th ed. 1811—16th ed. 1848. German: 1828.

Lacroix, *Traité élémentaire de trigonométrie.* 1st ed. 1798—7th ed. 1822—9th ed. 1837. German: 1814, 1822, 1837.

Lefebure de Fourcy, *Traité de géométrie descriptive.* 1829.

Legendre, *Élémens de géométrie.* 1st ed. 1794—4th ed. 1802—8th ed. 1809—12th ed. 1823—14th ed. 1839. Italian: 1802, 1822, 1834, 1847, 1854, 1858, 1871, 1896. German: 1822 (Crelle). English: 1824. Greek: 1829.

Leroy, *Traité de géométrie descriptive.* 1st ed. 1834—5th ed. 1859—14th ed. 1896†. German: 1st ed. 1846—3rd ed. 1873. Italian: 1837, 1869.

Leroy, *Traité de stéréotomie.* 1st ed. 1844—5th ed. 1870—14th ed. 1910†. Italian: 1844. German: 1846.

Leroy, *Analyse appliquée à la géométrie.* 1st ed. 1829—4th ed. 1854. German: 1840. Italian: 1848.

Monge, *Feuilles* [then] *Applications d'analyse appliquée à la géométrie.* 1st ed. 1795–3rd ed. 1817—5th ed. 1850†.

Monge, *Géométrie descriptive.* 1st ed. 1795—4th ed. 1820†—7th ed. 1847. Italian (part): 1805. Spanish: 1803. English: 1809. German: 1828–1829. Italian: 1838.

Olivier, *Cours de géométrie descriptive.* 1843–1844. German: 1857.

Olivier, *Mémoires de géométrie descriptive.* 1851.

Poncelet, *Traité des propriétés projectives.* 1st ed. 1822—2nd ed. 1865–1866.

Poncelet, *Applications d'analyse et de géométrie ...* 1862–1864.

Calculus and Analysis

Bossut, *Traités de calcul.* 1798.

Boucharlat, *Éléments de calcul.* 1st ed. 1813—5th ed. 1838. German: 1823. English: 1828. Spanish: 1830.

Briot and Bouquet, *Fonctions doublement périodiques.* 1st ed. 1859—2nd ed. 1875. German: 1862.

Carnot, *Réflexions ... calcul.* 1st ed. 1797—2nd ed. 1813—5th ed. 1881†. Portuguese: 1798. German: 1800. English: 1800–1801, 1832. Italian: 1803.

Cauchy, *Cours d'analyse.* 1821. German: 1828, 1885. Russian: 1864.

Cauchy, *Résumé ... calcul.* 1823. Russian (part): 1831.

Cauchy, *Calcul différentiel.* 1829. German: 1836 (Schnuse).

Cauchy, *Calcul ... géométrie.* 1826–1828. German: 1840, 1846(supp) (Schnuse).
Cauchy, *Sulla meccanica celeste e sopra un nuovo calcolo chiamato calcolo dei limiti.* Milan: 1835; and *De metodi analitici.* Rome: 1843 [original seemingly not published].
Cournot, *Traité élémentaire des fonctions ... calcul.* 1st ed. 1841—2nd ed. 1857. German: 1845 (Schnuse).
Duhamel, *Cours d'analyse.* 1st ed. 1840–1841—4th ed. 1886–1887. German: 1856.
Duhamel, *Éléments de calcul.* 1st ed. 1856—3rd ed. 1874–1876.
Lacroix, *Traité de calcul.* 1st ed. 1797–1800—2nd ed. 1814–1819. German (part): 1799–1800.
Lacroix, *Traité élémentaire ... calcul.* 1st ed. 1802—4th ed. 1828—7th ed. 1861–1862†—9th ed. 1881. English: 1816*. German: 1817, 1830–1831. Polish: 1824.
Lagrange, *Fonctions analytiques.* 1st ed. 1797—2nd ed. 1813. Portuguese: 1798. German: 1798–1799, 1823 (Crelle).
Lagrange, *Leçons ... calcul des fonctions.* 1804–1806. German: 1823 (Crelle).
Legendre, *Exercices de calcul intégral.* 1811–1816.
Legendre, *Traité des fonctions elliptiques.* 1825–1832.
Moigno, *Leçons de calcul.* 1840–1861. German (part): 1846 (Schnuse).
Navier, *Résumé ... analyse.* 1st ed. 1840†—2nd ed. 1856. German: 1st ed. 1848–1849—3rd ed. 1865. Spanish: 1851.
Puiseux, "Recherches sur les fonctions algèbriques" [paper]. 1850. German: 1861.
Sturm, *Cours d'analyse.* 1st ed. 1857–1859†—3rd ed. 1868—10th ed. 1895—15th ed. 1929. German: 1897–1898.

Mechanics

Biot, *Notions élémentaires de statique.* 1829. German: 1846 (Schnuse).
Bossut, *Traité d'hydrodynamique.* 1787. German: 1791–1792.
Boucharlat, *Éléments de mécanique.* 1st ed. 1815—2nd ed. 1827. American: 1833. German: 1846.
Carnot, *Principes ... équilibre ... mouvement.* 1803. German: 1805.
Delauney, *Cours élémentaire de mécanique.* 1857. Dutch [free]: 1851–1854. Italian: 1860. German: 1868. Spanish: 1864.
de Prony, *Mécanique philosophique.* 1800.
de Prony, *Leçons de mécanique.* 1810–1815.
Duhamel, *Leçons de mécanique.* [1838]. German: 1853.
Duhamel, *Cours de mécanique.* 1st ed. 1845–1846—3rd ed. 1886–1887. German: 1st ed. 1853–1854—2nd ed. 1857–1858.
Dupin, *Géométrie et méchanique.* 1825–1826. Spanish: 1827–1834. Polish (part): 1829. Italian: 1829–1830. German (part): 1832–1835.
Francoeur, *Géodésie.* 1st ed. 1835—3rd ed. 1855†—8th ed. 1895.
Francoeur, *Goniométrie.* 1820. English: 1824. American: 1830.
Francoeur, *Mécanique élémentaire.* 1st ed. 1800—3rd ed. 1807—5th ed. 1825. Spanish: 1803. Portuguese: 1812. German: 1825, 1850.
Francoeur, *Élémens de statique.* 1810.
Lacroix, *Géographie mathématique.* 1st ed. 1808—2nd ed. 1811. German: 1827. Spanish: ⊙.

Lagrange, *Méchanique analitique.* 1st ed. 1788—2nd ed. 1811–1815†. German: 1797, 1887.

Moigno, *Leçons de mécanique.* 1868 [statics].

Monge, *Traité élémentaire de statique.* 1st ed. 1798—4th ed. 1801—6th ed. 1826†—8th ed. 1846. Russian: 1st ed. 1803—2nd ed. 1825. German: 1806. American: 1851.

Morin, *Aide-mémoire ... mécanique.* 1st ed. 1837—4th ed. 1843—6th ed. 1871. German: 1st ed. 1838—3rd ed. 1851. German [Free]: 1856 (Austria-Hungary). Italian: 1842, 1868.

Morin, *Notions ... mécanique.* 1st ed. 1846—2nd ed. 1855. American: 1860.

Navier, *Résumé ... mécanique.* 1841.

Poinsot, *Éléments de statique.* 1st ed. 1803—4th ed. 1824—8th ed. 1842—12th ed. 1877†. German: 1828, 1831, 1887. English: 1847. Danish: 1853. Dutch: 1856.

Poinsot, *Théorie ... rotation des corps.* 1834 [then in *Éléments* 8th–10th eds.]. English: 1834. German: 1851.

Poisson, *Traité de mécanique.* 1st ed. 1811—2nd ed. 1833. German: 1825–1826, 1835–1836, 1888. English: 1842.

Poisson, *Nouvelle ... action capillaire.* 1831. Italian: 1833.

Sturm, *Cours de mécanique.* 1st ed. 1861†—5th ed. 1905. Italian: 1871. German: 1899–1900.

Astronomy

Biot, *Astronomie physique.* 1st ed. 1805—3rd ed. 1841–1857. American (part): 1827. Spanish (part): 1847. English (part): 1850.

Delambre, *Abrégé d'astronomie.* 1813.

Delambre, *Astronomie théorique.* 1814.

Delauney, *Cours élémentaire d'astronomie.* 1st ed. 1853—2nd ed. 1858. Italian: 1860.

Francoeur, *Uranographie.* 1st ed. 1812—6th ed. 1852.

Laplace, *Théorie du mouvement ... des planètes.* 1784 [not in *Oeuvres*]. German: 1800.

Laplace, *Exposition ... système du monde.* 1st ed. 1796—5th ed. 1824. German: 1797–1798. English: 1809, 1830. Italian: 1823.

Laplace, *Mécanique céleste.* 1799–1804, 1823–1825. German (part): 1800–1802. English (part): 1814, 1822–1827. American (part): 1829–1839 (Bowditch)*.

Pontécoulant, *Théorie élémentaire ... système du monde.* 1st ed. 1829–1846—2nd ed. 1856–1860. German (part): 1834.

Pontécoulant, *Physique céleste.* 1840. German: 1846.

Engineering

Brisson, *Essai ... navigation ... France.* 1829.

Carnot, *Essai ... machines.* 1783.

Carnot, *Défense ... places fortes.* 1st ed. 1810—3rd ed. 1812, 1823(supp). German: 1811, 1814, 1822. English: 1814.

Combes, *Traité de l'exploitation des mines.* 1844. German: 1844–1846.

Coriolis, *Calcul ... effet des machines.* 1829.

Coriolis, *Traité ... calcul de l'effet des machines.* 1844†. German: 1846 (Schnuse).

Delambre, ed., *Base du système métrique décimal.* 1806–1810.
de Prony, *Nouvelle architecture hydraulique.* 1790–1796. German: 1794–1800.
de Prony, *Eaux courantes.* 1804. German: 1812.
Dupin, *Voyages dans l'Angleterre.* 1820–1824. English: 1821–1825. German (part): 1825–1826.
Français, *Poussée des terres.* 1820. German (part): 1828.
Francoeur, *Élémens de technologie.* 1st ed. 1833—2nd ed. 1842.
Gauthey/Navier, *Oeuvres.* 1809–1816–1832. [Bridges and canals.]
Girard, *Traité ... résistance des solides.* 1798. German: 1811.
Hachette, *Traité élémentaire des machines.* 1st ed. 1811—4th ed. 1828.
Morin, *Expériences sur les roues hydrauliques.* 1836–1838. Italian: 1840.
Morin, *Des machines ... eaux.* 1863.
Morin, *Hydraulique.* 1st ed. 1849—3rd ed. 1865.
Morin, *Machines à vapeur.* 1863.
Morin, *Résistance des matériaux.* 1st ed. 1853—3rd ed. 1862. Italian: 1854, 1867.
Navier, *Chemins de fer.* 1826. English: 1836. German: 1839.
Navier, *Rapport ... ponts suspendus.* 1st ed. 1823—2nd ed. 1830. German: 1829.
Navier, *Résumé ... mécanique ... machines.* 1st ed. 1826—2nd ed. 1833–1838. Italian: 1836. German: 1851. [See also Saint-Venant.]
Navier/Bélidor, *Science de l'ingénieur.* 1st ed. 1813—2nd ed. 1830.
Navier/Bélidor, *Architecture hydraulique.* 1819.
Olivier, *Théorie géométrique des engrenages.* 1842. German: 1844 (Schnuse).
Pambour, *Théorie ... machine à vapeur.* 1st ed. 1839—2nd ed. 1844. German: 1839 (Schnuse), 1849 (Crelle). American: 1839.
Pambour, *Théorie ... machines locomotives.* 1st ed. 1835—2nd ed. 1840. English: 1840. German: 1841 (Schnuse).
Poisson, *Formules ... effets du canon.* 1826.
Poncelet, *Cours de mécanique industrielle.* 1st ed. 1829—3rd ed. 1870†. German: 1840–1845.
Poncelet, *Cours de mécanique ... machines.* 1st ed. 1836—2nd ed. 1874–1876. German: 1845–1848 (Schnuse).
Poncelet, *Mémoire sur les roues hydrauliques.* 1st ed. 1825—2nd ed. 1827. Italian: 1838.
Poncelet, *Stabilité des revêtements.* 1840. American: 1841. German: 1844.
Puissant, *Traité de géodésie.* 1st ed. 1805—3rd ed. 1842.
Puissant, *Traité de topographie.* 1st ed. 1807, 1810 (supp)—2nd ed. 1820.
Saint-Venant/Navier, *De la résistance des solides.* 1864. [3rd Navier *Résumé.*]

Mathematical Physics

Biot, *Traité de physique.* 1816. German: 1st ed. 1824–1825—2nd ed. 1828–1829.
Biot, *Précis de physique.* 1st ed. 1817—3rd ed. 1824. Italian: 1818. German: 1818–1819. American (part): 1826.
Biot/Fischer, *Physique.* 1st ed. 1806—4th ed. 1830. American (part): 1826.
Briot, *Essai ... lumière.* 1864. German: 1867.
Briot, *Théorie ... chaleur.* 1st ed. 1869—2nd ed. 1883. German: 1871.
Cauchy, *Lumière.* 1836 [Lithograph, Prague]. German: 1842 (Austria-Hungary).
Demonferrand, *Manuel de l'électricité dynamique.* 1823. German: 1824. Italian: 1824. English: 1827.

Fourier, *Théorie ... chaleur*. 1822. English: 1878. German: 1884.

Haüy, *Traité élémentaire de physique*. 1st ed. 1803—3rd ed. 1821. German: 1804. English: 1807 ⊙.

Lamé, *Traité de physique*. 1st ed. 1836–1837—2nd ed. 1840. German: 1838–1841 (Schnuse).

Moigno, *Repertoire d'optique moderne*. 1847–1850.

Moigno, *Traité de télégraphie électrique*. 1st ed. 1849—2nd ed. 1852.

Poisson, *Théorie ... chaleur*. 1835, 1837 (supp).

Collected and Selected Works (All †)

Arago, *Oeuvres*. 1854–1862. German: 1854–1860.

Cauchy, *Oeuvres*. 1882–1974.

Fourier, *Oeuvres*. 1888–1890.

Fresnel, *Oeuvres*. 1866–1870.

Lagrange, *Oeuvres*. 1867–1892.

Laplace, *Oeuvres*. 1st ed. 1843–1847 [books only]–2nd ed. 1878–1912.

CHAPTER 3

Spanish Initiatives to Bring Mathematics in Spain into the International Mainstream

Elena Ausejo and Mariano Hormigón
University of Zaragoza (Spain)

The International Mainstream: A Problem of Definition

Historiographical terminologies and concepts are not always as well-defined as mathematical ones. Such is the case with the concept of *international mainstream.* This phrase, understood primarily as opposed to *periphery*, has become fashionable over the last few years in an effort to name, in a politically correct fashion, any human group on the crest of the wave of a given activity. Thus, in national terms, the international mainstream is determined—in mathematics and science as in economics or politics—by a limited number of leading countries, the standards of which other countries try to adopt and adapt in an effort to secure the benefits enjoyed by the leaders. Historians of mathematics thus have the task of evaluating the level of modernity of a given country, that is, they must determine if, in that country, there were sufficient numbers of individuals, groups, or schools producing and teaching mathematics at a sufficiently advanced level. To do this, they must isolate a socially recognized mathematical community that shares internal characteristics (actual mathematical activity), external characteristics (organization and lines of communication), and teleological characteristics with its foreign contemporaries and successfully detect a group of individuals whose average mathematical yield in terms of these characteristics warrants its inclusion among "the moderns."[1] By this standard, Spain is one of the large set of mathematically peripheral countries.

Much has been written about Spanish intellectual isolation after Philip II's edict in 1559 forbidding Spanish scholars to study abroad and calling for the censorship of foreign books.[2] The history of Spanish mathematics shows, however, that even under the harshest political circumstances, there were always individuals and even groups of scholars in Spain who were well-aware of what was going on in mathematics beyond Spanish borders, at least at the level of normal (in the Kuhnian sense) mathematics. Still, the process of Spain's active involvement in the international mathematical mainstream could only start when there was a Spanish mathematical community large enough to insure the beginning of institutionalization and professionalization. By the end of the nineteenth century, the efforts of leaders like

[1]Mariano Hormigón, "Paradigms and Mathematics," in *Paradigms and Mathematics*, ed. Elena Ausejo and Mariano Hormigón (Zaragoza: Siglo XXI de España Editors, 1996), pp. 2-113.

[2]See, for example, José Mª López Piñero, *Ciencia y técnica en la sociedad española de los siglos XVI y XVII* (Barcelona: Labor Universitaria, 1979).

Zoel García de Galdeano (1846–1924), José Echegaray (1833–1916), and Eduardo Torroja (1845–1918) had left the Spanish mathematical community poised to make this transition. Two institutions—the Spanish Mathematical Society (SME) and the Mathematical Laboratory and Seminar (LSM) of the Council of Research—and one journal—the *Revista matemática hispano-americana*—were key agents in this process, which came abruptly to an end in the wake of the last Spanish Civil War (1936–1939). This chapter documents how Spain, one of the many countries on the mathematical periphery, worked its way toward the international mathematical mainstream and to what extent it succeeded in breaking into it.

The Enlightenment

On acceding to the Spanish crown in 1700, the Bourbons found a scholastic university system with numerous minor universities but three major ones, Salamanca, Valladolid, and Alcalá. The colleges existing within the latter effectively governed them, controlling both academic posts and the chair system. These colleges also functioned as what have been termed "invisible colleges" within the centralized, state administration in that they constituted effective networks of influence.[3] Moreover, each university had at most four faculties—theology, law, medicine, and arts—with the first two playing the major role. Given this system, not only was there no place for science (modern or not) within the university system, but an openly hostile attitude toward science pervaded the university intellectual environment.[4]

The Bourbons did not attempt to reform this system until the end of the reign (1759–1788) of Charles III. In 1786, they issued the Royal Charter of Minister Campomanes, which aimed to put an end to the power of the colleges by placing the Spanish universities under state control. At the same time, the charter attempted to rationalize the teaching structure by systematizing degrees, curricula, and textbooks and by centralizing the appointment of chairs. The reform had the positive effect of ending the profusion of minor universities; they were left devoid of content since they were unable to comply with the minimum requirement that they grant official, professionally recognized, academic titles. On the whole, however, the reform failed, mainly due to the fierce corporate resistance of the major universities, but also due to a lack of allocation of the financial resources that would at least have permitted—if not guaranteed—its launching.

More successful were efforts made in institutions parallel to the universities: military schools that trained engineers and artillery men, among others; marine academies that focused on navigation; and schools for the education of the nobility such as the *Seminario de nobles* and the *Reales estudios de San Isidro*. It was in these sorts of institutions, as usually happens during periods of non-revolutionary change, that scientific rejuvenation progressed as the century advanced. Notable changes also took place in newly created institutions like the academies for language, for history, and for medicine and, above all, in the so-called Economic Societies of Friends of the Country (*Sociedades económicas de amigos del país*). The latter were institutional innovations that attempted to carry out in Spain the enlightened

[3]Derek de Solla Price introduced the idea of "invisible colleges" as an analytic tool in *Big Science, Little Science* (New York: Columbia University Press, 1963). See also Diana Crane, *Invisible Colleges: Diffusion of Knowledge in Scientific Communities* (Chicago: University of Chicago Press, 1972).

[4]Mariano Peset and José Luis Peset, *La universidad española (siglos XVIII y XIX)* (Madrid: Taurus, 1974).

ideal of transforming society through productivity. Part of the general wave of utilitarian progress that ran through eighteenth-century Europe, these Economic Societies were concerned with issues related to agriculture, industry, trade, and political economy. In their attempt to instruct artisans, they created schools where the level of development of scientific disciplines such as mathematics and chemistry was quite high.[5]

Also at the end of the eighteenth century, virtually all Spanish scholars reinforced their knowledge by taking at least one European tour. In science, this was reflected in an active policy that aimed to bring Spain in line with European standards through education abroad. The *pensiones* (grants) given enabled a considerable number of young scholars to travel outside of Spain for their education with the goal of enabling them to establish, on their return, institutions similar to those they experienced in other nations. One institution that resulted from this strategy was the Civil Engineering Corps created in 1799. It played a key role in the development of mathematics in Spain, especially through its publication of the Spanish translation of Gaspard Monge's influential text, *Géométrie descriptive*.[6]

These diverse initiatives are reflected in the mathematical production of Spain; almost one third (71) of the 203 mathematical works published between 1700 and 1809 appeared during the reign of Charles III and almost half (100) during the successive reigns (1759–1808) of Charles III and Charles IV. This time period also witnessed the full introduction of differential calculus[7] and, even more importantly, the codification of then-modern mathematical knowledge in Benito Bails's seminal work, *Elementos de matemáticas*, which was published in ten volumes between 1772 and 1783.[8]

Benito Bails (1730–1797), the most influential Spanish mathematician in the late eighteenth and early nineteenth centuries, was the director of the Mathematics Section of the *Real academia de nobles artes de San Fernando* in Madrid. This academy charged him to write the complete course on mathematics that ultimately appeared as his ten-volume work. Bails then published a synthetic version of his treatise in 1776 in three volumes.[9] In the first, he dealt with pure mathematics (arithmetic, geometry, and plane trigonometry), in the latter two with so-called mixed mathematics (dynamics, hydrodynamics, optics, astronomy, and the calendar in the second, and geography, gnomonics, architecture, perspective, and tables of logarithms in the third). As a mark of the success of this work, it came out in a second edition in 1788–1790, a third in 1797–1799, and a fourth in 1805–1816. The third edition was substantially changed, with the first two volumes devoted to

---

[5]See, for instance, Víctor Arenzana, *La enseñanza de las matemáticas en España en el siglo XVIII: La Escuela de matemáticas de la Real sociedad económica aragonesa de amigos del país* (Zaragoza: Universidad de Zaragoza, 1987).

[6]Gaspard Monge, *Geometría descriptiva: Lecciones dadas en las Escuelas normales en el año tercero de la República, por Gaspard Monge, del Instituto nacional (Traducidas al castellano para el uso de los estudios de la Inspección general de caminos)* (Madrid: Imprenta Real, 1803); facsimile ed. (Madrid: Colegio de ingenieros de caminos, canales y puertos, 1996).

[7]See Norberto Cuesta Dutari, *La invención del cálculo infinitesimal y su introducción en España* (Salamanca: Universidad de Salamanca, 1985).

[8]Benito Bails, *Elementos de matemáticas*, 10 vols. (Madrid: Vda. de Joaquín Ibarra, 1772–1783). For a synthetic approach to mathematics in eighteenth-century Spain, see Mariano Hormigón, "Las matemáticas en la Ilustración española: Su desarrollo en el reinado de Carlos III," in *Ciencia, técnica y estado en la España ilustrada*, ed. Joaquín Fernandez Pérez and Ignacio Gonzalez Tascón (Zaragoza: MEC/SEHCYT, 1990), pp. 265-278.

[9]Benito Bails, *Principios de matemáticas*, 3 vols. (Madrid: Vda. de Joaquín Ibarra, 1776).

pure mathematics (arithmetic, tables of logarithms, geometry, plane trigonometry, and an appendix on probability in the first, and algebra, differential and integral calculus, and spherical trigonometry in the second) and the third treating mixed mathematics (dynamics, hydrodynamics, optics, and Copernican astronomy). The former third volume, more specifically adapted to the practical needs of the students of the Academy, was never reprinted, which seems to indicate that the *Principios* was reaching a much wider audience.

All these initiatives point to France as Spain's main source of foreign influence, in accordance with both the new ruling dynasty and the general intellectual framework of the Enlightenment. France remained the primary model and source of inspiration for Spanish mathematics until the last quarter of the nineteenth century, despite Spain's war against Napoleon I (1808–1814) in the opening years of the century.

The Nineteenth Century

Throughout the nineteenth century, civil and military engineers, artillery men and naval officers, secondary school teachers, and university professors gradually formed a mathematical community around the corporate links inherent in the different corps of civil and military officials.[10] Thus, during the first half of the century, three main groups formed within the Spanish mathematical community—military men,[11] secondary school teachers,[12] and civil engineers—although there were also isolated mathematicians, like José Mariano Vallejo (1779–1846) and Jacinto Feliú (1787–1867), who fell outside these groups.

Among the military men, engineers, artillery men, and naval officers were especially important. Their work, with its interesting results in the field of differential calculus[13] and higher geometry,[14] represented a certain institutional involvement by their respective academies and observatories in mathematical activity. They translated and prepared textbooks; they made tentative forays into research; they took certain initiatives in the field of scientific periodicals.[15]

Spain, like virtually the rest of Europe as well as the United States, followed the model of Lagrange and his colleagues of the French revolutionary period, especially in mathematics and physics. The popularity of Gaspard Monge (1746–1818) in scientific circles, especially among engineers, was remarkable. In 1819, for example, Colonel Mariano Zorraquín wrote *Geometry Analytic and Descriptive for Use by the*

---

[10] Mariano Hormigón, "The Formation of the Spanish Mathematical Community," *Istorico-matematicheskie issledovania* (1997):22-55.

[11] Angeles Velamazán, *La enseñanza de las matemáticas en las academias militares en España en el siglo XIX* (Zaragoza: Universidad de Zaragoza, 1994).

[12] Fernando Vea, *Las matemáticas en la enseñanza secundaria en España en el siglo XIX* (Zaragoza: Universidad de Zaragoza, 1995).

[13] Angeles Velamazán and Elena Ausejo, "De Lagrange a Cauchy: El cálculo diferencial en las academias militares en España en el siglo XIX," *LLULL: Revista de la Sociedad española de historia de las ciencias y de las técnicas* **16** (1993):327-370.

[14] Angeles Velamazán, "Nuevos datos sobre los estudios de geometría superior en España en el siglo XIX: La aportación militar," *LLULL: Revista de la Sociedad española de historia de las ciencias y de las técnicas* **16** (1993):587-620.

[15] Elena Ausejo, "Le *Periódico mensual de ciencias matemáticas y físicas* (Cádiz, 1848), premier journal scientifique espagnol: La constitution d'une communauté?" *Rivista di storia della scienza* **3** (1995):55-66.

Military Engineering Academy and explicitly arranged it following Monge's text.[16] Zorraquín's most brilliant student, Fernando García San Pedro (1796–1854), was influenced by his mentor's work and wrote a thesis on kinetic geometry while still a student. García San Pedro was also the first military professor in Spain to publish a textbook on differential and integral calculus.[17] This work became a mainstay in calculus teaching, owing to its rigorous approach to the gaps in Lagrangian analysis.[18] Similarly, the artillery lieutenant, José de Odriozola (1786–1863), wrote a *Complete Course on Mathematics* that clearly reflected his indebtedness to the curriculum of the *École polytechnique* in Paris.[19]

The work of secondary school teachers was even more relevant to textbook translations than that of the military professors. In Spain, the organization of secondary education began in 1836. The creation of high schools (*institutos*) in the provincial capitals meant permanent posts for civil servants whose job was science teaching. This thus represented a quantitative contribution to the formation of the scientific community, since it brought with it the creation of a layer of professionals who earned their living through their knowledge—however superficial—of a scientific discipline. These teachers also took on the task of preparing textbooks, ever under the influence of the French.

Civil engineers also played a key role in the development of mathematics in nineteenth-century Spain through their High Technical Schools. The first and ephemeral foundation of the Civil Engineering School dates to 1802 and was the work of one of the most interesting characters in the history of Europe in the first third of the nineteenth century, Agustín de Betancourt (1758–1824). The newborn school was closed because of the war against Napoleon I and remained closed during Fernando VII's reign except for the short period of the so-called Liberal Three Years (1820–1823). In 1834, after the death of Fernando, the school was reorganized.[20] The Mining Engineering School was founded in 1835, the Forestry Engineering School in 1846, the Industrial Engineering School in 1851, and the Agronomist Engineering School in 1855, all of them following essentially French models. Civil engineers, as well as mining and industrial engineers and agricultural and forestry experts introduced scientific content—especially mathematics—into their curricula as a means for differentiating the professions inherited from the Old Regime and those corresponding to the modern liberal spirit. Engineers—especially civil engineers—wanted to distinguish themselves as key to the moderate development

---

[16]Mariano Zorraquín, *Geometría analítica-descriptiva* (Alcalá de Henares: Imprenta Manuel Amigo, 1819), p. xiii.

[17]Fernando García San Pedro, *Teoría algebraica elemental de las cantidades que varían por incrementos positivos o negativos de sus variables componentes, o sea cálculo diferencial e integral* (Madrid: Imprenta que fue de García, 1828).

[18]Velamazán and Ausejo, pp. 328-366.

[19]José de Odriozola, *Curso completo de matemáticas puras*, 4 vols. (Madrid: Imprenta García, 1827–1829); vol. 1, *Aritmética y algebra elemental* (Madrid: Imprenta García, 1827); vol. 2, *Geometría elemental y trigonometría* (Madrid: Imprenta García, 1827); vol. 3, *Algebra sublime y geometría analítica* (Madrid: Imprenta García, 1829); and vol. 4, *Cálculo diferencial e integral* (Madrid: Imprenta García, 1829). The first volume of this work went through three editions, the second five; the fourth volume came out in partial reprints in 1833, 1843, 1844, 1850, 1852, and 1855. Cauchy's derivative was introduced later at the Artillery Academy by Francisco Sanchiz y Castillo. See Francisco Sanchiz y Castillo, *Tratado de cálculo diferencial* (Segovia: Imprenta Baeza, 1851).

[20]Antonio Rumeu de Armas, *Ciencia y tecnología en la España ilustrada: La Escuela de caminos y canales* (Madrid: Turner, 1980).

of the state. They tried—successfully in many cases—to combine economic growth and the reform of economic and social structures with individual advantage and personal fortune. For this reason, the power of the different engineering bodies in Spain was relevant and significant; it assured—although in a somewhat biased way—the stable development of Spanish scientific culture. Thanks to the gradual infiltration of the various engineering bodies into all aspects of Spanish life, the number of people at least potentially educated in scientific questions increased as did the number of people progressively interested in scientific publications. Moreover, the proliferation of schools aimed at preparing students for the extremely challenging entrance examination to the engineering schools represented a complement, if not a *modus vivendi*, for many people whose main intellectual and professional skills centered on science in general and mathematics in particular. These developments represent an unquestionable milestone in the scientific history of Spain; for the first time, mathematics and science were key to practicing a civilian profession with growing social prestige.

Contemporaneous with these changes, the Faculty of Philosophy became a High Faculty, with a section devoted to science, in 1843. The Law of Public Education of 1857—known as the Moyano Law after Minister of Education Claudio Moyano (1809–1890)—founded the Faculties of Science, which conferred two different degrees, one in physics and mathematics, the other in physics and chemistry. The science degree, however, was not accompanied by viable professional outlets. As a result, the first university groups that modestly tried to join the European mainstream did not appear until the closing two decades of the nineteenth century, and then in teaching rather than in research.

University professors did not actually join the Spanish mathematical community until the second half of the nineteenth century. Little by little, research groups became stable in different university centers of the country (mainly Madrid, Barcelona, and Zaragoza);[21] the first Spanish mathematical journal, *El Progreso matemático* (founded in 1891), was published;[22] and Spanish mathematicians began to take part in International Congresses of Mathematicians.[23] Even so, the problems of professionalization persisted due to the lack of occupational opportunities for science graduates. They were limited to careers in secondary or university education and denied more lucrative positions in institutions like the Ministry of Public Works, the Madrid Observatory, and the Geographic Institute. In a similar spirit, engineers largely managed to bypass the Science Faculties, despite the fact that the Moyano Law called for them to complete at least one preparatory course in the Science Faculties before proceeding to the High Technical Schools of Engineers. Finally, mathematicians, in particular, came into competition for jobs with unemployed engineers when the State stopped guaranteeing employment for engineers in the wake of the liberal revolution of 1868. This environment, characterized by

[21] Mariano Hormigón, "Las matemáticas en España en el primer tercio del siglo XX," in *Ciencia y sociedad en España: De la Ilustración a la Guerra Civil*, ed. José Manuel Sánchez Ron (Madrid: Ediciones El Arquero/CSIC, 1988), pp. 253-282.

[22] Mariano Hormigón, "García de Galdeano and *El Progreso matemático*," in *Messengers of Mathematics: European Mathematical Journals 1800–1946*, ed. Mariano Hormigón and Elena Ausejo (Zaragoza: Siglo XXI de España Editores, 1993), pp. 95-115.

[23] Mariano Hormigón, "El Affaire Cambridge: Nuevos datos sobre las matemáticas en España en el primer tercio del siglo XX," in *Actas del V Congreso de la Sociedad española de historia de las ciencias y de las técnicas*, ed. Manuel Valera and Carlos López Fernández, 3 vols. (Murcia: DM-PPU, 1991), 1:135-171.

the immense corporate weight of Spanish engineering, failed to be conducive to the professionalization of mathematics.[24]

The nineteenth century thus presented a complicated set of circumstances and tensions relative to the development of a mathematical research tradition. Spanish legislation on matters related to education was complicated; more than twenty-five study plans were in place in less than seventy-five years. This variety stemmed from the political instability resulting from the different liberal and conservative viewpoints; the liberals promoted scientific subjects, while the conservatives favored the humanities and came under the influence of the academic lobbies. Nevertheless, a Spanish mathematical community at the teaching level, if not at the research level, came into existence during the course of the century. This community drew its inspiration from France until the last quarter of the nineteenth century when it shifted its focus to Germany. This new orientation was largely the result of a private initiative, the *Institución libre de enseñanza* (Free Teaching Institution) (ILE),[25] as well as the personal initiatives of a number of key leaders. It affected virtually all of the professions in Spain.

The Role of Individuals in History

In 1919, the Italian mathematician, Gino Loria, reported on mathematics in Spain and singled out the three Spanish mathematicians—Echegaray, Torroja, and Galdeano—as the agents of mathematical modernization in Spain.[26] José Echegaray y Eizaguirre, an engineer, a politician, a writer, and a mathematician, was perhaps the most singular contributor relative to bringing mathematics in Spain into the international mainstream. He studied mathematics in Madrid and became a civil engineer. After a short period as a practicing engineer, he taught mathematics at the Civil Engineering School until 1868. Two years before, in 1866, he had become a member of the Royal Academy of Exact, Physical and Natural Sciences of Madrid and had given a famous lecture on the history of pure mathematics in Spain.

An intense political life as a staunch supporter of free trade followed. Echegaray became a member of Parliament with the 1868 Revolution and occupied several important posts in the monarchical administrations (1871–1873) of the six-year, revolutionary period (1868–1874), including that of minister on several occasions. He returned to public life again after General Pavía's *coup d'état* in 1874 and again in 1904. In between, he spent the period of the First Republic (1873–1874) in exile in Paris, became a member of the Spanish Academy in 1896, and received the Nobel Prize for Literature in 1904. He was appointed President of the Council of Public Education in that same year and served as the first President of the Mathematics Section of the *Junta para ampliación de estudios e investigaciones científicas* (Board

[24]Elena Ausejo and Angeles Velamazán, "Mathematics and Liberalism in 19th-century Spain," in *Paradigms and Mathematics*, ed. Elena Ausejo and Mariano Hormigón (Zaragoza: Siglo XXI de España Editores, 1996), pp. 237-264.

[25]The ILE followed the German philosopher, Karl Christian Friedrich Krause. The first German references in mathematics appear in secondary teaching with the translations of Richard Baltzer's *Elementos de matemáticas* between 1879 and 1881 and in university teaching with Karl von Staudt's geometry. See the next section.

[26]Gino Loria, "Le matematiche in Ispagna ieri ed oggio: I matematici moderni," *Scientia* **25** (1919):441-449.

of Scientific Research) (JAE) in 1907. He was also a founding member and the first President of the Spanish Mathematical Society.

Echegaray also wrote scientific articles for *El Imparcial*, the *Revista contemporánea*, *Ilustración española y americana*, the *Diario de la marina de la Habana*, *El Liberal*, and other newspapers and magazines, in addition to books and articles on mathematics and mathematical physics. In Spain, his position as a civil engineer, a successful playwright, a remarkable orator, and an influential moderate politician gave him corporate, professional, and social recognition shared by no one else of his day. In this sense, Echegaray was, for decades, the clearest reference of the taste for mathematics, the knowledge of which he valued highly. As he put it, "Time was when every cultured person knew Latin. Time will come—and it is not very far off—when every cultured person will have to know mathematics!"[27] His social prestige and influence meant that, in his day, no one in the field of science dared to make even the slightest criticism of him. (The situation in literature was different.) With historical perspective, however, it is clear that Echegaray, despite his unquestionable merits, was not very aware of the changes taking place in and already characterizing the new mathematics. This fact aside, his favorable position towards extending mathematics and his recommendations that mathematics be studied in order to practice the technical professions represented a vitalizing injection of optimism that helped consolidate the mathematical community in Spain.

Another significant personality, Eduardo Torroja Caballé, engendered a long line of scientists and technologists of outstanding presence and enormous relevance in the history of twentieth-century Spain, especially during Franco's regime. Torroja was a professor of geometry at the Central University of Madrid and introduced the work of Karl von Staudt into Spain. He became very influential both within and outside the mathematical community as head of the Madrid clan devoted to geometry—particularly projective geometry—of a rather obsolete style;[28] from 1876, when Torroja took over the second chair of geometry at the University of Madrid, the mathematics faculty evolved around him.

The third main character in the mathematical scene of the end of the nineteenth and into the beginning of the twentieth century was Zoel García de Galdeano. By way of mathematical didactics and criticism, García de Galdeano not only linked the perspectives of secondary and university-level education but also emphasized and supported research at both levels. The founder of the first Spanish mathematical periodical, he was the most internationally connected Spanish mathematician of his era and the perennial Spanish representative at international mathematical initiatives such as the international congresses. Due to his deep comprehension of then-current mathematical developments, García de Galdeano also played the role of the great importer of modern mathematics to Spain, devoting the decade of the

[27] "Hubo un tiempo en que toda persona culta sabía latín. ¡Tiempo llegará y no está muy lejano en que toda persona culta deba saber matemáticas!" José Echegaray, "La Escuela especial de ingenieros de caminos, canales y puertos y las ciencias matemáticas," *Revista de obras públicas* **44** (1897):2.

[28] Mariano Hormigón and Ana Millán, "Projective Geometry and Applications in the Second Half of the Nineteenth Century," *Archives internationales d'histoire des sciences* **42** (1992):269-289.

1880s to algebra,[29] the 1890s to geometry,[30] and the first decade of the twentieth century to mathematical analysis, differential equations, and a brief foray into the theory of numbers. In all of this work, the mathematics presented and produced achieved a level of modernity that was more than acceptable in comparison with that of the standard-bearing countries of Europe.[31]

Each of these three men made different contributions to bringing mathematics in Spain into the international mainstream. Echegaray worked for the social recognition both of Spain's unfavorable international position in mathematics and of the urgent need to import modern mathematics. Torroja marked the connection of Spanish mathematics to German sources especially in geometry. Finally, Galdeano, in addition to bringing modern mathematics to Spain, embodied the physical participation of Spanish mathematicians in the international mathematical community.

The First Third of the Twentieth Century

The twentieth century opened with the creation of two scientific institutions of capital importance in the process of bringing Spain in line with European standards: the Board of Scientific Research (JAE) (1907) and the Spanish Association for the Advancement of Science (AEPPC) (1908).[32] These, in turn, gave rise to the Mathematical Laboratory and Seminar (LSM) (1915) and the Spanish Mathematical Society (SME) (1911), respectively. The SME brought, despite the host of problems and tensions that marked its evolution during the first third of the twentieth century, both the stabilization of a centralizing forum that channeled professional relationships among Spanish mathematicians abroad and the consolidation of a specialized periodical, *Revista matemática hispano-americana*, as the most advanced representative of Spanish mathematical expression in the international context.[33] The LSM provided institutional recognition of research as a necessary and sufficient activity for the sociological definition of the professional category "mathematician." Moreover, the development of a line of work parallel to research in the field of secondary teacher training represented a very peculiar and idiosyncratic realization of the dual teaching and research imperative characteristic of the modern mathematical profession.[34] Still, while both of these initiatives were of fundamental importance in the institutionalization and professionalization of the Spanish mathematical community, they also reflected the problems associated with these processes.

---

[29]Mariano Hormigón, "García de Galdeano's Works on Algebra," *Historia Mathematica* **18** (1991):1-15.

[30]Mariano Hormigón, "García de Galdeano y la modernización de la geometría en España," *Dynamis* **3** (1983):199-229.

[31]Mariano Hormigón, "Una aproximación a la biografía científica de García de Galdeano," *El Basilisco* **16** (1984):38-47.

[32]Elena Ausejo, *Por la ciencia y por la patria: La institucionalización científica en España en el primer tercio del siglo XX* (Zaragoza: Siglo XXI de España Editores, 1993).

[33]Elena Ausejo and Ana Millán, "The Spanish Mathematical Society and Its Periodicals in the First Third of the 20th Century," in *Messengers of Mathematics: European Mathematical Journals 1800–1946*, ed. Mariano Hormigón and Elena Ausejo (Zaragoza: Siglo XXI de España Editores, 1993), pp. 159-187.

[34]Elena Ausejo and Ana Millán, "La organización de la investigación matemática en España en el primer tercio del siglo XX: El Laboratorio y Seminario matemático de la Junta para ampliación de estudios e investigaciones científicas (1915–1938)," *LLULL: Revista de la Sociedad española de historia de las ciencias y de las técnicas* **12** (1989):261-308.

The JAE was founded as the coordinating institution for research. It consisted of laboratories for the different disciplines and administered a generous grant policy in Spain and abroad with the goals of bringing Spanish research into line with European standards and of indirectly influencing curriculum renewal.[35] In fact, the opening decades of the twentieth century in Spanish culture and science were so positive that, following imperial traditions as well as the hierarchy of precious metals, it is usually referred to as the Silver Age of Spain.[36] Nevertheless, the JAE was an institution parallel to—and very often divergent from—the universities, which were not reformed. Young researchers from the JAE thus did not always have the chance to return the investment made in them, at least not at anything higher than the secondary level. As for the LSM, the strained relationship it had with the universities was a clear expression of the difficulties posed by the implementation of the research imperative in the long-established routine that prevailed mainly among the mathematical clan of the Central University of Madrid. The SME, on the other hand, actually focused its activity in Madrid, owing to a national centralist structure that reflected itself in the exclusive monopoly that the Central University of Madrid held (until 1954!) in granting the doctoral degree. This made the consolidation and development of research groups at provincial universities extremely difficult.

If belonging to the SME may be taken as a sign of attachment to the Spanish mathematical community, the Society's "List of Members" published in the sixth number of the first volume (February 1912) of the *Revista de la Sociedad matemática española*[37] enables a quantitative approach to the level of professionalization of this community, since it included reference to the professional situation of its members. The then recently created Society had at that time a total of 422 members distributed as follows: one protector member (the AEPPC, patron of the SME), one honorary member (Gomes Teixeira), one corresponding member (Henri Brocard), seven subscribing members, fifty-four institutional members, and 358 founding members. As regards the 358 founders, deducting the seven foreigners and the thirty whose profession is not given, the following professional profile emerges for the 89.7% of this block. (Compare Graph 1 and Table 1.)

These data reflect how the main historical characters defined themselves professionally. In round numbers, they may be reinterpreted as follows (see Graph 2 and Table 2): 38% teachers, 25% undecided, 18% technicians, and 9% miscellaneous (together with 10% foreigners and those of unknown profession). This classification, debatable in detail like almost all such classifications, nevertheless shows that, by the Prussian standards that came to define the standard internationally, the Spanish mathematical community had only reached a 56% level of professionalization in the most optimistic reading of the data, that is, by combining teachers and technicians, or a 38% level in a more restricted interpretation.

A second conclusion also emerges. Together with a discrete percentage of amateurs (9% "other"), a quarter of the society had no defined career trajectory. Clearly, the students, who represented an abnormally high 16% of the specialized scientific society, were not yet in a position to have chosen a career. More surprisingly, almost 9% of graduates and doctors of science were apparently either

[35] *1907–1987: La Junta para ampliación de estudios e investigaciones científicas 80 años después*, ed. José Manuel Sanchez Ron (Madrid: CSIC, 1988).

[36] Santiago Ramón y Cajal's Nobel Prize of 1908, the first for science awarded to a Spaniard, at least implied that something was improving.

[37] See *Revista de la Sociedad matemática española* **1** (1912):223-233.

GRAPH 1: SME FOUNDERS 1912

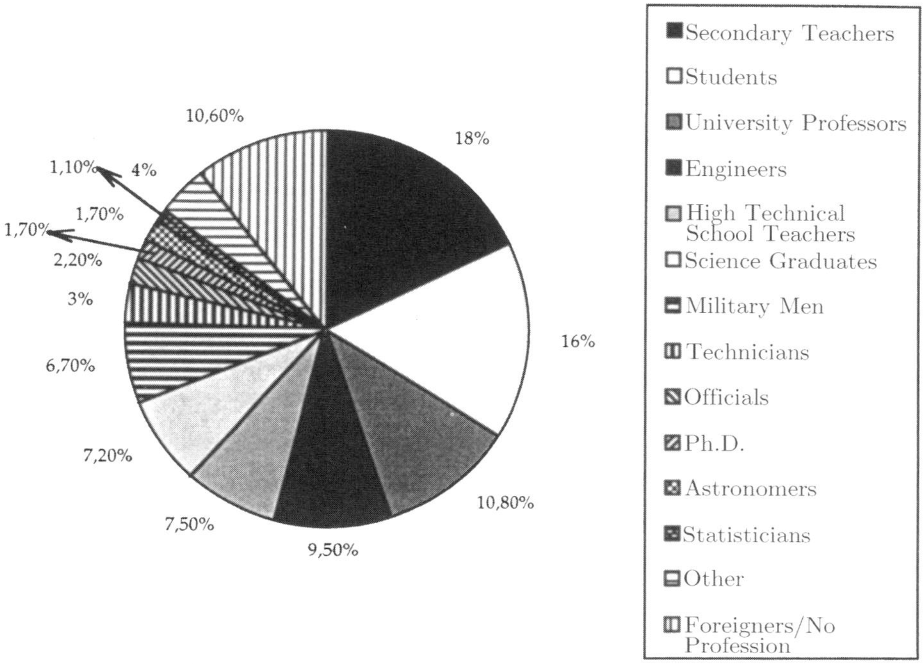

unemployed or employed in non-scientific jobs. Their association with the SME, however, suggests their interest in aligning themselves formally with science and mathematics.

TABLE 1: SME FOUNDERS 1912

Foreigners (7) / No profession (30)	37	10.3%
Secondary Teachers	65	18%
Students	57	16%
University Professors	39	10.8%
Engineers	34	9.5%
High Technical School Teachers	27	7.5%
Science graduates	26	7.2%
Military men	24	6.7%
Technicians	11	3%
Officials	8	2.2%
Ph.D.	6	1.7%
Astronomers	6	1.7%
Statisticians	4	1.1%
Other	14	4%
TOTAL	358	99.7%

Since France was the main source of mathematical influence in Spain throughout the nineteenth century, the evolution of the French mathematical community between 1874 and 1914 offers an appropriate benchmark for comparison of these quantitative data. Between 1874 and 1914, the number of teachers in the *Société mathématique de France* (SMF) increased from 49% to 72%, while engineers decreased from 46% to 23%. More specifically, university teachers increased from 10% to 33% of the SMF (from 20% to 47% of the teachers in the Society), while other categories remained constant in percentage.[38] Thus, the professional profile of the Spanish Mathematical Society in 1912 hardly approached that of the French Mathematical Society in 1874.

Another young mathematical society of that period, the American Mathematical Society (AMS), offers an example of the evolution of a mathematical community in a context of economic power and institutional support inconceivable in Spain. The AMS focused on research,[39] while in Spain this was the task of the LSM not the SME.[40] Moreover, whereas Germany served as the main source of influence for the AMS, it was Italian mathematics that defined a main point of reference for the LSM's modernization process of Spanish mathematics; the LSM sent research students abroad and engaged guest professors from different European universities. From 1915 to 1920, Spanish mathematical research focused on projective geometry, which still had an international network of researchers;[41] mathematical analysis, where Adolf Hurwitz, Karl Fueter, and George Polya in Zürich were among the international leaders; and nomography and numerical analysis, which reflected a timid introduction of the work of Carl Runge into Spain. From 1920 to 1930, the research wheel turned mainly around relativist mechanics and mathematical physics with Tullio Levi-Civita and Hermann Weyl as the primary foreign references.

GRAPH 2: SME FOUNDERS 1912

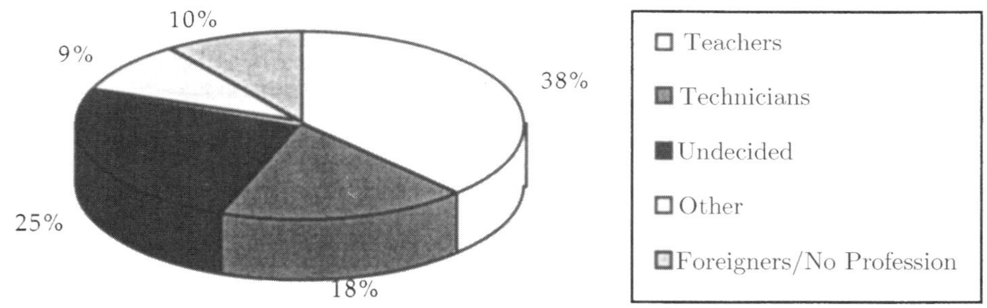

[38]Hélène Gispert, *La France mathématique: La Société mathématique de France (1870–1914)* (Paris: Cahiers d'histoire et de philosophie des sciences, Société mathématique de France, et Société française d'histoire des sciences et techniques, 1991), pp. 166-167.

[39]Karen Hunger Parshall and David E. Rowe, *The Emergence of the American Mathematical Research Community, 1876–1900: J. J. Sylvester, Felix Klein, and E. H. Moore*, HMATH, vol. 8 (Providence: American Mathematical Society and London: London Mathematical Society, 1994), pp. 418-419.

[40]Ausejo and Millán, "La organización de la investigación matemática en España."

[41]Hormigón and Millán, "Projective Geometry and Applications in the Second Half of the Nineteenth Century."

Table 2: SME Founders 1912		
Foreigners / No profession	37	10%
Teachers[i]	136	38%
Undecided[ii]	89	25%
Technicians[iii]	65	18%
Other[iv]	31	9%
TOTAL	358	100%

[i]Of these teachers, 65 were in secondary education, 39 taught at the university level, 27 were employed in high technical schools, 2 were school teachers, and 3 were mercantile teachers.
[ii]Of the "undecideds," 57 were students, 26 were science graduates, and 6 were Ph.D.s.
[iii]The technicians broke down this way: 34 engineers, 11 technicians, 8 officials, 6 astronomers, 4 statisticians, 1 seaman, and 1 employee of the Bank of Spain.
[iv]In the category of "other" were 24 military men, 2 presbyters, 2 pharmacists, 1 chemist, 1 mechanic, and 1 clockmaker.

In the analysis of the LSM and the Seminar of the JAE, some seventy-five individuals emerged as related in one way or another to the mathematical activity that developed throughout the twenty years from 1915 to 1935. Of these, thirty-six were candidates for positions in secondary teaching, that is, they were preparing to take the competitive examinations prefatory to work in secondary education; thirty-three were researchers; and ten, of whom four came from the group of researchers, were mentioned as research directors. Thirty-five of these seventy-five names appeared as authors of articles in Spanish mathematical periodicals. It can thus be said that the LSM researchers were producing Kuhn's sense of "normal" mathematics by international standards. However, no conclusions can be made regarding either their average yield or the modernity of their mathematical production. However, thirty-five is such a modest number in terms of human capital—especially when compared to the number of students, graduates, and doctors of science belonging to the SME in 1912—that the only possible conclusion is that mathematics was, during the first third of the twentieth century, a luxury that Spanish society could not afford. A true internationalization of mathematical knowledge may have been affected by similar economic and sociological circumstances in other countries.

Conclusion

Foreign mathematical influence on Spain shifted from France in the eighteenth century, to Germany in the last quarter of the nineteenth century, to Italy in the early twentieth century. Although there were always individuals and even groups of scholars in Spain who were perfectly aware of what was going on in mathematics outside Spanish borders, the establishment of a Spanish mathematical community—at the teaching level although not at the research level—took place throughout the nineteenth century. By 1900, the efforts of leaders such as Galdeano, Echegaray, and Torroja had resulted in the social recognition both of Spain's retrograde position in mathematics and of its urgent need to import modern mathematics. Their efforts had also spurred the participation of Spanish mathematicians in the international mathematical community. At the beginning of the twentieth century, the Spanish mathematical community was large enough to insure institutionalization, to begin the process of professionalization, and to encourage involvement in the international mainstream. The Spanish Mathematical Society and the Mathematical Laboratory and Seminar of the Board of Scientific Research were the two institutions involved in this development, but the low professional profile of the SME, and the small number of mathematicians engaged in full-time research reflected the kind of social and economic problems that peripheral countries encountered in their efforts to join the mathematical mainstream.

References

1907–1987: La Junta para ampliación de estudios e investigaciones científicas 80 años después. Ed. José Manuel Sánchez Ron. Madrid: CSIC, 1988.

Arenzana, Víctor. *La enseñanza de las matemáticas en España en el siglo XVIII: La Escuela de matemáticas de la Real sociedad económica aragonesa de amigos del país.* Zaragoza: Universidad de Zaragoza, 1987.

Ausejo, Elena. *Por la ciencia y por la patria: La institucionalización científica en España en el primer tercio del siglo XX.* Zaragoza: Siglo XXI de España Editores, 1993.

______. "Le *Periódico mensual de ciencias matemáticas y físicas* (Cádiz, 1848), premier journal scientifique espagnol: La constitution d'une communauté?" *Rivista di storia della scienza* **3** (1995):55-66.

Ausejo, Elena and Millán, Ana. "La organización de la investigación matemática en España en el primer tercio del siglo XX: El Laboratorio y Seminario matemático de la Junta para ampliación de estudios e investigaciones científicas (1915–1938)." *LLULL: Revista de la Sociedad española de historia de las ciencias y de las técnicas* **12** (1989):261-308.

______. "The Spanish Mathematical Society and Its Periodicals in the First Third of the 20th Century." In *Messengers of Mathematics: European Mathematical Journals 1800–1946.* Ed. Mariano Hormigón and Elena Ausejo. Zaragoza: Siglo XXI de España Editores, 1993, pp. 159-187.

Ausejo, Elena and Velamazán, Angeles. "Mathematics and Liberalism in 19th-century Spain." In *Paradigms and Mathematics.* Ed. Elena Ausejo and Mariano Hormigón. Zaragoza: Siglo XXI de España editores, 1996, pp. 237-264.

Bails, Benito. *Elementos de matemáticas.* 10 Vols. Madrid: Vda. de Joaquín Ibarra, 1772–1783.

______. *Principios de matemáticas.* 3 Vols. Madrid: Vda. de Joaquín Ibarra, 1776.

Cuesta Dutari, Norberto. *La invención del cálculo infinitesimal y su introducción en España.* Salamanca, Universidad de Salamanca, 1985.

De Solla Price, Derek J. *Big Science, Little Science.* New York: Columbia University Press, 1963.

Echegaray, José. "La Escuela especial de ingenieros de caminos, canales y puertos y las ciencias matemáticas." *Revista de obras públicas* **44** (1897):2.

García San Pedro, Fernando. *Teoría algebraica elemental de las cantidades que varían por incrementos positivos o negativos de sus variables componentes, o sea cálculo diferencial e integral.* Madrid: Imprenta que fue de García, 1828.

Gispert, Hélène. *La France mathématique: La Société mathématique de France (1870–1914).* Paris: Cahiers d'histoire et de philosophie des sciences, Société mathématique de France, et Société française d'histoire des sciences et techniques, 1991.

Hormigón, Mariano. "García de Galdeano y la modernización de la geometría en España." *Dynamis* **3** (1983):199-229.

———. "Una aproximación a la biografía científica de García de Galdeano." *El Basilisco* **16** (1984):38-47.

———. "Las matemáticas en España en el primer tercio del siglo XX." In *Ciencia y sociedad en España: De la Ilustración a la Guerra Civil.* Ed. José Manuel Sánchez Ron. Madrid: Ediciones El Arquero/CSIC, 1988, pp. 253-282.

———. "Las matemáticas en la Ilustración española: Su desarrollo en el reinado de Carlos III." In *Ciencia, técnica y estado en la España ilustrada.* Ed. Fernández Pérez and González Tascón. Zaragoza: MEC/SEHCYT, 1990, pp. 265-278.

———. "García de Galdeano's Works on Algebra." *Historia Mathematica* **18** (1991):1-15.

———. "El Affaire Cambridge: Nuevos datos sobre las matemáticas en España en el primer tercio del siglo XX." In *Actas del V Congreso de la Sociedad española de historia de las ciencias y de las técnicas.* Ed. Manuel Valera and Carlos López Fernández. 3 Vols. Murcia: DM-PPU, 1991, 1:135-171.

———. "García de Galdeano and *El Progreso matemático.*" In *Messengers of Mathematics: European Mathematical Journals 1800–1946.* Ed. Mariano Hormigón and Elena Ausejo. Zaragoza: Siglo XXI de España editores, 1993, pp. 95-115.

———. "Paradigms and Mathematics." In *Paradigms and Mathematics.* Ed. Elena Ausejo and Mariano Hormigón. Zaragoza: Siglo XXI de España editores, 1996, pp. 2-113.

———. "The Formation of the Spanish Mathematical Community." *Istorico-matematicheskie issledovania* (1997):22-55.

Hormigón, Mariano and Ausejo, Elena. *Messengers of Mathematics: European Mathematical Journals 1800–1946.* Zaragoza: Siglo XXI de España Editores, 1993

Hormigón, Mariano and Millán, Ana. "Projective Geometry and Applications in the Second Half of the Nineteenth Century." *Archives internationales d'histoire des sciences* **42** (1992):269-289.

López Piñero, José M^{a}. *Ciencia y técnica en la sociedad española de los siglos XVI y XVII.* Barcelona: Labor Universitaria, 1979.

Loria, Gino. "Le matematiche in Ispagna ieri ed oggio: I matematici moderni." *Scientia* **25** (1919):441-449.

Monge, Gaspard. *Geometría descriptiva: Lecciones dadas en las Escuelas normales en el año tercero de la República, por Gaspard Monge, del Instituto nacional. Traducidas al castellano para el uso de los estudios de la Inspección general de caminos.* Madrid: Imprenta Real, 1803; Facsimile Ed. Madrid: Colegio de ingenieros de caminos, canales y puertos, 1996.

Odriozola, José de. *Curso completo de matemáticas puras.* 4 Vols. Madrid: Imprenta García, 1827–1829.

Parshall, Karen Hunger and Rowe, David E. *The Emergence of the American Mathematical Research Community, 1876–1900: J. J. Sylvester, Felix Klein, and*

E. H. Moore. HMATH, Vol. 8. Providence: American Mathematical Society and London: London Mathematical Society, 1994.

Peset, Mariano and Peset, José Luis. *La universidad española (siglos XVIII y XIX)*. Madrid: Taurus, 1974.

Rumeu de Armas, Antonio. *Ciencia y tecnología en la España ilustrada: La Escuela de caminos y canales*. Madrid: Turner, 1980.

Sanchiz y Castillo, Francisco. *Tratado de cálculo diferencial*. Segovia: Imprenta Baeza, 1851.

Vea, Fernando. *Las matemáticas en la enseñanza secundaria en España en el siglo XIX*. Zaragoza: Universidad de Zaragoza, 1995.

Velamazán, Angeles. *La enseñanza de las matemáticas en las academias militares en España en el siglo XIX*. Zaragoza: Universidad de Zaragoza, 1994.

———. "Nuevos datos sobre los estudios de geometría superior en España en el siglo XIX: La Aportación militar." *LLULL: Revista de la Sociedad española de historia de las ciencias y de las técnicas* **16** (1993):587-620.

Velamazán, Angeles and Ausejo, Elena. "De Lagrange a Cauchy: El cálculo diferencial en las academias militares en España en el siglo XIX." *LLULL: Revista de la Sociedad española de historia de las ciencias y de las técnicas* **16** (1993):327-370.

Zorraquín, Mariano. *Geometría analtica-descriptiva*. Alcalá de Henares: Imprenta Manuel Amigo, 1819.

CHAPTER 4

International Mathematical Contributions to British Scientific Journals, 1800–1900

Sloan Evans Despeaux
University of Virginia (United States)

Introduction

Historians have often characterized British mathematics of the early nineteenth century as depressed and insular.[1] Several British mathematicians of the period shared this evaluation, but, among them, Charles Babbage gave perhaps the strongest critique in his 1830 work, *Reflections on the Decline of Science in England and on Some of Its Causes*.[2] In his "Introductory Remarks," Babbage stated that "[i]t cannot have escaped the attention of those, whose acquirements enable them to judge, and who have had opportunities of examining the state of science in other countries, that in England, particularly with respect to the more difficult and abstract sciences, we are much below other nations, not merely of equal rank, but below several even of inferior power."[3] To strengthen his claim, Babbage quoted John Herschel, who declared that in Britain "whole branches of continental discovery are unstudied, and indeed almost unknown, even by name. It is in vain to conceal the melancholy truth. We are fast dropping behind. In mathematics we have long since drawn the rein, and given over a hopeless race."[4]

The location of mathematical activity in nineteenth-century Britain is central to testing the validity of these perceptions of early nineteenth-century mathematics

---

[1]These views about mathematics at Oxford, Cambridge, and Britain can be seen in John Fauvel, "800 Years of Mathematical Traditions," in *Oxford Figures: 800 Years of the Mathematical Sciences*, ed. John Fauvel, Raymond Flood, and Robin Wilson (Oxford: University Press, 2000), p. 21; Ivor Grattan-Guinness, "Mathematics and Mathematical Physics from Cambridge, 1815-1840: A Survey of the Achievements and of the French Influences," in *Wranglers and Physicists*, ed. Peter M. Harman (Manchester: University Press, 1985), p. 84; Morris Kline, *Mathematical Thought from Ancient to Modern Times* (New York: Oxford University Press, 1972), p. 622.

[2]See Charles Babbage, *Reflections of the Decline of Science in England, and on Some of Its Causes* (London: B. Fellowes, 1830; reprint ed., New York: New York University Press, 1989), p. 1 (page citations are to the reprint edition); John F. W. Herschel, "Treatise on Sound," *Encyclopaedia Metropolitana* (1830) (quoted in Babbage, p. ix); John Playfair, "*Traité de méchanique céleste*," *Edinburgh Review* **11** (1807-1808):249-284; John Toplis, "On the Decline of Mathematical Studies, and the Sciences Dependent on Them," *Philosophical Magazine* **20** (1805):25-31; Charles William, Viscount Milton, *Report of the British Association for the Advancement of Science* **1** (1831):16; and "Preface," *Memoirs of the Analytical Society* (1813) (reprinted in *Science and Reform: Selected Works of Charles Babbage*, ed. Anthony Hyman (Cambridge: University Press, 1989), p. 15).

[3]Babbage, p. 1.

[4]Herschel, quoted in Babbage, p. ix.

and to tracking the progression of mathematics in Britain throughout the nineteenth century. While British mathematical outlets certainly included correspondence, monographs, and encyclopedia articles, journals increasingly became the primary venue for scientific research and communication. Thus, journal articles themselves provide important information about the absence or presence of international interaction in British mathematics.

While their opinions about the most effective metrics of science differ, sociologists have argued that scientific papers are valuable indicators of scientific activity. For instance, Derek de Solla Price asserted that

> [t]he act of creation in scientific research is incomplete without publication, for it is publication that provides the corrective process, the evaluation, and perhaps the assent of the relevant scientific community. ... Private property in science is established by open publication; the more open the publication and the more notice taken of it, the more is the title secure and valuable. ... For reasons such as these, strongly built into three centuries of development of norms of scientific behavior, we must regard the end product of scientific research as the openly published scientific paper or its functional equivalent.[5]

Viewing scientific papers contained in scientific journals as significant indicators of research, we can consider scientists who authored, read, and responded to papers in a given area within a given group of journals as a publication community.

In particular, this chapter traces international participation in the British mathematical publication community through the pages of British scientific journals. It reveals a small but important group of British mathematicians who actively promoted international interaction as well as the foreign mathematicians who responded to their encouragement. The reasons that compelled foreign mathematicians to contribute, the types of contributions they made, and the groups they represented all indicate the slow but steady international recognition of Britain as supporting a viable mathematical community.

Foreign Mathematics in British General Science Journals

British scientific periodicals experienced dramatic growth during the nineteenth century. Three-fourths of the periodicals studied in this chapter[6] began publication in the nineteenth century, and only one, the *Philosophical Transactions of the Royal Society of London*, began before 1780 (see Table 1). The supporters of these periodicals and the goals they brought to the enterprise shaped the scope of foreign mathematical interaction within this growing medium of scientific communication.

While the 6 May, 1665 appearance of the first British scientific periodical, the *Philosophical Transactions of the Royal Society*, was predated by the first scientific journal, the French *Journal des Sçavans*, by three months, it adopted the format emulated by future scientific journals.[7] Henry Oldenburg, Secretary of the Royal

[5]Derek de Solla Price, "Toward a Model for Science Indicators," in *Toward a Metric of Science: The Advent of Science Indicators*, ed. Yehuda Elkana *et al.* (New York: John Wiley and Sons, 1978), p. 80.

[6]The journals used here should be considered as a representative subset of nineteenth-century British scientific journals.

[7]Bernard Houghton, *Scientific Periodicals: Their Historical Development, Characteristics and Control* (London: Clive Bingley, 1975), pp. 13-14.

Table 1: British Scientific Periodicals, 1800-1900

Type		Journal	Years of Pub.
General Scientific Society Periodicals	Few* (if any) foreign math. contributions	*Manchester Literary and Philosophical Society*	
		Memoirs	1785 – 1887
		Proceedings	1857 – 1887
		Memoirs and Proceedings	1888 – present
		Transactions of the Cambridge Philosophical Society	1822 – 1928
		Transactions of the Royal Irish Academy	1787 – 1907
		Transactions of the Royal Society of Edinburgh - early (1783 - 1872)	1783 – present
	Consistent foreign math. contributions	*Transactions of the Royal Society of Edinburgh - later (1873-present)*	1783 – present
		Royal Society of London	
		Transactions	1665 – present
		Proceedings	1832 – present
		Report of the British Association for the Advancement of Science	1831 – 1938
Specialized Soc. Periodicals	Non-math. societies	*Royal Astronomical Society*	
		Memoirs	1822 – present
		Monthly Notices	1827 – present
	Math. Societies	*Proceedings of the London Mathematical Society*	1865 – present
		Proceedings of the Edinburgh Mathematical Society	1883 – present
Private Scientific Periodicals	Indirect foreign contributions	*Leybourne's Mathematical Repository*	1806 – 1835
		Cambridge Mathematical Journal	1837 – 1845
		Philosophical Magazine – early (1798 - 1865)	1798 – present
	Consistent direct foreign contributions	*Philosophical Magazine – later (1865 - present)*	1798 – present
		Cambridge and Dublin Mathematical Journal	1846 – 1854
		Quarterly Journal of Pure and Applied Mathematics	1855 – 1927
		Oxford, Cambridge, and Dublin Messenger of Mathematics	1862 – 1871
		Messenger of Mathematics	1871 – 1929

*A journal had "few contributions," if it had foreign contributions to only one volume. If it had foreign contributions to more than one volume, a journal had "consistent contributions."

Society, began the *Philosophical Transactions* as a private venture. The journal's foundation represented a natural extension of his duty to receive, announce, and reply to Society correspondence. While it was initially only affiliated with the Royal Society, the *Philosophical Transactions* finally gained official, financial support from the organization by the mid-eighteenth century.[8] Since that time, the periodicals of British general scientific societies have played a significant role in scientific and, more particularly, mathematical discourse.

With the economic prosperity and urban growth of the Industrial Revolution surfaced a demand for scientific knowledge "as a form of rational amusement, as theological edification, polite accomplishment, technical agent, social anodyne, and intellectual ratifier of the new industrial order."[9] In response to this growing interest in science, new general scientific societies and their journals joined the ranks of the Royal Society near the end of the eighteenth century. The Royal Society of Edinburgh and the Royal Irish Academy, which began publishing their *Transactions* in the 1780s, emerged in established seats of the United Kingdom, while "Literary and Philosophical" societies and their journals were founded in burgeoning industrial towns.[10]

Societies with primarily local concerns perhaps naturally had few foreign contributions to their periodicals. For example, the existence of only one nineteenth-century foreign mathematical contribution to the Royal Irish Academy's *Transactions* reflects its initially national interests: "though the members who compose it [the Academy] are not entirely without hopes that their efforts may hereafter become perhaps extensively useful and respected, yet the original intent of their institution must be considered as confining their views for the present more immediately to Ireland."[11] Similarly, the Manchester Literary and Philosophical Society aimed to encourage scientific activity among the inhabitants of its city, while the Cambridge Philosophical Society wanted to present in its periodical the efforts of the present and former students of the university. The presence of foreign mathematicians in these societies was almost exclusively limited to formal honorary memberships.

While the foreign presence in locally focused societies was usually nominal, more broadly based British scientific societies made international outreach a priority. For instance, during the first meeting of the British Association for the Advancement of Science (BAAS), the President, Charles William, Viscount Milton, underscored the importance of adopting international goals: "In our insular and insulated country, we have few opportunities of communicating with the cultivators of science in other parts of the world. It is the more necessary, therefore, to adopt means for opening new channels of communication with them, and at the same time of promoting a greater degree of scientific intercourse among ourselves."[12] Founded in 1831, the BAAS distinguished itself by migratory annual meetings in metropolitan, academic, industrial, provincial, and colonial centers, as well as by offering opportunities for participation at different levels of scientific skill and social

---

[8] A. Jack Meadows, *Communication in Science* (London: Butterworth & Co., 1974), pp. 66-67.

[9] Jack Morrell and Arnold Thackray, *Gentlemen of Science: Early Years of the British Association for the Advancement of Science* (Oxford: Clarendon Press, 1981), p. 12.

[10] These provincial literary and philosophical societies included: Manchester (f. 1781, *Memoirs* f. 1785), Newcastle-upon-Tyne (f. 1793, *Report* f. 1793), Birmingham (f. 1800, *Proceedings* f. 1876), Glasgow (f. 1802, *Proceedings* f. 1841), Liverpool (f. 1812, *Report and Proceedings* f. 1844), Plymouth (f. 1812, *Report and Transactions* f. 1862), Leeds (f. 1818, *Transactions* f. 1837), Cork (f. 1819), York (f. 1822, *Annual Report* f. 1824), Sheffield (f. 1822, *Annual Report* f. 1824), Whitby (f. 1822, *Report* f. 1824), Hull (f. 1822, *Report* f. 1864), and Bristol (f. 1823).

[11] "Preface," *Transactions of the Royal Irish Academy* **1** (1787):xi.

[12] Charles William, Viscount Milton, p. 16.

standing.[13] The Association induced international interaction in a variety of ways, including the naming of corresponding members, the direct solicitation of papers from foreign scientists, and monetary grants for research.[14]

Similarly, the Royal Society of London encouraged foreign participation through fellowships and medals. Its most prestigious honor, the Copley Medal, was established in 1736 and awarded to the British *savant* who had made what was deemed the most significant scientific contribution. In 1753, the Society internationalized the incentive of the award by including foreigners as candidates;[15] after 1831, it involved its periodicals in the adjudication of the Copley Medal by requiring that "the Copley Medal shall be awarded to the living author of such philosophical research, either published or communicated to the Society."[16]

In addition to general scientific societies, individuals provided support for general science periodicals. In 1798, British entrepreneur, Alexander Tilloch, began the *Philosophical Magazine*. While not a scientist, Tilloch successfully applied his journalistic skills to the endeavor. In fact, while other independent general science journals struggled during the nineteenth century, the *Philosophical Magazine* prevailed.[17] By 1832, it had absorbed William Nicholson's *A Journal of Natural Philosophy, Chemistry and the Arts*, Thomas Thomson's *Annals of Philosophy*, and David Brewster's *Edinburgh Journal of Science*.[18] Tilloch placed great importance on informing his readers of continental scientific happenings. To this end, he printed abstracts of the proceedings of European learned societies and published short biographies of European scientists. In mathematics, Tilloch published translations of memoirs from foreign publication outlets.[19] Eventually, the *Philosophical Magazine* replaced these translations with direct foreign mathematical contributions.

---

[13] Morrell and Thackray, pp. 108 and 135. During the nineteenth-century, the BAAS met twice in Canada (in Montreal in 1884 and in Toronto in 1897). For the Toronto meeting, the foreign (which here includes colonial) contributions came exclusively from Canada and the United States. However, foreign contributors to the Montreal meeting also included the Germans Ferdinand Lindemann and Walther Dyck. With the exception of these two meetings, foreign mathematical participation in the *BAAS Report* was mostly European.

[14] Corresponding members were expected to attend meetings, transmit meeting invitations to other foreign scientists, and report on the state of science in their homelands; for their efforts they were to be paid £10. These expectations enjoyed only limited success. See Morrell and Thackray, pp. 380-381.

[15] Sir Henry Lyons, *The Royal Society, 1660-1940* (Cambridge: University Press, 1944), pp. 172-173. The first foreigner to win the medal was Benjamin Franklin; in 1753, he received it for his *Philosophical Transactions* article, "Curious Experiments and Observations on Electricity." *Ibid.*, p. 182.

[16] *Record of the Royal Society of London* (London: Harrison and Sons, 1897), p. 125. The phrase "or communicated" meant that Copley medal winners were not required to publish in the *Philosophical Transactions*. In fact, the limitation that a Royal Medal winner publish in the *Philosophical Transactions* (in effect until 1850) was negatively viewed by several fellows because it "invited and encouraged [fellows] to enter only into the petty contests of a single society." George Airy, quoted in Roy M. MacLeod, "Of Medals and Men: A Reward System in Victorian Science 1826-1914," *Notes and Records of the Royal Society of London* **25** (1971):81-105 on p. 91.

[17] In 1825, upon Tilloch's death, the *Philosophical Magazine*'s editorship was assumed by Richard Taylor, Tilloch's printer. Taylor led the journal until 1851, when his son William Francis took over. William H. Brock and A. Jack Meadows, *The Lamp of Learning: Taylor & Francis and the Development of Science Publishing* (London: Taylor & Francis, 1984), p. 83.

[18] Robert M. Gascoigne, *A Historical Catalogue of Scientific Periodicals, 1665-1900* (New York: Garland Publishing, Inc., 1985), pp. 123-124.

[19] From 1800 to 1865, only nine of the thirty-three foreign mathematical articles were direct contributions.

British Specialized Science Journals as a Venue for Foreign Mathematics

Throughout the nineteenth century, foreign and domestic mathematicians increasingly participated in British specialized scientific periodicals supported by societies and individuals. This participation reflects the increasing specialization of publication outlets during the nineteenth century. Throughout Europe during the first half of the nineteenth century, general science periodicals yielded to specialized science journals. In fact, by 1844, there were ten specialized scientific societies publishing journals in London alone.[20] With the growth of the journals of these societies as well as commercial specialized scientific journals, the supremacy of general science society journals waned. General science journals grew from seven in 1800 to twenty-two in 1900, while British specialized science journals grew from three to at least sixty-seven during the nineteenth century.[21]

This process of specialization in British journals did not immediately limit mathematics publications to exclusively mathematical journals. For example, the *Monthly Notices* and *Memoirs of the Royal Astronomical Society* represented viable avenues for applied mathematics from their establishment in the 1820s to the end of the nineteenth century. From the outset, this society viewed international communication as vital to more comprehensive observational astronomy, better-made instruments, and theoretical innovations.[22] In this environment of international openness, foreign astronomers and mathematicians recognized the publications of the society as an appropriate place for their results.[23]

During the second half of the nineteenth century, the journals of British mathematical societies provided a new publication outlet for foreign mathematics. Both the London Mathematical Society (LMS) and the Edinburgh Mathematical Society (EMS) included foreigners among their ordinary members, and the LMS began to elect Honorary Foreign Members soon after its establishment in 1865.[24] Although

---

[20] Gascoigne, pp. 3-89. The disciplines of these societies and the dates of the foundation of their journals were natural history (1788), geology (1807), horticulture (1810), astronomy (1821), geography (1832), zoology (1833), entomology (1834), statistics (1835) (listed here although Gascoigne did not include statistics in his study), chemistry (1841), and microscopy (1844). Beyond the metropolis, Gascoigne lists eight British specialized societies with journals during this period. The disciplines of these societies were geology, natural history, and botany. Morrell and Thackray count five general and twenty-five specialized science societies in the metropolis by 1850 (up from two general and two specialized in 1780); in the provinces for 1850, they count thirty general and forty-two specialized science societies (up from none of either category in 1780). Morrell and Thackray, p. 546.

[21] Gascoigne, pp. 132-170. The specialized journals listed concerned mathematics, astronomy, physics, chemistry, geology, geography, natural history, botany, zoology, and experimental biology. This trend of specialization occurred throughout Europe. According to Gascoigne, "while the numbers of general science periodicals grew more or less steadily from the seventeenth century to 1900 their proportion among scientific periodicals of all types decreased from 100% in 1770 and earlier to 27% in 1900." *Ibid.*, p. 131.

[22] "Address of the Society, Explanatory of Their Views and Objects," *Memoirs of the Royal Astronomical Society* **1** (1822):1-7.

[23] The *Journal of the Institute of Actuaries and Assurance Magazine* made a similar international commitment. The mathematical articles in the *Journal* are, however, primarily actuarial and are not analyzed here.

[24] Election to this position anticipated, in some cases, activity with the Society. Of the twenty-two nineteenth-century Honorary Foreign Members, ten contributed to the Society's *Proceedings*

the EMS emphasized pedagogy while the LMS focused on theoretical research, both regularly received foreign contributions.[25]

British mathematical journals supported by individuals rather than societies also provided a forum for the publication of foreign mathematics. However, as in the case of the *Philosophical Magazine*, this representation did not immediately include direct foreign contributions. Duncan Gregory, founding editor of the *Cambridge Mathematical Journal*, identified goals for his journal characteristic of those outlined for other early nineteenth-century scientific journals. He believed his journal could prove useful "by publishing abstracts of important and interesting papers that have appeared in the Memoirs of foreign Academies, and in works not easily accessible to the generality of students."[26] He expressed the motivation behind this explicitly: "We hope in this way to keep our readers, as it were, on a level with the progressive state of Mathematical science, and so lead them to feel a greater interest in the study of it."[27]

Like the *Cambridge Mathematical Journal*, the *Oxford, Cambridge, and Dublin Messenger of Mathematics* was "a journal supported by junior mathematical students."[28] The "board of editors composed of members of the three universities"[29] considered their journal "as an improved guide to the old Mathematical heights, but we also offer abundant room for those who can tell us anything of heights or depths hitherto unexplored."[30] Three foreign mathematicians provided improvements on older mathematical explanations as well as original research. The most notable of this group, Luigi Cremona, made two contributions on conics that were communicated by Thomas Archer Hirst.[31]

In 1871, nine years after the journal began, James W. L. Glaisher joined the editorial staff. He soon took over the journal's editorship and changed its name to

---

in or before 1900. Moreover, these ten represent almost one-third of the Society's nineteenth-century foreign contributors, and all but one, Amédée Mannheim, were elected to this position *before* making a contribution.

[25] The EMS began in 1883 with the goal of "the mutual improvement of its members in the Mathematical Sciences ... [through] [r]eviews of works both British and Foreign, historical notes, discussion of new problems or new solutions, and comparison of the various systems of teaching in different countries, or any other means tending to the promotion of mathematical Education." See Gargill G. Knott, Andrew J. G. Barclay, and Alexander Y. Fraser, "Circular, January 23, 1883," in Robert A. Rankin, "The First Hundred Years," *Proceedings of the Edinburgh Mathematical Society* **26** (1983):135-150 on p. 136. On the LMS, see Adrian C. Rice and Robin J. Wilson, "From National to International Society: The London Mathematical Society, 1867–1900," *Historia Mathematica* **25** (1998):185-217 on p. 194. A British society concerned with mathematics education, the Association for the Improvement of Geometrical Teaching (later known as the Mathematical Association), began publishing its *Report* soon after its 1871 foundation. (This publication was renamed the *Mathematical Gazette* in 1894.) The latter publications have not been used for this study, but deserve mention.

[26] Duncan Gregory, *Cambridge Mathematical Journal* **1** (1837):1.

[27] *Ibid.*

[28] Title page, *Oxford, Cambridge, and Dublin Messenger of Mathematics* **1** (1862). This statement appeared on the title page of each of this journal's five volumes.

[29] *Ibid.* This board consisted of Henry William Challis of Oxford, James McDowell, Henry John Purkiss, Charles Taylor, William Peverill Turnbull, and William Allen Whitworth of Cambridge, and John Casey of Dublin.

[30] "Preface," *Oxford, Cambridge, and Dublin Messenger of Mathematics* **3** (1866):v.

[31] Luigi Cremona, "On the Fourteen Point Conic," *Oxford, Cambridge, and Dublin Messenger of Mathematics* **3** (1866):13-14; and "On Normals to Conics, a New Treatment of the Subject," *Oxford, Cambridge, and Dublin Messenger of Mathematics* **3** (1866):88-91.

the *Messenger of Mathematics*. In the same year, Glaisher graduated from Cambridge as Second Wrangler on the Mathematical Tripos and obtained a fellowship at Trinity College, where he resided the rest of his life.[32] Under his guidance, foreign contributions soared; in fact, the *Messenger* had more nineteenth-century foreign mathematical contributions than any other journal considered here.[33]

A change of editorship for the *Cambridge Mathematical Journal* also positively affected its foreign contributions. When William Thomson, who would later become Lord Kelvin, took over its editorship in 1846 and renamed it the *Cambridge and Dublin Mathematical Journal*, the contributors to the journal, as well as its title, broadened geographically. The most notable direct foreign contribution to the journal was Charles Hermite's "Sur la théorie des fonctions homogènes à deux indéterminées."[34] In this, his third article to Thomson's journal, Hermite presented his law of reciprocity for the first time. This law states that the number of covariants of the pth order in the coefficients of a binary form of degree m is equal to the number of covariants of the mth order in the coefficients of a binary form of degree p. In the early 1850s, invariant theory was focused on systematically determining the covariants and invariants for binary forms of each successive degree. Thus, Hermite's surprising result, presented in the *Cambridge and Dublin Mathematical Journal*, effectively cut the work of invariant theorists in half by doubling the number of known invariants and covariants.[35]

The editors of independent mathematical journals expressed a desire to communicate international developments such as Hermite's law of reciprocity and a need to lead British mathematics out of isolation. For example, the editors of the *Quarterly Journal of Pure and Applied Mathematics*,[36] which included James Joseph Sylvester, Arthur Cayley, George Gabriel Stokes, and Charles Hermite, wrote that "it would be little credible to English Mathematicians that they should stand aloof

[32]Andrew Russell Forsyth, "James Whitbread Lee Glaisher," *Journal of the London Mathematical Society* **4** (1929):101-112 on p. 101.

[33]The mathematical questions that appeared in the *Educational Times* from 1849 and that later formed a separate journal bear mentioning here, although they were not used in this study. Despite its pedagogical orientation and the "question and answer" nature of its mathematical contributions, the journal received many questions that later sparked original research. The *Educational Times* and *Mathematical Questions with Their Solutions From the 'Educational Times'* included among their contributors Cayley, William Kingdon Clifford, Augustus De Morgan, Hirst, Charlotte Angas Scott, and Sylvester in Britain, and Carl Borchardt, Cremona, and Ernst Kummer from abroad. Ivor Grattan-Guinness, "Contributing to The Educational Times: Letters to W. J. C. Miller," *Historia Mathematica* **21** (1994):204-205.

[34]Charles Hermite, "Sur la théorie des fonctions homogènes à deux indéterminées," *Cambridge and Dublin Mathematical Journal* **9** (1854):172-217.

[35]*Ibid.* On invariant theory in the nineteenth century, see Tony Crilly, "The Rise of Cayley's Invariant Theory (1841–1862)," *Historia Mathematica* **13** (1986):241-254; Karen Hunger Parshall, "Toward a History of Nineteenth-Century Invariant Theory," in *The History of Modern Mathematics*, ed. David E. Rowe and John McCleary, 2 vols. (Boston: Academic Press, Inc., 1989), 1:157-206; and Karen Hunger Parshall, *James Joseph Sylvester: Life and Work in Letters* (Oxford: Clarendon Press, 1998).

[36]The *Quarterly Journal of Pure and Applied Mathematics* began in 1855, the year after the *Cambridge and Dublin Mathematical Journal* folded. Glaisher joined its editorial staff in 1878 and edited both the *Messenger of Mathematics* and the *Quarterly Journal* until his death in 1928. The *Messenger* died with Glaisher, while the *Quarterly Journal* survived as a new series in Oxford. Godfrey Harold Hardy, "Dr. Glaisher and the 'Messenger of Mathematics'," *Messenger of Mathematics* **58** (1929):159-160 on p. 159.

from the general movement, or else remain indebted to the courtesy of the editors of foreign journals for the means of taking a part in the rapid circulation and interchange of ideas by which the present era is characterised."[37]

Changes in Foreign Participation through the Nineteenth Century

While their goals and sources of support varied, over half of the journals listed in Table 1 contain consistent direct contributions from foreign mathematicians.[38] The overall increase in the number of foreign contributors and contributions to these journals throughout the nineteenth century reflects a growing international presence in this publication medium (see Table 2).[39] However, consistent growth was not the rule for individual journals; the exclusively mathematical journals fueled the general trend of growth, while the majority of general science journals in Table 2 experienced no significant increase in the number of foreign mathematical contributions over the century.[40] The stagnation of foreign mathematical contributions to these general scientific society journals, simultaneous with the growth of such contributions to mathematical journals, indicates a trend of specialization in mathematics publication in Britain.

Even among the exclusively mathematical journals, this growth was uncertain and sometimes checked. For example, while the *Quarterly Journal of Pure and Applied Mathematics* contained thirteen foreign contributions in its first volume (1855–1856), it would take the next five volumes (1857–1862) to match this number of foreign articles; in fact, no other single nineteenth-century volume of the *Quarterly Journal* contained thirteen or more foreign contributions. Several of the foreign contributions to the journal's first volume were in the form of correspondence to its editors or responses to the editors' work.

---

[37] James Joseph Sylvester, *et al.*, "Address to the Reader," *Quarterly Journal of Pure and Applied Mathematics* **1** (1855):i-ii.

[38] Table 2 contains thirteen of the original twenty journals studied. The seven journals not included either had no direct foreign contributions (such as the journals of the Manchester Literary and Philosophical Society) or had such contributions only in one volume (such as the *Transactions of the Royal Irish Academy* or the *Transactions of the Cambridge Philosophical Society* whose 1899 volume carried several foreign contributions resulting from the jubilee held in honor of G. G. Stokes's fiftieth year in the Lucasian Professorship). A journal received "consistent direct contributions" if it received original foreign contributions for more than one volume.

[39] While counts of scientific papers have been advocated by Derek de Solla Price as a good gauge of scientific activity, other sociologists have argued that citation counts, co-citation counts, or citation typing represent better measures. For reviews and criticisms of these techniques, see Henry G. Small, "The Lives of a Scientific Paper," in *Selectivity in Information Systems: Survival of the Fittest*, ed. Kenneth S. Warren (New York: Praeger Science Publishers, 1985), pp. 83-97; Daryl E. Chubin and Soumyo D. Moitra, "Content Analysis of References: Adjunct or Alternative to Citation Counting?" *Social Studies of Science* **5** (1975):423-441; and David Edge, "Quantitative Measures of Communication in Science: A Critical Review," *History of Science* **17** (1979):102-134. Because of the absence of a citation database such as the *Science Citation Index* for the nineteenth century, I do not use citation counting here. However, I have looked for references to British mathematicians in the citations in the foreign contributions considered. See Table 5.

[40] Contributions to both the *Philosophical Magazine* and the *Report of the British Association for the Advancement of Science* grew over the century.

Table 2: Foreign Contributors and Contributions by Decade: 1800-1900

Journal	**Foreign Direct Contributors/Contributions (# : #)**										
	1800-1809	1810-1819	1820-1829	1830-1839	1840-1849	1850-1859	1860-1869	1870-1879	1880-1889	1890-1900	**1800-1900**
Philosophical Transactions of the Royal Society of London							2:3			2:2	**4:5**
Philosophical Magazine			2:3			1:1	4:8	9:16	10:13	12:15	**35:56**
Transactions of the Royal Society of Edinburgh									2:2	1:2	**2:4**
Memoirs of the Royal Astronomical Society			3:3	1:1	1:1	2:2	1:3	2:2	1:1	1:2	**10:15**
Monthly Notices of the Royal Astronomical Society				2:3	2:4	9:11	6:16	6:6	3:3	4:9	**27:52**
BAAS: *Report*				1:1	1:3	1:1		3:2*	3:3	2:2	**10:12**
BAAS: *Transactions*					2:2	3:5	5:5	7:12	12:15	9:10	**33:49**
Proceedings of the Royal Society of London				2:3	1:1	2:3	2:3	3:3	4:5	3:3	**16:21**
Cambridge and Dublin Mathematical Journal					5:6	8:10					**10:16**
Quarterly Journal of Pure and Applied Mathematics						9:21	7:19	11:25	8:12	17:35	**40: 112**
Proceedings of the London Mathematical Society							1:1	6:10	12:15	16:31	**32:57**
Oxford, Cambridge, and Dublin Messenger of Mathematics							3:8	2:3			**3:11**
Messenger of Mathematics								14:64	17:37	13:41	**34: 142**
Proceedings of the Edinburgh Mathematical Society									4:7	13:19	**15:26**
Total for all journals above	0:0	0:0	5:6	5:8	10:17	29:54	26:66	49: 143	56: 113	66: 171	**192: 578**

* In 1872, A. Erman (1806-1877) and H. Petersen (1815- ?) wrote "Report on the Gaussian Constants for the year 1829" together.

As an editor of the *Quarterly Journal*, Hermite naturally supported the journal with articles.[41] Two of his contributions to the first volume were extracts of letters to Sylvester and Cayley. In the letter to Sylvester, Hermite proposed a method with which to find a complete system of positive integer solutions to the equation $ax +$

[41] Besides his above-mentioned contribution to the *Cambridge and Dublin Mathematical Journal* and his work with the *Quarterly Journal*, Hermite published in the *Proceedings of the Royal Society of Edinburgh*, the *Proceedings of the LMS*, the *Report of the BAAS*, and the *Messenger of Mathematics*.

$by = n$, where a and b are positive integers with no common divisors.[42] The letter to Cayley concerned cubic forms; to the same number of the *Quarterly Journal*, Hermite included his article, "Sur les formes cubiques à deux indéterminées," on the same subject.[43]

Enrico Betti also presented results to the *Quarterly Journal* through correspondence. In a letter to Sylvester, Betti extended the solution of a problem first posed by Niels Henrik Abel, then posed again by Leopold Kronecker to the Berlin Academy in 1853. The problem, finding the most general algebraic function that satisfies a given equation of a given degree, was solved by Kronecker for prime degrees. In his letter, Betti announced his solution of the problem for degrees that are powers of primes and provided a sketch of the proof.[44] However, he chose the inaugural volume of the *Annali di matematica pura ed applicata*, of which he was an editor, for the complete demonstration of his solution.[45]

Besides letter extracts, the first volume of the *Quarterly Journal* also contained original articles from foreign mathematicians. Francesco Brioschi, a professor of applied mathematics at the University of Pavia, had previously contributed to the *Cambridge and Dublin Mathematical Journal* and had actively read and responded to articles in the *Quarterly Journal*. In his first contribution, he used the determinant of functions to solve a differential equation that he had found in an article in the *Cambridge and Dublin Mathematical Journal* by University of Upsala professor, Carl Johann Malmsten.[46] In his second article, Brioschi used determinants to prove two theorems in analytic geometry indicated by Cayley in the first number of the *Quarterly Journal*.[47] Brioschi's support of communication via periodicals is reflected in his role in the 1858 foundation of the *Annali di matematica pura ed applicata* and in his place on the journal's editorial staff for the rest of his life.[48]

Although Brioschi's participation represented active communication through reading and responding to articles in the *Quarterly Journal*, his contributions were not extensive expositions. The majority of the foreign contributions to the inaugural volume of the *Quarterly Journal* were short notes; in fact, only two exceeded six pages in length. Reinhold Hoppe, *Privatdozent* at the University of Berlin, provided one of these exceptions. In his article, "Determination of the Motion of Conoidal Bodies through an Incompressible Fluid," Hoppe contributed the only applied mathematical article to the first volume of the *Quarterly Journal*. In this paper, he considered a body moved by a given force through an incompressible fluid. He deemed this case as one "where the agreement between natural and analytic functions is so perfect, that the latter seem to exist exactly for the purpose

[42]Charles Hermite, "Extract of a Letter from M. Hermite to Professor Sylvester," *Quarterly Journal of Pure and Applied Mathematics* **1** (1855):370-373.

[43]Charles Hermite, "Correspondence: Mr. Cayley and M. Hermite on Cubic Forms," *Quarterly Journal of Pure and Applied Mathematics* **1** (1855):88-89, and "Sur les formes cubiques à deux indéterminées," *Quarterly Journal of Pure and Applied Mathematics* **1** (1855):20-22.

[44]Enrico Betti, "Extract from a Letter of Signor Enrico Betti to Mr. Sylvester," *Quarterly Journal of Pure and Applied Mathematics* **1** (1855):91-92.

[45]Enrico Betti, "Sopra l'equazioni algebriche con più incognite," *Annali di matematica pura ed applicata* **1** (1858):1-8.

[46]Francesco Brioschi, "Sur une propriété d'un déterminant fonctionnel," *Quarterly Journal of Pure and Applied Mathematics* **1** (1855):365-367.

[47]Francesco Brioschi, "Note sur deux théorèmes de géométrie," *Quarterly Journal of Pure and Applied Mathematics* **1** (1855):368-370.

[48]Charles C. Gillispie, ed., *Dictionary of Scientific Biography*, 16 vols. and 2 supps. (New York: Charles Scribner's Sons, 1970–1990), s.v. "Brioschi, Francesco" by Joseph Pogrebyssky.

to answer every question that can be proposed with regard to a certain motion of a fluid."[49]

In the other extended article in the *Quarterly Journal*'s inaugural volume, Alfred Enneper, at the time a doctoral student at Göttingen University, considered the function $\Pi(z) = \int_0^\infty u^z e^{-u} du$ for imaginary and complex domain values.[50] Enneper was the most frequent foreign contributor to the first volume of the *Quarterly Journal*; besides the contribution just mentioned, he presented a derivation of a formula in the theory of elliptic functions, and contributed two short discussions concerning integral calculus.[51] With the exception of the two articles by Hoppe and Enneper, the foreign contributions to the inaugural volume of the *Quarterly Journal* have the appearance of mere wishes for the journal's success rather than full-fledged research contributions.[52]

While editorial influence spawned foreign mathematical contributions, so too did controversy. For example, almost half of the foreign mathematical contributions to the *Monthly Notices of the Royal Astronomical Society* from 1850 to 1869 consisted of a series of articles from 1859 to 1861 debating the secular acceleration of the mean motion of the moon.[53] The catalyst for this debate, John Couch Adams, began his career in astronomy by co-discovering Neptune with Urbain Le Verrier, was president of the Royal Astronomical Society from 1851 to 1853, and soon after turned his attention to lunar theory.[54] In 1853, in the *Philosophical Transactions of the Royal Society of London*, he presented his calculation of the series for the acceleration of the moon's mean motion. The series yielded a value almost half as small as those previously calculated by the Royal Astronomer to the Observatory of Turin, Giovanni Plana,[55] and Peter Andreas Hansen, head astronomer at the Observatory of Seeberg.[56] In 1856, Plana accepted Adams's results but then soon refuted them with calculations that also differed from Plana's original values; in 1859, Charles-Eugène Delaunay corroborated Adams's values through an independent method and communicated his findings to the Paris Academy.[57] Delaunay,

---

[49]Reinhold Hoppe, "Determination of the Motion of Conoidal Bodies through an incompressible Fluid," *Quarterly Journal of Pure and Applied Mathematics* **1** (1855):301-315 on p. 301.

[50]Alfred Enneper, "On the Function $\Pi(z)$ with Imaginary and Complex Variable," *Quarterly Journal of Pure and Applied Mathematics* **1** (1855):393-402.

[51]Alfred Enneper, "Elementary Demonstration of an Equation Between Two Transcendental Functions," "On the Definite Integral $\int_0^\infty \frac{\sin^q u}{u^m} du$," and "A General Theorem Relating to Multiple Periodic Series," *Quarterly Journal of Pure and Applied Mathematics* **1** (1855):272-276, 276-279, and 280-285, respectively.

[52] The remaining two contributions came from Francesco Faà de Bruno, at the time a student in Paris. In his two articles, Faà de Bruno presented a formula using the determinant to find the derivative of any given order of a function and recorded an invariant of the twelfth degree.

[53]These twelve articles are from three foreign contributors; foreign contributions on lunar theory continued to appear in the *Monthly Notices* throughout the rest of the nineteenth century.

[54]Gillispie, ed., *Dictionary of Scientific Biography*, s.v., "Adams, John Couch" by Morton Grosser. Adams was again president of the Royal Astronomical Society from 1874 to 1876 and became Lowndean Professor of Astronomy and Geometry at Cambridge in 1858.

[55]Gillispie, ed., *Dictionary of Scientific Biography*, s.v. "Plana, Giovanni" by F. G. Tricomi.

[56]"Obituary; Peter Andreas Hansen," *Monthly Notices of the Royal Astronomical Society* **35** (1875):168-170 on pp. 169-170. Hansen became an Associate of the Royal Astronomical Society in 1837 and received the Society's Gold Medal in 1842 and 1860.

[57]"Obituary; John Couch Adams," *Monthly Notices of the Royal Astronomical Society* **53** (1893):184-209 on p. 196.

a chief engineer at the *École des mines* who would later succeed Le Verrier as director of the Paris Observatory, produced these results as part of a comprehensive investigation of lunar theory that he began in 1846.[58]

This lunar theory controversy did not end with Delaunay's corroborating results. One of the most persistent opponents of these results, Phillipe Gustave Doulcet, Comte de Pontécoulant, repeatedly attacked these findings from 1859 to 1861. Described by Ivor Grattan-Guinness "as an outsider algebraist of the French mathematical community,"[59] Pontécoulant disagreed with the results of Adams and Delaunay because they disagreed with observations. In an 1859 article to the *Monthly Notices of the Royal Astronomical Society*, he claimed this disparity made "the existence of the new terms introduced by M. Adams *very problematic.*"[60] Other astronomers shared Pontécoulant's misgivings. In the same number of the *Monthly Notices*, Hansen wrote that while he could not at that instant find the error in Delaunay's calculations, "I must determine to deem [them] as incorrect."[61] Through a grant from the British government, Hansen had published his lunar tables in London in 1857. In 1860, during a presentation to the French Academy, Le Verrier asserted that these tables refuted Delaunay's conclusions: "Now the theory of M. Hansen agrees with them all, and one demonstrates to M. Delaunay that, with his formulas, one does not reach such agreement. Thus, doubts remain, and more than doubts relative to M. Delaunay's formulas. Most certainly, truth is on the side of M. Hansen."[62]

In a reply to those who disagreed with his results, Adams emphasized "that the question is a purely mathematical one, with the decision of which observation has nothing whatever to do."[63] Moreover, he argued that causes not accounted for in the theory of the mean motion of the moon explained differences between his findings and observations.[64] By 1861, Adams's British colleagues, Cayley, William

---

[58] "Obituary; Charles Eugène Delaunay," *Monthly Notices of the Royal Astronomical Society* **33** (1873):203-209 on pp. 204-207. These investigations appeared in the two-volume work, *La théorie de la lune*, published in 1860 and 1867. On the basis of this work, Delaunay was elected an Associate of the Royal Astronomical Society in 1862 and awarded the Society's Gold Medal in 1870.

[59] Ivor Grattan-Guinness, *Convolutions in French Mathematics, 1800–1840*, vol. 2, *The Turns* (Basel: Birkhäuser Verlag and Berlin: Deutscher Verlag der Wissenschaften, 1990), p. 1204.

[60] "Cette conséquence seule paraîtrait déjà rendre *très problématique* l'existence des nouveaux termes introduits par M. Adams" (his emphasis). Phillipe Gustave Doulcet, Comte de Pontécoulant, "Sur la variation séculaire du moyen mouvement de la lune," *Monthly Notices of the Royal Astronomical Society* **19** (1859):235-236 on p. 236.

[61] "Delaunay's Säcularänderung der mittleren Mondlänge muss ich entschieden für unrichtig halten." Peter Andreas Hansen, "Extract of a Letter from Prof. Hansen to the Astronomer Royal, dated Gotha, May 31, 1859," *Monthly Notices of the Royal Astronomical Society* **19** (1859):236-237 on p. 236.

[62] "Pour un astronome, la première condition est que ses théories satisfassent aux observations. Or la théorie de M. Hansen les représente toutes, et l'on prouve à M. Delaunay qu'avec ses formules on ne saurait y parvenir. Nous conservons donc des doutes et plus que des doutes sur les formules de M. Delaunay. Très certainement la vérité est du côté de M. Hansen." Urbain Le Verrier, quoted in "Obituary; John Couch Adams," p. 197.

[63] John Couch Adams, "Reply to Various Objections Which Have Been Brought against His Theory of the Secular Acceleration of the Moon's Mean Motion," *Monthly Notices of the Royal Astronomical Society* **20** (1860):225-240 on p. 226.

[64] In 1865, Delaunay correctly suggested that the slowing of the earth's rotation due to tidal friction represented the unrecognized cause of the discrepancy between theory and observation. A year before, the American, William Ferrel, who would later contribute to British journals, was the first to treat this tidal friction quantitatively in a paper to the American Association for

Table 3: Foreign Contributions Considered by Page Length: 1800-1900

Journal		Foreign Math. Contributions/All Math. Contributions (in pages)*										
		1800-1809	1810-1819	1820-1829	1830-1839	1840-1849	1850-1859	1860-1869	1870-1879	1880-1889	1890-1900	1800-1900
BAAS	*Repts.*				25/297	11/85	4/132	0/322	27/388	7/236	46/525	**120/1985**
					8.4%	12.9%	3.0%	0.0%	7.0%	3.0%	8.8%	**6.0%**
	Abs.				0/83	7/76	11/63	6/63	26/165	20/106	14/114	**84/670**
					0.0%	9.2%	17.5%	9.5%	15.8%	18.9%	12.3%	**12.5%**
Cambridge and Dublin Mathematical Journal						16/1152	111/1451					**127/2603**
						1.4%	7.6%					**4.9%**
Quarterly Journal of P&A Mathematics							164/1137	169/2309	358/2220	173/3000	703/2661	**1567/11,327**
							14.4%	7.3%	16.1%	5.8%	26.4%	**13.8%**
Proceedings of the London Math. Society								6/679	82/1850	64/3909	462/5419	**614/11,857**
								0.1%	4.4%	1.6%	8.5%	**5.2%**
Oxford, Cambridge, and Dublin Messenger of Mathematics								43/1018	17/252			**60/1270**
								4.2%	6.7%			**4.7%**
Messenger of Mathematics									178/1728	117/1920	204/2033	**499/5681**
									10.3%	6.1%	10.0%	**8.8%**
Proceedings of the Edinburgh Math. Society										35/539	106/1515	**141/2054**
										6.5%	7.0%	**6.9%**
Total for all journals above					**25/380**	**34/1313**	**290/2783**	**224/4391**	**688/6603**	**416/9710**	**1535/12,267**	**3212/37,447**
					6.6%	**2.6%**	**10.4%**	**5.1%**	**10.4%**	**4.3%**	**12.5%**	**8.6%**

* Percentages are rounded to nearest 0.1%.

Donkin, and Sir John Lubbock, had arrived at Adams's results by a variety of methods;[65] moreover, by this time Plana had reached the same conclusions and Hansen had acknowledged his agreement with them in the *Monthly Notices*.[66] For his part, however, Pontécoulant continued to attack these results in the *Monthly*

the Advancement of Science. Gillispie, ed., *Dictionary of Scientific Biography*, s.v. "Delaunay, Charles-Eugéne" by Jean Kovalevshky and s.v. "Ferrel, William" by Harold L. Burstyn.

[65] "Obituary; John Couch Adams," pp. 197-198.

[66] Peter Andreas Hansen, "From a Letter from Professor Hansen to the Astronomer Royal, dated Gotha, 1861, Feb. 2," *Monthly Notices of the Royal Astronomical Society* **21** (1861):152-155 on p. 154.

Notices into 1863, calling them "steeped in inacurracy,"[67] and he never accepted them.[68]

Through the multiple arguments concerning this controversy that they published in the *Monthly Notices of the Royal Astronomical Society*, Pontécoulant and Hansen became two of the most prolific foreign mathematical contributors in this study.[69] While certainly not as prolific as Pontécoulant and Hansen, foreign contributors generally published multiple mathematical contributions to British journals; on average, these journals contain three articles per foreign contributor.[70] In their study of nineteenth-century mathematics through the *Catalogue of Scientific Papers of the Royal Society of London*, Roland Wagner-Döbler and Jan Berg calculated that each mathematical contributor published in total an average of 6.48 articles in all nineteenth-century journal outlets covered by the *Catalogue*.[71] Presumably, authors who contributed to foreign journals wrote more extensively than those who published only in local journals; however, these two averages indicate that, in general, the number of foreign mathematical papers these authors contributed to British journals was not trivial.

The small but increasing numbers of foreign mathematical contributions to British journals over the nineteenth century imply a growing international presence within the British publication community. However, when viewed by the proportion of pages these contributions occupied among the total number of mathematical pages in British journals, the international presence does not seem to have grown steadily (see Table 3). Of the six exclusively mathematical journals with consistent foreign contributions and the *Report of the British Association for the Advancement of Science*, which had a separate Mathematical and Physical Sciences Section, only three of these journals attained foreign page proportions of over 10%.[72] Of these,

[67] "[P]lusiers des résultats qu'il [M. Delaunay] présente peuvent être avec raison regardés comme entachés d'inéxactitude." Phillipe Gustave Doulcet, Comte de Pontécoulant, "Méchanique céleste: Observations sur la comparison établie par M. Delaunay entre les expressions des coordonnées de la lune déduites de sa théorie avec celles qui avaient été obtenues antérieurement," *Monthly Notices of the Royal Astronomical Society* **23** (1863):259-266 on p. 260.

[68] "Obituary; John Couch Adams," p. 198.

[69] Thirteen out of the 192 foreign contributors in this study made nine or more nineteenth-century contributions. These contributors were: from Australia, Edward J. Nanson (29), James Cockle (25), and John H. Michell (15); from the United States, George A. Miller (19), William Woolsey Johnson (12), Leonard E. Dickson (11), and J. J. Sylvester (9; all contributions he made to British journals while at Johns Hopkins University from 1876 to 1883); from France, V. M. Amédée Mannheim (24), Hermite (13), and Pontécoulant (9); from Belgium, Paul Mansion (29) and Hansen (10); and from Switzerland, Ludwig Schläfli (13).

[70] The *Messenger of Mathematics* was a repository for multiple foreign contributions. The average number of articles per foreign contributor to this journal was 4.1. This average for all journals except the *Messenger* was 2.4. Many of the prolific contributors above made most of their contributions to the *Messenger*. For example, Nanson made all of his contributions to it, while it contained sixteen of Mannheim's contributions and twenty-five of Mansion's.

[71] Roland Wagner-Döbler and Jan Berg, "Nineteenth-Century Mathematics in the Mirror of Its Literature: A Quantitative Approach," *Historia Mathematica* **23** (1996):288-318 on p. 297. Wagner-Döbler and Berg considered only journal articles in pure mathematics while I include pure and applied mathematics here, so this comparison of productivity can only be considered as approximate.

[72] Of the societies not listed in Table 3, only the Royal Astronomical Society had a substantial proportion of foreign contributions (considered by page length) to its *Memoirs* and *Monthly Notices*. Page statistics for the *Philosophical Magazine* have not yet been calculated.

the *Quarterly Journal for Pure and Applied Mathematics* achieved the highest percentage, 26.4%, from 1890 to 1900. Considered over the entire nineteenth century, foreign contributions represented from 4.7% of all pages of mathematics for the *Oxford, Cambridge, and Dublin Messenger of Mathematics* to 13.8% for the *Quarterly Journal of Pure and Applied Mathematics*. Considering the totals for all journals, the proportion of foreign pages fluctuates through the century but ends at a high of over 12%. Three journals achieved their highest foreign page percentages during the 1890s, and for two journals, the *Quarterly Journal* and the *Proceedings of the LMS*, the 1890s percentage was at least 63% higher than any previous percentage. While slow but steady growth characterizes the number of foreign contributors and their contributions to British scientific journals, growth for the proportion of the journal literature in Britain occupied by foreign mathematics was erratic.

A Geographical Profile of the Publication Community

While primarily British, the geographical distribution of the British mathematical publication community reveals support given to British mathematics by established centers of mathematics as well as publication opportunities taken by emerging groups of mathematicians (see Table 4).[73] The comparatively high French mathematical presence in British journals to the middle of the nineteenth century reflects the strength of French mathematics during this period.[74] Likewise, the growth of German mathematical contributions coincides with the emergence of Germany as a mathematical power later in the nineteenth century.

Mathematical contributions from the British colonies increased considerably during the last decade of the nineteenth century.[75] This late growth in colonial contributions agrees with the later, twentieth-century foundation of mathematical societies of Canada, Australia, New Zealand, and Calcutta.[76] Over 40% of these contributions concentrated on one journal, the *Messenger of Mathematics*. Moreover, these articles represent over 35% of the total foreign contributions to the journal.[77]

---

[73]Here, contributors are identified with the countries from which they are writing. For example, the contributions Sylvester made to British journals while at the Johns Hopkins University are considered to be from the United States.

[74]The first nineteenth-century mathematical contribution to British scientific journals not written in English was Adrien-Quentin Buée's 1806 "Mémoire sur les quantités imaginaires" in the *Philosophical Transactions of the Royal Society of London*. Buée, however, was at the time a political refugee in England and so his contribution is not counted as foreign.

[75]A mathematician is termed "colonial" here for the period he resided in a British colony. For example, the contributions of Horace Lamb, a frequent contributor to the *Messenger of Mathematics*, are considered as Australian for the period from 1876 to 1885, when Lamb was a professor of mathematics at Adelaide University in Australia.

[76]This trend also agrees with the colonial mathematical presence in June Barrow-Green's BRITMATH database. This database, which covers British mathematics from 1860 to 1940, contained the biographical data for seventy-one of the colonial mathematicians. Of these mathematicians, only six held university appointments that began before 1871, twenty-four began their appointments from 1871 to 1900, thirty-five held twentieth-century appointments, and six were undetermined.

[77]Nine out of the twenty-six colonial contributors (35%) contributed to the *Messenger of Mathematics*. From the United States, thirteen out of forty-seven (27.7%) of its contributors were published in the *Messenger*; these account for 24.4% of all U.S. contributions and 22.5% of all foreign contributions to the *Messenger*.

Table 4: Foreign Contributors by Region and Decade: 1800-1900

Region or Country*		Number of Contributors : Number of Contributions										
		1800-1809	1810-1819	1820-1829	1830-1839	1840-1849	1850-1859	1860-1869	1870-1879	1880-1889	1890-1900	**1800-1900**
United States							1:1	1:2	10:31	11:17	28:80	**47:131**
France				2:3	2:4	3:3	8:15	3:10	6:16	8:30	11:20	**33:101**
British Colonies	Australia							2:12	4:32	5:17	4:30	**8:91**
	Other			1:1				3:7	5:7	5:9	9:15	**18:39**
Germany						2:4	4:9	3:8	8:11	7:10	4:4	**23:46**
Belgium					1:1	1:4	1:2	1:4	1:24	1:2	1:3	**3:40**
Northern Europe						1:2	5:8	5:8	5:8	4:4	2:6	**17:36**
Italy				1:1			5:7	3:5	5:7	6:9	1:1	**12:30**
Eastern Europe	Austria			1:1	1:1				1:1		2:4	**4:7**
	Other							2:2	2:4	4:5	1:1	**9:12**
Russia						3:4	2:3	1:1	1:1	2:6	2:2	**9:17**
Switzerland					1:2		1:6	1:7				**2:15**
Far East										2:2	3:4	**4:6**
Southern Europe									1:1	2:2	1:1	**3:4**
South America							2:3					**2:3**

* Region and countries are by modern borders.
British Colonies: Australia, Canada, India, South Africa.
Eastern Europe: Austria, Czech Republic, Hungary, Poland, Rumania.
Far East: Japan
Northern Europe: Denmark, Netherlands, Sweden.
South America: Brazil.
Southern Europe: Portugal, Spain, Greece.

The most remarkable growth in foreign mathematical contributions came from the United States. While it produced no mathematical contributions before 1854, the United States provided a substantial number of contributions during the 1870s and 1880s and an unrivaled number during the 1890s.[78] In fact, with the exception of France and Australia, the United States from 1890 to 1900 outnumbered every other foreign country for the entire century in contributors and contributions. This rapid increase followed the general trend of nineteenth-century graduate education

[78]All but nine of the thirty-one American contributions made during the 1870s were to the *Messenger of Mathematics* or its predecessor, the *Oxford, Cambridge, and Dublin Messenger of Mathematics*. However, the *Messenger*'s American contributions waned during the nineteenth century's last two decades: four of the seventeen 1880s contributions and only eight of the eighty 1890s contributions were to this journal.

in the United States; in 1870, there were forty-four graduate students in America, while by 1900 there were as many as 5,668.[79]

Almost one-quarter of the American contributions came from just two men: L. E. Dickson and G. A. Miller. In their "Profile of the American Mathematical Research Community: 1891–1906," Della Fenster and Karen Parshall found that Dickson and Miller were highly active at home as well as in Britain. For the years studied, Dickson published sixty-seven papers and Miller published sixty-two in the four mathematical journals extant in the United States.[80] Moreover, Dickson and Miller were the most prolific contributors to these journals among the sixty-two "most active" members of the American mathematical community at that time.[81]

After earning his doctorate in 1893 from Cumberland College in Kentucky and serving as an instructor for two years at the University of Michigan, Miller augmented his mathematical training in Leipzig and Paris. Throughout this European tour and afterward back in America, Miller focused his energies on group theory. The sheer number of his results often outweighed their profundity, and he "literally made a career out of dissecting finite groups in virtually every way conceivable."[82] His contributions to British journals reflected these interests; in these, Miller listed primitive, transitive, and intransitive substitution groups, or all groups in general, for certain orders. Miller's numerous results sometimes found unconventional homes. The appearance of his articles, "The Substitution Groups Whose Order Is Four," "The Operation Groups of order $8p$, p Being Any Prime Number," "The Transitive Substitution Groups of Order $8p$, p Being Any Prime Number," and "On the Simple Isomorphisms of a Substitution-Group to Itself," to the *Philosophical Magazine* formed a sharp contrast to the applied mathematics articles that usually appeared there.[83]

Dickson also contributed heavily to group theory but attained results with more depth and coherence than Miller's. With an 1896 University of Chicago Ph.D. in hand, Dickson, like Miller, pursued postdoctoral studies in Leipzig and

---

[79]Karen Hunger Parshall and David E. Rowe, *The Emergence of the American Mathematical Research Community (1876-1900): J. J. Sylvester, Felix Klein, and E. H. Moore*, HMATH, vol. 8 (Providence: American Mathematical Society and London: London Mathematical Society, 1994), pp. 262-263.

[80]Della Dumbaugh Fenster and Karen Hunger Parshall, "A Profile of the American Mathematical Research Community: 1891–1906," in *The History of Modern Mathematics*, vol. 3, ed. Eberhard Knobloch and David E. Rowe (Boston: Academic Press, Inc., 1994), pp. 179-227 on p. 186. The four extant journals were the *Bulletin of the American Mathematical Society*, the *Transactions of the American Mathematical Society*, the *Annals of Mathematics*, and the *American Journal of Mathematics*.

[81]"Most active" mathematicians contributed, on average, at least one publication (to the four extant U.S. journals), talk, or year of service (to the American Mathematical Society or the American Association for the Advancement of Science) per year over the period from 1891–1906. See Fenster and Parshall, pp. 184-186. Out of the forty-six American contributors to British journals, sixteen were considered "most active," and six were considered "active" by Fenster and Parshall in the American mathematical community.

[82]*Ibid.*, pp. 209-210.

[83]These appear in the fifth series of the *Philosophical Magazine* in volumes **41** (1896):431-437; **42** (1896):195-200; **43** (1897):117-125; and **45** (1898):234-242, respectively. Brock and Meadows have estimated that, "by the last quarter of the nineteenth century, eighty to ninety percent of all [*Philosophical Magazine*] articles dealt with physics. By the beginning of the twentieth century, the *Philosophical Magazine* was recognised as being essentially a physics journal." Brock and Meadows, p. 206.

Paris. The research Dickson accomplished while on the faculties of the Universities of California, Texas, and, ultimately, Chicago, strengthened the foundation of algebraic excellence in the United States.[84] To the *Quarterly Journal of Pure and Applied Mathematics*, Dickson contributed a series of papers that generalized group-theoretic results from Camille Jordan's *Traité des substitutions*.[85] Dickson also investigated linear groups in several of his nineteenth-century British contributions.[86] He soon codified his research in linear groups in the book, *Linear Groups with an Exposition of the Galois Field Theory*, which appeared in 1901 and inspired American interest in finite group theory.[87]

The mathematical training of Dickson, Miller, and the other forty-five American authors published in British journals provides a characteristic example of the educational paths taken by American mathematics students during the nineteenth century. Almost half of these contributors pursued mathematical studies abroad;[88] this fact illustrates the American reliance on European universities for advanced mathematics training after Sylvester left the Johns Hopkins University in 1883, following his establishment of America's first research-level mathematics program there.[89] However, since over half of these contributors received graduate mathematical training at home,[90] Europe was not the only option for advanced study.

Society Involvement and Personal Influence: Factors in Foreign Participation

Beyond geographical factors, the personal influence of a few British mathematicians as well as the efforts of British scientific societies induced many foreign mathematicians to contribute. Biographical information about the foreign contributors, and references made by the authors themselves in their contributions, indicated personal or societal factors for almost sixty percent of the articles (see Table 5).[91]

---

[84]Parshall and Rowe, p. 381.

[85]Leonard E. Dickson, "A Triply Infinite System of Simple Groups," "The First Hypoabelian Group Generalized," and "Simplicity of the Abelian Group on Two Pairs of Indices in the Galois Field of Order $2n$, $n > 1$," in *Quarterly Journal of Pure and Applied Mathematics* **29** (1898):169-178; **30** (1899):1-16; and **30** (1899):383-384, respectively.

[86]Leonard E. Dickson, "A Class of Linear Groups including the Abelian Group," *Quarterly Journal of Pure and Applied Mathematics* **31** (1900):60-65; "The Structure of Certain Linear Groups with Quadratic Invariants," *Proceedings of the London Mathematical Society* **30** (1899):70-98; and "Concerning the Four Known Simple Linear Groups of Order 25920, with an Introduction to the Hyper-Abelian Linear Groups," *Proceedings of the London Mathematical Society* **31** (1900):30-68.

[87]Karen Hunger Parshall, "A Study in Group Theory: Leonard Eugene Dickson's *Linear Groups*," *The Mathematical Intelligencer* **13** (1) (1991):7-11, and Parshall and Rowe, p. 381.

[88]At least twenty out of forty-seven of these contributors received training abroad, and at least six were trained in Britain. For a discussion of the foreign education of the "most active" American mathematicians, see Fenster and Parshall, pp. 202-203.

[89]For Sylvester's foundation of a research-level mathematics program at the Johns Hopkins, see Karen Hunger Parshall, "America's First School of Mathematical Research: James Joseph Sylvester at the Johns Hopkins University, 1876-1883," *Archive for History of Exact Sciences* **38** (1988):153-196, and Parshall and Rowe, pp. 53-97.

[90]At least twenty-six out of the forty-seven contributors received American graduate training (in addition, possibly, to graduate training abroad); fifteen of these received their training exclusively in the U.S. The training of six of the American contributors has not been determined.

[91]These factors were not disjoint. Thus, the percentages in Table 5 do not add to 100%.

Over 8% of foreign contributors responded to work they had read in British scientific journals. For these responses, British journals represented a forum for constructive argument and criticism between foreign and domestic mathematicians. In his *Convolutions in French Mathematics, 1800–1840*, Ivor Grattan-Guinness described one of these arguments as exemplifying "the growing internationalisation of science."[92] In 1826, Siméon-Denis Poisson published "Sur l'attraction des sphéroïdes" in *Connaissance des temps*, in which he codified his previous work as well as the work of Laplace in this area.[93] James Ivory, a retired Royal Military College professor, disagreed with Poisson's treatment of an integral concerning surface harmonics. The next year, he addressed his concerns with "Some Remarks" in the *Philosophical Magazine*: "[w]ill this pass for demonstration? It is a mere assertion. It is one of those curt and imperative attempts at proof, of which too many occur in the modern mathematics, which are none of its improvements, and which ought never to be admitted without scrupulous examination."[94] Poisson submitted a defense against Ivory's criticisms that was published in the next volume of the *Philosophical Magazine*. While he clarified some of Ivory's objections, Poisson did not definitively settle the argument. Grattan-Guinness described the debate between Poisson and Ivory as "another example of a problem in *multi*-variate analysis which surpassed the powers of the time."[95] The publication of this debate in the *Philosophical Magazine* opened an international dialogue between the two mathematicians.

While quite small, the group of British mathematicians who made international interaction a priority influenced foreign contributors in a variety of ways. Thomas Archer Hirst's efforts in this direction illustrate the variety of forms international encouragement could take.[96] In 1850, Hirst began a lifelong commitment to international interaction with graduate study at the University of Marburg. After earning his Ph.D. there in 1852 for a thesis "On Conjugate Diameters of the Triaxial Ellipsoid," he traveled to Göttingen, where he met Carl Friedrich Gauss. He then visited Berlin, where he began friendships with Peter Lejeune-Dirichlet and Jakob Steiner; the latter's approach to synthetic geometry especially appealed to Hirst. This continental tour, taken so early in his life, ended in Paris where he attended the lectures of Joseph Liouville and Gabriel Lamé.

During a return visit to Paris in 1857, Hirst spent much of his time translating significant mathematical works. In particular, he translated into English the work of Louis Poinsot on the percussion of bodies, which appeared in installments from 1858 to 1859 in the *Philosophical Magazine*.[97] Hirst completed the translation "with the consent of the author and with the advantage of occasional suggestions

[92]Ivor Grattan-Guinness, *Convolutions*, p. 1190.

[93]*Ibid.*, p. 1192.

[94]James Ivory, "Some Remarks on a Memoir by M. Poisson, Read to the Academy of Sciences at Paris, Nov. 20, 1826, and Inserted in the *Connaissance des temps* 1829," *Philosophical Magazine*, new ser., **1** (1827):324-331 on pp. 326-327.

[95]Grattan-Guinness, *Convolutions*, p. 1194 (his emphasis).

[96]The following account of Hirst's life is based on J. Helen Gardner and Robin J. Wilson, "Thomas Archer Hirst—Mathematician Xtravagant, I-VI," *American Mathematical Monthly* **100** (1993):435-441, 531-538, 619-625, 723-731, 827-834, and 907-915.

[97]Louis Poinsot, "On the Percussion of Bodies," *Philosophical Magazine*, ser. 4, **15** (1858):161-180, 263-290, 348-359; and **18** (1859):241-259. On the impact of efforts at translations from the French, compare Ivor Grattan-Guinness's chapter in this volume.

Table 5: Factors for Foreign Contributions to British Journals: 1800-1900		
Factor	**Number**	**Percent***
Asked to contribute by a British mathematician.	8	1.4%
Had an article translated by a British mathematician.	13	2.2%
Contribution was a letter extract to a British mathematician.	15	2.6%
Contribution was communicated by a British mathematician.	20	3.5%
Responded to work read in a British scientific journal.	47	8.1%
Spent time in Britain.†	104	18.0%
Cited a British mathematician in contribution.	82	14.2%
Awarded medals, memberships, or grants by British scientific societies.§	262	45.3%
Undetermined.	248	42.9%

* Percentage out of the 578 contributions. Since these factors are not disjoint, the percentages do not add to 100%.

† The contributor was born in, educated in, worked in, or visited Britain. One hundred of these contributions (96.2%) are from contributors who spent time in Britain *before* making their mathematical contribution.

§ This category includes contributors who received medals, memberships, or grants from British scientific societies *at any time* during their careers.

from him."[98] In his journal, Hirst recounted a meeting with Poinsot, during which the French mathematician remarked, "[w]e cast our seed upon the waters knowing not where it may fall, but it is nevertheless pleasant after long years of labour to find that these seeds have taken root."[99] In his translator's footnote, Hirst extolled the virtues of this work; he then painted a mixed picture of British reception of French mathematical works. Hirst asserted that "[m]athematicians at this day, too, are so well acquainted with the current mathematical literature of our continental neighbours, that for them even an announcement of the publication, in Liouville's *Journal* for September 1857, of Poinsot's most recent memoir is superfluous; for their use, an English version of this memoir is certainly not called for."[100] Hirst did not extend his positive assessment of British mathematicians to all British mathematical practitioners: "[n]evertheless the works of the able author ... are far from being so familiar to Englishmen generally as they deserve to be."[101]

Hirst continued to visit the continent and maintained relationships with several of the foreign mathematical contributors to British journals. Besides translation, he communicated foreign articles and was cited several times in foreign contributions. For example, Rudolf Sturm, a professor at Darmstadt's *Technische Hochschule*,

[98]Thomas Archer Hirst, in *ibid.*, p. 162.

[99]Louis Poinsot, as quoted in Gardner and Wilson, p. 727.

[100]Thomas Archer Hirst, in Poinsot, p. 162.

[101]*Ibid.*

cited Hirst's work in projective geometry[102] as the inspiration for his 1876 article "On Correlative Pencils" in the *Proceedings of the London Mathematical Society*. "The present paper," he acknowledged, "originated from a proposition of my friend, Dr. Hirst, whose own investigations on the correlation of two planes and of two spaces are closely connected with mine."[103]

Besides individual contacts with foreign mathematicians, Hirst exercised his influence within British scientific societies. In 1865, as a member of the Council of the Royal Society of London, he successfully proposed Michel Chasles for the Society's Copley Medal; in 1867, he again successfully recommended Chasles for another society honor, the London Mathematical Society's first foreign membership.[104]

Henry J. S. Smith also encouraged British interaction with international mathematics. An Oxford graduate and holder of the Savilian Chair of Geometry at Oxford from 1860 until his death in 1883, Smith regularly visited foreign mathematicians on the continent and entertained those who visited England. While not in the mainstream of English mathematics, his work in number theory, especially his *Report on the Theory of Numbers* was "well received by continental mathematicians."[105] Smith, like Hirst, used his position in a scientific society to encourage international participation. As President of the Mathematical and Physical Sciences Section of the BAAS, Smith invited Hermite and Felix Klein to the 1873 meeting, where Hermite spoke on the irrationality of e.[106]

British scientific societies often used inducements such as medals or memberships to promote both international scientists in Britain and British societies internationally. In fact, the authors of over 45% of the nineteenth-century foreign mathematical contributions received a medal from, membership in, or grant to one of these societies. Thus, for a significant proportion of the foreign members of the British mathematical publication community, these society rewards either predicted future foreign contributions or rewarded the authors of contributions already made.

---

[102]Hirst made investigations in projective geometry from the 1860s to the end of his life, a period when projective geometry particularly captivated British mathematicians. See Joan L. Richards, *Mathematical Visions: The Pursuit of Geometry in Victorian England* (Boston: Academic Press, Inc., 1988), pp. 131-143.

[103]Rudolf Sturm, "On Correlative Pencils," *Proceedings of the London Mathematical Society* **7** (1876):175-194 on p. 176.

[104]Rice and Wilson, pp. 187-189.

[105]Keith Hannabuss, "Henry Smith," in *Oxford Figures: 800 Years of the Mathematical Sciences*, ed. John Fauvel, Raymond Flood, and Robin Wilson (Oxford: University Press, 2000), pp. 203-217 on p. 206. One of Smith's number-theoretical results was missed by these continental mathematicians and highlighted the fact that mathematics in British journals at that time had not completely entered the international arena. In his "On the Orders and Genera of Quadratic Forms Containing More Than Three Indeterminates," published in the *Proceedings of the Royal Society* in 1867, Smith gave a formula expressing the number of ways an integer can be expressed as a sum of five squares. However, fifteen years later, Smith noticed an announcement in the *Comptes rendus* that posed the same problem for the *Grand Prix* of the Paris Academy of Sciences. After inquiring about this oversight in a letter to Hermite, Smith learned that no one on the committee that posed the problem knew of Smith's earlier solution. The *Grand Prix* eventually went to both Smith and University of Königsberg student, Hermann Minkowski; Smith, however did not live to receive the prize. *Ibid.*, pp. 214-216.

[106]*Ibid.*, p. 212. "[P]resumably his newly discovered proof of the transcendence of e would have been too technical for such a general audience." *Ibid.*

Conclusions

During the nineteenth century, the venues, goals and support of scientific publication in Britain expanded. This expansion reflected the changes in nineteenth-century British mathematics. The journal articles that mathematicians contributed in order to communicate their work now serve as an important record of their mathematical activity. Through the study of these journals, the contours of the nineteenth-century British mathematical community become more clearly defined. In particular, outlets for the publication of research-level mathematics, initially dominated by general scientific journals, increased to include the periodicals of specialized societies and independent mathematics journals. The editors and supporters of these new periodicals shared an increased awareness of the importance of international interaction. As these new outlets opened, international participation in the British mathematical publication community experienced slow but steady growth. However, the lack of steady growth in the percentage of mathematical pages occupied by foreign contributions indicates that the bulk of mathematical journal literature in Britain continued to be domestic.

The geographical diversity of the foreign contributors shows that mathematicians from a variety of countries with established, as well as emerging, mathematical programs recognized British journals as a viable channel for mathematics. Furthermore, the encouragement by influential British mathematicians and British societies of international participation reveals that these individuals and organizations recognized the benefits and need of fostering foreign participation in their journals. Thus, while the British mathematical publication community had not reached a state of internationalization by the end of the nineteenth century, an analysis of British scientific journals confirms that it was most definitely departing from insularity.

References

Adams, John Couch. "Reply to Various Objections Which Have Been Brought against His Theory of the Secular Acceleration of the Moon's Mean Motion." *Monthly Notices of the Royal Astronomical Society* **20** (1860):225-240.

"Address of the Society, Explanatory of Their Views and Objects." *Memoirs of the Royal Astronomical Society* **1** (1822):1-7.

Babbage, Charles. *Reflections of the Decline of Science in England, and on Some of its Causes.* London: B. Fellowes, 1830. Reprint Ed. New York: New York University Press, 1989.

Barrow-Green, June. *BRITMATH* [database] (to appear on the World Wide Web through a Java Web Applet).

Betti, Enrico. "Extract from a Letter of Signor Enrico Betti to Mr. Sylvester." *Quarterly Journal of Pure and Applied Mathematics* **1** (1855):91-92.

———. "Sopra l'equazioni algebriche con più incognite." *Annali di matematica pura ed applicata* **1** (1858):1-8.

Brioschi, Francesco. "Note sur deux théorèmes de géométrie." *Quarterly Journal of Pure and Applied Mathematics* **1** (1855):368-370.

———. "Sur une propriété d'un déterminant fonctionnel." *Quarterly Journal of Pure and Applied Mathematics* **1** (1855):365-367.

Brock, William H. and Meadows, A. Jack. *The Lamp of Learning: Taylor & Francis and the Development of Science Publishing.* London: Taylor & Francis, 1984.

Buée, Adrien-Quentin. "Mémoire sur les quantités imaginaires." *Philosophical Transactions of the Royal Society of London* **46** (1806):23-88.

Chubin, Daryl E. and Moitra, Soumyo D. "Content Analysis of References: Adjunct or Alternative to Citation Counting?" *Social Studies of Science* **5** (1975):423-441.

Cremona, Luigi. "On the Fourteen Point Conic." *Oxford, Cambridge, and Dublin Messenger of Mathematics* **3** (1866):13-14.

______. "On Normals to Conics, a New Treatment of the Subject." *Oxford, Cambridge, and Dublin Messenger of Mathematics* **3** (1866):88-91.

Crilly, Tony. "The Rise of Cayley's Invariant Theory (1841–1862)." *Historia Mathematica* **13** (1986):241-254.

De Solla Price, Derek. "Toward a Model for Science Indicators." In *Toward a Metric of Science: The Advent of Science Indicators*, Ed. Yehuda Elkana *et al.* New York: John Wiley and Sons, 1978, pp. 69-95.

Dickson, Leonard E. "A Class of Linear Groups including the Abelian Group." *Quarterly Journal of Pure and Applied Mathematics* **31** (1900):60-66.

______. "Concerning the Four Known Simple Linear Groups of Order 25920, with an Introduction to the Hyper-Abelian Linear Groups." *Proceedings of the London Mathematical Society* **31** (1900):30-68.

______. "The First Hypoabelian Group Generalized." *Quarterly Journal of Pure and Applied Mathematics* **30** (1899):1-16.

______. "Simplicity of the Abelian Group on Two Pairs of Indices in the Galois Field of Order $2n$, $n > 1$." *Quarterly Journal of Pure and Applied Mathematics* **30** (1899):383-384.

______. "The Structure of Certain Linear Groups with Quadratic Invariants." *Proceedings of the London Mathematical Society* **30** (1899):70-98.

______. "A Triply Infinite System of Simple Groups." *Quarterly Journal of Pure and Applied Mathematics* **29** (1898):169-178.

Edge, David. "Quantitative Measures of Communication in Science: A Critical Review." *History of Science* **17** (1979):103-134.

Enneper, Alfred. "Elementary Demonstration of an Equation Between Two Transcendental Functions." *Quarterly Journal of Pure and Applied Mathematics* **1** (1855):272-276.

______. "A General Theorem Relating to Multiple Periodic Series." *Quarterly Journal of Pure and Applied Mathematics* **1** (1855):280-285.

______. "On the Definite Integral $\int_0^\infty \frac{\sin^q u}{u^m} du$." *Quarterly Journal of Pure and Applied Mathematics* **1** (1855):276-279.

______. "On the Function $\Pi(z)$ with Imaginary and Complex Variable." *Quarterly Journal of Pure and Applied Mathematics* **1** (1855):393-402.

Faà de Bruno, Francesco. "Invariant of the Twelfth Degree of the Quintic (a, b, c, d, e) $(x, y)^5$." *Quarterly Journal of Pure and Applied Mathematics* **1** (1855):361-363.

______. "Note sur une nouvelle formule de calcul différentiel." *Quarterly Journal of Pure and Applied Mathematics* **1** (1855):359-360.

Fauvel, John. "800 Years of Mathematical Traditions." In *Oxford Figures: 800 Years of the Mathematical Sciences.* Ed. John Fauvel, Raymond Flood, and Robin J. Wilson. Oxford: University Press, 2000, pp. 1-27.

Fenster, Della Dumbaugh and Parshall, Karen Hunger. "A Profile of the American Research Community: 1891-1906." In *The History of Modern Mathematics.*

Vol. 3. Ed. Eberhard Knobloch and David E. Rowe. Boston: Academic Press, Inc., 1994, pp. 179-227.

Forsyth, Andrew Russell. "James Whitbread Lee Glaisher." *Journal of the London Mathematical Society* **4** (1929):101-112.

Gardner, J. Helen and Wilson, Robin J. "Thomas Archer Hirst—Mathematician Xtravagant, I-VI." *American Mathematical Monthly* **100** (1993):435-441, 561-538, 619-625, 723-731, 827-834, and 907-915.

Gascoigne, Robert M. *A Historical Catalogue of Scientific Periodicals, 1665-1900.* New York: Garland Publishing, Inc, 1985.

Gillispie, Charles C., Ed. *Dictionary of Scientific Biography.* 16 Vols. and 2 Supps. New York: Charles Scribner's Sons, 1970–1990.

Grattan-Guinness, Ivor. "Contributing to *The Educational Times*: Letters to W. J. C. Miller." *Historia Mathematica* **21** (1994):204-205.

———. *Convolutions in French Mathematics, 1800–1840.* Vol. 2. *The Turns.* Basel: Birkhäuser Verlag and Berlin: Deutscher Verlag der Wissenschaften, 1990.

———. "Mathematics and Mathematical Physics from Cambridge, 1815–1840: A Survey of the Achievements and of the French Influences." In *Wranglers and Physicists.* Ed. Peter M. Harman. Manchester: University Press, 1985, pp. 84-110.

Gregory, Duncan. "Preface." *Cambridge Mathematical Journal* **1** (1837):1-2.

Hannabuss, Keith. "Henry Smith." In *Oxford Figures: 800 Years of the Mathematical Sciences.* Ed. John Fauvel, Raymond Flood, and Robin J. Wilson. Oxford: University Press, 2000, pp. 203-217.

Hansen, Peter Andreas. "Extract of a Letter from Prof. Hansen to the Astronomer Royal, Dated Gotha, May 31, 1859." *Monthly Notices of the Royal Astronomical Society* **19** (1859):236-237.

———. "From a Letter from Professor Hansen to the Astronomer Royal, Dated Gotha, 1861, Feb. 2." *Monthly Notices of the Royal Astronomical Society* **21** (1861):152-155.

Hardy, Godfrey Harold. "Dr. Glaisher and the 'Messenger of Mathematics'." *Messenger of Mathematics* **58** (1929):159-160.

Hermite, Charles. "Correspondence: Mr. Cayley and M. Hermite on Cubic Forms." *Quarterly Journal of Pure and Applied Mathematics* **1** (1855):88-89.

———. "Extract of a Letter from M. Hermite to Professor Sylvester." *Quarterly Journal of Pure and Applied Mathematics* **1** (1855):370-373.

———. "Sur la théorie des fonctions homogènes à deux indéterminées." *Cambridge and Dublin Mathematical Journal* **9** (1854):172-217.

———. "Sur les formes cubiques à deux indéterminées." *Quarterly Journal of Pure and Applied Mathematics* **1** (1855):20-22.

Hoppe, Reinhold. "Determination of the Motion of Conoidal Bodies through an Incompressible Fluid." *Quarterly Journal of Pure and Applied Mathematics* **1** (1855):301-315.

Hyman, Anthony. *Science and Reform: Selected Works of Charles Babbage.* Cambridge: University Press, 1989.

Ivory, James. "Some Remarks on a Memoir by M. Poisson, Read to the Academy of Sciences at Paris, Nov. 20, 1826, and Inserted in the *Connaissance des temps* 1829." *Philosophical Magazine.* New Ser. **1** (1827):324-331.

Kline, Morris. *Mathematical Thought from Ancient to Modern Times.* New York: Oxford University Press, 1972.

Lyons, Sir Henry. *The Royal Society, 1660-1940.* Cambridge: Cambridge University Press, 1944.

MacLeod, Roy M. "Of Medals and Men: A Reward System in Victorian Science 1826-1914." *Notes and Records of the Royal Society of London* **25** (1971):81-105.

Meadows, A. Jack. *Communication in Science.* London: Butterworth & Co., 1974.

Miller, George A. "The Operation Groups of Order $8p$, p Being Any Prime Number." *Philosophical Magazine.* Ser. 5. **42** (1896):195-200.

———. "On the Simple Isomorphisms of a Substitution-Group to Itself." *Philosophical Magazine.* Ser. 5. **45** (1898):234-242.

———. "The Substitution Groups Whose Order is Four." *Philosophical Magazine.* Ser. 5. **41** (1896):431-437.

———. "The Transitive Substitution Groups of Order $8p$, p Being Any Prime Number." *Philosophical Magazine.* Ser. 5. **43** (1897):117-125.

Milton, Viscount (Charles William). "Address." *Report of the British Association for the Advancement of Science* **1** (1831):15-17.

Morrell, Jack and Thackray, Arnold. *Gentlemen of Science: Early Years of the British Association for the Advancement of Science.* Oxford: Clarendon Press, 1981.

"Obituary; Charles Eugène Delaunay." *Monthly Notices of the Royal Astronomical Society* **33** (1873):203-209.

"Obituary; John Couch Adams." *Monthly Notices of the Royal Astronomical Society* **53** (1893):184-209.

"Obituary; Peter Andreas Hansen." *Monthly Notices of the Royal Astronomical Society* **35** (1875):168-170.

Parshall, Karen Hunger. "America's First School of Mathematical Research: James Joseph Sylvester at the Johns Hopkins University, 1876-1883." *Archive for History of Exact Sciences* **38** (1988):153-196.

———. *James Joseph Sylvester: Life and Work in Letters.* Oxford: Clarendon Press, 1998.

———. "A Study in Group Theory: Leonard Eugene Dickson's *Linear Groups.*" *The Mathematical Intelligencer* **13** (1) (1991):7-11.

———. "Toward a History of Nineteenth-Century Invariant Theory." In *The History of Modern Mathematics.* Ed. David E. Rowe and John McCleary. 2 Vols. Boston: Academic Press, Inc., 1989, 1:157-206.

Parshall, Karen Hunger and Rowe, David E. *The Emergence of the Americn Mathematical Research Community, 1876–1900: J. J. Sylvester, Felix Klein, and E. H. Moore.* HMATH. Vol. 8. Providence: American Mathematical Society and London: London Mathematical Society, 1994.

Playfair, John. "*Traité de méchanique céleste.*" *Edinburgh Review* **11** (1807-1808): 249-284.

Poinsot, Louis. "On the Percussion of Bodies." *Philosophical Magazine.* Ser. 4. **15** (1858):161-180, 263-290, 348-359; **18** (1859):241-259.

Pontécoulant, Comte de (Phillipe Gustave Doulcet). "Mécanique céleste: Observations sur la comparison établie par M. Delaunay entre les expressions des coordonnées de la lune déduites de sa théorie avec celles qui avaient été obtenues antérieurement." *Monthly Notices of the Royal Astronomical Society* **23** (1863):259-266.

———. "Sur la variation séculaire du moyen mouvement de la lune." *Monthly Notices of the Royal Astronomical Society* **19** (1859):235-236.

"Preface." *Oxford, Cambridge, and Dublin Messenger of Mathematics* **3** (1866):v-vi.

"Preface." *Transactions of the Royal Irish Academy* **1** (1787):ix-xvii.

Rankin, Robert A. "The First Hundred Years." *Proceedings of the Edinburgh Mathematical Society* **26** (1983):135-150.

Record of the Royal Society of London. London: Harrison and Sons, 1897.

Rice, Adrian C. and Wilson, Robin J. "From National to International Society: The London Mathematical Society, 1867-1900." *Historia Mathematica* **25** (1998):185-217.

Richards, Joan L. *Mathematical Visions: The Pursuit of Geometry in Victorian England.* Boston: Academic Press, Inc., 1988.

Small, Henry G. "The Lives of a Scientific Paper." In *Selectivity in Information Systems: Survival of the Fittest.* Ed. Kenneth S. Warren. New York: Praeger Science Publishers, 1985, pp. 83-97.

Sturm, Rudolf. "On Correlative Pencils." *Proceedings of the London Mathematical Society* **7** (1876):175-194.

Sylvester, James Joseph; Cayley, Arthur; Stokes, George Gabriel; Ferrers, Norman Macleod; and Hermite, Charles. "Address to the Reader." *Quarterly Journal of Pure and Applied Mathematics* **1** (1855):i-ii.

Toplis, John. "On the Decline of Mathematical Studies, and the Sciences Dependent on Them." *Philosophical Magazine* **20** (1805):25-31.

Wagner-Döbler, Roland and Berg, Jan. "Nineteenth-Century Mathematics in the Mirror of Its Literature: A Quantitative Approach." *Historia Mathematica* **23** (1996):288-318.

CHAPTER 5

International Participation in Liouville's *Journal de mathématiques pures et appliquées*

Jesper Lützen
Copenhagen University (Denmark)

Introduction

Since the seventeenth century, journals have been an important vehicle for the exchange of scientific knowledge in general and of mathematics in particular. Although most journals had a national base, they were often distributed internationally and so tended to advance internationalization. This was certainly true of Liouville's *Journal de mathématiques pures et appliquées*, which, after Crelle's *Journal für die reine und angewandte Mathematik* (f. 1826), is the oldest specialized mathematics journal to enjoy continuous publication from its founding to the present. Begun in 1836 as a national French enterprise, Liouville's *Journal* was widely read abroad and helped to spread French mathematical ideas throughout the Western world. It further facilitated internationalization through the significant number of foreign contributions it published. These helped both to acquaint the French mathematical community with foreign mathematical ideas and to overcome the self-centered attitude that had prevailed in France throughout the first third of the nineteenth century. The present chapter focuses on the foreign contributions to Liouville's *Journal* and on the contributions by its editor, Joseph Liouville, that were instrumental in bringing foreign, in particular German, results and approaches to France in such areas as mechanics, potential theory, and differential geometry.

The National Enterprise

In the 1830s, Paris was the center of mathematics in the West. French mathematicians could look back on a glorious past with "revolutionary" masters such as Joseph-Louis Lagrange, Pierre Simon Laplace, Adrien-Marie Legendre, Gaspard Monge, Joseph Fourier, and André-Marie Ampère. Although this old guard was dying out, plenty of excellent younger people were waiting in the wings to replace them: Siméon-Denis Poisson; Louis Poinsot; Victor Poncelet; Augustin-Louis Cauchy; Joseph Liouville; Jean-Claude Bouquet; Michel Chasles; Jean-Marie-Constant Duhamel; Gabriel Lamé; Louis Navier; Adhémar Barré, Comte de Saint-Venant; and Charles Sturm were among the most prominent of this next generation working in Paris in the 1830s. With hindsight, it is clear that after this time Germany gradually overtook France as the world's leading power in mathematics. However in the 1830s, there was no reason to believe that the future would be less glorious for French mathematics than its past and present. In fact, with new generations of mathematicians such as Joseph Bertrand, Pierre-Ossian Bonnet, Charles

Hermite, Victor Puiseux, Joseph Serret and later Gaston Darboux, Camille Jordan, Émile Mathieu, and Edmond Laguerre, not to mention Henri Poincaré and later mathematicians, Paris probably continued to outmatch any one German city be it Göttingen, Berlin, or Königsberg.[1]

Yet, after 1831, France did not have a specialized mathematical journal in which these mathematicians could publish their papers. The only journal of that type was Crelle's Berlin-based *Journal für die reine und angewandte Mathematik*. It was a twenty-six-year-old assistant at the *École polytechnique*, Joseph Liouville (1809-1882), who took up the challenge to produce and sustain a new French mathematics journal.

Liouville had already felt personally the need for such a journal earlier in his career. His first mathematical paper was published in 1830 in the *Annales de mathématiques pures et appliquées*, founded in 1810 by Joseph Diaz Gergonne as the world's first, widely circulated, specialized mathematics journal. However, the year after Liouville's paper appeared, Gergonne discontinued the publication of the *Annales*, owing to his appointment as rector of the *Académie de Montpellier* and to the time pressures of that post. The same year, Ferussac's *Bulletin des sciences mathématiques, physiques et chimiques*, in which Liouville had also published a short note, was likewise discontinued. Liouville thus turned to the less specialized *Journal de l'École polytechnique* and to the *Mémoires des savants étrangers* of the *Académie des sciences* as outlets for his productivity. The latter, unfortunately, had a backlog of roughly five years, so to make his ideas known more quickly and more widely Liouville sent five papers to August Leopold Crelle for inclusion in what Liouville himself called his "excellent journal."[2] In 1835, on the initiative of the physicist and astronomer, François Arago, the *Académie des sciences* did begin to publish a weekly *Comptes rendus des séances de l'Académie des sciences*, which contained short notes written or presented by members of the *Académie*, but Liouville still felt that the mathematical capital of the world ought to have a mathematics journal of its own.

Liouville got Gergonne's permission to declare his new journal a continuation of the *Annales*, and he decided to give it a similar title, adopting the same slight variation that Crelle had chosen. Liouville ended the "Avertissement" of the first volume of the *Journal de mathématiques pures et appliquées* with the following plea to the French community of mathematicians:

> Now, it is up to the geometers, in particular, the French geometers, to make this enterprise prosper. The most distinguished among them have already promised us some articles, and doubtless they will keep their promise. We dare say that it is in the interest of their reputation: the failure of a useful journal which they have refused to support would be honorable neither for them nor for France.[3]

---

[1]Compare the argument on the supposed decline of French mathematics in Ivor Grattan-Guinness's chapter above.

[2]See Erwin Neuenschwander, "Die Edition mathematischer Zeitschriften im 19. Jahrhundert und ihr Beitrag zum wissenschaftligen Austausch zwischen Frankreich und Deutschland," Mathematisches Institut der Universität Göttingen, Preprint 4 (1984), p. 58.

[3]"C'est maintenant aux géomètres, et surtout aux géomètres français, qu'il appartient de faire prospérer cette entreprise. Les plus distingués d'entre eux nous ont déjà promis des articles, et, sans aucun doute, ils tiendront leur promesse. Nous osons dire que leur réputation y est

As this statement makes clear, Liouville's creation of his journal had definite national overtones and led, in fact, to publication patterns along more nationalistic lines. Indeed, in the period after the discontinuation of Gergonne's *Annales*, not only Liouville but many other French mathematicians had published rather widely in Crelle's *Journal*;[4] this cross-frontier activity almost stopped in 1836 when French mathematicians began to publish in Liouville's new French journal instead. In one sense, then, Liouville's *Journal* hindered rather than promoted international collaboration. This national tendency was counterbalanced, however, by the international contributions to the *Journal*, which informed French mathematicians about the progress outside France.

Foreign Contributions to Liouville's *Journal*

From the start, Liouville succeeded in persuading most of the prominent French mathematicians to publish their work in his journal. The list of foreign contributors is also impressive, including such names as:[5] Arthur Cayley (17,0), Pafnuti Chebyshev (17,10), Rudolf Clausius (6,5), Peter Lejeune-Dirichlet (21,15), Gotthold Eisenstein (2,1), Carl Friedrich Gauss (3,3), Carl G. J. Jacobi (22,18), Leopold Kronecker (8,2), Ernst Kummer (5,3), James MacCullagh (1,1), Ferdinand Minding (2,1), Franz Neumann (2,2), Julius Plücker (3,1), Ernst Schering (1,0), Ludwig Schläfli (1,0), Hermann Schwarz (1,0), Jakob Steiner (4,2), Lord Kelvin (7,1), and Karl Weierstraß (1,1).

In the first series (the first twenty volumes) of the *Journal*, 25% of the papers were written by foreign authors. Their distribution according to country was: Germany 8.6%, Great Britain 9.4% (of which 5.3% were Irish and 1% Scottish), Belgium 2.9%, Italy 1.8%, Russia 1%, and Scandinavia 0.4%.[6] This percentage of foreign contributions is not significantly lower than the corresponding percentage, 30%, of foreign contributions to the first fifty volumes of Crelle's *Journal* (1826–1855).[7] The figures are somewhat deceptive, however. In Crelle's *Journal*, the vast majority of papers were original in the sense that they were published there for the first time. This was also the case for the French contributions to Liouville's *Journal*, but, as is apparent from the numbers in the previous paragraph, a majority of the foreign

intéressée: la chute d'un Journal utile qu'ils auraient refusé de soutenir ne serait honorable ni pour eux ni pour la France." Joseph Liouville, "Avertissement," *Journal de mathématiques pures et appliquées* **1** (1836):1-4 on p. 4.

[4]Neuenschwander, p. 28.

[5]This list is drawn from *ibid.*, pp. 31-35. The numbers in parenthesis are the number of papers by the author in Liouville's *Journal* during the period 1836–1880, followed by the number of these papers that had already appeared elsewhere.

[6]These numbers are somewhat uncertain. I have been unable to find information about several of the insignificant contributors to Liouville's *Journal*. Most of these are clearly French (students at the Parisian schools, etc.). I have distributed these authors according to their names; authors have been classified as French if they have French-sounding surnames. This means that some Belgian authors may have been misclassified. Moreover, problems arise when assigning a nationality to a foreigner who lived more or less permanently in Paris. I have, for example, classified Guglielmo Libri and Sturm as French, although they were born in Italy and Switzerland, respectively, but I have classified Catalan as Belgian. Franz Woepcke's thirteen papers have not been counted at all because it seems equally wrong to classify them as German or French. A more careful search for the minor contributors and different choices in cases of doubt would change the numbers slightly, but this would certainly not change the overall picture.

[7]Neuenschwander, p. 28. Compare also the figures for Great Britain in Sloan Despeaux's chapter above.

contributions, in particular those from German writers, were French translations of papers that had originally been published in other languages (primarily in German and Latin) in other journals (such as Crelle's). In fact, in one essential respect Liouville's *Journal* was decidedly less international than its German counterpart. Liouville only published papers in French (disregarding a few short republications of papers in Latin). Crelle, on the other hand, published 38% of the papers in the first fifty volumes of his *Journal* in languages other than German. Moreover, as Neuenschwander has noted,[8] the first eleven volumes of Liouville's *Journal* did not contain any original contributions by German mathematicians based outside of Paris, except short letters to the editor.

In his "Avertissement," Liouville had explicitly stated his intention that, for the most part, the *Journal* would contain original papers submitted directly by their authors. He planned only occasionally to reprint papers or summaries of papers that had already appeared elsewhere. As far as the French contributions were concerned, he stuck to this decision, but he clearly changed his mind regarding foreign papers. He may have discovered that foreigners were more reluctant than he had expected to send him papers. In order to increase the international participation in his journal, he asked his friends and colleagues such as Victor Amédée Lebesgue and Hervé Faye to translate foreign papers for his journal. This activity, already in evidence in the journal's second volume, indicated that Liouville worked diligently for both an internationalization of mathematics and an exchange of ideas across national borders. However, his "Avertissement" equally reflected his feeling that this international collaboration should be centered in Paris and that it should take place in French. As noted above, Liouville succeeded in establishing an international presence in the first series of his journal; he was less successful in the second series, the nineteen volumes from 1856 to 1874. Only 14% of the papers in the second series were written by foreign authors, and these were distributed among the following countries: Germany 7.0%, England 1.9%, Ireland 0.2%, Italy 0.5%, Russia 2.3%, and Scandinavia 0.5%.

These percentages require comment. The high percentage of French contributions in the second series resulted partly from a change in Liouville's own personal publishing patterns. In the first series, he wrote 16.5% of the papers himself. Many of these were long or medium-length papers; others were short notes. In

TABLE 1:
BREAKDOWN OF CONTRIBUTORS TO
LIOUVILLE'S *Journal* BY SERIES

Papers by French Authors (Series 1)	391	68.5%
Papers by Foreign Authors (Series 1)	180	31.5%
Papers by French Authors (Series 2)	259	74.6%
Papers by Foreign Authors (Series 2)	88	25.4%

the second series, however, he flooded the volumes with rather insignificant short notes on number theory. They made up 44% of all the papers in the second series! Quite naturally, this skews the statistics and suggests a French domination of the *Journal*. Yet, if we disregard Liouville's contributions to the two series as has been

[8]Neuenschwander, pp. 28-29.

done in generating the data in Table 1 above, a general downward trend relative to foreign contributions is still apparent. In particular, the last eight volumes of the second series contained few foreign contributions and hardly any German ones. The latter fact was probably due, at least in part, to the Franco-Prussian War,[9] but it was no doubt equally a result of the general decrease in the journal's quality; Liouville had lost both his mathematical and his organizational power. In fact, when Liouville finally left the editorship to Henri Résal in 1875, the *Journal* almost died. It was rescued at the last minute by Camille Jordan who became the editor in 1885.

Countering the Sense of French Self-sufficiency

During its period of glory, 1770–1830, Parisian mathematics had become virtually self-sufficient. It cannot be denied that French mathematicians recognized a *princeps mathematicorum* like Gauss, but they did not appreciate Gauss's works as highly as they should have.[10] Moreover, although foreigners like Lagrange and Sturm, who lived in Paris, were accepted, knowledge of and interest in foreign mathematics was minimal. As Niels Henrik Abel put it, somewhat harshly, in 1826: "I do not like the Frenchman as well as the German. The Frenchman is very reserved towards foreigners. ... Everybody [in Paris] works for himself without caring about others. Everybody wants to teach, nobody wants to learn. The most absolute egotism prevails everywhere. The only thing the Frenchman seeks in a foreigner is the practical. Nobody can think except him. He is the only one who can produce something theoretical. These are his thoughts."[11]

After the inauguration of Liouville's *Journal*, French mathematicians gradually became more aware of mathematical progress outside of France. It is naturally hard to estimate how much the foreign papers in Liouville's *Journal* contributed to this and how much was simply due to the fact that there was relatively more, important, foreign (in particular German) mathematics to be aware of. Nevertheless, since many French mathematicians did not read German, the translations and original French contributions by German (and other non-French) authors made these foreign ideas more readily available to the French mathematical community.

This said, it would be misleading to conclude that, just because Liouville promoted and was influenced by (see below) foreign mathematics, his French contemporaries were equally open to foreign ideas. Liouville was probably the most internationally oriented, nineteenth-century French mathematician before Hermite, Darboux, and Guillaume-Jules Hoüel. He had many personal contacts with foreign mathematicians, some of his own age like Dirichlet, Jacobi, and Steiner, and many of a younger generation like William Thomson (later Lord Kelvin), Michael and William Roberts (from Ireland), Chebyshev, Erik Holmgren, and Carl Anton Bjerknes. The members of this younger generation had met Liouville during stays

[9]On the impact of the Franco-Prussian War on mathematics in France, see the chapters by Hélène Gispert and Thomas Archibald below.

[10]On the reception of Gauss's work in France, see Ivor Grattan-Guinness's chapter above.

[11]"Ellers synes jeg ikke saa godt om Franskmanden som Tyskeren. Franskmanden er umaadelig tilbageholden mod Fremmede. ... Enhver arbeider for sig selv uden at bryde sig om andre. Alle ville belære og ingen vil lære. Den meest absolute Egoisme herser [sic] overalt. Det eneste Franskmanden søger hos fremmede er det Practiske; Ingen kan tænke uden ham. Han er den eneste som kan frembringe noget theoretisk. Saaledes er hans Tanker." Niels Henrik Abel to Berndt Holmboe, 24 October, 1826, in *Festskrift ved hundredaarsjubilæet for Niels Henrik Abels fødsel*, ed. Elling Holst, Carl Størmer, and Ludvig Sylow (Kristiania: Jacob Dybwad, 1902), p. 44.

in Paris and had benefited from his hospitality as well as from his mathematical instruction and guidance. Through these friends (Dirichlet in particular) and through the papers sent to his *Journal*, Liouville—unlike most of his colleagues—was well-informed about mathematical developments outside France. The following incident illustrates the point.

In 1847, Gabriel Lamé believed that he had found a proof of Fermat's last theorem based on factorization in cyclotomic integers. When he reported on his proof at a meeting of the *Académie des sciences*, Liouville rose and pointed out that the proof relied on the (unproved) assumption that—as in the ordinary integers—factorization in prime numbers was unique in the cyclotomic integers. "Isn't there a gap to be filled here?" Liouville asked.[12] (Indeed there is, and the gap cannot be filled.) Liouville's prompt reaction was made possible by the international work he had done for his *Journal*.

In 1843, Liouville had asked Faye to translate a paper by Jacobi for the journal.[13] In his paper, Jacobi had pointed out that "if λ is a divisor of $p-1$, the prime p can be represented in several ways as a product of two complex numbers made up of λth roots of unity."[14] As Harold Edwards has convincingly argued, it was this remark that alerted Liouville to the problematic point in Lamé's "proof." Indeed, in his objection, Liouville specifically pointed out that "it is especially in a paper by Mr. Jacobi ... that one can find useful information."[15]

A few months later, Ernst Kummer informed Liouville that the unique factorization property does not hold, and he sent Liouville a paper on the matter. Liouville published this, as well as Lamé's purported proof, in his *Journal*. It was symptomatic of the situation in Paris that, even after Kummer had given this final blow to Lamé's method, Cauchy tried to get part of the credit for the proof by referring to some earlier papers containing ideas similar to Lamé's.[16] It was equally characteristic of Liouville's outward-looking mathematical persona that, in 1857, he worked successfully within the *Académie des sciences* to secure a prize for Kummer for his contributions to the solution of Fermat's last theorem. Moreover, in connection with this award,[17] Liouville's foreign contacts, and especially Dirichlet, helped him fend off the criticisms Cauchy put forward regarding Kummer's ideas.

---

[12] "N'y a-t-il pas là une lacune à remplir." Joseph Liouville, "Remarques à l'occasion d'une communication de M. Lamé sur un théorème de Fermat," *Comptes rendus des séances de l'Académie des sciences de Paris* **24** (1847):315-316 on p. 316. For a full account of the incident, see Harold M. Edwards, "The Background of Kummer's Proof of Fermat's Last Theorem for Regular Primes," *Archive for History of Exact Sciences* **14** (1975):219-236.

[13] Carl G. J. Jacobi, "Über die complexen Primzahlen, welche in der Theorie der Reste der 5ten, 8ten und 12ten Potenzen zu betrachten sind," *Monatsbericht der Akademie der Wissenschaften zu Berlin 1839* and *Journal für die reine und angewandte Mathematik* **19** (1839):314-318, or Carl G. J. Jacobi, *C. G. J. Jacobi's Gesammelte Werke*, ed. Carl W. Borchardt and Karl Weierstraß, 7 vols. (Berlin: Verlag von G. Reimer, 1881–1891), 6:275-280. The French translation appeared as "Sur les nombres premiers complexes que l'on doit considérer dans la théorie des résidus," *Journal de mathématiques pures et appliquées* **8** (1843):268–272.

[14] "dass, wenn λ ein Theiler von $p-1$ ist, sich die Primzahl p, und in der Regel auf mehrere verschiedene Arten, als Product zweier complexen Zahlen darstellen läfst, welche aus λ^{ten} Wurzeln der Einheit zusammengesetzt sind." Jacobi, *Werke*, 6:279.

[15] "c'est surtout dans un article de M. Jacobi (*Journal de mathématiques*, tome VIII, page 268), que l'on pourra trouver des renseignements utiles." Liouville, "Remarques à l'occasion d'une communication de M. Lamé sur un théorème de Fermat," p. 316.

[16] Edwards has shown that Cauchy never took the trouble to read Kummer's work carefully.

[17] On the utility of prizes in the internationalization of mathematics, compare Thomas Archibald's chapter in the present volume.

International Inspirations for Liouville's Work: Mechanics

During the first ten years of his career as a mathematician (1830–1840), Liouville drew his mathematical inspiration almost exclusively from his French peers. From then on, however, his research was increasingly inspired by foreign works. In particular, Jacobi and, to a somewhat lesser extent, William Rowan Hamilton, sparked his interest in and shaped his approach to mechanics.[18]

The first foreign work that provoked Liouville to write a paper in mechanics was a short note that Jacobi had sent to the *Académie des sciences* in 1834.[19] There, Jacobi announced his discovery that certain ellipsoids with three different axes could serve as figures of equlibrium of rotating masses of fluid (that is, fluid planets). This was a great surprise because, in the *Mécanique céleste*, Laplace had analyzed only those ellipsoids of revolution (i.e., ellipsoids with two equal equatorial axes) that Colin Maclaurin had already shown to be in equilibrium. Moreover, Lagrange had "proved" that such equilibrium figures must possess rotational symmetry. According to Liouville, in his communication to the *Académie*, Jacobi "promised even more by announcing that he would take up the *Mécanique céleste* again in, what he called, the pitiable state where Laplace had left it."[20] Liouville immediately took up this challenge to French mathematics. Within a week, he had proved that Jacobi's result was a relatively easy corollary of formulas that could be found in the *Mécanique céleste*, thereby restoring Laplace's reputation.[21]

It is interesting that the first time Liouville was directly inspired by a foreign work, his aim was to defend the French colors. This nationalist attitude soon gave way, however, to an appreciation of foreign contributions. In his original paper, Liouville had cast doubt on the importance of Jacobi's result, but in 1839, he described Jacobi's theorem as "remarkable,"[22] and in an unpublished note in 1843, he termed it "Mr. Jacobi's beautiful discovery."[23] In 1842, Liouville continued to highlight the importance of Jacobi's work. He showed that, for those values of the angular momentum where there exist both a Maclaurin and a Jacobi ellipsoid in equilibrium, the Jacobi ellipsoid is in stable equilibrium. He published this

---

[18]This and the following two sections focus specifically on international influences on Liouville's work in mechanics, potential theory, and differential geometry. It is equally the case that his work on elliptic functions was inspired by Abel and Jacobi and that his number-theoretic research drew crucially from Dirichlet.

[19]Carl G. J. Jacobi, "Über die Figur des Gleichgewichts," *Annalen der Physik und Chemie* **33** (1834):229-233, or Jacobi, *Werke*, 2:17-22.

[20]"et certes M. Jacobi promettait davantage, en annonçant qu'il allait reprendre la Mécanique céleste dans l'état pitoyable, disait'il, où Laplace l'avait laissée." This quote is taken from Liouville's paper, "Note sur la figure d'une masse fluide homogène, en équilibre, et douée d'un mouvement de rotation," *Journal de l'École polytechnique* **14** (1834):289-296. See the footnote on pp. 290-291 there. I have not been able to locate Jacobi's letter to the *Académie*. The phrase does not occur in the printed German version of Jacobi's paper, namely, "Über die Figur des Gleichgewichts."

[21]Liouville, "Note sur la figure d'une masse fluide homogène."

[22]Joseph Liouville, "Observations sur un mémoire de M. Ivory, sur l'équilibre des ellipsoïdes homogènes," *Journal de mathématiques pures et appliquées* **4** (1839):169-174 on p. 169.

[23]Joseph Liouville, "Recherches sur la stabilité de l'équilibre des fluides" (1843). This note was published in Appendix B of Jesper Lützen "Joseph Liouville's Work on the Figures of Equilibrium of a Rotating Mass of Fluid," *Archive for History of Exact Sciences* **30** (1984):113-166. The quote is on p. 161.

surprising result without proof, although enough evidence remains to reconstruct his reasoning.[24]

In his later works on rational mechanics, Liouville followed Jacobi even more closely. In 1838, Liouville published translations of two of Jacobi's major papers on mechanics and differential equations that had appeared the previous year in Crelle's *Journal*.[25] Moreover, in 1840, he reprinted a note that Jacobi had published in the *Comptes rendus* concerning Poisson brackets.[26] These papers, as well as others by Poisson, led Liouville to his famous theorem to the effect that if one knows half the integrals of the equations of motion of a mechanical system and if these are in involution, then one can find the remaining integrals by quadrature. It is even possible that Liouville had discussed related matters with Jacobi when Jacobi visited Paris in 1842. Indeed, in Jacobi's lecture the following semester (published in 1866), he explicitly formulated this *theorema gravissimum*.[27] Liouville's geometrization of the principle of least action was likewise a direct continuation of Jacobi's researches on this matter.[28]

Finally, Liouville's papers on so-called Liouvillian integrable mechanical systems sprang directly from a geometrical paper by Jacobi that Liouville had printed in translation in his *Journal* in 1841.[29] In this paper, Jacobi showed how ellipsoidal coordinates could be used to determine the equation of geodesics on ellipsoids. His paper, however, did not contain any proofs; these were supplied by Liouville in 1844. Liouville's idea was to consider the geodesics as paths of particles moving on the surface of the ellipsoid when no forces act on them (so-called geodesic motion). He then showed that, when elliptical coordinates are used, one can easily separate variables in Lagrange's equations of motion. This had also been Jacobi's route of discovery, but Liouville did not know that. Liouville then realized that this method could be generalized to many other surfaces and to mechanical systems influenced by conservative forces, as long as the potential energy had a simple

---

[24]See Lützen, "Liouville's Work on the Figures of Equilibrium," and Jesper Lützen, *Joseph Liouville: Master of Pure and Applied Mathematics* (New York: Springer-Verlag, 1990), chapter 11. These also contain a more thorough discussion of Liouville's work on rotating masses of fluid as a whole than can be given here.

[25]Carl G. J. Jacobi, "Zur Theorie der Variations-Rechnung und der Differential-Gleichungen," *Journal für die reine und angewandte Mathematik* **17** (1837):68-82, or Jacobi, *Werke*, 4:37-55; and "Über die Reduction der Integration der partiellen Differentialgleichungen erster Ordnung zwischen irgend einer Zahl Variabeln auf die Integration eines einzigen Systems gewöhnlicher Differentialgleichungen," *Journal für die reine und angewandte Mathematik* **17** (1837):97-162, or Jacobi, *Werke*, 4:57-127. The French translations appeared as "Sur le calcul des variations et sur la théorie des équations différentielles," *Journal de mathématiques pures et appliquées* **3** (1838):44-59, and "Sur la réduction de l'intégration des équations différentielles partielles du premier ordre entre un nombre quelconque de variables à l'intégration d'un seul système d'équations différentielles ordinaires," *Journal de mathématiques pures et appliquées* **3** (1838):60-96 and 161-201, respectively.

[26]Carl G. J. Jacobi, "Sur un théoreme de Poisson," *Comptes rendus des séances de l'Académie des sciences de Paris* **11** (1840):529; reprinted in *Journal de mathématiques pures et appliquées* **5** (1840):350–351, or Jacobi, *Werke*, 4:143-146.

[27]Lützen, *Liouville*, pp. 670-679.

[28]*Ibid.*, pp. 680-686.

[29]Carl G. J. Jacobi, "Note von der geodätischen Linie auf einem Ellipsoid und den verschiedenen Anwendungen einer merkwürdigen analytischen Substitution," *Journal für die reine und angewandte Mathematik* **19** (1839):309-313, or Jacobi, *Werke*, 2:59-63. For the French translation, see "De la ligne géodésique sur un ellipsoïde, et des différents usages d'une transformation analytique remarquable," *Journal de mathématiques pures et appliquées* **6** (1841):267-272.

form that allowed for the separation of variables. He published these ideas in three major papers in 1846 and 1847.[30]

As these various examples make clear, Liouville's works in mechanics were, for the most part, inspired directly by those works of Jacobi published in Liouville's *Journal*. Other French mathematicians who followed Jacobi's lead include Bertrand and Edmond Bour.[31] Their papers, but especially Liouville's, contributed much to the spread of Jacobi's ideas on mechanics in France.

International Inspirations for Liouville's Work: Potential Theory

Potential theory was another main focus of Liouville's research, and, here again, he drew critically from the work of foreign mathematicians. In 1839, the French geometer, Michel Chasles, had been misled by an analogy with heat conduction to announce that a given conducting surface could carry several equilibrium charge distributions. Owing to his reading of Gauss's work, however, Liouville soon set Chasles straight. In 1842, Liouville published a translation of a then-recent paper of Gauss that detailed theorems pertinent to Chasles's claim.[32] Gauss's work, as well as Liouville's own ongoing research on ellipsoids of revolution, induced Liouville to embark on a study of potential theory for ellipsoids. Nothing much came of this until 1845 when the young William Thomson visited Paris and gave Liouville a copy of George Green's privately published and virtually unknown *Essay on the Application of Mathematical Analysis to the Theories of Electricity and Magnetism*.[33] Strangely enough, Liouville did not print a translation of Green's work in his journal,[34] but he was very enthusiastic about its content. This new potential-theoretic input prompted Liouville to return to Gauss's problem, namely, given a function on a

---

[30] Joseph Liouville, "Sur quelques cas particuliers où les équations du mouvement d'un point matériel peuvent s'intégrer: Premier mémoire," *Journal de mathématiques pures et appliquées* **11** (1846):345-378; "Sur quelques cas particuliers ou les équations du mouvement d'un point matériel peuvent s'intégrer: Second mémoire," *Journal de mathématiques pures et appliquées* **12** (1847):410-444; and "Mémoire sur l'intégration des équations différentielles du mouvement d'un nombre quelconque de points matériels," *Connaissance des temps pour 1850* (1847):1-40, and *Journal de mathématiques pures et appliquées* **14** (1849):257-299.

[31] Bertrand was chosen as the editor of the third edition of Lagrange's *Mécanique analytique* (1853) and, probably in preparation, he devoted his course of 1852–1853 at the *Collège de France* to mechanics. It is unclear if his resulting interest in Jacobi's work was influenced directly by Liouville or only indirectly through Liouville's earlier inclusion of Jacobi's papers in his *Journal*. Bour, on the other hand, was primarily inspired by Bertrand, secondarily by Liouville.

[32] Carl Friedrich Gauss, "Allgemeine Lehrsätze in Beziehung auf die im verkehrten Verhältnisse des Quadrats der Entfernung wirkenden Anziehungs- und Abstossungs-Kräfte," *Resultate aus den Beobachtungen des magnetischen Vereins im Jahre 1839*, ed. Carl Friedrich Gauss and Wilhelm Weber (Leipzig: n.p., 1840), or Carl Friedrich Gauss, *Werke*, ed. Königlichen Gesellschaft der Wissenschaften zu Göttingen, 12 vols. (Göttingen: Dieterischen Universitätsdruckerei, 1863–1929), 5:197-242. The French translation appeared as "Théorèmes généraux sur les forces attractives et répulsives qui agissent en raison inverse du carré des distances," *Journal de mathématiques pures et appliquées* **7** (1842):273-324.

[33] George Green, *An Essay on the Application of Mathematical Analysis to the Theories of Electricity and Magnetism* (Nottingham, 1828). This was eventually published in installments in the *Journal für die reine und angewandte Mathematik* **39** (1850):73-89; **44** (1852):356-374; and **47** (1854):161-221. See also George Green, *Mathematical Papers of the Late George Green*, ed. Norman M. Ferrers (London: Macmillan, 1871), pp. 3-115.

[34] He probably found it too long. In the publication of the *Essay* in Crelle's *Journal* in 1850, 1852, and 1854, it takes up ninety-four pages.

closed surface S, show that there exists a charge distribution on the surface that gives rise to the potential U on the surface.

In a series of papers from 1845 to 1846, Liouville showed how some important theorems he had deduced for Lamé functions (ellipsoidal harmonics) in connection with his work on fluid ellipsoids could be used to solve Gauss's problem analytically for an ellipsoid. Moreover, he hinted at a generalization of these results to arbitrary surfaces, in which case the Lamé functions must be replaced by what would today be called eigenfunctions of a certain integral operator related to the surface and the equilibrium measure on it. The only published evidence of this generalization is a short note from 1845 in which Liouville proved, in modern terms, the "orthogonality" of the eigenfunctions corresponding to different eigenvalues of the integral operator and the reality of the eigenvalues.[35] However, Liouville's notebooks contain enough calculations to allow a reconstruction of a large part of his theory, a theory remarkable in many respects. Liouville found the eigenvalues and eigenfunctions by a variational method named after Lord Rayleigh and Walter Ritz, who used it more than forty years later. Liouville also pointed out the problematic nature of the Dirichlet principle. Above all, the general path Liouville followed in his theory is strikingly similar to a modern treatment of the spectral theory of compact operators and unparallelled in the published literature for the next four decades.[36]

As with the examples drawn from Liouville's research in mechanics, the influence of foreign mathematics on Liouville's work in potential theory is clear. Liouville intimately knew the work of French authors like Chasles and Lamé, but his major source of inspiration came from two foreigners, Gauss and Green, the work of the former made available to francophone mathematicians thanks to Liouville's *Journal*.

International Inspirations for Liouville's Work: Differential Geometry

Liouville's work in differential geometry can be divided into four somewhat overlapping periods. Between 1841 and 1844, he gave analytic treatments of theorems having their roots in synthetic geometry. Much of this research was inspired primarily by Chasles in France and Steiner in Germany, but Liouville's work on geodesics and other curves on ellipsoids drew additionally from the ideas of the Dublin mathematicians, Michael Roberts and James MacCullagh, as well as from those of the German mathematician, Ferdinand Joachimsthal.[37] From 1844 to 1845, Liouville shifted his focus. In work closely related to research of Jacobi, he used mechanics, as well as elliptic and abelian functions, to treat geodesics and especially those on ellipsoids. In 1845, however, William Thomson diverted Liouville's attention from these studies when he brought the geometric transformation

[35] Joseph Liouville, "Sur une propriété générale d'une classe de fonctions," *Journal de mathématiques pures et appliquées* **10** (1845):327-328.

[36] See Lützen, *Liouville*, chapter fifteen for Liouville's work on potential theory in general and for this unpublished theory in particular. Compare, too, Jesper Lützen "The Birth of Spectral Theory—Joseph Liouville's Contributions," *Proceedings of the International Congress of Mathematicians, Kyoto 1990*, ed. Ichiro Satake (Tokyo: Springer-Verlag, 1991), pp. 1651-1663, and Christian Berg and Jesper Lützen, "J. Liouville's Unpublished Work on an Integral Operator in Potential Theory: A Historical and Mathematical Analysis," *Expositiones Mathematicae* **8** (1990):97-136.

[37] Lützen, *Liouville*, pp. 710-715.

called inversion in spheres under Liouville's view. Liouville published several papers on this over the five years from 1845 to 1850 before turning to a series of important papers on the Gaussian differential geometry of surfaces from 1847 to around 1852. In all these works, except those from the first period, Liouville's main inspiration came from foreign papers partly published in his *Journal*. These papers, like the journal that contained them, were instrumental in bringing these new ideas to France. In order to illuminate this transfer of knowledge, consider Liouville's final two differential-geometric periods.

During a stay in Paris in 1845, William Thomson became friendly with Liouville, who encouraged his work on electricity and magnetism and who invited him to write papers for his *Journal*. This was a method that Liouville often used to get foreign contributions. While in Paris, Thomson discussed his ideas about inversions in spheres (electrical images, as Thomson called them) with Liouville, and after his return to Cambridge, he sent Liouville three letters on the subject, which were promptly printed in the *Journal* together with a long note appended by Liouville.[38] In his note, Liouville discussed a formula given by Thomson expressing the distance between two points in terms of the distance between their images under the transformation. He noted that this formula showed that the transformation was conformal (angle-preserving). In the paper that followed, he further showed that, in the plane, the holomorphic functions and their conjugates are the only conformal maps,[39] a fact that he made more explicit in a footnote to a paper by Roberts.[40] This result had been implied, but nowhere explicitly stated, in a prize essay by Gauss to the Royal Danish Academy in 1822,[41] but it is uncertain if Liouville knew of this paper. Liouville went on to state the problem of finding all the conformal mappings of three dimensional space to itself, but in 1848 he could not solve it. Two years later, however, he announced his solution in his *Journal*[42] and included a proof of it as "Note 6" in his new edition of Monge's *Application de l'analyse à la géométrie.*[43] The result, now usually called Liouville's theorem, stated that the only conformal mappings of $\mathbb{R}^3$ into itself are those composed of an inversion in a

[38]See William Thomson (Lord Kelvin), "Extrait d'une lettre de M. William Thomson à M. Liouville," *Journal de mathématiques pures et appliquées* **10** (1845):364-367, and "Extraits de deux lettres adressées à M. Liouville," *Journal de mathématiques pures et appliquées* **12** (1847):256-264 as well as Joseph Liouville, "Note sur deux lettres de M. Thomson relatives à l'emploi d'un système nouveau de coordonnées orthogonales dans quelques problèmes des théories de la chaleur et de l'électricité, et au problème de la distribution d'électricité sur le segment d'une couche sphérique infiniment mince," *Journal de mathématiques pures et appliquées* **12** (1847):265-290.

[39]Joseph Liouville, "Sur un théorème de M. Gauss concernant le produit des deux rayons de courbure principaux en chaque point d'une surface," *Journal de mathématiques pures et appliquées* **12** (1847):291-304, and *Comptes rendus des séances de l'Académie des sciences de Paris* **25** (1847):707.

[40]Michael Roberts, "Extrait d'une lettre adressée à M. Liouville," *Journal de mathématiques pures et appliquées* **13** (1848):209-220.

[41]Carl Friedrich Gauss, "Allgemeine Auflösung der Aufgabe: Die Theile einer gegebenen Fläche auf einer andern gegebenen Fläche so abzubilden, dass die Abbildung dem Abgebildeten in den kleinsten Theilen ähnlich wird (Als Beantwortung der von der königlichen Societät der Wissenschaften in Copenhagen für 1822 aufgegebenen Preisfrage)," *Astronomische Abhandlungen herausgegeben von H. C. Schumacher* **3** (1825):1-30, or Gauss, *Werke*, 4:189-216.

[42]Joseph Liouville "Théorème sur l'équation $dx^2+dy^2+dz^2=\lambda(d\alpha^2+d\beta^2+d\gamma^2)$," *Journal de mathématiques pures et appliquées* **15** (1850):103.

[43]Gaspard Monge, *Application de l'analyse à la géométrie*, 5th ed., ed. Joseph Liouville (Paris: Bachelier, 1850).

sphere and a similitude. This famous result thus traces its roots directly back to Thomson's papers on electrical images published in Liouville's *Journal.*

While an Englishman's work drove this discovery during the third period of Liouville's research in differential geometry, it was Gauss's pivotal 1828 tract, "Disquisitiones generales circa superficies curvas," that informed Liouville's ideas on the theory of surfaces from 1847 to 1852.[44] Gauss's paper was, however, unknown in France prior to the 1840s. In 1843, for example, Bertrand admitted in print that "after having written this memoir, I have learned about a memoir by Mr. Gauss entitled *Disquisitiones generales.*"[45] This early French ignorance of Gauss's masterpiece was replaced by a keen interest in 1847, and, as pointed out by Karin Reich, this change was due to Liouville.[46] Indeed, in that year, Liouville published a new proof of the central *theorema egregium* in Gauss's "Disquisitiones generales" and that inspired other mathematicians to other proofs.[47] Liouville himself contributed many other important papers in this field. They were characterized by a more conscious, intrinsic approach to differential geometry than any other work before Riemann's *Habilitationsschrift.*[48] Liouville also made the "Disquisitiones generales" more readily available to French mathematicians by republishing it in his new edition of Monge's *Application.* Many of his own results were also appended to this work with the explicit aim of drawing the attention of young mathematicians to Gauss's approach.

Conclusion

During the years following the discontinuation of Gergonne's *Annales*, French mathematicians published a large part of their works in Crelle's *Journal.* This cross-frontier activity ended in 1836 with the founding of Liouville's *Journal des mathématiques pures et appliquées.* The immediate consequence of the appearance of this journal was thus a more national publication pattern. However, this national tendency was counterbalanced by Liouville's conscious attempt to internationalize the journal. He had several friends abroad (some of them former students) who sent him original papers, and he arranged for his French colleagues to prepare French translations for his journal of interesting papers published in foreign journals. These international contributions helped keep French mathematicians informed about mathematical developments abroad.

At a more individual level, Liouville himself often took these foreign ideas as the point of departure for his own work, thereby pushing internationalization even

[44]Carl Friedrich Gauss, "Disquisitiones generales circa superficies curvas," *Commentationes societatis regiae scientiarum Gottengensis recentiores* (Math. Classe) **6** (1828):99-146, or Gauss, *Werke*, 4:217-258. The text was also included later in Monge, *Application de l'analyse à la géométrie*, ed. Joseph Liouville, pp. 505-546.

[45]"Après avoir rédigé ce Mémoire, j'ai eu connaissance d'un Mémoire de M. Gauss, intitulé: *Disquisitiones generales.*" Joseph Bertrand, "Démonstration de quelques théorèmes sur les surfaces orthogonales," *Journal de l'École polytechnique* **17** (1843):157-173 on p. 158.

[46]Karin Reich, "Die Geschichte der Differentialgeometrie von Gauss bis Riemann (1828–1868)," *Archive for History of Exact Sciences* **11** (1973):273-382.

[47]Liouville, "Sur un théorème de M. Gauss." Compare *Comptes rendus des séances de l'Académie des sciences de Paris* **25** (1847):707.

[48]Bernhard Riemann "Ueber die Hypothesen welche der Geometrie zu Grunde liegen" (Habilitationsvortrag, 1854), *Abhandlungen der königlichen Gesellschaft der Wissenschaften zu Göttingen* (Math. Classe) **13** (1867):1-20, or Bernhard Riemann, *Bernhard Riemann's Gesammelte Mathematische Werke*, ed. Heinrich Weber (Leipzig: B. G. Teubner Verlag, 1892), pp. 272-287. Compare Lützen, *Liouville*, pp. 739-755.

further. His republication of Jacobi's works, for example, and his own works on mechanics introduced Jacobi's ideas to the French mathematical community, while his republication of Gauss's "Disquisitiones generales" and his many contributions to the intrinsic geometry of surfaces did a similar service to Gaussian differential geometry. Despite all of his publication efforts in the French context, Liouville did not discourage French mathematicians from publishing abroad. He himself published two papers in foreign journals after the founding of his *Journal.* In 1846, he supported the initiative of his young friend, William Thomson, by publishing a paper in the first volume of the *Cambridge and Dublin Mathematical Journal*, a continuation of the *Cambridge Mathematical Journal* that changed its name when Thomson became its editor. In 1856, he made a similar gesture in support of the first volume of Oscar Xaver Schlömilch's *Zeitschrift für Mathematik und Physik.* The established editor with the nationalistic as well as internationalistic outlook seemed consciously to encourage similar outlooks in his counterparts abroad.

References

Abel, Niels Henrik. *Festskrift ved hundredaarsjubilet for Niels Henrik Abels fødsel.* Ed. E. Holst, C. Størmer, and L. Sylow. Kristiania: Jacob Dybwad, 1902.

Berg, Christian and Lützen, Jesper. "J. Liouville's Unpublished Work on an Integral Operator in Potential Theory: A Historical and Mathematical Analysis." *Expositiones Mathematicae* **8** (1990):97-136.

Bertrand, Joseph. "Démonstration de quelques théorèmes sur les surfaces orthogonales." *Journal de l'École polytechnique* **17** (1843):157-173.

Edwards, Harold M. "The Background of Kummer's Proof of Fermat's Last Theorem for Regular Primes." *Archive for History of Exact Sciences* **14** (1975): 219-236.

Gauss, Carl Friedrich. "Allgemeine Auflösung der Aufgabe: Die Theile einer gegebenen Fläche auf einer andern gegebenen Fläche so abzubilden, dass die Abbildung dem Abgebildeten in den kleinsten Theilen ähnlich wird (Als Beantwortung der von der königlichen Societät der Wissenschaften in Copenhagen für 1822 aufgegebenen Preisfrage)." *Astronomische Abhandlungen herausgegeben von H. C. Schumacher* **3** (1825):1-30. In *Werke*, 4:189-216.

———. "Allgemeine Lehrsätze in Beziehung auf die im verkehrten Verhältnisse des Quadrats der Entfernung wirkenden Anziehungs- und Abstossungs-Kräfte." *Resultate aus den Beobachtungen des magnetischen Vereins im Jahre 1839.* Ed. Carl F. Gauss and Wilhelm Weber. Leipzig: N.p., 1840. In *Werke*, 5:197-242. French translation as "Théorèmes généraux sur les forces attractives et répulsives qui agissent en raison inverse du carré des distances." *Journal de mathématiques pures et appliquées* **7** (1842):273-324.

———. "Disquisitiones generales circa superficies curvas." *Commentationes societatis regiae scientiarum Gottengensis recentiores* (Math. Classe) **6** (1828): 99-146. In *Werke*, 4:217–258.

———. *Werke.* Ed. Königlichen Gesellschaft der Wissenschaften zu Göttingen. 12 Vols. Göttingen: Dieterischen Universitätsdruckerei, 1863–1929.

Green, George. *An Essay on the Application of Mathematical Analysis to the Theories of Electricity and Magnetism.* Nottingham, 1828. In *Journal für die reine und angewandte Mathematik* **39** (1850):73-89; **44** (1852):356-374; and

47 (1854):161–221. In *Mathematical Papers of the Late George Green.* Ed. Norman M. Ferrers. London: Macmillan, 1871, pp. 3-115.

Jacobi, Carl Gustav Jacob. *C. J. G. Jacobi's Gesammelte Werke.* Ed. Carl W. Borchardt and Karl Weierstraß. 7 Vols. Berlin: Verlag von G. Reimer, 1881–1891.

———. "Note von der geodätischen Linie auf einem Ellipsoid und den verschiedenen Anwendungen einer merkwürdigen analytischen Substitution." *Journal für die reine und angewandte Mathematik* **19** (1839):309-313. In *Werke,* 2:59-63. French translation as "De la ligne géodésique sur un ellipsoïde, et des différents usages d'une transformation analytique remarquable." *Journal de mathématiques pures et appliquées* **6** (1841):267-272.

———. "Sur un théorème de Poisson." *Comptes rendus des séances de l'Académie des sciences de Paris* **11** (1840):529, and *Journal de mathématiques pures et appliquées* **5** (1840):350-351. In *Werke,* 4:143-146.

———. "Über die complexen Primzahlen, welche in der Theorie der Reste der 5ten, 8ten und 12ten Potenzen zu betrachten sind." *Monatsbericht der Akademie der Wissenschaften zu Berlin* (1839):186-191, and *Journal für die reine und angewandte Mathematik* **19** (1839):314-318. In *Werke,* 6:275-280. French translation as "Sur les nombres premiers complexes que l'on doit considérer dans la théorie des résidus." *Journal de mathématiques pures et appliquées* **8** (1843):268-272.

———. "Über die Figur des Gleichgewichts." *Annalen der Physik und Chemie* **33** (1834):229-233. In *Werke,* 4:17-22.

———. *Vorlesungen über Dynamik gehalten an der Universität zu Königsberg in Wintersemester 1842–1843 und nach einem von C. W. Borchardt ausgearbeiteten Hefte, Berlin 1866.* In *Werke,* Supplementband.

———. "Zur Theorie der Variations-Rechnung und der Differential-Gleichungen." *Journal für die reine und angewandte Mathematik* **17** (1837):68-82. In *Werke,* 4:37-55. French translation as "Sur le calcul des variations et sur la théorie des équations différentielles." *Journal de mathématiques pures et appliquées* **3** (1838):44-59.

Liouville, Joseph. "Avertissement." *Journal de mathématiques pures et appliquées* **1** (1836):1-4.

———. "De la ligne géodésique sur un ellipsoïde quelconque." *Journal de mathématiques pures et appliquées* **9** (1844):401-408.

———. "Mémoire sur l'intégration des équations différentielles du mouvement d'un nombre quelconque de points matériels." *Connaissance des temps pour 1850* (1847):1-40, and *Journal de mathématiques pures et appliquées* **14** (1849):257-299.

———. "Note sur deux lettres de M. Thomson relatives à l'emploi d'un système nouveau de coordonnées orthogonales dans quelques problèmes des théories de la chaleur et de l'électricité, et au problème de la distribution d'électricité sur le segment d'une couche sphérique infiniment mince." *Journal de mathématiques pures et appliquées* **12** (1847):265-290.

———. "Note sur la figure d'une masse fluide homogène, en équilibre, et douée d'un mouvement de rotation." *Journal de l'École polytechnique* **14** (1834):289-296.

———. "Observations sur un mémoire de M. Ivory, sur l'équilibre des ellipsoïdes homogènes." *Journal de mathématiques pures et appliquées* **4** (1839):169-174.

———. "Recherches sur la stabilité de l'équilibre des fluids." (1843) Manuscript 3640 (1880-). Bibliothèque de l'Institut de France. In Lützen, Jesper. "Joseph Liouville's Work on the Figures of Equilibrium of a Rotating Mass of Fluid." *Archive for History of Exact Sciences* **30** (1984):113-166. Appendix B.

———. "Remarques à l'occasion d'une communication de M. Lamé sur un théorème de Fermat." *Comptes rendus des séances de l'Académie des sciences de Paris* **24** (1847):315-316.

———. "Solution d'un problème relatif à l'ellipsoïde." *Comptes rendus des séances de l'Académie des sciences de Paris* **20** (1845):1609-1612.

———. "Sur quelques cas particuliers où les équations du mouvement d'un point matériel peuvent s'intégrer: Premier mémoire." *Journal de mathématiques pures et appliquées* **11** (1846):345-378.

———. "Sur quelques cas particuliers ou les équations du mouvement d'un point matériel peuvent s'intégrer: Second Mémoire." *Journal de mathématiques pures et appliquées* **12** (1847):410-444.

———. "Sur un théorème de M. Gauss concernant le produit des deux rayons de courbure principaux en chaque point d'une surface." *Journal de mathématiques pures et appliquées* **12** (1847):291-304, and *Comptes rendus des séances de l'Académie des sciences de Paris* **25** (1847):707.

———. "Sur une propriété générale d'une classe de functions." *Journal de mathématiques pures et appliquées* **10** (1845):327-328.

———. "Théorème sur l'équation $dx^2 + dy^2 + dz^2 = \lambda(d\alpha^2 + d\beta^2 + d\gamma^2)$." *Journal de mathématiques pures et appliquées* **15** (1850):103.

Lützen, Jesper. "The Birth of Spectral Theory: Joseph Liouville's Contributions." *Proceedings of the International Congress of Mathematicians, Kyoto 1990.* Ed. Ichiro Satake. Tokyo: Springer-Verlag, 1991, pp. 1651–1663.

———. *Joseph Liouville: Master of Pure and Applied Mathematics.* New York: Springer-Verlag, 1990.

———. "Joseph Liouville's Work on the Figures of Equilibrium of a Rotating Mass of Fluid." *Archive for History of Exact Sciences* **30** (1984):113-166.

Monge, Gaspard. *Application de l'analyse à la géometrie.* 5th Ed. Ed. Joseph Liouville. Paris: Bachelier, 1850.

Neuenschwander, Erwin. "Die Edition mathematischer Zeitschriften im 19. Jahrhundert und ihr Beitrag zum Wissenschaftligen Austausch zwischen Frankreich und Deutschland." Mathematisches Institut der Universität Göttingen. Preprint 4. 1984.

Reich, Karin. "Die Geschichte der Differentialgeometrie von Gauss bis Riemann (1828-1868)." *Archive for History of Exact Sciences* **11** (1973):273-382.

Riemann, Bernhard. "Ueber die Hypothesen welche der Geometrie zu Grunde liegen." (Habilitationsvortrag 1854). *Abhandlungen der königlichen Gesellschaft der Wissenschaften zu Göttingen.* Math. Classe **13** (1867):1-20. In *Bernhard G. F. Riemann's Gesammelte Mathematische Werke.* Ed. Heinrich Weber. Leipzig: B. G. Teubner Verlag, 1892, pp. 272-287.

Roberts, Michael. "Extrait d'une lettre adressée à M. Liouville." *Journal de mathématiques pures et appliquées* **13** (1848):209-220.

Thomson, William (Lord Kelvin). "Extrait d'une lettre de M. William Thomson à M. Liouville." *Journal de mathématiques pures et appliquées* **10** (1845):364-367.

———. "Extraits de deux lettres adressées à M. Liouville." *Journal de mathématiques pures et appliquées* **12** (1847):256-264.

CHAPTER 6

The Effects of War on France's International Role in Mathematics, 1870–1914

Hélène Gispert
Université de Paris Sud (France)

Introduction

On the surface, the internationalization of mathematical activity seems to have developed irrespective of national boundaries and characteristics. But how can such a process have occurred during periods when national identity was high on the political agenda? In the case of France, the interval between 1870 and 1914—from its entry and swift defeat in the Franco-Prussian War to the outbreak of World War I—was a period when nationalistic rhetoric was at its height. One of the major effects of the French defeat in 1871 was a desire for sweeping national reforms in science and higher education. The French case is thus of particular interest for examining the question of the construction of an international community.

Yet, during the period under consideration, the explicit—or even implicit—idea of the desirability of an international mathematical community is not easily found in mathematical France. Moreover, the belief that there was mathematical activity abroad with which French mathematicians should be acquainted and involved was very much a minority view during the 1860s and 1870s. Indeed, the work of men like Gaston Darboux and Charles Hermite was quite exceptional. In 1869, Darboux founded the *Bulletin des sciences mathématiques et astronomiques*, a journal specifically designed to increase French awareness of foreign mathematical research. Contemporaneously, Hermite worked to these same ends, utilizing his rich store of international contacts and awareness.[1]

If one French mathematical area could be described as international at this time, it was mathematical pedagogy. Its proponents tried, at the very end of the century, to attract foreign students to the Sorbonne; Charles-Ange Laisant was one of the two founders of the international review, *L'Enseignement mathématique*; and French mathematicians such as Carlo Bourlet and others were active in the International Commission on Mathematical Instruction (ICMI). In mathematical pedagogy, but only in this field, it seems possible and relevant to pose the question of the role—and not just the place—of French mathematicians or mathematics on the international mathematical stage.

Trying to access France's international role raises yet another difficulty, however: it cannot be viewed only from within. Any analysis needs also to take into account the real effects of French initiatives and mathematical life as seen from

[1]On Hermite's activities to bring foreign mathematics to France, see the chapter by Thomas Archibald that follows.

abroad. Since this would necessarily draw from a very large work as yet unwritten, the present chapter can only serve as an initial attempt to deal with this vast topic. It will stress a number of historical factors at work in the internationalization of mathematics that may merit future exploration. It takes the following as its point of departure: one of the major effects of France's defeat was the desire for a national recovery in science and in higher education. To this end, national societies were created to organize and support this nationalistic will, of which two, partly or exclusively, involved mathematical fields: the *Société mathématique de France* (SMF) and the *Association française pour l'avancement des sciences* (AFAS). Both defined mathematical networks on a national as well as an international scale. Since foreign mathematicians attended their meetings and congresses, and since their publications included foreign papers, these bodies both should have had some form of international standing.[2] Taking into account the variety of participants and the differences between the mathematical standards and research values of the two bodies—on a national and international scale—thus sheds light on the complexity of internationalization processes in the decades around the turn of the twentieth century.

Part I: The Effects of War on French Mathematical Standing

The Creation of a National Mathematical Society: Nationalism and Professionalization

Just before the war, in an official report on the progress of geometry in France, Michel Chasles had expressed publicly for the first time the need for a learned society exclusively devoted to the mathematical sciences. He was conscious of the limitations both of the old élitist societies then extant in France and of the publications that French mathematicians had at their disposal. He suggested the creation of a society targeting all *géomètres*, which would aim to make its members' works known through regular publication. Chasles took as an example the recently created London Mathematical Society and its *Proceedings*, expressing concern that the existence of such a society and journal put the British at a definite mathematical advantage:

> Mathematics is making considerable advances abroad. A simple fact would suffice to show everyone how much we should fear being left behind in this part of the sciences .... We have in our *Société philomatique* a mathematical section, with a small number of members, whose communications only appear after long intervals [and then] together with [those] on other topics in a very limited, trimestrial *Bulletin*; however, a mathematical society was founded in London in 1865 with a membership of one hundred, and this number is increasing; a society whose

[2]For more on these two societies, see Hélène Gispert, "Réseaux mathématiques en France dans les débuts de la Troisième République," *Archives internationales d'histoire des sciences* **49** (1999):122-149.

French activity in the International Commission on Mathematical Instruction (ICMI) and the international dimension and role of the *Comptes rendus de l'Académie des sciences de Paris* also warrant analysis in the context of France's international role in mathematics, but are not considered here.

> *Proceedings*, like those of the Royal Society of London, ... publishes abstracts, more or less extended, of many papers. Is not [the existence of the *Proceedings*], which we applaud, an ingredient of future superiority in mathematical culture that should worry us? It worries us all the more because for twenty years the state of our classical mathematical studies has experienced a weakening that we cannot ignore.[3]

In criticizing the institutional features of the practice and diffusion of knowledge in France, Chasles joined numerous French scientists who had tried increasingly in the 1860s to warn their countrymen of the decline of French scientific expertise. The French defeat in the Franco-Prussian War amplified the criticisms of scientists like Louis Pasteur, who, in a newspaper article in March 1871, claimed that it was German science that had won the war:

> I propose to prove in this piece that if, at its moment of greatest peril, France did not find superior men to utilize the resourcefulness and the courage of its children, we must attribute this, I believe, to France's lack of interest, over the past half-century, in the great works of thought, particularly in the exact sciences. ... Whereas Germany has increased its universities, established a more beneficial rivalry between them, bestowed its teachers and its professors with honor and consideration, created vast laboratories endowed with the finest instruments, France ... has given only sporadic attention to its establishments of higher education.[4]

With the war over and the revolutionary *Commune de Paris* violently suppressed, French mathematicians quickly moved to create their new organization. The *Société mathématique de France*, officially founded in 1872, was open, without restriction, to all those with an interest in the advancement and dissemination of

---

[3] "Les mathématiques prennent à l'étranger des développements considérables. Un simple fait suffirait pour montrer aux yeux de tous combien nous devons craindre de nous laisser arriérer dans cette partie des sciences .... Nous possédons dans notre société philomatique une section des mathématiques, d'un nombre de membres limité, dont les communications ne paraissent que de loin en loin avec d'autres matières dans un *Bulletin* trimestriel fort restreint; or il s'est formé à Londres, en 1865, une Société mathématique d'une centaine de membres, et le nombre s'en accroît encore; société dont les *Proceedings*, à l'instar de la Société royale de Londres ... , font connaître les travaux par des analyses plus ou moins étendues. Ce fait [l'existence des *Proceedings*] auquel nous applaudissons n'est-il pas dans la culture des Mathématiques un élément de supériorité future qui doit nous préoccuper? Il nous préoccupe d'autant plus que l'état de nos études classiques des Mathématiques a éprouvé, depuis une vingtaine d'années, un affaiblissement que l'on ne peut se dissimuler." Michel Chasles, *Rapport sur les progrès de la géométrie* (Paris: Imprimerie nationale, 1870), pp. 378-379.

[4] "Je me propose de démontrer dans cet écrit que si, au moment du péril suprême, la France n'a pas trouvé des hommes supérieurs pour mettre en oeuvre ses ressources et le courage de ses enfants, il faut l'attribuer, j'en ai la conviction, à ce que la France s'est désintéressée, depuis un demi-siècle, des grands travaux de la pensée, particulièrement dans les sciences exactes. ... Tandis que l'Allemagne multipliait ses universités, qu'elle établissait entre elles la plus salutaire émulation, qu'elle entourait ses maîtres et ses docteurs d'honneur et de considération, qu'elle créait de vastes laboratoires dotés des meilleurs instruments de travail, la France ... ne donnait qu'une attention distraite à ses établissements d'instruction supérieure." Louis Pasteur in *Le Salut Public* (Lyon) (March 1871); reprint ed., *Pour l'avenir de la science française* (Paris: Éditions "Raison d'être," Collection "À la lumière des textes oubliés," 1947).

mathematics. Chasles's comments, however, reveal two separate issues underlying its formation. It was created not only to develop mathematics in France but also in reaction to the development of mathematics abroad. When viewed in this light, it is far from obvious that a national society should be the first step towards an international community.

In the same year, two other societies were also created: the *Société française de physique* and the *Association française pour l'avancement des sciences.* These three societies shared the nationalistic intention of working to promote French scientific research. Émile Bouty, speaking of the creation of the *Société française de physique*, might just as well have had the SMF and the AFAS in mind when he remarked: "If they were led to seek out one another by the love of science, another sentiment further strengthens their union: the love of country. As far as their impact may extend, they want, for their part, to contribute to the development of the intellectual and moral strength of France."[5]

The *Société mathématique de France* and its *Bulletin*, the latter precisely the sort of publication venue that Chasles had found lacking in France in 1870, had the precise aim of organizing and reinforcing the French mathematical community by promoting the production of its members. As far as the SMF was concerned, however, the word "production" applied only to the publication of members' original research and to its diffusion within the society. A comparison with the intentions of another journal, the *Bulletin des sciences mathématiques et astronomiques*, provides for a clearer understanding of the standards adopted by the *Bulletin* of the SMF.

Unlike Darboux's *Bulletin des sciences mathématiques et astronomiques*, which aimed to increase awareness and availability in France of papers published abroad, the *Bulletin* of the SMF openly declared its exclusivity and its aim to become the showcase for the research of its members. This stance seems peculiar, considering the more open publication policies of the journals of other national societies at the end of the nineteenth century such as those of the *Deutsche Mathematiker-Vereinigung* (DMV) or of the American Mathematical Society.[6] The effects of such a position in France, during a period of strong professionalization of mathematical research activity, are extremely important. They conditioned the process of structuralization of the French national mathematical community and, consequently, of France's roles on the international scene.

Concurrent with the development of the SMF and its *Bulletin*, France also witnessed the development of higher education, which French scientists had been demanding for years. Until the Franco-Prussian War, the leading institution for

---

[5] "S'ils ont été conduits à se rechercher par l'amour de la Science, un autre sentiment vient encore fortifier leur union: l'amour du pays. Aussi loin que peut s'étendre leur action, ils veulent, pour leur part contribuer au développement des forces intellectuelles et morales de la France." Émile Bouty, "Notice sur la vie et les travaux de J. C. Almeida," *Journal de physique* **9** (1880):430.

[6] On the roles and philosophies of journals in French and other contexts, see Hélène Gispert, "Le *Bulletin de la Société mathḿatique de France*, le journal de recherche d'une communauté (1873–1914)," *Rivista di storia della scienza*, 2nd ser. **4** (2) (1996):1-11 on pp. 4-6; Hélène Gispert and Renate Tobies, "A Comparative Study of the French and German Mathematical Societies before 1914," in *L'Europe mathématique–Mathematical Europe*, ed. Catherine Goldstein, Jeremy J. Gray, and Jim Ritter (Paris: Édition de la Maison des sciences de l'homme, 1996), pp. 408-430; and Della Dumbaugh Fenster and Karen Hunger Parshall, "A Profile of the American Mathematical Research Community: 1891–1906," in *The History of Modern Mathematics*, vol. 3, ed. Eberhard Knobloch and David E. Rowe (Boston: Academic Press, Inc., 1994), pp. 179-227, as well as the chapters by Sloan Despeaux and June Barrow-Green in the present volume.

higher mathematics had been the prestigious *École polytechnique*, created during the French Revolution for the training of engineers. From the beginning of the nineteenth century, virtually every important French mathematician had studied or taught—and had often done both—at this school. Mathematics also defined the main topic of the entrance examination in the famous system of *classes préparatoires* taken at the end of secondary school. Mathematical studies were virtually absent from the university curriculum, however, apart from the Parisian Sorbonne, where a few students attended the lectures given by famous mathematicians. Most of these students were from the *École normale supérieure* (ENS)—the other school founded during the Revolution—and intended to become secondary school teachers.

In the thirty years after the war, from 1870 to 1900, this educational profile changed dramatically. The number of university positions in mathematics doubled, reaching a total of sixty. The university became a locus of professionalization for mathematics and competed in this arena with the *École polytechnique*. (See Table 1.) But there was also a second, qualitative change, linked to these quantitative changes. The creation of these positions implied a real intention on the part of these institutions to foster research as well as teaching. Holders of these new positions, mostly young and brilliant mathematicians, were thus able to avoid the traditional positions in *lycées* or *classes préparatoires* once they had obtained their doctorates. A distinction gradually arose between careers in secondary and in higher education and also, to some extent, between careers at the university and at the *École polytechnique*.

TABLE 1:
UNIVERSITY EXPANSION IN MATHEMATICAL FRANCE, 1860–1910

	1860	**1870**	**1880**	**1890**	**1900**	**1910**
Math. Chairs	32	32	41	43	43	51
All Math. Posts[i]	32	32	44	51	64	77
% Math./Science Positions	32%	31%	32%	27%	29%	27%

[i] The positions counted here are based on the *Annuaires de l'instruction publique*. They only take into account chairs, *maîtres de conférences*, and adjunct professors. They do not correspond to the number of teachers on the faculties; teachers at the *lycées* sometimes taught complementary courses. There may also be—for the sciences—differences between the results presented here and in other researches; laboratory personnel were most often not taken into account in this table.

This expansion of higher mathematical teaching, a direct consequence of the new political ambitions after the war, symbolized the birth of a new mathematical sphere—and with it new interests, new standards, and a new ethos—the sphere of the *École normale supérieure*. The development of the *Société mathématique de France*, which was both the arena and the tool for the structuralization of the French national community during these years, further amplified the effects of the processes of professionalization.

Structuralization of the SMF: Establishing an Academic Center

One important evolutionary trend in the composition of the SMF during this period is particularly notable. On its creation in 1872, the SMF's strongest constituency was primarily in the sphere of the *École polytechnique*. By the beginning of the twentieth century, however, the *École polytechnique* had lost its position of superiority, its dominance usurped by the *École normale supérieure*. (See Table 2, top half.) Similarly, the proportion of (mostly non-teaching) engineers was greatly diminished, whereas the relative number of mathematics teachers had rapidly increased. In the latter membership category, the number of university teachers grew fourfold. (See Table 2, bottom half.) The SMF thus progressed in the teaching milieu (particularly at the university level) but regressed with regard to that of engineering. This provides one explanation for the meager growth in the number of French members of the society during the period, which is indeed quite peculiar to the SMF: from an initial number of 141, they had only grown to 178 by 1914.

TABLE 2:
STUDIES AND PROFESSIONS OF THE
FRENCH MEMBERS OF THE SMF

Studies	**1874**	**1880**	**1890**	**1900**	**1914**
Polytechniciens	103	80	76	86	52
% in SMF[i]	58%	53%	48%	46%	29%
Normaliens	25	21	29	44	63
% in SMF	14%	14%	18%	24%	35%
Professions	**1874**	**1880**	**1890**	**1900**	**1914**
Teachers[ii]	87	72	84	113	129
% in SMF	49%	47%	54%	61%	72%
Engineers	81	58	56	68	41
% in SMF[i]	46%	38%	36%	38%	23%
% Teachers	23%	28%	36%	38%	52%
% Mil. Officers	35%	33%	29%	27%	22%

[i]The percentages are calculated relative to the French members of the SMF.
[ii]Included in this category are all of the teachers in the SMF who are French as well as teachers of engineering.

With the structuralization of the national mathematical community around the SMF went an implicit process of selection. This eventually resulted in the exclusion of certain categories of members who had initially felt attracted to the new organization. In spite of its essentially open membership policies, the élitist publication philosophy of the *Bulletin* of the SMF symbolized, as well as reinforced, this phenomenon. The evolution of the members' activities highlights this trend.

At the inception of the SMF, only one member in four was an author, that is, was publishing papers regularly. By 1914, this ratio was nearly one half. The percentage of teachers among the authors, which was 70% in the early 1870s, was overwhelming by 1914, with university teachers rising from 20% in the first few years to 50%. (See Table 3.) By 1914, the national mathematical community had become an academic one. Moreover, focusing on the output of SMF authors reveals three further features. First, the French mathematical milieu was clearly divided into two groups: authors (could we say mathematicians?) who published papers only in periodicals devoted to mathematics education and authors who published in any sort of journal, research or otherwise. Second, this division was analogous to others relating to sources of education (the *École polytechnique* as opposed to the *École normale supérieure*), academic employment (the *École polytechnique* and *classes préparatoires* versus the university), and degrees of qualification (doctorate or lesser degrees). Third and finally, the production of French mathematicians, at least as seen through the eyes of the *Société mathématique de France*, completely ignored the field of applied mathematics.[7]

TABLE 3:
PRODUCTION OF THE FRENCH AUTHORS OF THE SMF

	1874	**1880**	**1890**	**1900**	**1914**
Authors	47	45	54	73	70
% in SMF	26%	30%	34%	40%	40%
Research Authors	32	30	37	51	49
% in SMF	18%	20%	24%	27%	28%
% Teachers/Authors	72%	71%	77%	82%	90%
% Univ. Teachers/Authors	21%	27%	33%	34%	56%

A Non-academic Periphery: Actors in and Values of the AFAS

The period from 1870 to 1914 also witnessed a differentiation or stratification of the different components of mathematical activity and its participants. Just prior to World War I, not all participants in the mathematical life of France participated in the *Société mathématique de France*. Moreover, not all SMF members necessarily published papers in its *Bulletin*. Analyzing participants both within and outside the SMF thus reveals the following profile of mathematical activity in France. Academic mathematical research focused mainly on specific domains such as analysis (differential equations and the theory of functions) and was published in the *Comptes rendus de l'Académie des sciences*, the *Bulletin* of the SMF, Liouville's *Journal*, or similarly spirited foreign journals. More elementary research

[7]On the two first points, precise data are given in Hélène Gispert, "Le milieu mathématique français et ses journaux en France et en Europe," in *Messengers of Mathematics: European Mathematical Journals (1800–1946)*, ed. Mariano Hormigón and Elena Ausejo (Zaragoza: Siglo XXI de España Editores, 1993), pp. 133-158. On the third point, see Hélène Gispert, *La France mathématique: La Société mathématique de France (1870–1914)* (Paris: Cahiers d'histoire et de philosophie des sciences, Société mathématique de France, et Société française d'histoire des sciences et techniques, 1991), pp. 85-86 and 128-131.

devoted primarily to geometry or number theory was published in specific intermediate journals.[8] The latter were often educational publications, which developed dramatically in the 1890s. Research on more applied, less academic topics like actuarial mathematics, ballistics, (non-rational) mechanics, and geodesy appeared, but in more specifically targeted journals, and progressively disappeared from the so-called "mathematical" academic milieu. Finally, there was also the diffusion of mathematical results on a scale larger than that defined by the SMF or particular journals, namely, through publication by specific presses. As these phenomena suggest, mathematical results were being diffused beyond the boundaries of the SMF. At the national level, then, a scale of distinctions and disciplinary values had developed within the mathematical community that had specific outlets. It seems that the SMF, which at this time could have been seen as representative of the French mathematical community, failed to appreciate some of the values that were emphasized elsewhere. As far as overall values are concerned, it is interesting to note in this context that those of the SMF evolved over time. Linked in the 1870s to the engineering culture of the *École polytechnique*, the SMF later took its lead from the more academic *École normale supérieure.*

The history of nineteenth-century mathematics, and particularly the history of mathematics in its final third, has usually focused on the participants and mathematical interests of those in academic research. It has thus set up a hierarchy in the mathematical landscape that has resulted in the creation of a privileged "center" and of a so-called "mathematical periphery."[9] In France, the "center" has been traditionally included in the SMF, while the "periphery" consisted of the activities and participants in which the SMF became less and less concerned. This mathematical periphery represents a *terra incognita* that histories of mathematics seldom approach owing to its relative obscurity. In the case of France, however, the *Association française pour l'avancement des sciences*, created after the Franco-Prussian War, provided an institutional structure through which to study the periphery. An analysis of the AFAS thus provides a wider perspective on the evolution of an international mathematical community. In some ways, however, it also slightly complicates things, since the SMF was not the only French organization with a claim to international stature in the mathematical world of the turn of the twentieth century. In other words, the SMF did not constitute the only potential French component of an international mathematical community.

The AFAS, even more than the SMF, was an outcome of the effects of the war and the subsequent defeat. It was a joint creation of industrial, banking, and commercial interests, on the one hand, and of the scientific milieux, on the other. It was "inspired by the patriotic desire to contribute to the intellectual recovery of a

---

[8]The most successful intermediate journal in France was *Les nouvelles annales de mathématiques: Journal des candidats à l'École polytechnique*, where solutions of typical *classes préparatoires* school problems were published. These dealt, for example, with conics and the geometry of the triangle as well as with elementary algebra and non-transcendental analysis.

[9]For the term "mathematical periphery," see Eduardo Ortiz, "The Nineteenth Century International Mathematical Community and Its Connection with Those on the Iberian Periphery," in *L'Europe mathématique–Mathematical Europe*, ed. Catherine Goldstein, Jeremy J. Gray, and Jim Ritter (Paris: Éditions de la Maison des sciences de l'homme, 1996), pp. 323-346, and Jaroslav Folta, "'Local' and 'General' Developments in Mathematics: The Czech Lands," in *ibid.*, pp. 271-290. Compare also the chapters by Karen Parshall and Adrian Rice and by Elena Ausejo and Mariano Hormigón above.

country laid low by so many blows."[10] This political agenda came through clearly in speeches delivered at the AFAS's first congress in Bordeaux in 1872. "Today more than ever," declared one AFAS founder,

> the domain of intelligence [and] the scientific world also have their battles, their victories, and their laurels. ... It is there that one must go first to seek revenge. To the rich and idle classes, it (the Association) wants to show what the study of nature and of its forces has [to offer] that is engaging and grand; it wants to open to their industry, which is too often led astray, avenues where it would find satisfaction in a manner attractive and honorable to them, and glorious for the country. To the working classes, even to those who spend their life in proletarian labors, [the Association] wants to make known what science has [to offer] that is useful, what it has [to offer] for the good of all, for the prosperity of the nation.[11]

The creation of the AFAS, like its subsequent activity, was the fruit of a real geopolitical seachange provoked by the defeat. It questioned the suitability of the humanities to form the cultural élite and highlighted the role of the sciences, and so of mathematics, in French society. But which science? Which mathematics to promote? The AFAS symbolized a new Republic that had forged a fresh alliance between the political, the scientific, and the industrial milieux, a Republic that attempted to develop itself by relying on a science that was not opposed to applications. The AFAS participated in this new Republic's rising cult. Science, however, had several functions. It was thought of and promoted as a factor of social and industrial progress (hence, its immense utility), but it also required a political function. In other words, it needed to be a factor of political and social consensus.

In the field of mathematics, the AFAS, with an entirely open policy towards paper submission, received slightly more than 1,200 contributions from more than 350 congress participants during the period from 1870 to 1914.[12] From a quantitative point of view, it can thus be compared to the SMF, whose *Bulletin* published over 1,000 papers in the same period. From a qualitative point of view, however, the comparison becomes more complicated and cannot be limited to the observation that most of the communications to the AFAS fall outside the category of academic research defined above.

What was the intellectual and institutional identity of this important part of French mathematical activity? At the intellectual level, the AFAS's statutes—like

---

[10] "Inspirée par le désir patriotique de contribuer au relèvement moral du pays abattu par tant de secousses." Alfred Cornu, "Histoire de l'Association française," in *Compte rendu du Congrès de l'AFAS* (Bordeaux: AFAS, 1872), pp. 44-49 on p. 44.

[11] "De nos jours plus que jamais, le domaine de l'intelligence, le terrain de la science ont aussi leurs batailles, leurs victoires et leurs lauriers. ... C'est là qu'il faut d'abord aller chercher la revanche. Aux classes riches et oiseuses elle [l'Association] veut montrer ce que l'étude de la nature et de ses forces a d'aimable et de grand; elle veut ouvrir à leur activité, qui trop souvent s'égare, des voies où elle trouverait à se satisfaire d'une manière attrayante et honorable pour elles, glorieuses pour le pays. Aux classes laborieuses, à celles mêmes dont la vie passe dans les labeurs du prolétariat, elle veut faire comprendre ce que la science a d'utile, ce qu'elle fait pour le bien de chacun, pour la prospérité de la patrie." J. L. A. De Quatrefages, "La science et la patrie," in *ibid.*, pp. 36-41.

[12] A complete study of the mathematical section of the AFAS is given in Gispert, "Réseaux mathématiques."

those of the SMF—expressed the will to favor the progress and diffusion of science from the mutual point of view of pure theory and the development of its practical applications. Nevertheless, the two societies interpreted this intention very differently, as evidenced by the two different mathematical publics they came to target. The SMF focused on academic research, and by the end of the century, promoted the ENS's standards of research in its *Bulletin*, namely, pure and increasingly theoretical and abstract research. The AFAS, on the other hand, laid particular stress on applications such as in astronomy, industry, engineering, and education.[13] Themes, levels, interests, links with society: everything seemed increasingly different during this period. What, then, were the networks that participated in or furthered the mathematical initiatives of the AFAS? What was the AFAS's institutional identity?

From the beginning, the proportion of the mathematical academic milieu among the advocates and promoters of the AFAS was very low, a situation at variance with that of the Association's other disciplinary fields.[14] With the exception of Michel Chasles, all founders of the mathematical sciences section were astronomers. Moreover, in its first decade, the officers and the principal speakers in its mathematical sections belonged to the *polytechnique* component of the SMF. These were military men, engineers, or teachers at the *École polytechnique*, who published papers in both mathematical research journals and intermediate publications. During the following decades, mirroring the evolution of the SMF summarized above, a break arose between AFAS networks and the academic research center. Prominent AFAS members still belonged to the SMF, with many of them teaching at the *École polytechnique* or in the *classes préparatoires*, but most of them now published almost exclusively in intermediate journals. There were some noticeable exceptions, however. Charles Laisant and Maurice d'Ocagne were both prolific authors in the *Bulletin* of the SMF and active AFAS members. Still, few prominent mathematical AFAS members were associated with the university system and few published in the *Comptes rendus de l'Académie des sciences* or even in the SMF's *Bulletin*.

Broadening the analysis beyond the main speakers, the differences become still more prominent and expose a major effect of the evolution under way during this

---

[13]In 1905, Ernest Lebon explictly stated the objectives of the AFAS relative to the mathematical sciences: "The communications made at the sessions meet all the objectives that the Association pursues: to publicize improvements applicable in astronomy, industry, engineering and education [les communications faites en séance répondent toutes au but que poursuit l'association: faire connaître les perfectionnements applicables dans l'astronomie, l'industrie, l'art de l'ingénieur et l'enseignement]." See Ernest Lebon, "Compte rendu du Congrès de Cherbourg de l'AFAS," *L'Enseignement mathématique* **7** (1905):406-409 on p. 406. In the same spirit, in 1894, the mathematical sections of the AFAS had expressed the following wish, which had been unanimously approved: "The first and second sections, regretting that a tighter link has not been established between those who devote themselves to mathematical science and those who, every day, have to apply its results, express the wish that publications, written for engineers, have as their specific aim to extract from pure mathematical works, and above all from analytical mechanics, everything that seems to be of interest from the point of view of the sciences of application ... [La 1ème et 2ème Section, regrettant qu'il ne s'établisse pas un lien plus étroit entre les personnes qui cultivent la science mathématique et celles qui ont chaque jour à en appliquer les résultats; expriment le voeu que des publications, rédigées en vue des ingénieurs, prennent pour tâche spéciale d'extraire des travaux de mathématique pure et surtout de mécanique analytique, tout ce qui semble présenter un intérêt du point de vue des sciences d'application ...]." Charles-Ange Laisant, "Voeu de la 1er et 2e sections," *Compte rendu du Congrès de l'AFAS* (Caen: AFAS, 1894), pp. 103-104.

[14]For the analysis here and in the following paragraph, compare Gispert, "Le *Bulletin de la Société mathématique de France*," pp. 3-4, and Gispert, "Réseaux mathématiques," p. 137.

period. Not all of the authors of the "center"—those, for example, whose papers were published in the *Bulletin*—were prominent; but most of them—as noted above—taught higher-level mathematics. Of the authors belonging to the AFAS, on the other hand, the proportion of retired military men, non-*polytechnicien* engineers, and amateurs without any stated profession increased throughout the entire period. The process of professionalization outlined above thus seems to have pushed to the sidelines—or to the periphery—those who were not specifically involved in a professional mathematical activity.

Part II: International Connections of French Mathematical Worlds

Foreign Contributions to the Journals of the SMF and the AFAS

Given this profile of the structuralization of the French national community, what are the international implications of the analysis? What was the status, stature, and role at the international level of the two distinct French mathematical worlds under discussion here? Pursuing these questions will lead to issues, first, of the existence and organization of several international mathematical communities as identified by specific ideas and projects; second, of the eventual connection of these communities; and, third, of their place in the history of mathematics.

Some preliminary answers are provided by looking at the network developed outside of France by the AFAS in comparison with the network of foreign authors who published in the *Bulletin* of the SMF.[15] The proportion of foreign contributions to the proceedings of AFAS congresses and the volumes of the *Bulletin* are nearly the same for the period from the 1870s through 1914: around 12% and 14%, respectively.[16] There are, however, very important differences.

Initially, the number of foreign contributors and the average frequency of their contributions to each journal are not the same. The SMF had more authors: seventy foreigners published in the *Bulletin*, more than one in three of *Bulletin* authors. Their links to the *Bulletin*, and probably also to the SMF, were still very tenuous, however. The average number of papers by foreign authors for the entire period was less than two per author.

The situation for the AFAS was the exact opposite: about forty foreigners attended their congresses during the two last decades of the nineteenth century and gave lectures in the mathematical science sessions, representing a little more than one author in ten. More significant is the fact that fifteen of these were either an Honorary President or Vice-President of the mathematical sessions, gave an average of four papers each, and, interestingly, were all university teachers. Hence, at the AFAS, the profile of foreign speakers was very different from that of their French counterparts. The French periphery, neglected by the French academic center, on the one hand, was well-catered-for by the foreign university world, on the other.

This foreign academic community contributed heavily to the AFAS's formative years. Prominent mathematicians—such as Pafnuti Chebychev, James Joseph Sylvester, Francesco Brioschi, Luigi Cremona, and Eugène Catalan, the latter formerly in Paris but by this point a professor in Liège and registered as a *savant*

---

[15]For a more detailed analysis, see Gispert, "Réseaux mathématiques."

[16]For data on international contributions to mathematics journals in Great Britain and Sweden, see the chapters by Sloan Despeaux and June Barrow-Green, respectively, in the present volume.

étranger—attended and lectured several times during the first decade. Durable relationships had been established with French officials, particularly by several scholars from the United Kingdom, Belgium, Russia, and to a lesser extent Italy, and these links lasted until the mid-1890s.[17] For its part, the SMF did not attract foreigners from entirely the same countries. Germany was the most common native country of foreign *Bulletin* authors, just ahead of Russia, the central European countries, Belgium, and Italy.[18]

What particularly distinguished the United Kingdom and Germany from the other contributing countries was that each participated in only one of these two French mathematical arenas: the British in the AFAS and the Germans in the SMF. The non-participation of Germany at the congresses of the AFAS can be explained, initially at least, by the explicit refusal of the founders to invite German scientists to the postwar congresses. This stance was relaxed over time, and by the 1880s, several sections were inviting German scholars to AFAS meetings. However, the only German lecturer to contribute to a mathematical science section was Georg Cantor, who lectured at the 1894 congress in Rouen. He was not, however, invited to be one of the section's Honorary Presidents. The fact that German mathematicians were increasingly active in the SMF, and the importance of their contributions in its *Bulletin*, would thus seem to suggest that political and nationalistic feelings were not the only cause for this almost complete absence of Germany at the AFAS.

The case of Britain, unaffected by the political considerations caused by the 1870 war, offers another explanation of the selective presence of these two countries in French mathematical life. Compare, for example, British attendance at the mathematical sections of the AFAS, on the one hand, and at meetings of the SMF and the early International Congresses of Mathematicians (ICM), on the other. The fact that the British were among the most active foreign participants in the AFAS contrasts sharply with the fact that very few British mathematicians were members of the SMF or attended the early ICMs, with the obvious exception of the Cambridge ICM in 1912. The British Association for the Advancement of Science, founded in 1831, had actually been the model for the AFAS, and some of its prominent members had participated in AFAS activities. For example, three distinguished presidents of the Mathematical and Physical Sciences Section of the BAAS, Arthur Cayley, Sylvester, and James W. Glaisher, had been regular attendees of the AFAS's mathematical section in its first two decades. During these same years, no French mathematicians of such stature were similarly involved in the AFAS's activities. The BAAS thus seemed to be linked far more closely to the British learned community than the AFAS was with the French.

The analysis of the relative roles in France of Germany, which participated actively in the SMF and its *Bulletin*, and the United Kingdom, which so greatly contributed to the AFAS and its *Comptes rendus*, confirms other distributions present in the international mathematical landscape that historians have only recently begun to notice. The countries at the forefront of mathematical research at this time

---

[17]The success of the similar British Association for the Advancement of Science (BAAS) might have positively influenced the international attendance at AFAS congresses (see below). The BAAS enjoyed great international prestige and encouraged foreign attendance through invitation and through the awarding of prizes. Again, Sloan Despeaux's chapter above provides more detail on the BAAS and its international activities.

[18]The number of German and Russian authors in the *Bulletin* was nine and eight, respectively.

were Germany, followed by France, then Great Britain and Italy. But what was the role—and, again, not only the place—of each of these countries on the international stage?

In the case of France, it is worth pointing out a characteristic attitude of many prominent French mathematicians: irrespective of age, they rarely published in foreign research journals. Édouard Goursat, Paul Painlevé, Émile Picard—with only twenty "foreign" papers out of the four hundred he wrote—and Émile Borel—with nine out of ninety—are notable examples. Indeed, top-level research occupied the least significant portion of French SMF members' foreign publications: only a quarter of their foreign papers were published in such journals.[19] By far the greater part of their international mathematical work was on other levels. In the intermediate mathematical fields—such as school mathematics (preparatory level), mathematics for engineers, the diffusion or popularization of science, pedagogy, and mathematical recreations—French authors, many of them members of the AFAS, regularly published in English, Belgian, Italian, Spanish, and Portuguese reviews.[20] But there was no such German journal available to them.

Center and Periphery: From National Contexts to the International Scale

Thus, different strata of international mathematical life corresponded to the two mathematical networks that developed in France around the SMF and the AFAS. In the same way, they also corresponded to the two groups of French participants in the mathematical life of France, those at the academic research center and those at the periphery. These strata, as shown above, can be identified through specific foreign journals which, despite their national nature, fostered international collaborations. Still, questions arise. To what extent did these strata correspond to distinct components of the national mathematical milieux of different countries? And, to what extent did they each take part in the structuralization of an (or *the*) international mathematical community?

Regarding the first question, it seems that the French mathematical landscape was somewhat peculiar, when compared to that of England, for example. As Adrian Rice and Robin Wilson have shown relative to the London Mathematical Society,[21] the growing gap between the different strata of activity and, above all, between the participants in the French mathematical scene, does not seem to exist to the same extent—or even at all—in late-nineteenth-century England. In Italy, too, the mathematical activities or careers of men like Cremona, Angelo Genocchi, Francesco Brioschi, the latter the editor of the *Annali di matematica pura ed applicata* and of the engineering journal, *Il Politecnico*, might suggest that there was no such gap

[19]A list of such research journals would include titles such as *Acta Mathematica*, *Rendiconti del Circolo matematico di Palermo*, *Mathematische Annalen*, *Annali di matematica pura ed applicata*, and Crelle's *Journal*. For a more detailed analysis of "foreign" research papers of the French mathematical community, see Gispert, "Le milieu mathématique français et ses journaux en France et en Europe," pp. 140-141. Note that this analysis in terms of "papering" in no way excludes contacts with foreign mathematicians and their mathematical works. As noted above and in Thomas Archibald's chapter to follow, Charles Hermite's role was very important in making German works known in France.

[20]For example, *Mathesis*, *The Educational Times*, *Messenger of Mathematics*, *El progresso matematico*, and *Les nouvelles correspondances mathématiques*. See *ibid.*, pp. 143-144.

[21]Adrian C. Rice and Robin J. Wilson, "From National to International Society: The London Mathematical Society, 1867–1900," *Historia Mathematica* **25** (1998):185-217.

there either, even though it must be said, in the case of Brioschi's journals, that they targeted separate fields and adopted different standards.

Germany exhibited a somewhat different pattern. By the beginning of the twentieth century, it supported a society devoted to applied mathematics or, more properly, to the applications of mathematics. Thus, for the first time, we find a separate organization for mathematicians working in applied and even industrial fields; in France, the *Société française de statistiques de Paris*, created in the 1860s, was much more specific and attracted mostly those who were not *géomètres*. The coexistence in Germany of two societies, the *Deutsche Mathematiker-Vereinigung* and the society for applied mathematics just mentioned, may lead, once more, to the question of the structuralization of the different strata on the national and international levels. Was it only the DMV that was involved at the international level, for example, or with the International Congresses?

In the case of the United States, the American Mathematical Society and its *Bulletin* appear to have had quite a different orientation from that of the *Société mathématique de France* and its *Bulletin*. The former organization was even less élitist than the latter, despite the progressively research-oriented policy and steadfast avoidance of pedagogical issues that eventually prompted the founding of the Mathematical Association of America in 1915.[22]

With these brief profiles of the stratification, or lack thereof, in a variety of national venues as a backdrop, partial answers to the second question now emerge. First, the model of the *École polytechnique* and its geometrical legacy in, for example, projective geometry, played a very important role not only in France but also in many European countries throughout the nineteenth century.[23] This geometric *polytechnicien* model—its participants and its ethos—was, however, largely ignored in the process of structuralization of the French national community during the closing decades of the century. Moreover, it is probable, although not yet established, that it was equally neglected during the evolution of the international community of what were called "mathematicians" at the beginning of the twentieth century, despite the strong representation of its participants at some levels within mathematics internationally. The final years of the nineteenth century also appear to have been the very time when the international mathematical network linked to the AFAS broke up, since no more foreign mathematicians attended its congresses after this point.

Still, this was similarly the period when the International Congresses of Mathematicians came into being. It would be of both interest and relevance to examine

[22]On these institutional developments in the United States, see Karen Hunger Parshall and David E. Rowe, *The Emergence of the American Mathematical Research Community, 1876–1900: J. J. Sylvester, Felix Klein, and E. H. Moore*, HMATH, vol. 8 (Providence: American Mathematical Society and London: London Mathematical Society, 1994), pp. 418-419. On the issue of stratification within the early American mathematical community, see David L. Roberts, "Albert Harry Wheeler (1873–1950): A Case Study in the Stratification of American Mathematical Activity," *Historia Mathematica* **23** (1996):269-287.

[23]See, for example, Mariano Hormigón and Ana Millán, "Projective Geometry and Applications in the Second Half of the Nineteenth Century," *Archives internationales d'histoire des sciences* **42** (1992):269-289 on the state of research and teaching in this field in Europe in the second half of the century, and Luboš Nový, "Les mathématiques et l'évolution de la nation tchèque 1860–1918," in *L'Europe mathematique—Mathematical Europe*, pp. 269-289 on the Czech school of geometry at the end of the century.

in depth not only who attended such meetings but also from which countries or regions and professions they came and on what subjects they spoke. The example of central Europe merits mention, since these mathematicians had consistently strong representation at the early ICMs. Concurrently, the Czech Society of Mathematicians had more than 1,000 members, more than three times that of the SMF or the London Mathematical Society. But, to what extent did those thousand members belong to the different strata analyzed above? To what extent did they join this newly institutionalized international community of mathematicians? Or did they remain outside? Answers to such questions—for the case of central Europe as for any nation—would shed light on the issue of the internationalization of mathematical activity at the beginning of the twentieth century.

Conclusion

Studies of national communities have begun to pay greater attention to a wider set of participants and fields. They have thus enlarged and enriched our understanding of what may be called a "mathematical community" and its production in a given time, while exposing the different strata comprising such milieux. Such recent work includes studies of nineteenth-century France, Britain, and the United States.[24] As the case of France between the Franco-Prussian War and World War I shows, whole strata representing diverse participants and fields no longer contributed to what had become, around the turn of the twentieth century, a community of mathematicians institutionalized around the SMF. The latter seems, however, to have been construed as "the" French mathematical community by other mathematical institutions of the period—the ICMs and the mathematical societies of other countries, for example—and this false perspective has sometimes been taken by historians of mathematics as well. An international community thus seems—in the history of mathematics as most often considered up until now—to have imposed itself to the exclusion of all other communities. Once again, the study of French mathematical activity between 1870 and 1914 illustrates that the situation was more complicated than this. Other international relations existed and developed outside that mathematical arena, for example, through the Associations for the Advancement of Science and through the different venues (institutional or disciplinary) of instruction, whether they be secondary, the earliest years of higher education, or in the engineering schools with their development of periodicals and their translation and promotion of numerous manuals. Although partly eclipsed by the ICMs, the ICMI and its journal, *L'Enseignement mathématique*—which are

[24]See, for example, the bibliographies of Michael Price, *Mathematics for the Multitude? A History of the Mathematical Association* (Leicester: Mathematical Association, 1994) and June Barrow-Green, "Mathematics in Britain, 1860–1940: The Creation of a Source-oriented Database," in *Computing Techniques and the History of Universities*, ed. Peter Denley (Göttingen: Max-Planck-Institut für Geschichte & Scripta Mercaturæ Verlag, 1996) for Great Britain; Fenster and Parshall, and David L. Roberts, "Albert Harry Wheeler (1873–1950): A Case Study in the Stratification of American Mathematical Activity," *Historia Mathematica* **23** (1996):269-287 for the United States; and Ivor Grattan-Guinness, "*L'ingénieur savant* 1800–1830: A Neglected Figure in the History of French Mathematics and Science," *Science in Context* **6** (1993):405-433 and Jean Dhombres and Mario H. Otero, "Les *Annales de mathématiques pures et appliquées*: Le journal d'un homme seul au profit d'une communauté enseignante," in *Messengers of Mathematics: European Mathematical Journals (1800-1946)*, ed. Mariano Hormigón and Elena Ausejo (Zaragoza: Siglo XXI de España Editores, 1993), pp. 3-70 for France in the first half of the nineteenth century.

beginning to be studied—reflect this little-known aspect of turn-of-the-twentieth-century mathematical relations today.[25] It would be a valuable contribution to the history of mathematics, as well as to the area of international studies, for future works to focus on the effects of marginalization and even exclusion of countries, topics, ethos, and professions in the complex of processes that led to the international mathematical community in the twentieth century.

References

Barrow-Green, June. "Mathematics in Britain, 1860–1940: The Creation of a Source-oriented Database." In *Computing Techniques and the History of Universities*. Ed. Peter Denley. Göttingen: Max-Planck-Institut für Geschichte & Scripta Mercaturæ Verlag, 1996.

Chasles, Michel. *Rapport sur les progrès de la géométrie*. Paris: Imprimerie nationale, 1870.

Cornu, Alfred. "Histoire de l'Association française." *Compte rendu du Congrès de l'AFAS*. Bordeaux: AFAS, 1872, pp. 44-49.

Dhombres, Jean and Otero, Mario, H. "Les *Annales de mathématiques pures et appliquées*: Le journal d'un homme seul au profit d'une communauté enseignante." In *Messengers of Mathematics: European Mathematical Journals (1800–1946)*. Ed. Mariano Hormigón and Elena Ausejo. Zaragoza: Siglo XXI de España Editores, 1993, pp. 3-70.

Fenster, Della Dumbaugh and Parshall, Karen Hunger. "A Profile of the American Mathematical Research Community: 1891–1906." In *The History of Modern Mathematics*. Vol. 3. Ed. Eberhard Knobloch and David E. Rowe. Boston: Academic Press, Inc., 1994, pp. 179-227.

Folta, Jaroslav. " 'Local' and 'General' Developments in Mathematics: The Czech Lands." In *L'Europe mathématique–Mathematical Europe*. Ed. Catherine Goldstein, Jeremy J. Gray, and Jim Ritter. Paris: Éditions de la Maison des sciences de l'homme, 1996, pp. 271-290.

Gispert, Hélène. "Le *Bulletin de la Société mathématique de France*, le journal de recherche d'une communauté (1873–1914)." *Rivista di storia della scienza*. 2nd Ser. **4** (2) (1996):1-11.

______. *La France mathématique: La Société mathématique de France (1870–1914)*. Paris: Cahiers d'histoire et de philosophie des sciences, Société mathématique de France, et Société française d'histoire des sciences et techniques, 1991.

______. "Le milieu mathématique français et ses journaux en France et en Europe." In *Messengers of Mathematics: European Mathematical Journals (1800–1946)*. Ed. Mariano Hormigón and Elena Ausejo. Zaragoza: Siglo XXI de España Editores, 1993, pp. 133-158.

______. "Réseaux mathématiques en France dans les débuts de la Troisième République." *Archives internationales d'histoire des sciences* **49** (1999):122-149.

[25]In October 2000, an international symposium on 100 years of *L'Enseignement mathématique* was held in Geneva organized by the journal as well as by the ICMI. The proceedings of this conference will appear in *L'Enseignement mathématique* in 2002.

Gispert, Hélène and Tobies, Renate. "A Comparative Study of the French and German Mathematical Societies before 1914." In *L'Europe mathématique–Mathematical Europe.* Ed. Catherine Goldstein, Jeremy J. Gray, and Jim Ritter. Paris: Éditions de la Maison des sciences de l'homme, 1996, pp. 408-430.

Goldstein, Catherine; Gray, Jeremy J.; and Ritter, Jim, Ed. *L'Europe mathématique–Mathematical Europe.* Paris: Éditions de la Maison des sciences de l'homme, 1996.

Grattan-Guinness, Ivor. "*L'ingénieur savant* 1800–1830: A Neglected Figure in the History of French Mathematics and Science." *Science in Context* **6** (1993):405-433.

Hormigón, Mariano and Ausejo, Elena, Ed. *Messengers of Mathematics: European Mathematical Journals (1800–1946).* Zaragoza: Siglo XXI de España Editores, 1993.

Hormigón, Mariano and Millán, Ana. "Projective Geometry and Applications in the Second Half of the Nineteenth Century." *Archives internationales d'histoire des sciences* **42** (1992):269-289.

Laisant, Charles-Ange. "Voeu de la 1er et 2e sections." *Compte rendu du Congrès de l'AFAS.* Caen: AFAS, 1894, pp. 103-104.

Lebon, Ernest. "Compte rendu du Congrès de Cherbourg de l'AFAS." *L'Enseignement mathématique* **7** (1905):406-409.

Nový, Luboš. "Les mathématiques et l'évolution de la nation tchèque 1860–1918." In *L'Europe mathématique–Mathematical Europe.* Ed. Catherine Goldstein, Jeremy J. Gray, and Jim Ritter. Paris: Éditions de la Maison des sciences de l'homme, 1996, pp. 269-289.

Ortiz, Eduardo. "The Nineteenth Century International Mathematical Community and Its Connection with Those on the Iberian Periphery." In *L'Europe mathématique—Mathematical Europe.* Ed. Catherine Goldstein, Jeremy J. Gray, and Jim Ritter. Paris: Éditions de la Maison des sciences de l'homme, 1996, pp. 323-346.

Parshall, Karen Hunger and Rowe, David E. *The Emergence of the American Mathematical Research Community, 1876–1900: J. J. Sylvester, Felix Klein, and E. H. Moore.* HMATH. Vol. 8. Providence: American Mathematical Society and London: London Mathematical Society, 1994.

Pasteur, Louis. In *Le Salut Public* (Lyon) (March 1871). Reprint Ed. *Pour l'avenir de la science française.* Paris: Éditions "Raison d'être," Collection "À la lumière des textes oubliés," 1947.

Price, Michael. *Mathematics for the Multitude? A History of the Mathematical Association.* Leicester: Mathematical Association, 1994.

Rice, Adrian C. and Wilson, Robin J. "From National to International Society: The London Mathematical Society, 1867–1900." *Historia Mathematica* **25** (1998):185-217.

Roberts, David L. "Albert Harry Wheeler (1873–1950): A Case Study in the Stratification of American Mathematical Activity." *Historia Mathematica* **23** (1996):269-287.

Charles Hermite (1822–1901)

CHAPTER 7

Charles Hermite and German Mathematics in France

Thomas Archibald*
Acadia University (Canada)

Introduction

The period 1848–1918 is so strongly constitutive of the "modern" mathematical world of the mid- to late-twentieth century that it is very easy to overlook differences between the basic attitudes of European mathematicians of that time and those of today. Yet, these differences are historically important, in particular for our understanding of the way in which a set of quite distinct metaphysical frameworks, cultures of taste, and sets of problems were blended to become the international mathematics that existed by around 1900. Academics were heavily implicated in the pervasive social changes taking place in the last half of the nineteenth century; the class interests of university professors were closely allied with those of the aristocracy. Thus, despite individual cases in which university professors took a stand for constitutional monarchy or republicanism (as in the case of the "Göttingen Seven" in 1837[1] or Jacobi in 1848) or the extensive role played by intellectuals in the Dreyfus affair (among them, the mathematicians, Henri Poincaré, Paul Appell, and Émile Picard), there is frequently a strong identification of professors with the upper classes in both France and Germany, an identification that was expressed in a variety of ways: social position, participation in certain kinds of reward systems, and political attitudes and stances.

These tendencies are very nicely expressed in the correspondence of Charles Hermite (1822–1901) with his colleagues, particularly with two mathematicians outside France: Hermite's former student, the Swede Gösta Mittag-Leffler (1846–1917), and Paul Du Bois-Reymond (1831–1889), a German. The surviving correspondence is one-sided and authored by Hermite. It is on his views and activities that this chapter will concentrate.

---

*I wish to thank the Centro internazionale della storia delle università e della scienza, Università di Bologna for the invitation to present an earlier version of this paper in March 1998 as well as Karen Parshall and Adrian Rice for the invitation to participate in the conference, "Mathematics Unbound: The Evolution of an International Mathematical Research Community (1800–1945)," in May 1999. The research for this paper was supported in part by the Social Sciences and Humanities Research Council of Canada, which support I gratefully acknowledge.

[1]Seven Göttingen University professors, among them Wilhelm Weber, were removed from their positions in 1837 when they protested the annulment of Hanover's constitution by the new King, Ernst August.

The correspondence with Du Bois-Reymond, published by Emil Lampe in 1916, has been described by mathematician and historian of mathematics, Hans Freudenthal, as "a valuable human document."[2] These letters afford not only a unique picture of the life of the mathematics professor in late-nineteenth-century France but also indicate clearly the tensions between France and Germany in the period immediately following the Franco-Prussian War of 1870–1871. They are thus extremely helpful for understanding the formation of the international community of professional mathematicians and for grasping the part played by national differences in the reception and development of key mathematical research areas during this period. The Mittag-Leffler correspondence was edited and published by Pierre Dugac in 1984 and serves both to expand on the picture provided by the Du Bois-Reymond correspondence and to corroborate the views Hermite expressed there.[3]

Perhaps surprisingly for a patriot in a period following military defeat by an enemy power, Hermite was a strong apologist for German mathematics in the France of his day. He worked hard to foster an understanding of what he viewed as the most important work emanating from German centers, most particularly that of Leopold Kronecker and Karl Weierstraß. By virtue of his position as a mentor to the next generation of French mathematicians, his efforts to interest students in German work had a profound effect on the development of the research programs of French mathematicians. Thus, research programs and mathematical approaches that had first matured in Germany were transplanted to the markedly different French environment with an astonishing and lasting success that touched all aspects of the pure mathematical endeavor. In the context following the Franco-Prussian War, it might be tempting, given this quickening of interest in German material, to make the analogy with the post-World-War-II ascendancy of American cultural styles in, for example, Germany. Yet, looking at France in the 1870s and 1880s does not reveal German dominance in other cultural realms such as literature, art, and music, indeed far from it. Thus, the situation in mathematics calls for the identification of some specific factors in that field that gave unique importance to the value of understanding and appreciating German work and that facilitated its assimilation and transmission.[4] Charles Hermite was one of these factors.

Hermite and German Mathematical Values

The urgency that Hermite felt to engage in his role as interpreter of German work was intimately linked to his view of social conditions in France following the Franco-Prussian War. In particular, his efforts were connected, on the one hand, to his ideal of an international, quasi-aristocratic élite of scientists and *savants* living in gentlemanly harmony and, on the other, to his concerns about the consequences for France of continued repudiation of the Germans.

---

[2]Charles C. Gillispie, *Dictionary of Scientific Biography*, 16 vols. and 2 supps. (New York: Charles Scribner's Sons, 1970–1990), s.v. "Hermite, Charles" by Hans Freudenthal.

[3]Pierre Dugac, "Lettres de Charles Hermite à Gösta Mittag-Leffler (1874–1883)," *Cahiers du Séminaire d'histoire des mathématiques* **5** (1984):49-285. The original correspondence is at the Mittag-Leffler Institute, Djursholm, Sweden.

[4]The situation in mathematics is paralleled in some of the other sciences, most notably in physics and medicine. See Robert Fox, "The View over the Rhine: Perceptions of German Science and Technology in France 1860–1914," in *Frankreich und Deutschland: Forschung, Technologie und industrielle Entwicklung im 19. und 20. Jahrhundert,* ed. Yves Cohen and Klaus Manfrass (Munich: C. H. Beck, 1990), pp. 14-24.

Many European mathematicians at the end of the nineteenth century saw themselves almost as aristocrats. Recent studies, in particular those of Christophe Charle and Martin Zerner, provide a sociological picture of this group and make it clear that, indeed, successful professors were among the *haute bourgeoisie*.[5] Charle clearly identifies the aristocratic aspirations of most of the professoriate in the realm of *lettres*, and Zerner reflects that a busy senior professor could earn up to 50,000 francs annually in this period. This was a considerable sum, permitting a house or apartment in a very good neighborhood, servants, a country home, and vacations abroad. Such earnings placed the professoriate among an élite that was more than intellectual, despite the liberal associations of academic protest in the Dreyfus affair and other public debates of the time.

Hermite was born in a small town in Lorraine. He was a hereditary anti-radical, his grandfather's fortune having been destroyed by the Revolution and his great-uncle guillotined. Hermite had a rather unorthodox career and was largely self-taught in mathematics, although his teacher at the Louis-le-Grand preparatory school, Louis Richard, had also taught Galois. Hermite's affinity for German mathematical work dated from his earliest years, when he successfully extended some important theorems of Jacobi on elliptic functions. This was the foundation of his career, achieved at the early age of twenty-one.[6] He thus may be associated with the Jacobian ideal of pure mathematics and was doubtless aware of Jacobi's statement in response to Fourier that mathematics was undertaken not for application but for the honor of the human spirit. Hermite was appointed to the *Académie des sciences* in 1856 and held various teaching positions, notably at the *École polytechnique* and at the *Collège de France*, before his appointment to the *Faculté des sciences* of the *Université de Paris* in 1869.[7]

Of key importance for Hermite's influence in the transmission of German mathematics were Hermite's dynastic connections, in what Martin Zerner has called the Kingdom of Mathusia, ruled rather surprisingly by Joseph Bertrand.[8] Bertrand's mathematical work is now little recalled, but his unique position of power in French mathematics and science in the 1870s and 1880s was guaranteed by his role as one of the two perpetual secretaries of the Paris *Académie des sciences*. As Bertrand's brother-in-law, Hermite had unique access to Bertrand at times, although their relations were sufficiently difficult that they sometimes did not speak. Hermite was father-in-law to the mathematician, Émile Picard, and was likewise related by marriage to another important, young mathematician of the period, Paul Appell. Indeed, his extended family also had links to Jean-Marie-Constant Duhamel and Olinde Rodrigues (in earlier generations) and to Émile Borel (later).[9] One final

---

[5]See, for example, Christophe Charle, *La république des universitaires, 1870–1940* (Paris: Seuil, 1994). A different picture is provided by Fritz Ringer in *Fields of Knowledge: French Academic Culture in Comparative Perspective, 1890–1920* (Cambridge and Paris: Éditions de la Maison de sciences de l'homme, 1991). Mathematics, in particular, is discussed in Martin Zerner, "Le regne de Joseph Bertrand (1874-1900)," *Cahiers d'histoire et de philosophie des sciences* **34** (1991):296-322.

[6]Curiously, he never learned to read German; Jacobi's most relevant papers, for example, were in French or Latin.

[7]Further details may be found in Claude Brezinski, *Charles Hermite, père de l'analyse mathématique moderne*, *Cahiers d'histoire et de philosophie des sciences* **32** (1990) and in the references therein.

[8]Zerner, p. 301.

[9]*Ibid.*

biographical detail of importance: Hermite had been very closely touched by the Franco-Prussian War. His home town, where many of his relatives still lived, had been occupied by Prussian officers during the siege of Metz. A portion of his family was in territory ceded to Germany at the conclusion of hostilities,[10] and he felt the changes that had occurred very deeply.

The information that survives about Hermite's efforts rests, as noted above, on his correspondence with two people who were unusually well-positioned to assist him in grasping current trends in the German mathematical communities. The first of the two correspondents, Paul Du Bois-Reymond, was born in Berlin and studied there. His father was a self-made Prussian diplomat from Neuchâtel in Switzerland (a Prussian fiefdom until the mid-nineteenth century), who ensured his fluency in both French and German. Paul's brother Emil, the physiologist, was at the top of the Berlin scientific world in 1870 and was a well-known and rather chauvinistic critic of the French at the time of the war.[11] During the period of the correspondence, Du Bois-Reymond was an *Ordinarius* first at Tübingen, and, after 1885, at the *Technische Hochschule Charlottenberg* back in Berlin. He had close ties to Weierstraß but had difficult relations with many members of the Weierstraß school, most notably Hermann Amandus Schwarz.[12] Du Bois-Reymond's correspondence with Hermite revolved around mathematical questions in analysis, concerning primarily the convergence of trigonometric series and improper integrals. However, Hermite's candid discussion of his life as a professor and his efforts to promote Du Bois-Reymond's work provide a valuable picture of his role as a vector of German ideas in France. As Du Bois-Reymond's portion of the correspondence is lost, except for one letter, his responses must be inferred.

Mittag-Leffler, certainly one of the most intriguing figures in international mathematics of the late-nineteenth century, was also a pivotal figure in the transmission of information between Paris and Berlin. Visiting Paris on a postdoctoral travel grant in 1874, he studied elliptic functions with Hermite; then, acting on Hermite's advice, he went to Berlin to attend the lectures of Weierstraß and Kronecker. He subsequently held positions in Helsinki (then Helsingfors) and in Stockholm, where, in 1882, he founded his journal, *Acta Mathematica*, with royal support. He kept in very regular contact with Paris and Berlin, both via travel and in correspondence, and *Acta* became an important vehicle for the transmission of mathematical information from one country to another.[13]

Hermite's Critique of Radicalism

Hermite's efforts took place against a turbulent political background, and understanding his position with respect to contemporary political issues is useful in grasping his strategies. He shared his views freely with his correspondents and

---

[10] Emil Lampe, "Briefe von Ch. Hermite zu P. du Bois-Reymond aus den Jahren 1875–1888," *Archiv der Mathematik und Physik* **24** (3) (1916):193-220 and 289-310 on p. 197.

[11] Emil du Bois-Reymond, "Über den deutschen Krieg," Rectoral address given 3 August, 1870 at the University of Berlin and translated in *Revue des cours littéraires de la France et de l'étranger* **7** (1870):658-668.

[12] Gillispie, ed. *Dictionary of Scientific Biography*, s.v. "Du Bois-Reymond, Paul David Gustav" by Luboš Nový and the references therein.

[13] For biographical details on Mittag-Leffler, see Lars Gårding, *Mathematics and Mathematicians: Mathematics in Sweden before 1950*, HMATH, vol. 13 (Providence: American Mathematical Society and London: London Mathematical Society, 1998), chapter seven. On Mittag-Leffler and *Acta*, see June Barrow-Green's chapter that follows.

was often able to assume their agreement with his attitudes. Hermite favored the restoration of the monarchy as well as the retention and, indeed, expansion of religious involvement in education. He was strongly opposed to democratic measures of various kinds, as is amply attested in the letters. He was particularly caustic about universal suffrage, which he found deeply wrong-headed. In a letter to Mittag-Leffler of 1883, he described his horror at picking up his voter card at the *mairie*: " ... I savored the impression of contact with the sovereign people, Our Lords the laborers, tailors, second-hand-shop keepers, wine merchants, tripe merchants, etc., who demanded their cards while harassing the employees."[14] Beyond a mere quasi-aristocratic distaste for the common people, however, Hermite showed a genuine fear of the consequences of radical republicanism. He frequently mentioned the possibility of German armed intervention in an unstable or radical French state and went so far as to investigate acquiring a property in Brittany for the express purpose of having a place to go in the event of another invasion.[15]

This needs a little explanation. The aftermath of the war had left a complicated political situation in France, in which the positions of Hermite were those of a substantial minority on the right. The notion of *rapprochement* with and even appeasement toward Germany was also a common right-wing view, while elements of the left were much more belligerent towards the Germans. The right was, however, insufficiently organized, divided, for example, between Bourbonists and Orleanists; the superior leadership of the left led to its dominance, especially from about 1879. For Hermite, this had important implications, and not just on general political grounds. The left favored and, indeed, carried out much educational reform, a key element of which focused attention on the Sorbonne and the *grandes écoles* because of the proposal to eliminate members of religious orders from teaching.[16]

This focus had various consequences. Members of some orders, notably the Jesuits, were banned from teaching in public institutions. The question of the content of education and its appropriateness for republican life also arose. In discussions of the latter set of issues, the question of foreign influence had an ambiguous status. While Germany was generally seen as bad, the organization of higher education there had various elements that were politically correct for at least some of the left, for example, the distribution of universities across the country and the well-developed system of higher technical schools. Moreover, the obvious rise of German science and technology compared to the relative stagnation of French efforts was apparent. Thus, those on both the right and the left expressed an interest in German models.[17] This atmosphere was one in which Hermite could pursue his efforts at building German mathematics into the research community and the curriculum, although it was not without obstacles. Likewise, it afforded various opportunities for integration of the mathematical communities in the two countries.

[14] "...j'ai savouré l'impression de contact avec le peuple souverain, Nos Seigneurs les ouvriers, tailleurs, brocanteurs, marchands de vin, marchands de crépines, etc. qui recherchent leurs cartes en gouaillant les employés." Dugac, "Lettres (1874–1883)," letter of 3 February, 1883, p. 195.

[15] See, for example, Lampe, letters of 27 June, 1879 and 21 July, 1879, pp. 201-202.

[16] Dugac, "Lettres (1874–1883)," letter of 5 April, 1880, p. 70 and note 38.

[17] For a discussion of this and related matters, see the paper by Robert Fox cited above. The Education Ministry in the new Third Republic sent a commission of inquiry to the German universities in 1881, a fact noted by Hermite in a letter to Mittag-Leffler of 24 September, 1881. See Dugac, "Lettres (1874–1883)," p. 130.

The Promotion of Franco-German Relations in Mathematics

Hermite's expressed view of international relations was essentially this: that the scientific confraternity transcends national boundaries. Writing to Du Bois-Reymond in 1875, he asked:

> Do you believe that just because we belong to countries with different languages, between which the war has made a bloody trench, that there is no longer any resemblance in the work and the professional duties, between those who have devoted their entire lives to the same studies? ... I do not at all hesitate to tell you that at the end of a tiring lecture ... I take great pleasure in conversing with a colleague from Germany, and deploring just as energetically as he would the impossibility of uninterrupted reflection.[18]

Hermite continued that "[t]he analogy of our situations creates a natural and legitimate sympathy, of which, above and against the resentments of politics and the war, I make myself the medium."[19] As for the then-current view of Germany within the French scientific community: " ... certain signs announce that a real and general peace will be concluded, to restore to science its most precious privilege, to create a bond of esteem and personal affection between all those who dedicate themselves to it."[20] Hermite insisted here, and frequently throughout the course of the correspondence, that the hopes and fears he expressed were not uniquely his own. In fact, he insisted so much that it raises the question whether he was really correct. He frequently mentioned, however, the agreement of his colleague, Jean-Claude Bouquet, and also sometimes of Bertrand.

As mentioned earlier, Hermite's admiration of German mathematics dated to the beginning of his career. His admiration of the German institutional setting, academic life, and cultural values was strongly reinforced by a visit to Göttingen in 1877 for the centenary of Gauss's birth. Hermite was deeply moved there, as he explained to Du Bois-Reymond, by the fact that such a large celebration would be given in honor of a man whose life's work had been dedicated to science alone. Such honors would, he felt, be impossible in France.[21] Subsequent to this, his comments comparing France unfavorably to Germany seem to become more frequent, particularly as the French political situation became more unstable and more republican.[22]

---

[18] "Pensez-vous donc que pour appartenir à des pays de langues différentes, et entre lesquelles la guerre a fait un sillon sanglant, il n'existe plus aucune ressemblance, dans les travaux et les devoirs personnels, de ceux qui ont voué toute leur vie aux mêmes études? ... je n'hesite point à vous dire qu'au sortir d'une leçon fatiguante ... je me mets avec bonheur à m'entretenir familièrement avec un collègue d'Allemagne, et à déplorer aussi vivement qu'il peut le déplorer pour son compte, de n'avoir point la possibilité de méditer sans interruption." Lampe, undated, pp. 195-196.

[19] "L'analogie des situations crée une sympathie naturelle et si légitime, qu'envers et contre les ressentiments de la politique et de la guerre, je m'en fais l'organe." *Ibid.*

[20] " ... certains signes annoncent qu'une paix réelle et générale va se conclure, pour restituer à la science son plus précieux privilège, de créer un lien d'estime et d'affection personelles entre tous ceux qui s'y consacrent." *Ibid.*, p. 196.

[21] Dugac, "Lettres (1874–1883)," letter of 23 April, 1878, p. 52.

[22] See, for example, Dugac, "Lettres (1874–1883)," letter of 11 October, 1881, p. 133 and letter of 20 March, 1882, p. 154. Such views are also mentioned in Lampe, letter of 8 August, 1881, p. 125.

Hermite's methods for the promotion of German mathematics took two main forms: attempts to get official French recognition for German mathematical achievements and efforts to promote the teaching and learning of current German mathematics.[23] Already around 1875, Hermite's efforts in creating this "real and general peace" in the scholarly community may be seen in the appointment of Carl Wilhelm Borchardt as corresponding member of the *Académie des sciences*.[24] One of the standard methods of recognition of foreign *savants*, a corresponding membership conferred publication rights in the *Comptes rendus*, an important platform internationally. For the corresponding member, it was also a form of recognition that could prove useful in a variety of ways, such as for professional advancement at home, and that could thus be both a material and a symbolic reward.[25]

Borchardt, the man who had succeeded August Leopold Crelle in 1856 as the editor of Crelle's *Journal für die reine und angewandte Mathematik*, joined a number of his compatriots who had been nominated before the war: among mathematicians, Ernst Kummer (as foreign associate) and Weierstraß and Kronecker (as corresponding members). Borchardt was both a colleague and a friend of Hermite, whom he had met during a trip to Paris in 1846.[26] The two had worked on closely related subjects, and Hermite had frequently published papers in Borchardt's journal. Borchardt, an Academy member in Berlin, was also at the heart of the Berlin mathematical establishment. The editorial board of his journal included Weierstraß, Kronecker, and Hermann von Helmholtz.

Getting someone into the *Académie*, as a corresponding member or as a regular member, was a complex task. In the case of corresponding members, the *secrétaire perpetuel*—that is, Hermite's brother-in-law, Bertrand—had a good deal to do with starting the process that would lead to the position being filled. A committee of members from related domains met in secret to rank nominees; following that a vote was taken in the *Académie* as a whole, in awareness of the ranking but not necessarily in obedience to it. There was thus room for much behind-the-scenes maneuvering at all stages of the process. The result in this case was positive. Borchardt had good relations in France and, since the war, had remained in correspondence with Hermite, Bertrand, and Michel Chasles. It was not a simple matter, however: he failed to obtain a majority in the first round, and in the second round received twenty-nine votes, with twenty-three opposed and four abstaining.[27]

Bad feeling towards Germany, in the *Académie*, the scientific community, and the community at large, was certainly not a myth at this time, despite Borchardt's election. The republican parties of the left, increasingly powerful in the mid-1870s

---

[23]Similar efforts at building international connections were taking place even earlier in Britain. For example, James Joseph Sylvester proposed Jakob Steiner for foreign membership in the Royal Society in 1862, while he was later a supporter of Michel Chasles, Julius Plücker, and Jean Victor Poncelet for the Copley Medal. See J. J. Sylvester to Thomas Archer Hirst, 27 February, 1863 and J. J. Sylvester to Thomas Archer Hirst, 1 November, 1865 in Karen Hunger Parshall, *James Joseph Sylvester: Life and Work in Letters* (Oxford: Clarendon Press, 1998), pp. 118-119 and 127-129.

[24]Lampe, letter of 17 April, 1876, p. 196.

[25]This kind of function for a corresponding membership is likewise illustrated by J. J. Sylvester. See Karen Hunger Parshall and Eugene Seneta, "Building a Mathematical Reputation: The Case of J. J. Sylvester (1814–1897)," *American Mathematical Monthly* **104** (1997):210-222.

[26]Dugac, "Lettres (1874–1883)," p. 77.

[27]The tallies are in the *Comptes rendus des séances de l'Académie des sciences de Paris* **82** (1876):814.

and ascendant to government in 1879, contained a strongly anti-German component, with a rhetoric that harkened back to the days of the Revolution and leaders who had bitterly opposed surrender. Since these parties had a close eye on the educational system, advancing a pro-German cause in the context of their leadership was not without risk of reprisal. There was nevertheless an element of the left that shared the rather widely held view that French science (in particular) had fallen behind that of the Germans, and that the higher educational system was particularly weak in this regard.[28] Louis Pasteur was a famous proponent of this view already in the 1860s, and it was taken up as a theme in criticizing the educational system, especially its more conservative features and its religious elements.[29]

In the mathematical community, anti-German feeling displayed itself through a variety of mostly rather minor acts. Seen from the German standpoint, namely Borchardt's, these included the failure to cite German work on which French work was based (Camille Jordan); pettily critical reviews in Darboux's *Bulletin des sciences mathématiques* (by Maurice Levy of a paper by Borchardt); resignation from membership in scientific societies (Jordan again); and outright plagiarism. Writing to Rudolf Lipschitz, Borchardt made these charges specific, drawing an analogy between the ordinary plunder of war and this kind of plunder in the scientific sphere, and citing, in particular, a work of Pierre Alphonse Laurent that he felt had been stolen directly from Eduard Heine.[30]

In addition to society membership, another form of distinction that could be accorded to Germans was the award of prizes and medals. For the established French scientist, recognition by the state came in one principal form: admission to a national order accompanied by a medal. In France, the different degrees of the Legion of Honor were the standard awards to meritorious servants of the public good.

A medal is a form of recognition accorded not merely to the *savant*, but to the military man, public servant, or private citizen. It therefore permits comparisons with others in different walks of life whose efforts may be thus recognized. It may be worn or displayed in a prominent place in the home and so is a portable and easily understood expression of distinction. A medal in an aristocratic society may carry with it the sense of inclusion in the aristocracy, a feeling that one is quasi-royal. It is, moreover, an honor to the family as well as to the recipient.

Some academics were inveterate collectors of such honors; Mittag-Leffler is one example. An anecdote still told recounts how he appeared at an official function in a fine jacket, covered with medals all of whose ribbons matched its blue color. When asked how it happened that the ribbons should all match, Mittag-Leffler supposedly gave the obvious answer: "These are the blue ones."[31]

Hermite knew well the interest of academics in receiving such honors. Unhappy about his own lack of advancement, in April of 1882, he seized a political opportunity in order to seek awards for foreign colleagues: first Weierstraß, then

[28]The complexities of the French political situation in the 1870s and 1880s are outlined well in Dominique Barjot, Jean-Pierre Chaline and André Encrevé, *La France au XIXe siècle, 1814–1914* (Paris: Presses universitaires de France, 1995), pp. 447-502.

[29]Fox, pp. 14-18. Compare also Hélène Gispert's chapter above.

[30]See Rudolf Lipschitz, *Briefwechsel mit Cantor, Dedekind, Helmholtz, Kronecker, Weierstraß und anderen*, ed. Winfried Scharlau (Braunschweig: Vieweg Verlag, 1986), letters of 17 and 21 December, 1871, pp. 21-23 as well as the accompanying notes.

[31]Efforts to verify the truth of this anecdote have so far been unsuccessful.

Kronecker. The procedure was to ask one of the *secrétaires perpetuels*, in this case the chemist, Jean-Baptiste-André Dumas, to approach the President of the *Conseil des ministres*, concerning the possibility of an award. In the Third Republic, this individual was the ceremonial head of state, rather like the French President today. The decision appears to have been essentially up to this man, Charles de Saulces de Freycinet, a *polytechnicien* and engineer. Freycinet had recently sought Hermite's support for a position as a free member of the *Académie*, and Hermite grabbed his chance, writing immediately to Dumas requesting an award for Weierstraß. Almost at once, he realized that Kronecker would be jealous, and learning that Weierstraß would receive an award, he quickly sought one for Kronecker as well.[32]

Hermite had to provide the basis for the honor. For Weierstraß, the arguments were interesting, since they insisted on the importance of his work for the French State. For example, Hermite cited the fact that Weierstraß's work was so fundamental that it was being taught to students at the *École polytechnique*. For Kronecker, he was obliged to fall back on his own prestige as the main basis for the award, since his argument was, in essence, that Kronecker had worked in the same area as Hermite and that they had simultaneously obtained closely related results using different methods. These results were very famous at the time they were produced (in the mid-1850s), at least in scientific circles, since they concerned a long-open problem that had defied generations of researchers: the solution of equations of the fifth degree (impossible algebraically), achieved by Hermite and Kronecker using elliptic functions.[33] The medals to Weierstraß and Kronecker followed closely similar (actually slightly better) awards to Helmholtz and Gustav Robert Kirchhoff. These were the first Germans to have been honored in this way following the Franco-Prussian War.[34]

Bringing German Mathematics to French Students: A Thankless Task

If medals and academy membership were important in establishing Franco-German links at the highest professional levels, much more important relative to long-term developments was Hermite's effort both to make the work of Germans known in France and to publicize French work in Germany. This took several forms. One involved teaching mathematics that would allow students access to German work. Another was oriented toward research and included encouraging German publication in French journals, arranging for translations of selected works, and ensuring that German mathematicians were well-aware of French efforts.

The role of the universities in French and German society differed somewhat at this time, and this was reflected in marked differences between the systems in the two countries, particularly as regards teaching. In Germany, the professor typically lectured on his research, which was necessarily somewhat current. By contrast, in France, there was a greater emphasis on lower-level pedagogical lecturing, and in some institutions (notably the *École polytechnique*), committees kept a close watch on the curriculum. This had led to a textbook tradition, and published French mathematical *cours* (or lecture courses) often had long lives, running many editions

[32]Dugac, "Lettres (1874–1883)," letter of 18 April, 1882, pp. 156-157.

[33]These results appeared in a variety of papers, but brief accounts may be found in Charles Hermite, "Résolution de l'équation du cinquième degré," *Comptes rendus de l'Académie des sciences de Paris* **46** (1858):715-722, and Leopold Kronecker, "Sur la résolution de l'équation du cinquième degré," *ibid.*, pp. 1150-1152.

[34]Dugac, "Lettres (1874–1883)," letter of 18 April, 1882, p. 156.

over decades with only minor modifications. For example, the calculus textbook of Sylvestre-François Lacroix, first published in 1797, had a sixth edition in 1862, with an appendix by Hermite on elliptic functions. In Germany, such books were a late development, and, indeed, French books were much used there by students who likewise paid tutors to help them get the basics.[35] Hermite's conviction about the superiority of the German system was reinforced once again by his 1877 visit to Göttingen, where he saw with admiration and amazement professors and students sitting, drinking beer together, and singing.[36] This may have encouraged him in subsequent efforts to follow a more German pattern in his lecturing by the inclusion of current work.

This emulation of Germany aside, Hermite frequently complained to Du Bois-Reymond about the amount of effort he devoted to teaching. Beginning in 1882, he had incorporated a good deal of recent German work into his analysis course at the *Faculté des sciences*, notably results due to Weierstraß on primary factorization of functions of a complex variable. These results were brand-new. They had appeared for the first time in 1874 in Weierstraß's Berlin lectures, but Hermite had learned about them three years later by virtue of his attendance at the Gauss memorial symposium. They became widely accessible only in 1878 and were, to some degree, completed by Mittag-Leffler's work of 1880 through 1882. Thus, the students were lucky enough (in Hermite's view) to be getting German-style lectures on the most recent work. In fact, these theorems were introduced quite early in the course, since they permitted certain economies in the treatment and allowed Hermite to devote more time to his favorite topic, elliptic functions. It should be noted that these theorems are not that easy, the more so because the treatment uses Weierstrassian arguments that were unknown to all but a handful of professional mathematicians in France (and to relatively few people in Germany). Thus, students could not rely on the traditional method of paying a tutor to get them through the difficult spots.

Hermite learned from his colleague, Charles Biehler, then director of preparatory studies at the *Collège Stanislas*, a post-secondary institution preparing students for the entry competitions of the *grandes écoles*, that most of the students could not follow him.[37] It seems rather characteristic of Hermite that this was a real surprise. Indeed, his letters of this period to Mittag-Leffler note with what enthusiasm the students heard of these theorems.[38] Nevertheless, Hermite responded to the classroom situation like a highly responsible pedagogue; he asked for Bouquet (whose students at the *École normale supérieure* came to the lectures) to arrange for notes to be taken, lithographed, and sold at a bookstore for around 0.25 francs a lecture (a then-common system in France).[39]

Negative student reaction was not turned aside by these efforts, however, and achieved expression in a modern form. In July 1881, the republican government had passed legislation declaring the freedom of the press. In June 1882, just at the end of the university year, the editors of a new journal called *Le Passant* had

---

[35]Recall Ivor Grattan-Guinness's discussion of the translation of French texts and textbooks above.

[36]Dugac, "Lettres (1874–1883)," letter of 23 April, 1878, p. 52.

[37]*Ibid.*, letter of 6 April, 1882, pp. 153-154.

[38]*Ibid.*, letter of 24 December, 1881, p. 138.

[39]*Ibid.*, letter of 20 March, 1882, p. 154. These lecture notes, transcribed by Marie-André Andoyer, a student at the *École normale supérieure*, were revised by Hermite and published in several editions. See Charles Hermite, *Cours d'analyse de la Faculté des sciences*, ed. Marie-André Andoyer (Paris: Hermann, 1882). For the material on elliptic functions, see pp. 211-265.

the idea that they should include some critical investigations of the educational community and commissioned one A. Dunoiset to make reports. He began with the science faculty, a particularly attractive target. Staffed by what Dunoiset termed "grizzled Methusalahs" deviously standing in the way of promised reforms, the faculty had serious space and equipment problems. His complaints have a ring remarkably reminiscent of student complaints today. For example, the chemist, Charles-Adolphe Wurtz, "the most distinguished scientist among the chemists is, you guessed it, the most detestable of teachers."[40] The question of why researchers were needed to teach first- and second-year courses was brought up repeatedly by the critic.

Mathematics fared a little better at the hands of *Le Passant*, but Hermite was singled out on two grounds: his teaching and his role in obtaining a position for his son-in-law, Picard. In the journalist's view, Picard would still be in Toulouse if he had not been the son-in-law of Hermite, and his nomination to the chair of kinematics was "a real scandal."[41] Hermite himself was criticized in these terms: "You come to learn how to integrate and M. Hermite presumes that you have been integrating all your life. His course is not a course, it is a dazzling conversation, broken up into bits, with interminable digressions and meanderings all across Europe [a reference to the frequent inclusion of discussions of recent foreign research] ... he is detested by the students."[42] His references to religion and a belief in providence likewise come in for criticism.

Hermite's response was shock. In letters to both Du Bois-Reymond and Mittag-Leffler, he expressed several times his dismay that his students really might detest him.[43] However, the shock to his dignity at being treated this way in public was clearly as much in his mind as his concern for his students. His ultimate diagnosis? Freedom of the press is an evil: "Without doubt, Sir, you are liberal and subscribe to the freedom of the press. Alas the use of this liberty under the government of the republic disgusts me profoundly."[44]

German Mathematics and the French Mathematical Research Community

Research communication between France and Germany was not in a good state before 1870, but Hermite made a concerted effort to improve the situation. At this time, for a French academic to read German was atypical, particularly in the older generations that dominated the French university community.[45] Furthermore, the library situation in Paris (and *a fortiori* everywhere else in France) was not good so that, in general, one would actually have had to subscribe personally to many journals in order to see them. This was expensive, although it was the route

[40] "C'est le savant le plus illustre des trois: vous avez deviné que c'est le plus détestable des maîtres." A. Dunoiset, "Gazette de l'Université," *Le Passant: Journal illustré* **1** (4) (1882):5.

[41] *Ibid.*

[42] "Vous venez pour apprendre à intégrer, M. Hermite suppose que vous avez intégré toute votre vie. Son cours n'est pas un cours, c'est une conversation abracadabrante, faites à bâtons rompus, avec des digressions interminables, des chevauchées perpetuelles à travers l'Europe entière ... il est ... execré des étudiants." A. Dunoiset, "Gazette de l'Université," *Le Passant: Journal illustré* **1** (5) (1882):4-5.

[43] For example, see Lampe, letter of 5 July, 1882, p. 211.

[44] "Sans doute, Monsieur, vous êtes libéral et partisan de la liberté de la presse; hélas, l'usage de cette liberté, sous le gouvernement de la République, m'en dégoûte profondément." *Ibid.*

[45] On the languages of mathematics, see Jeremy Gray's chapter in the present volume.

followed, for example, by Darboux.[46] Translations of a few articles did appear in Liouville's *Journal*, although this was erratic, and a German practice earlier in the century of trying to announce key results in letters (to Liouville or others in France) intended for publication had waned as the French mathematical community weakened and the German one attained independence and maturity.[47] There were no regular international meetings and no exchanges as such.

The situation began to change in 1870, when Gaston Darboux and Guillaume-Jules Hoüel brought out the first volume of their *Bulletin des sciences mathématiques et astronomiques*.[48] Darboux presented summaries, or at least titles, of papers in most of the international journals and presented translations of some articles of critical importance, a noteworthy example being Riemann's thesis on trigonometric series. However, there were various problems: reliable translators were in short supply, and the mathematics in some of the papers was incomprehensible even when in French—not only the notoriously difficult Riemann but also the work of Weierstraß and his students.

Hermite worked to have longer pieces translated. The Mittag-Leffler correspondence provides detailed accounts of Hermite's efforts to produce translations of fundamental papers of Weierstraß and Cantor, while the correspondence with Du Bois-Reymond documents his desire to translate one of the latter's books.[49] Hermite's chosen translators were initially his best students. Émile Picard, at that time only a potential son-in-law, was drafted to translate a Weierstraß paper, and Paul Appell and Henri Poincaré assisted with the work of Cantor.[50] Such papers could be published in the *Annales* of the *École normale supérieure*, run by Louis Pasteur, or they could appear in *Acta Mathematica*, Mittag-Leffler's new journal, once that was begun. Such efforts at translation were not without their problems. Poincaré, in particular, felt the French readers would balk at the combination of mathematics and philosophy in Cantor's researches.[51]

Such efforts of translation were obviously very important in assisting the young mathematicians engaged in them to assimilate German concepts and the German language. At times, these junior researchers were immediately able to produce work in the same vein; Picard is an example.[52] Expository accounts of German work became common thesis topics at the Sorbonne during this period, with Hermite figuring prominently in the dedications and acknowledgments. Students were encouraged to go beyond the original, but understanding was all that was required

---

[46]Hélène Gispert, "La correspondence de G. Darboux avec J. Hoüel: Chronique d'un rédacteur (déc. 1869—nov. 1871)," *Cahiers du Séminaire d'histoire des mathématiques* **8** (1987): 67-202 on p. 89.

[47]Jesper Lützen discusses these practices in some detail in his chapter on Liouville and his *Journal* in the present volume.

[48]See Gispert, "Correspondence de Darboux." Beginning in 1885, the journal became simply the *Bulletin des sciences mathématiques*.

[49]Dugac, "Lettres (1874–1883)," pp. 53-54 and 192-193 as well as the footnote on p. 272. See also Lampe, letter of 22 June, 1886, p. 297 and the chapter on Mittag-Leffler and *Acta Mathematica* by June Barrow-Green below.

[50]Dugac, "Lettres (1874–1883)," pp. 53-54 and 192-193.

[51]*Ibid.*, letter of 5 March, 1883, p. 199.

[52]See, for example, Émile Picard, "Sur une propriété des fonctions entières," *Comptes rendus des séances de l'Académie des sciences* **88** (1879):1024-1029.

for the doctorate. German work that was treated in this way included work by Schwarz, Otto Hölder, Riemann, and Lazarus Fuchs.[53]

Also useful were efforts to have German mathematicians publish in French periodicals. Darboux's journal was clearly a pioneer in this regard.[54] Similarly, Hermite attempted to get accounts of German work published in the *Comptes rendus*, which were much more prestigious, widely available, and universally read. His letters to Du Bois-Reymond, in particular, contain various accounts of his attempts to place German research there. One of his first letters explained the system: a paper in the *Comptes rendus* could be no more than three pages and had to be forwarded by an academician to the *secrétaire perpetuel.*[55] Hermite subsequently served as a vehicle for work by Du Bois-Reymond and several of his students.[56] Since Du Bois-Reymond was essentially a native French speaker, there was no obstacle to publication, which had to be in French.

Hermite did face difficulites in this effort, however. His redoubtable brother-in-law, Bertrand, as one of the two perpetual secretaries, decided what went into the *Comptes rendus* and what did not. Bertrand's opposition to the newer mathematics became obvious and was not restricted to German work. For example, he set up barriers to the appearance in the *Comptes rendus* of a note by Darboux inspired by Riemann.[57] According to Hermite, Bertrand consciously determined to obstruct the numerous publications of young mathematicians, ostensibly because he wanted works that were riper, more complete, more profound.[58] Bertrand may well have failed both to grasp the details of these works and to understand their interest. This was certainly a question that Hermite posed from time to time in his correspondence, especially regarding Cantor's work.[59] At one point, Hermite feared that Bertrand would go so far as to exclude mathematical work from the *Comptes rendus* altogether.

Concluding Remarks

While Hermite stressed international cooperation, there was also a strong feeling of rivalry with the Germans, particularly among the younger men who were his students. It is likewise clear that Hermite did not like to be the target of criticism coming from Germany, for example, by Schwarz who, like Giuseppe Peano, criticized the definition Hermite had given of surface area in his lectures.[60] This

---

[53]See Hélène Gispert, "La France mathématique," *Cahiers d'histoire et de philosophie des sciences* **34** (1991):11-180 for a discussion of many such theses.

[54]For example, one could mention the appearance of the paper in which Lipschitz announced the result in the existence theory of differential equations employing what is now known as a Lipschitz condition. Rudolf Lipschitz, "Sur la possibilité d'intégrer complètement un système donné d'équations différentielles," *Bulletin des sciences mathématiques et astronomiques* **10** (1879):177-179.

[55]Lampe, undated letter, p. 194-195.

[56]See, for example, *ibid.*, pp. 205 and 209.

[57]Gaston Darboux, "Sur une équation linéaire aux dérivées partielles," *Comptes rendus des séances de l'Académie des sciences de Paris* **95** (1882):69-72

[58]Dugac, "Lettres (1874–1883)," letter of 25 July, 1882, p. 162.

[59]*Ibid.*, letter of 13 April, 1883, pp. 209-210.

[60]Hermite used a definition involving inscribed polyhedra. Schwarz had pointed out in a letter to Hermite that such polyhedra could have a surface area exceeding any given bound. The letter appears in Hermann Amandus Schwarz, *Gesammelte mathematische Abhandlungen*, 2 vols. (Berlin: Akademie der Wissenschaften, 1890), 2:309-311. Hermite was aware that the definition was not rigorous, as may be seen in Dugac, "Lettres (1874–1883)," p. 77.

rivalry was felt in both directions, and Weierstraß was very impressed by the current French crop of young mathematicians, including Poincaré, Picard, and Appell. Apparently speaking to an assembled group Weierstraß stated that "[w]e are going to have to pull together like the devil, gentlemen, if Paris is not to become once again the mathematical capital."[61] Hermite's program was enormously successful, and although it was certainly not his alone, his leadership was central to the development of a mastery of German concepts. This was displayed all the more clearly in the next generation, when individuals such as Maurice Fréchet and Émile Borel took the theory of real-valued functions in important new directions.

Yet, to return to the theme of the formation of the international mathematical community, such rivalries did not consitute an overwhelming obstacle for Hermite and those who, like him, saw the importance of French participation in an international community of mathematicians. Instead, they fueled interest in the work of their rivals and stimulated efforts on both sides of the border to be fully aware of the mathematical production of their competitors. Hermite's dream of an international fraternity of scientists sharing mathematical interests and values as well as social attitudes was, of course, realized only partially, and mathematicians were to become embroiled once again in Franco-German hostilities during the Great War of 1914–1918.[62] Yet even during the war years, Hermite's collegial voice spoke posthumously to his colleagues: Lampe's edition of the correspondence with Du Bois Reymond appeared in 1916, at the depth of hostilities.

References

Barjot, Dominique; Chaline, Jean-Pierre; and Encrevé, André. *La France au XIXe siècle, 1814–1914*. Paris: Presses universitaires de France, 1995.

Brezinski, Claude. *Charles Hermite, père de l'analyse mathématique moderne. Cahiers d'histoire et de philosophie des sciences* **32** (1990).

Charle, Christophe. "Intellectuels, Bildungsbürgertum et professions au XIXème siècle: Essai de bilan historiographique comparé (France, Allemagne)." *Actes de la recherche en sciences sociales* (1995):85-95 and 106-107.

———. *Les intellectuels en Europe au XIXème siècle: Essai d'histoire comparée.* Paris: Seuil, 1996.

———. *La république des universitaires, 1870–1940.* Paris: Seuil, 1994.

Darboux, Gaston. "Sur une équation linéaire aux dérivées partielles." *Comptes rendus des séances de l'Académie des sciences de Paris* **95** (1882):69-72.

Du Bois-Reymond, Emil. "Über den deutschen Krieg." Rectoral address given 3 August, 1870 at the University of Berlin. Translated in *Revue des cours littéraires de la France et de l'étranger* **7** (17 September, 1870):658-668.

Dugac, Pierre. "Éléments d'analyse de Karl Weierstraß." *Archive for History of Exact Sciences* **10** (1973):41-176.

———. "Lettres de Charles Hermite à Gösta Mittag-Leffler (1874–1883)." *Cahiers du Séminaire d'histoire des mathématiques* **5** (1984):49-285 (the footnotes to this edition assume the pagination starts at page 1).

[61] "Wir müssen uns teuflich zusammenhangen wenn nicht Paris noch einmal der Hauptsitz der Mathematik werden wird." Gösta Mittag-Leffler to Charles Hermite, 3 August, 1882, in Pierre Dugac, "Elements d'analyse de Karl Weierstraß," *Archive for History of Exact Sciences* **10** (1973):41-176 on pp. 158-159.

[62] Compare the discussion in the chapter by Sanford Segal in the present volume.

———. "Lettres de Charles Hermite à Gösta Mittag-Leffler (1884–1891)." *Cahiers du Séminaire d'histoire des mathématiques* **6** (1986):79-217.

Dunoiset, A. "Gazette de l'Université." *Le Passant: Journal illustré* **1** (4) (1882):5 and **1** (5) (1882):4-5.

Fox, Robert. "The View over the Rhine: Perceptions of German Science and Technology in France 1860–1914." In *Frankreich und Deutschland: Forschung, Technologie und industrielle Entwicklung im 19. und 20. Jahrhundert.* Ed. Yves Cohen and Klaus Manfrass. Munich: C. H. Beck, 1990, pp. 14-24.

Gårding, Lars. *Mathematics and Mathematicians: Mathematics in Sweden before 1950.* HMATH. Vol. 13. Providence: American Mathematical Society and London: London Mathematical Society, 1998.

Gillispie, Charles C., Ed. *Dictionary of Scientific Biography.* 16 Vols. and 2 Supps. New York: Charles Scribner's Sons, 1970–1990.

Gispert, Hélène. "La correspondence de G. Darboux avec J. Hoüel: Chronique d'un rédacteur (déc. 1869–nov. 1871)." *Cahiers du Séminaire d'histoire des mathématiques* **8** (1987):67-202.

———. "La France mathématique." *Cahiers d'histoire et de philosophie des sciences* **34** (1991):11-180.

Hermite, Charles. "Résolution de l'équation de cinquième degré." *Comptes rendus des séances de l'Académie des sciences de Paris* **46** (1858):715-722.

———. *Cours d'analyse de la Faculté des sciences.* Ed. Marie-André Andoyer. Paris: Hermann, 1882. Lithographed notes.

Kronecker, Leopold. "Sur la résolution de l'équation du cinquième degré." *Comptes rendus des séances de l'Académie des sciences de Paris* **46** (1858):1150-1152.

Lampe, Emil. "Briefe von Ch. Hermite zu P. du Bois-Reymond aus den Jahren 1875–1888." *Archiv der Mathematik und Physik* **24** (3) (1916):193-220 and 289-310.

Lipschitz, Rudolf. *Briefwechsel mit Cantor, Dedekind, Helmholtz, Kronecker, Weierstraß und anderen.* Ed. Winfried Scharlau. Braunschweig: Vieweg Verlag, 1986.

———. "Sur la possibilité d'intégrer complètement un système donné d'équations différentielles." *Bulletin des sciences mathématiques et astronomiques* **10** (1879):177-179.

Parshall, Karen Hunger. *James Joseph Sylvester: Life and Work in Letters.* Oxford: Clarendon Press, 1998.

Parshall, Karen Hunger and Seneta, Eugene. "Building a Mathematical Reputation: The Case of J. J. Sylvester (1814–1897)." *American Mathematical Monthly* **104** (1997):210-222.

Picard, Émile. "Sur une propriété des fonctions entières." *Comptes rendus des séances de l'Académie des sciences de Paris* **88** (1879):1024-1029.

Ringer, Fritz. *Fields of Knowledge: French Academic Culture in Comparative Perspective, 1890–1920.* Cambridge: University Press and Paris: Éditions de la Maison des sciences de l'homme, 1992.

Schwarz, Hermann Amandus. *Gesammelte mathematische Abhandlungen.* 2 Vols. Berlin: 1890. Reprint Ed. New York: Chelsea Publishing Co., 1972.

Zerner, Martin. "Le regne de Joseph Bertrand (1874–1900)." *Cahiers d'histoire et de philosophie des sciences* **34** (1991):296-322.

Gösta Mittag-Leffler (1846–1927)

CHAPTER 8

Gösta Mittag-Leffler and the Foundation and Administration of *Acta Mathematica*

June E. Barrow-Green
Open University (United Kingdom)

Introduction

The name of Gösta Mittag-Leffler (1846-1927) is familiar to many mathematicians through his contributions to complex function theory, in particular, the Mittag-Leffler theorem that enters into many university courses in the subject.[1] However, although he was an accomplished mathematician, his primary contribution to mathematics lies elsewhere. It was as founder and editor-in-chief of the journal, *Acta Mathematica*, that Mittag-Leffler found his true *métier*. For forty-five years, his duties as editor of *Acta* were his main preoccupation, and he excelled in the role. The journal has remained in the front rank of mathematical periodicals from its inaugural volume up to the present day, and credit for establishing its enduring reputation lies firmly with Mittag-Leffler. André Weil, who in 1927 spent a month visiting Mittag-Leffler at his home in Djursholm, said (taking a cue from Oscar Wilde) that "the *Acta Mathematica* were the product of his [Mittag-Leffler's] genius, while nothing more than talent went into his mathematical contributions."[2] G. H. Hardy echoed Weil's opinion observing that "to the outside world [Mittag-Leffler] was, above everything, the editor of the *Acta Mathematica*."[3]

It was due to Mittag-Leffler that *Acta* became the first truly international mathematical journal with both its contributors and readers coming from across the globe. Mittag-Leffler traveled widely on the journal's behalf, meeting the leading mathematicians of the day, and he became a familiar figure in mathematical circles across Europe. He attended each of the first five International Congresses of Mathematicians, from Zürich in 1897 to Cambridge in 1912, and was a Vice-President of the Congress on three out of the five occasions. On his invitation,

[1]The Mittag-Leffler theorem is an extension to meromorphic functions of the Casorati-Weierstraß theorem. The Casorati-Weierstraß theorem—which states that in the neighborhood of an essential singularity c, a function $f(z)$ can come arbitrarily close to any given value and, consequently, has no given value for $z = c$—was published by Casorati in 1868 and by Weierstraß in 1876. See Umberto Bottazzini, *The Higher Calculus: A History of Real and Complex Analysis from Euler to Weierstraß* (New York: Springer-Verlag, 1986), pp. 284-285. Mittag-Leffler first published his theorem in 1877 in *The Proceedings of the Swedish Academy of Sciences* and then republished it with various generalizations as his first contribution to *Acta Mathematica*. Gösta Mittag-Leffler, "Sur la répresentation analytique des fonctions monogènes uniformes d'une variable indépendante," *Acta Mathematica* **4** (1884):1-79.

[2]André Weil, "Mittag-Leffler As I Remember Him," *Acta Mathematica* **148** (1982):9-13.

[3]Godfrey H. Hardy, "Gösta Mittag-Leffler, 1846-1927," *Proceedings of the Royal Society of London* A **119** (1928):v-viii.

Stockholm was chosen as the venue for the 1916 Congress, although due to the outbreak of the First World War it never took place.[4] He was also on the editorial board of the *Rendiconti del Circolo matematico di Palermo* and a member of the International Commission for the Bolyai Prize (1910). His numerous honorary memberships of foreign societies and honorary degrees from foreign universities are a further testament to his standing in the international community.

The Founding of *Acta Mathematica*

The idea for the journal, which emerged at the beginning of the 1880s, did not originate with Mittag-Leffler but came from the Norwegian mathematician, Sophus Lie (1842–1899).[5] Lie, who was professor of mathematics at the University of Christiana, had been one of the co-founders of the Norwegian journal, *Archiv för mathematik og naturvidenskab* (1876). It is not surprising that Lie should have been thinking about starting another journal at this time, for it was a period of unprecedented Scandinavian mathematical activity. Apart from a brief flowering in the 1820s encompassing the work of Abel, it was not until the 1870s that an appreciable number of Scandinavian mathematicians of international caliber emerged. A decade later, they had begun to make their mark. In addition to Lie and Mittag-Leffler, they included Albert Victor Bäcklund, Karl Anton Bjerknes, Julius Petersen, Ludvig Sylow, Hieronymous Zeuthen, and the mathematical astronomers, Hugo Gyldén and Anders Lindstedt.[6]

Lie initially raised the idea of a new journal on his visit to Stockholm in the spring of 1881 when he met Mittag-Leffler, who had just been appointed as mathematics professor at the newly founded Stockholm Högskola (later to become Stockholm University). On this occasion, Lie proposed the idea of an international journal to be edited from Sweden and suggested that Mittag-Leffler serve as the editor. It was exactly the kind of scheme to appeal to Mittag-Leffler. Not only was he a proficient mathematician and a capable organizer, but he was also a skilled communicator with well-cultivated international contacts. He had built up an extensive network of correspondents and was committed to promoting international communication between mathematicians.[7] He was both well-traveled and a staunch

[4]After the hostilities ceased, Mittag-Leffler expected the Stockholm Congress to be rescheduled, but he was disappointed. The first postwar Congress took place in Strasbourg in 1920. See Olli Lehto, *Mathematics Without Borders: A History of the International Mathematical Union* (New York: Springer-Verlag, 1998), pp. 15, 28, 331-332, and 335. Compare also Lehto's chapter in the present volume.

[5]Gösta Mittag-Leffler, "Sophus Lie," *Acta Mathematica* **22** (1899):i-ii, and Yngve Domar, "On the Foundation of *Acta Mathematica*," *Acta Mathematica* **148** (1982):3-8.

[6]For a history of Swedish mathematics, see Lars Gårding, *Mathematics and Mathematicians: Mathematics in Sweden before 1950*, HMATH, vol. 13 (Providence: American Mathematical Society and London: London Mathematical Society, 1998).

For information on Ludvig Sylow and his work, see Bent Birkeland, "Ludvig Sylow's Lectures on Algebraic Equations and Substitutions, Christiania (Oslo), 1862: An Introduction and a Summary," *Historia Mathematica* **23** (1996):182-199.

[7]Mittag-Leffler's political skills were widely acknowledged, although occasionally with reservation. Felix Klein described Mittag-Leffler as "a courtier and a diplomat." Ganesh Prasad, *Some Great Mathematicians of the Nineteenth Century* (Benares: Benares Mathematical Society, 1934), p. 239. Lie considered Mittag-Leffler to be an "excellent socializer" but for his own taste found him "a much too intriguing diplomat." David E. Rowe, "Three Letters from Sophus Lie to Felix Klein on Parisian Mathematics During the Early 1880's," *The Mathematical Intelligencer* **7** (3) (1985):74-77.

REDACTION

SVERIGE:

A. V. BÄCKLUND,	Lund.
H. TH. DAUG,	Upsala.
H. GYLDÉN,	Stockholm.
HJ. HOLMGREN,	"
C. J. MALMSTEN,	Upsala.
G. MITTAG-LEFFLER,	Stockholm.

NORGE:

C. V. BJERKNESS,	Christiania.
O. J. BROCH,	"
S. LIE,	"
S. SYLOW,	Fredrikshald.

DANMARK:

L. LORENZ,	Kjöbenhavn.
J. PETERSEN,	"
H. G. ZEUTHEN,	"

FINLAND:

L. LINDELÖF,	Helsingfors.

FIGURE 1. Editorial Board of *Acta Mathematica*, 1882

supporter of the Scandinavian scientific enterprise. In short, he was ideally qualified to take on such an editorial role.

Moreover, Lie's idea was timely. There was a real need for an international journal. In the aftermath of the Franco-Prussian War, the two leading mathematical journals, *Journal für die reine und angewandte Mathematik* (Crelle's *Journal*) (1826) and *Journal de mathématiques pures et appliquées* (Liouville's *Journal*) (1836), edited in Berlin and Paris, respectively, had both ceased to have a genuinely international character. The more recently founded *American Journal of Mathematics* (1878) carried articles from outside the United States but was predominantly a showcase for the mathematical research of the Johns Hopkins University.[8] In addition, there was no Scandinavian journal that specialized in advanced mathematical research, and a journal such as Lie had suggested would provide a welcome means to advance Scandinavian mathematics at the international level.

With the aim of promoting Scandinavian mathematics high on their agenda, Lie and Mittag-Leffler decided upon an editorial board made up of the leading Scandinavian mathematicians. (See Figure 1.) The idea was that these mathematicians would undertake not just to make regular contributions to the journal but also to submit their best papers to it. Moreover, to give the journal a high degree of accessibility, French or German, or in exceptional cases English or Latin, were to be the languages of publication.[9]

Mittag-Leffler felt optimistic about the future quality of Scandinavian mathematical research, although Lie and some of his other Scandinavian colleagues were not quite as sanguine. As Yngve Domar has pointed out, Mittag-Leffler's confidence stemmed from his previous success in building up a research department in Helsinki—he had held the chair of mathematics there for four years before his return to Sweden—and his belief that he could do likewise in Stockholm.[10] However, if the journal were to have, as Mittag-Leffler envisaged, a genuinely international character, then it would need more than the best of Scandinavian mathematics; mathematicians from outside Scandinavia, and, in particular, from both Germany and France, would have to be wooed. Mittag-Leffler was the ideal person to do such courting. In the 1870s, he had studied under Karl Weierstraß in Berlin and Charles Hermite in Paris. He was anxious to promote mathematical harmony between the two countries and had continued to maintain strong ties with both mathematicians. As it transpired, it was his relationship with Hermite that proved crucial for the initial success of the journal.[11]

Hermite had kept Mittag-Leffler informed of mathematical life in Paris, and it was through him that Mittag-Leffler had learned about three extremely talented young French mathematicians: Paul Appell, Émile Picard, and Henri Poincaré. All three were former students of Hermite. Mittag-Leffler felt sure that if he could engage them, and especially Poincaré, the most brilliant of the three, in the start of the journal, then success would surely follow. As he observed in a letter to Carl Johan Malmsten, a retired mathematics professor from the University of Uppsala and ex-cabinet minister: "According to my firm belief, we now find ourselves in a

---

[8]Karen Hunger Parshall and David E. Rowe, *The Emergence of the American Mathematical Research Community, 1876-1900: J. J. Sylvester, Felix Klein, and E. H. Moore*, HMATH, vol. 8 (Providence: American Mathematical Society and London: London Mathematical Society, 1994), pp. 88-94.

[9]French and German were given equal status in the publicity for the journal. However, Mittag-Leffler did reveal to Poincaré that French would be the preferred language, and indeed approximately 60% of the papers (and pages) of the first ten volumes were in French. Philippe Nabonnand, ed., *La correspondence entre Henri Poincaré et Gösta Mittag-Leffler* (Basel: Birkhäuser Verlag, 1999), p. 87.

Latin had been rarely used in mathematical publication since the mid-nineteenth century, but it was still an integral part of a general European education and was therefore widely understood. This was important for a journal with international aspirations. Nevertheless, no papers in Latin were ever published in *Acta*. For more on the issue of a language for mathematics, see Jeremy Gray's chapter below.

[10]Domar, p. 4.

[11]On the relationship between Hermite and Mittag-Leffler from the French side, recall Thomas Archibald's chapter above.

period quite similar to that of the discovery of elliptic functions. Then Abel made the success of Crelle's German journal. In the same way Poincaré will make a success of our Swedish journal."[12]

Mittag-Leffler was already in correspondence with Poincaré and, prompted by his knowledge of Poincaré's recent remarkable discovery of Fuchsian functions, wasted no time in writing to him with details of the enterprise.[13] He pointed out the potential parallel with Abel, again drawing attention to the nationalities involved: "You know that it was Abel, a Norwegian, who was responsible for the success of *Crelle's Journal*. Now it has occurred to Gyldén and me that you, a Frenchman, might be as generous and make a success of our journal."[14] He then took the bold step of asking Poincaré for the manuscript of his substantial paper on Fuchsian groups, notwithstanding that he knew it was supposedly destined for the *Journal de l'École polytechnique*. He also asked him for four others, luring him with the promise of fast and efficient editing. Mittag-Leffler considered the capture of Poincaré's manuscripts for the launch of the journal of such importance that he implied to Poincaré that the project's immediate feasibility was contingent on his participation—without it, he said, the journal would have to be delayed two to three years. Mittag-Leffler's initiative was rewarded. Poincaré agreed,[15] and in so doing dispelled from the minds of Mittag-Leffler's colleagues all remaining doubts about the journal's viability.

With Poincaré's support for the journal secured, Mittag-Leffler began work on financial matters. He needed to raise sufficient funds to cover the journal's launch as well as its expenses in the period before it could become self-financing. Since the journal was to be independent of any institution, there was no natural source of funding to tap, and his net would have to be cast wide. To help with the task, he engaged his friend Malmsten's assistance. Malmsten's background made him ideally suited for the job, and his help and advice proved invaluable. Together, they secured the support of King Oscar II of Sweden and Norway, a known patron of mathematics.[16] With the King's sponsorship—he promised 1,500 crowns from

[12]Domar, p. 5. The first four volumes of Crelle's *Journal* contained seventeen papers by Abel, of which six were published in the first volume.

[13]Gösta Mittag-Leffler to Henri Poincaré, 29 March, 1882, Institut Mittag-Leffler, Djursholm, Sweden, in Nabonnand, ed., pp. 87-93.

[14]"Vous savez que c'est Abel, un norvégien, qui a fait surtout le succès du *journal de Crelle*. Maintenant nous avons pensé M. Gyldén et moi que vous, un français, serez peut-être assez généreux pour vouloir faire le succès de notre journal." See *ibid.*, p. 87.

[15]Poincaré's reply to Mittag-Leffler is lost but, as Philippe Nabonnand has argued, it seems likely that Poincaré agreed to publish his papers in *Acta Mathematica* in order to establish priority for his work and to ensure a rapid dissemination of his theory in Germany. See Philippe Nabonnand, "The Poincaré-Mittag-Leffler Relationship," *The Mathematical Intelligencer* **21** (2) (1999):58-64.

Mittag-Leffler's success at enticing French mathematicians to *Acta* did, of course, mean that French mathematicians had less work to submit to French journals. At the time of *Acta*'s launch, Liouville's *Journal* was going through a particularly difficult period, and the following year, Hermite remarked to Mittag-Leffler that he had been told that Liouville's *Journal* was approaching an end due to the fact that all the interesting works were going to *Acta*. But as Jesper Lützen has argued, the near demise of Liouville's *Journal* was probably as much to do with the then editor, Henri Résal, as with the success of *Acta*. Jesper Lützen, *Joseph Liouville 1809–1882: Master of Pure and Applied Mathematics* (New York: Springer-Verlag, 1990), pp. 236-237.

[16]King Oscar, who had studied mathematics at the University of Uppsala, maintained an active interest in the subject and provided individual awards to mathematicians across Europe.

his own private purse[17]—the enterprise gained immediate credibility, and other sponsors were quickly found, including Hermite who promised 720 crowns.[18] The financial situation was assisted further by the offer of support from the governments of Denmark, Norway, and Sweden, each of which was prepared to make an annual donation of 1,000 crowns. In the meantime, Mittag-Leffler's own financial prospects were looking increasingly healthy—he was about to marry the heiress to a substantial fortune. So the economic future of the journal, for the early years at least, looked assured.

The project was well on its way. Mittag-Leffler had the promises of sufficient funds, he had engaged the backing of the appropriate Scandinavian mathematicians, and French cooperation was assured in the shape of Poincaré; only one element was missing. Mittag-Leffler wanted and needed German support for the plan. Weierstraß was the obvious person to approach, but there was a complication. In 1880, Weierstraß, together with Leopold Kronecker, had jointly taken charge of Crelle's *Journal*, and Mittag-Leffler was conscious that his own efforts to found a journal might be construed as acting against the interests of his former teacher. He therefore decided to keep the project secret from the Germans until it was further advanced, with the hope that he could win them over later by surprise. When the time came, the patronage of King Oscar once again proved invaluable. With the King's authority, Malmsten sent a letter to Ernst Kummer, Kronecker, Ernst Schering, and Weierstraß—each of whom had previously been honored by the King—asking for their support. They all replied positively, and papers were promised from Kronecker, Schering, and Weierstraß (although, in the event, they were not all forthcoming).

With the first volume of the journal organized, Mittag-Leffler devoted himself to publicity and promotional work. Even while on honeymoon in Germany and France during the summer of 1882, he found time to make personal contacts and successfully to solicit contributions—Lazarus Fuchs in Heidelberg and Eugen Netto and Theodor Reye in Strasbourg, were among those who succumbed to his charm. Meanwhile, he orchestrated a comprehensive advertising campaign sending information to mathematicians, learned societies, and other scientific journals across the globe.[19]

As far as the format and the price of the new journal were concerned, Crelle's *Journal*, which sold in Sweden at nine crowns per volume, provided the model. The question of the journal's name was a little harder to resolve. Up until October 1882, it was going to be called *Acta Mathematica Eruditorum*, but, at the last minute, it was decided to abbreviate it to simply *Acta Mathematica*. In December 1882, the first issue was published, and the contents page of the first volume makes impressive reading. (See Figure 2.) Three members of the editorial board contributed articles: Gyldén on the three-body problem, Malmsten on life annuities, and Hieronymus Zeuthen on geometry. From Germany, there were papers by Fuchs on ordinary differential equations, Reye on geometry, and Netto and Schering, both on algebra. While the French contribution, crowned by Poincaré's memoirs on Fuchsian groups and Fuchsian functions, also included papers from Appell and Picard on function

---

[17]The money was to be paid over a period of three years. By way of comparison, Mittag-Leffler's annual salary as a professor was 7,000 crowns.

[18]The sponsors are listed in the front of *Acta Mathematica* **1** (1882).

[19]See, for example, the notice published in the *Quarterly Journal of Pure and Applied Mathematics* **19** (1881–1882):189.

ACTA MATHEMATICA, 1. 1882/1883

INHALT. TABLE DES MATIÈRES.

FIGURE 2. Table of contents of the first volume of *Acta Mathematica*, 1882–1883

theory, as well as a short note on elliptic functions from Hermite. A communiqué (written in both French and German) from the editorial board made it clear that the main focus of the journal would be mathematical analysis—not surprising given the mathematical interests of Mittag-Leffler—and that papers containing either new results or original methods would be equally welcome. Moreover, the frontispiece, a portrait of Abel, not only hinted at a parallel with the launch of Crelle's *Journal* but also provided an apposite icon for the quality and type of mathematics being sought.

Subsequent volumes, with papers from Italy, Russia, and the United States, reenforced further the international character of the journal. The list of authors contained in these early volumes is striking and, apart from those already mentioned, includes the French mathematicians Gaston Darboux and Eduoard Goursat; the Germans Georg Cantor, Leo Koenigsberger, Rudolph Lipschitz, Hermann Minkowski, Heinrich Weber, and Weierstraß; the Italians Eugenio Beltrami, Felice Casorati, Gino Loria, and Salvatore Pincherle; the Russians Pafnuti Chebyshev and Sonya Kovalevskaya; the American George William Hill; as well as several Scandinavians, including Mittag-Leffler himself.

As editor Mittag-Leffler did not confine himself to seeking out new articles, he also arranged for the translation and/or republication of papers that had appeared elsewhere. Generally, these were papers that had been published in rather more obscure journals or had been published privately.[20] Although sometimes, as was the case with a paper of Weierstraß's and also, more notably, with some of Cantor's papers (discussed in more detail below), he arranged the translation of papers that had appeared in well-known publications, but which he considered warranted further exposure.

Thus, from its inception, it is clear that *Acta* was genuinely international in terms of its authors, but what of its readers? In 1885, after the publication of Volume 5, Mittag-Leffler was very happy to receive a letter from Weierstraß congratulating him on the success of the journal and emphasizing its international character.

> It gives me great pleasure to say that I have been pleased with the success of your undertaking. Perhaps it would be forgivable if I as editor-in-chief of the oldest currently existing mathematical journal have had a twinge of jealousy. From the beginning *Acta* has obtained so many old masters and young vigorous talents as authors, not only from Scandinavia but also from Germany, France, and Italy. It is my pleasure and hope that *Acta* will continue as an international organ for the development of our science with the same success as it has had up to now and that it may remain the most cosmopolitan of them.[21]

[20] A typical example is George W. Hill, "On the Part of the Motion of the Lunar Perigee Which Is a Function of the Mean Motions of the Sun and Moon," *Acta Mathematica* **8** (1886):1-36, which was published privately in the United States by John Wilson & Son, Cambridge, Massachusetts in 1877.

[21] "Es drängt mich zugleich der Befriedigung Ausdruck zu geben, mit der ich den erfreulichen Fortgang Ihres Unternehmens begleite. Vielleicht wäre es verzeihlich, wenn ich als Mitherausgeber der ältesten von den gegenwärtig existierenden mathematischen Zeitschriften eine Anwandlung von Neid darüber empfände, daß es Ihnen gelungen ist, von Anfang an für die *Acta* so viele altbewährte Meister und junge, aufstrebende Talente als Mitarbeiter zu gewinnen, und zwar nicht bloß aus den skandinavischen Ländern, sondern auch aus Deutschland, Frankreich, Italien. Es ist mein Wunsch und meine Hoffnung, daß die *Acta* auch fernerhin mit ebenso glänzenden Erfolge, wie bis jetzt, ein internationales Organ für die Fortentwicklung unserer Wissenschaft, der am meisten kosmopolitsichen von allen, bleiben mögen." See Gösta Mittag-Leffler, "Procès-verbaux," in *Compte rendu du Congrès des mathématiciens tenu à Stockholm, 22-25 septembre 1909* (Leipzig: n.p., 1910):3.

Further confirmation of international readership (and proof that Mittag-Leffler's efforts had achieved tangible recognition well beyond the confines of mathematical analysis) is displayed in a notice published in *Nature* in June 1884:

> The new Scandinavian journal, *Acta Mathematica*, has already gained such a reputation that the French government has decided to subscribe for 15 copies for the Facultés des Sciences. In his note to the Swedish Ambassador in Paris on the subject, M. Jules Ferry pointed out that it is the first time his Government has supported a foreign publication, which he trusts will be an acknowledgement to the high international position the *Acta Mathematica* has gained and of the value it has become to French science. This journal is already supported by three Scandinavian governments.[22]

Moreover, the fact that the notice was published in a British journal, despite having no obvious British connection, supports the idea that *Acta* had already become well known in British scientific circles. Additional verification of *Acta*'s rapid international circulation is provided by the widespread coverage it received in other publications in the years immediately after its launch. Enthusiastic notices appeared in a number of mathematical journals written by mathematicians from France, Germany, and Italy, as well as Scandinavia,[23] and a detailed analysis of the contents of the first three volumes was published in the *Bulletin des sciences mathématiques et astronomiques*.[24]

Throughout his career, Mittag-Leffler continued to work energetically on the journal's behalf, traveling extensively, strengthening contacts, and obtaining manuscripts. He was warmly received—on a visit to London in 1887 Thomas Archer Hirst described him as "a bright, intelligent, vigorous Swede"[25]—and his journeys were productive. He was an excellent judge of the quality of work submitted to him for publication and notably receptive to modern developments, even when, as in the case of Cantor's work, others were slow to understand or appreciate the significance of new results.

In contrast, Lie's active involvement with the project was largely confined to the initial stages of *Acta*'s development. He did become a member of the editorial board, but he never published in the journal. Perhaps he never intended to do more than act as a catalyst for Mittag-Leffler with regard to the founding of the journal, but it may also be the case that a clash of styles and/or personalities diminished his appetite for the enterprise. By the beginning of 1883, he was admitting to feeling rather uncomfortable with the grandeur of Mittag-Leffler's plans and was referring to *Acta* as "Mittag-Leffler's journal."[26] At all events, it is clear that, once publication began, Mittag-Leffler became the driving force, and Lie effectively dropped out of the picture. And Lie was not the only member of the editorial

---

[22] *Nature* **30** (12 June, 1884):153.

[23] A list of the notices is published in "Table des matières des tomes 1-10 composée par G. Eneström," *Acta Mathematica* **10** (1887):349-397 on p. 351.

[24] Jules Tannery, "Revue," *Bulletin des sciences mathématiques et astronomiques* **8** (2) (1884):136-171.

[25] William H. Brock and Roy M. McLeod, ed., *Natural Knowledge in a Scientific Context: The Journals of Thomas Archer Hirst, F.R.S.* (London: Mansell 1980): Folio 2417, dated 7 June, 1887.

[26] Rowe, p. 77.

board not to play an active part in the journal's early years. Of the fourteen original members of the editorial board, only Malmsten and Zeuthen made any significant contribution. It was Mittag-Leffler who shouldered the responsibilities and Mittag-Leffler who ensured the journal's survival.

Poincaré, Cantor, Kovalevskaya, and *Acta Mathematica*

Mittag-Leffler's ability to attract papers from the élite of Europe's mathematicians set *Acta* quickly on the path to success. However, the relationship between *Acta* and its contributors was not entirely one-sided. Poincaré, Cantor, and Kovalevskaya all featured prominently in *Acta*'s pages and did much to enhance the journal's reputation, but *Acta*, in its turn, played a role in each of their lives. An examination of their involvement with *Acta*, and hence with Mittag-Leffler, serves to illustrate the symbiosis that can exist between a journal and the mathematicians whose work it serves to promote.

Of all the mathematicians associated with the early years of *Acta Mathematica*, the person who more than anyone else ensured its initial success was Poincaré. Poincaré not only made a substantial contribution to the very first volume of *Acta*, but he contributed to nine out of the first ten volumes and went on to publish a total of twenty-five papers in the journal, the last one appearing in 1911, the year before he died. Moreover, his association with the journal did not stop with his death. Volume 38 (published in 1921) was dedicated to him and was made up entirely of articles about his work, and volume 39 (published in 1923)—which was devoted to Weierstraß, Kovalevskaya, and Poincaré—contained a previously unpublished paper by him. Mittag-Leffler, in his acknowledgment of Poincaré's contribution to the journal, described him as the most eminent and the most faithful of its authors.[27]

Poincaré's many papers in *Acta* covered a wide variety of topics and included several important memoirs. In terms of their effect on the success of the journal, his original series of five memoirs on Fuchsian functions were undoubtedly the most significant. These memoirs, which were concerned with the construction of Fuchsian and Kleinian groups and their corresponding automorphic functions, were notable not only for their mathematical content but also for the fact that they were at the heart of a renowned priority and naming dispute with Klein.[28] Another exceptional paper was his memoir on stability theory, in which he discussed the equilibrium of a rotating fluid mass employing approximative methods using Lamé series expansions.[29] This appeared in 1885 and, although independent, almost coincided with a similar attack on the problem by the Russian mathematician, Alexander Liapunov.[30] A paper that also ranks high in his *oeuvre* is his paper of 1886 on divergent series, in which he was led to the construction of a general theory

---

[27]Gösta Mittag-Leffler, *Acta Mathematica* **39** (1923):iii.

[28]The Poincaré-Klein correspondence relating to this controversy is published in "Correspondance d'Henri Poincaré et de Félix Klein," *Acta Mathematica* **39** (1923):94-132, or *Oeuvres de Henri Poincaré*, 11 vols. (Paris: Gauthier-Villars, 1916–1956), 11:26-65 (hereinafter cited *Oeuvres de HP*), or *Cahiers du Séminaire d'histoire des mathématiques* **10** (1989):89-140.

[29]Henri Poincaré, "Sur l'équilibre d'une masse fluide animée d'un mouvement de rotation," *Acta Mathematica* **7** (1885):259-380, or *Oeuvres de HP*, 7:40-140.

[30]Alexander Liapunov, *On the Stability of Ellipsoidal Forms of Equilibrium of a Rotating Liquid* (St. Petersburg: n.p., 1884). (His master's thesis translated into French as "Sur la stabilité des figures ellipsoidales d'équilibre d'un liquide animé d'un mouvement de rotation," *Annales de la Faculté des sciences de l'Université de Toulouse* (2) **6** (1904):5-116.)

of what are today called asymptotic expansions.[31] The paper is notable too for the fact that it generated a public disagreement between him and Ludwig Thomé.[32]

Three years later, Poincaré's famous memoir on the three-body problem was published in *Acta*. It was the prize-winning essay in the international competition organized by Mittag-Leffler through the auspices of *Acta* and on behalf of King Oscar II to celebrate the King's sixtieth birthday in 1889.[33] The international character of the competition and the fact that it was associated with a journal rather than a national academy put it in sharp contrast to other mathematical prize competitions of the time. This further emphasized Mittag-Leffler's commitment to mathematical internationalization. Poincaré's memoir appeared in a special competition issue of the journal and, as described elsewhere,[34] there is an intriguing history attached both to the memoir and to the competition itself. Poincaré's success in the competition and the publication of his memoir were naturally of great benefit to the journal. However, there were some, notably Kronecker, who questioned the link between the competition and the journal, and who argued that the Swedish Academy of Sciences would have provided a more appropriate forum from which to launch the competition. King Oscar's openly acknowledged support for *Acta*, however, did make it a natural and legitimate vehicle in which to promote the competition and publish the winning entries, and it seems that the criticism was founded more on professional jealousy than on genuine feelings. Nevertheless, it was clearly in Mittag-Leffler's interests to do all that he could to ensure that the competition attracted the highest caliber entrants. Thus, the fact that all four of the topics set for the competition were ones on which Poincaré could have submitted an entry may not have been a complete coincidence.[35]

Although it is certain that Poincaré's early involvement with *Acta* was critical for the journal's success, the journal was also helpful to Poincaré. When Mittag-Leffler made his first approach, Poincaré had only recently returned to Paris from Caen, and his career in the capital was in its infancy. He was young, and publishing in French journals provided no guarantee of international readership. From the point of view of priority, it was important for his ideas to be disseminated across international borders, especially into Germany, as swiftly as possible. A journal such as *Acta* that published quickly and had a circulation in both France and Germany catered precisely to his needs.

In terms of personality, Poincaré and Mittag-Leffler were very different characters—Poincaré, shy and retiring; Mittag-Leffler, gregarious and outgoing. Poincaré, however, respected Mittag-Leffler as a mathematician and had confidence in his

---

[31] Henri Poincaré, "Sur les intégrales irrégulières des équations linéaires," *Acta Mathematica* **8** (1886):295-344, or *Oeuvres de HP*, 1:290-332.

[32] Ludwig W. Thomé, "Bemerkung zur Theorie der linearen Differentialgleichen," *Journal für die reine und angewandte Mathematik* **101** (1887):203-208, and Henri Poincaré, "Remarques sur les intégrales irrégulières des équations linéaires: Réponse à M. Thomé," *Acta Mathematica* **10** (1887):310-312, or *Oeuvres de HP*, 1:333-335.

[33] Henri Poincaré, "Sur le problème des trois corps et les équations de la dynamique," *Acta Mathematica* **13** (1890):1-270, or *Oeuvres de HP*, 7:262-490.

[34] June E. Barrow-Green, *Poincaré and the Three Body Problem*, HMATH, vol. 11 (Providence: American Mathematical Society and London: London Mathematical Society, 1997).

[35] Although there is no hard evidence to show that all of the questions were set with Poincaré in mind, there is no doubt that Mittag-Leffler looked for opportunities to promote Poincaré's genius. This is clearly demonstrated in his later extensive (but ultimately unsuccessful) attempts to secure a Nobel Prize for Poincaré. See Elisabeth Crawford, *The Beginnings of the Nobel Institution* (Cambridge: University Press, 1984), pp. 136-149.

ability as an editor. They maintained a strong relationship. That Poincaré continued to publish in *Acta Mathematica* long after his reputation was made is a tribute to Mittag-Leffler's editorial skills as well as a concrete acknowledgment of the ongoing success of the journal.

Unfortunately, the good relationship between Mittag-Leffler and Poincaré was not mirrored in all of Mittag-Leffler's relationships with his authors. In particular, the roles played by Mittag-Leffler and *Acta* in the life of Georg Cantor turned out very differently to the ones they played in Poincaré's.[36] Despite an auspicious beginning, a disagreement over the publishing of a certain article resulted in the collapse of relations between Cantor and *Acta* after only a couple of years and caused essentially irreparable damage to the professional relationship between Cantor and Mittag-Leffler.[37] Nevertheless, Cantor retained warm feelings towards Mittag-Leffler, and several years later in a letter to Poincaré, he explicitly declared his regard for Mittag-Leffler and for his work with *Acta*.[38]

Cantor's association with *Acta* was initially triggered in January 1883 by a note from A. Hermann, the French publisher of *Acta*, to Mittag-Leffler. The note, which came via Hermite, suggested that *Acta* should publish French translations of some important papers by German mathematicians.[39] The suggestion was timely since Mittag-Leffler himself had been thinking along similar lines. He gave a positive response and, a few days later, put forward Cantor's name. (No similar plan was put forward for translating French papers into German since, as Mittag-Leffler himself remarked, German mathematicians had a good understanding of French.[40])

Cantor's papers on set theory had been appearing in German in the *Mathematische Annalen* and in Crelle's *Journal* since the beginning of the 1870s, but not without opposition—opposition that had left him feeling increasingly isolated at Halle University. However, Mittag-Leffler, who had been one of the first to understand the significance of Cantor's work and to recognize its links with recent results in function theory, was very supportive of Cantor's ideas and keen to make them more widely accessible. From Mittag-Leffler's point of view, republishing Cantor's papers in *Acta* would be good for the journal and for Cantor. With Hermite's cooperation, he arranged for the translation of a selected number of papers for publication, the translations being done by Father Dargent, a relative of Hermite, and then checked by Poincaré.[41] When volume 2 of *Acta* appeared in 1883, it contained the translations of nine of Cantor's papers, together with an extract of a letter from Cantor to Mittag-Leffler and an extract of a letter from Ivar Bendixson (one of Mittag-Leffler's students) to Cantor. Altogether the papers and letters filled more than one quarter of the journal.

---

[36]For a detailed biography of Cantor, see Joseph W. Dauben, *Georg Cantor: His Mathematics and Philosophy of the Infinite* (Cambridge: Harvard University Press, 1979).

[37]For a complete chronology of this incident, see Ivor Grattan-Guinness, "An Unpublished Paper by Georg Cantor," *Acta Mathematica* **124** (1970):65-107.

[38]Georg Cantor to Henri Poincaré, 22 January, 1896, in *Cahiers de Séminaire d'histoire des mathématiques* **7** (1986):103-105. See also Dauben, p. 138.

[39]Nabonnand, ed., pp. 120-121. The idea of carrying French translations of previously published works was not new. Joseph Liouville had used it to good effect in the early volumes of his *Journal de mathématiques pures et appliquées*. Compare the chapter by Jesper Lützen in the present volume.

[40]Nabonnand, ed., p. 121.

[41]*Ibid.*, p. 121.

The first four of Cantor's papers were on the foundations of set theory and contained some of his most celebrated results, including the discovery of the non-denumerability of the real numbers and the proof of the possibility of a one-to-one correspondence between the line and n-dimensional space.[42] The remaining five, which were on the development of set theory, concluded with part of his celebrated *Grundlagen* on transfinite numbers.[43] These five formed part of the series of six on infinite linear point sets—the *Punktmannigfaltigkeitslehre*—that were published in the *Mathematische Annalen* between 1879 and 1884. Mittag-Leffler did not publish the *Grundlagen* in its entirety due to his unease about the philosophical aspects of Cantor's arguments, although he did manage to persuade Cantor to rework his original paper so as to eliminate all the strictly philosophical sections. However, his concern in this respect remained with him and resurfaced more strongly later on.

Cantor's final contribution to volume 2, which was in the form of an extract from a letter to Mittag-Leffler and subtitled "First communication," concerned various properties of n-dimensional point sets.[44] His next paper, which appeared in volume 4 and concerned the powers of perfect sets, was also in the form of an extract from a letter to Mittag-Leffler, and in it, he announced the forthcoming publication of a solution to the continuum problem.[45] However, his next paper in *Acta* (which turned out to be his last) appeared the following year in volume 7 and contained no mention of this solution.[46] This paper, although it was entitled "Second communication" and was intended as the sequel to the "First communication" published in volume 2, contained the continuation of results from the sixth paper in the *Mathematische Annalen* series and is acknowledged as the concluding paper in that series.[47]

Cantor's ideas had already provoked considerable criticism in some quarters, notably Berlin. One of his strongest critics was Kronecker, who was in the vanguard of a campaign to discredit Cantor's work on transfinite set theory.[48] By way of retaliation, Cantor attempted to aggravate Kronecker by applying for a position in Berlin. Kronecker responded by writing to Mittag-Leffler to ask if he could publish a short paper in *Acta* in which he would demonstrate "that the results of modern function theory and set theory are of no significance."[49] Cantor, having been

---

[42]Georg Cantor, "Sur une propriété du système de tous les nombres algébriques réels," *Acta Mathematica* **2** (1883):305-310; "Une contribution à la théorie des ensembles," *op. cit.*, pp. 311-328; "Sur les séries trigonométriques," *op. cit.*, pp. 329-335; and "Extension d'un théorème de la théorie des séries trigonométriques," *op. cit.*, pp. 336-348.

[43]Georg Cantor, "Sur les ensembles infinis et linéaires de points: I-IV," *Acta Mathematica* **2** (1883):349-380, and "Fondements d'une théorie générale des ensembles," *op. cit.*, pp. 381-408.

[44]Georg Cantor, "Sur divers théorèmes de la théorie des ensembles des points situés dans un espace continu à N dimensions," *Acta Mathematica* **2** (1883):409-414.

[45]Georg Cantor, "De la puissance des ensembles parfaits de points," *Acta Mathematica* **4** (1884):381-392.

[46]Georg Cantor, "Über verschiedene Theoreme aus der Theorie der Punctmengen in einem n-fach ausgedehnten stetigen Raume G_n," *Acta Mathematica* **7** (1885):105-124.

[47]Dauben, pp. 149-150.

[48]*Ibid.*, p. 134.

[49]"dass due Ergebnisse der modernen Funktionentheorie und Mengenlehre von keiner realen Bedeutung sind." Quoted in a letter from Georg Cantor to Gösta Mittag-Leffler, 26 January, 1884. Arthur M. Schoenflies, "Die Krisis in Cantor's mathematisichem Schaffen," *Acta Mathematica* **50** (1927):5. For a full account of the differences between Cantor and Kronecker, see Dauben, pp. 66-69 and 133-138.

informed of the idea by Mittag-Leffler, initially did not object. He thought that if Kronecker's views became public, they would be opposed and hence rejected. But old fears reemerged, and he changed his mind. Some years previously, Kronecker had tried to prevent him from publishing in Crelle's *Journal*, and he now thought that Kronecker was attempting to keep him out of *Acta*. He warned Mittag-Leffler that if he published Kronecker's paper, then he could no longer be associated with the journal. In the event, Kronecker's paper was never published in *Acta*, if indeed it was ever submitted. The crisis was averted, but it was a portent for things to come.

In 1884, Cantor had been working on the problem of the continuum hypothesis, but without success. However, his attempts to find a proof had led him to a number of fresh ideas concerning the decomposition of point sets, and in November of that year, he sent Mittag-Leffler the first part of a paper on ordered sets together with some other manuscripts. In January, he received the page-proofs of the first eight pages of the paper. In February, with the aim of having the work published as two articles—a paper on ordered sets and a paper that was later published in volume 7—he sent off the rest of the manuscript to Mittag-Leffler, requesting the remaining page-proofs as soon as possible. Mittag-Leffler agreed to his request but then on 9 March wrote to him suggesting that the first of these articles, the one on the theory of ordered sets, should be withdrawn from the press:

> I am convinced that the publication of your new work, before you have been able to explain new positive results, will greatly damage your reputation among mathematicians. I know very well that basically this is all the same to you. But if once your theory is discredited in this way it will be a long time before it will again command the attention of the mathematical world. It may well be that you and your theory will never be given the justice you deserve in our lifetime. Then the theory will be rediscovered in a hundred years or so by someone else, and then it will subsequently be found that you already had it all. Then, at least, you will be given justice. But in this way [i.e., by publishing the article] you will exercise no significant influence, which you naturally desire as does everyone who carries out scientific research.[50]

Although Cantor initially agreed to the idea, and requested the return of the manuscript (which he received in part), as time wore on he felt increasingly aggrieved by Mittag-Leffler's response and before long decided to turn his back on *Acta* completely. Matters were worsened by the fact that the letter arrived at a

---

[50] "Aber ich bin auch davon wohl bewusst dass sehr wenige Mathematiker meinen Geschmack theilen, und ich bin davon über zeugt dass die Veröffentlichung Ihrer neuen Arbei, früher als Sie neue positive Resultate darlegen können, Ihr Ansehen bei den Mathematikern sehr viel schaden wird. Ich weiss wohl, dies ist Ihnen im Grunde einerlei. Aber wenn Ihre Theorien einmal auf diese Weise der mathematischen Welt an sich ziehen. Ja es kann wohl sein dass man Ihnen und Ihre[!] Theorien nie in unserer Lebenszeit Gerechtigkeit zu Theil kommen lässt. So werden die Theorien wieder einmal nach 100 Jahren oder mehr von Jemand entdeckt und dann findet man wohl nachträglich aus, dass Sie doch schon das alles hatten und dann thut man Ihnen zuletzt Gerechtigkeit, aber auf diese Weise werden Sie keinen bedeutenden Einfluss auf die Entwicklung unserer Wissenschaft ausgeübt haben." Gösta Mittag-Leffler to Georg Cantor, 9 March, 1885, in Dauben, p. 138 (in translation). See also Grattan-Guinness, pp. 101-102.

critical moment in Cantor's life—he had recently suffered a nervous breakdown and was becoming progressively discouraged by his inability to resolve the continuum hypothesis. Although Mittag-Leffler's caution that few mathematicians were ready for Cantor's new terminology and increasingly philosophical approach was undoubtedly genuine (he had expressed similar reservations in the past), Cantor became convinced that Mittag-Leffler was really only concerned about the reputation of *Acta*. More than a decade later, Cantor confided to Poincaré that he thought the antipathy to his work emanating from Berlin was behind Mittag-Leffler's decision.[51] Cantor believed that, at the time, Mittag-Leffler was still essentially dependent on the goodwill of the Berlin mathematicians for the success of the journal and so to publish his work would endanger the relationship. Furthermore, since Mittag-Leffler had been one of the few mathematicians who up until that time had shown unwavering support for his mathematical ideas, Cantor was especially hurt by Mittag-Leffler's apparently negative attitude. His rejection appears to have been the final blow; their relationship foundered, and Cantor never again published in *Acta Mathematica*. From that moment on, however, Cantor gave up mathematics almost entirely.[52]

Thus, Cantor owed much to Mittag-Leffler and *Acta* for their part in making his work more widely available. Indeed, Mittag-Leffler himself was one of the first mathematicians to use Cantor's work to prove results of his own. But, tragically, it was also Mittag-Leffler and *Acta* who, albeit unintentionally, were in part to blame for Cantor's disenchantment with mathematics. Mittag-Leffler's decision to err on the side of caution with respect to Cantor's later publications would not have been taken lightly; he was certainly well-qualified to assess the merits of Cantor's work. And Cantor's charge that Mittag-Leffler was protecting himself by not publishing these papers appears harsh in the light of Mittag-Leffler's earlier support.

As far as the journal itself was concerned, the inclusion of Cantor's papers in its second volume undoubtedly brought it added publicity. Nevertheless, at the time of their publication, the journal was at a critical phase in its development, and, given the opinions that had been aired about Cantor's work, their publication was not entirely risk-free. Mittag-Leffler's readiness to run the gauntlet of possible academic opposition indicates his courage and strength as an editor. Although the affair ended badly for Cantor, it did *Acta* no harm. Building on the success of Poincaré's papers in the first volume, the publication of Cantor's papers in the second helped secure the reputation of the journal as one in the forefront of mathematical developments.

In his dealings with *Acta*, Cantor not only corresponded with Mittag-Leffler, but he also discussed his ideas with one of the other editors, the Russian mathematician, Sonya Kovalevskaya.[53] Kovalevskaya, one of Weierstraß's most famous students, was not only the first female editor of the journal (or indeed of any major scientific journal), she was also the first woman to publish in it. Her success

---

[51]Georg Cantor to Henri Poincaré, 22 January, 1896, in *Cahiers de Séminaire d'histoire des mathématiques* **7** (1986):103-105. See also Dauben, p. 138.

[52]The complex circumstances leading up to Cantor's eventual disillusionment with mathematics are described well in Dauben, pp. 133-140 and 280-282.

[53]See, for example, the letter from Georg Cantor to Sonya Kovalevskaya, 7 December, 1884, in which Cantor discusses his memoir on the theory of order types. The letter, which is in the archives of the Institut Mittag-Leffler, is reproduced in Dauben, pp. 310-311.

with *Acta*, both as an editor and as an author, was a remarkable achievement for a woman of her time, and she owed much to Mittag-Leffler for his unfailing support.

Mittag-Leffler heard Kovalevskaya present a paper at a conference in St. Petersburg in 1880 and from then on devoted considerable energy to championing her as a mathematician. Eventually overcoming substantial opposition from many of his colleagues, he arranged a teaching position for her in Stockholm.[54] She arrived there at the end of 1883, and shortly afterwards, Mittag-Leffler invited her to join the editorial board of *Acta*. Time spent in both Berlin and Paris meant that she was well-placed to communicate with the mathematicians there, as well as those in her native Russia. Her main responsibility at *Acta* was to act as a sort of international liaison officer, although she also took over some of the fund-raising activities.

One of Kovalevskaya's assignments was to try to persuade the Russian Academy of Sciences to give both institutional and financial support to the journal. Although she made strenuous efforts on this account during her visits to Russia, she never managed to succeed. It seems that the lack of support was essentially political and stemmed from *Acta*'s endorsement of the Finnish mathematical establishment as a separate entity. This endorsement, which derived from the fact that Mittag-Leffler had spent four years as a professor at the University in Helsinki prior to his return to Stockholm in 1881, was construed by the Russians as support for the Finnish movement for independence from Russia. As a result, Russian mathematicians thought it wiser to avoid a formal association with the journal.

Kovalevskaya had rather more success with her task of obtaining Russian manuscripts for *Acta*. Chebyshev, in particular, was keen to contribute and was grateful to her for translating the first of his five papers to appear in the journal. She was also popular with several other Russian colleagues, who felt that she would treat their work fairly. To a great extent, they considered her as their own representative in the West and relied on her to publicize their results. This led to a large amount of correspondence—mainly in the form of requests of one sort or another (reprints, preprints, technical queries, etc.)—and kept a channel of communication open between the two countries. As a result of Kovalevskaya's efforts on behalf of *Acta*, Russian mathematicians were provided both with a means to make their work known in Europe and a conduit for contact with European mathematical developments. Conversely, just as Mittag-Leffler had hoped when he originally involved Kovalevskaya with the journal, *Acta* itself benefited from having such a direct Russian connection. Kovalevskaya's presence on the editorial board not only provided the journal with an *entrée* into Russian mathematical circles but also helped to widen the journal's international circulation.

Five of Kovalevskaya's papers were published in *Acta*.[55] The first was a paper on Abelian integrals.[56] Although it had originally formed part of her doctoral dissertation of 1874, she had also used it in 1880 to relaunch her career in mathematics after an interval of six years by presenting it at a conference in St. Petersburg, the same conference at which Mittag-Leffler had made up his mind to try to find a position for her in Stockholm. Her second paper, which was a response to a problem

---

[54] For details of Kovalevskaya's life, see Ann Hibner Koblitz, *A Convergence of Lives: Sofia Kovalevskaia: Scientist, Writer, Revolutionary* (Boston: Birkhäuser Verlag, 1983).

[55] For an account of Kovalevskaya's mathematics, see Roger Cooke, *The Mathematics of Sonya Kovalevskaya* (New York: Springer-Verlag, 1984).

[56] Sonya Kovalevskaya, "Über die Reduction einer bestimmten Classe Abel'scher Integrale 3ten Ranges auf elliptische Integrale," *Acta Mathematica* **4** (1884):393-414.

on the wave theory of light posed by Weierstraß, centered on finding solutions to Lamé's equations.[57] Her next two papers were related to her work on the rotation of a solid body, for which she had earlier won the prestigious *Prix Bordin* of the Paris Academy in 1888.[58] Her final paper in the journal, which was published posthumously, concerns a theorem of Bruns in potential theory and was, as an application of the Cauchy-Kovalevskaya theorem, related to her doctoral research in partial differential equations.[59]

Over the longer term, Kovalevskaya's association with *Acta* served to her advantage, both heightening and broadening her reputation. However, in the years following her death, her reputation declined rapidly, due, in part at least, to her paper on the wave theory of light. Only shortly after her death, Vito Volterra discovered a mistake in the paper—a mistake that had, in fact, originated with Gabriel Lamé—which he explained in an article published in *Acta* in the following year.[60] Opponents of women mathematicians gleefully seized upon the discovery of the mistake, and Kovalevskaya's mathematical credibility suffered as a result. Naturally, the fact that Weierstraß and Mittag-Leffler had also failed to spot the mistake was overlooked. It was a blow from which her reputation took some time to recover.

Mittag-Leffler, nevertheless, continued to hold her memory in good faith. In addition to an obituary that appeared in *Acta* the year after she died,[61] he also published a long article on her and Weierstraß in volume 39,[62] dedicating the volume jointly to her. In his willingness to promote Kovalevskaya and her work, Mittag-Leffler displayed his concern for young and disadvantaged mathematicians, and her example affirms his ability as an editor to recognize talent regardless of circumstances. No doubt Kovalevskaya's connection with the Russian mathematical community and the endorsement of her work by Weierstraß worked solidly in her favor, but given the very real prejudice that prevailed against female mathematicians, Mittag-Leffler deserves credit for his readiness to campaign so strongly on her behalf.[63]

Acta Mathematica Volumes 1-20

The preceding history of the foundation of *Acta* and the discussion of some of its early contributors reveal an international journal in both conception and realization. But a close examination of the composition of the first twenty volumes brings into focus an even more precise sense of the extent of its international coverage.

---

[57] Sonya Kovalevskaya, "Über die Brechung des Lichtes in cristallinischen Mitteln," *Acta Mathematica* **6** (1885):249-304.

[58] Sonya Kovalevskaya, "Sur le problème de la rotation d'un corps solide autour d'un point fixe," *Acta Mathematica* **12** (1889):177-232, and "Sur une propriété du système d'équations différentielles qui définit la rotation d'un corps solide autour d'un point fixe," *Acta Mathematica* **14** (1890):81-94.

[59] Sonya Kovalevskaya, "Sur un théorème de M. Bruns," *Acta Mathematica* **15** (1891):45-52.

[60] Vito Volterra, "Sur les vibrations lumineuses dans les milieux biréfringents," *Acta Mathematica* **16** (1892):153-206.

[61] Gösta Mittag-Leffler, "Sophie Kovalevsky, Notice Biographique," *Acta Mathematica* **16** (1892):385-390.

[62] Gösta Mittag-Leffler, "Weierstraß and Sonja Kowalewsky," *Acta Mathematica* **39** (1923): 133-198.

[63] Mittag-Leffler's support for female scientists also extended to Marie Curie, whose cause he assisted in his dealings with the Nobel Prize Committee. See Crawford, p. 141.

Initially, *Acta* was published almost biannually with volumes 1-10 appearing between 1882 and 1887. These first ten volumes contained papers by seventy-seven authors from thirteen countries, with France (16) and Germany (28) being responsible for over 57% of the contributors. Scandinavian authors, i.e., those from Sweden (11), Norway (2), Denmark (4), and Finland (3), accounted for just over 25% of the total number. If the number of articles and the number of pages are counted, then the Scandinavian total reduces to less than 21% and 18%, respectively.

The second ten volumes, which were published on an annual basis, appeared between 1887 and 1897. Analyzing the content of these volumes produces results almost identical to those of the first group, with Scandinavian contributions in this second group accounting for just under 25% of the journal, irrespective of the method of counting. The most striking difference between the two sets of volumes is in the total number of papers published; this fell from 162 to 125 despite a slight increase in the number of pages. Although there was a trend towards slightly longer papers, the reduction in the number of papers can be largely explained by the special nature of volume 13—the Oscar competition volume—which contained only two long papers.

Given the existing state of research in mathematics, particularly in analysis, it was to be expected that the majority of authors would come from France and Germany. Nevertheless, there were distinctions between their types of contributions. The first ten volumes contained almost twice as many German as French authors, but in terms of the number of articles, the ratio was reduced to just over three to two, i.e., the French authors contributed more articles per person than their German counterparts. If just the numbers of pages are considered, the French mathematicians again contributed more than the German, by a ratio of four to three. A similar situation occurred in the second ten volumes. In this case, the number of German authors exceeded the number of French by a ratio of five to two, and the number of German articles exceeded the number of French by a ratio of just under two to one. With regard to the numbers of pages, France again contributed more than Germany, although this time only by a ratio of seven to six.

CONTRIBUTIONS TO *Acta Mathematica* 1882-1897
1882-1887: VOLUMES 1-10
1887-1897: VOLUMES 11-20

COUNTRY	NUMBER OF AUTHORS		NUMBER OF MEMOIRS/NOTES		NUMBER OF PAGES		NUMBER OF PAGES %	
	82-87	**87-97**	**82-87**	**87-97**	**82-87**	**87-97**	**82-87**	**87-97**
France	16	12	42.5	25	1594	1234	40	29
Germany	28	29	64.5	47	1220	1027	31	24
Scandinavia	20	19	34	26	932	958	21	22
(Sweden)	(10)	(8)	(19)	(14)	(510)	(419)	(13)	(10)
Italy	4	4	6	7	92	323	2	8
Russia	3	6	4	10	45	554	1	13
Others*	6	8	11	10	192	173	5	4
TOTAL	77	78	162	125	3979	4269	100	

*1882-87: Austria, Belgium, Holland, Switzerland, United States of America
1887-97: Austria, Belgium, England, Holland

Subject of Memoir	1882-1887	1887-1897
Biography	0	2
Set Theory	9	0
Algebra	12	14
Number Theory	18	16
Probability	2	2
Series	1	3
Analysis	89	50
Geometry	18	20
Mechanics & Mathematical Physics	11	18
Other	2	0

Language of Memoir	1882-1887	1887-1897
French	95	69
German	66	54
English	1	2

Thus, it appears that Mittag-Leffler was able to strike a remarkably even balance between France and Germany with regard to the acceptance of papers for the journal, and such a balance was just what he had hoped for from the outset. Nevertheless, the results from such analyses should be interpreted with caution, since they take no account of any distortion that might arise from the contributions of a single author responsible for a large number of long papers. For example, Poincaré contributed 684 pages to the first ten volumes, approximately 43% of the French total, or 17% of the overall total, and in the second ten volumes, he contributed 440 pages, 36% of the French total, or 10% of the overall total. Of the latter, more than half was accounted for by Poincaré's prize-winning memoir, i.e., one paper from Poincaré accounted for 21% of the French total or 6% of the overall total. Without Poincaré's contribution, the figures would look quite different, with Germany ahead of France on each count. As far as *Acta* was concerned, there was no German equivalent of Poincaré, a fact that further emphasizes Poincaré's significance for the journal.

Although the number of published papers that originated from outside Scandinavia, France, or Germany was comparatively small, the papers did come from several different countries—Austria, Belgium, England, Holland, Italy, Russia, Switzerland, and the United States—showing that the journal was indeed reaching a wide audience. Naturally, the restriction on language would have posed a problem for mathematicians from some countries but not for others. For example, the Russians had the advantage of having Kovalevskaya to act as a translator for them. But language difficulties cannot account for the low number of British or American articles in the journal—only three in the first twenty volumes.[64] Although it is possible that there was some reluctance on Mittag-Leffler's part to accept articles written

[64]George W. Hill, "On the Part of the Motion of the Lunar Perigee Which Is a Function of the Mean Motions of the Sun and Moon," *Acta Mathematica* **8** (1886): 1-36; William Thomson, "On the Division of Space with Minimum Partition Area," *Acta Mathematica* **11** (1887-1888):121-134; and James J. Sylvester, "On a Funicular Solution of Buffon's 'Problem of the Needle' in Its Most General Form," *Acta Mathematica* **14** (1890-1891):185-206.

in English, it is more likely that he received few from which to choose. This was a period in which the academic concerns of the majority of British mathematicians lay predominantly outside mathematical analysis. It was also a period in which American mathematical research in general was in its infancy with relatively few American mathematicians ready to find a place on the international stage.[65] In addition, British and American mathematicians had their own journals in which to publish their research. In Britain, mathematicians could choose from publications such as the *Proceedings of the London Mathematical Society* (1865), the *Proceedings of the Edinburgh Mathematical Society* (1883), and the *Quarterly Journal of Pure and Applied Mathematics* (1855), as well as from some of the more general scientific journals such as the *Philosophical Transactions of the Royal Society* (1665) and the *Transactions of the Cambridge Philosophical Society* (1822).[66] In contrast (and indicative of the scale of the country's research enterprise), America supported only one research-based mathematics journal, the *American Journal of Mathematics*.

Conclusion

During the first half of the nineteenth century when August Leopold Crelle and Joseph Liouville began publishing their journals, there was almost a complete dearth of journals devoted exclusively to mathematics,[67] but by 1882, the year of *Acta*'s launch, the practice of publishing mathematical journals was becoming well established. Many of the new journals were imbued with a national identity, being associated with a local or national mathematical society or with a particular academic institution, and were produced primarily for the benefit of the members and their colleagues, providing them with a natural venue for publication. These included journals such as *Matematicheskii Sbornik* (1866),[68] the journal of the Moscow Mathematical Society, the *Bulletin de la Société mathématique de France* (1873), and the *American Journal of Mathematics* mentioned above. Some were restricted to carrying only articles written by members, but others could and did contain articles by foreign mathematicians. A broadly similar situation existed in the case of journals with no specific affiliation, such as the Italian *Annali di matematica pura ed applicata* (1858).[69] Although such journals published translations of foreign papers and disseminated information about international mathematical

[65] For a detailed analysis of the growth of the American mathematical research community, see Parshall and Rowe.

[66] Although the majority of these journals were associated with a national society and thus a natural medium for their members, they were not all—as Sloan Despeaux shows in her chapter in the present volume—the exclusive preserve of British mathematicians.

[67] Prior to the founding of Crelle's *Journal* and Liouville's *Journal*, the most significant mathematical journal was the rather short-lived *Annales de mathématiques pures et appliquées* (Gergonne's *Journal*), which flourished between 1810 and 1832 and which provided the model for Liouville's *Journal*. See Jean Dhombres and Mario H. Otero, "Les *Annales de mathématiques pures et appliquées*: Le journal d'un homme seul au profit d'une communauté enseignante," in *Messengers of Mathematics: European Mathematical Journals (1800–1946)*, ed. Mariano Hormigón and Elena Ausejo (Zaragoza: Siglo XXI de España Editores, 1993), pp. 3-70.

[68] See Serguei S. Demidov, "La Revue *Matematicheskii Sbornik* dans les années 1866–1935," in *Messengers of Mathematics: European Mathematical Journals (1800–1946)*, ed. Mariano Hormigón and Elena Ausejo (Zaragoza: Siglo XXI de España Editores, 1993), pp. 235-256.

[69] For a discussion of the *Annali di matematica pura ed applicata*, see Hélène Gispert, "Une comparison des journaux français et italiens dans les années 1860–1875," *L'Europe mathématique–Mathematical Europe*, ed. Catherine Goldstein, Jeremy J. Gray, and Jim Ritter (Paris: Éditions de la Maison des sciences de l'homme, 1996), pp. 389-406.

activity (as well as publishing national research), they too targeted a domestic market. *Acta* was different. Mittag-Leffler set out to produce a journal that would be recognized as international both inside and outside its country of origin, and he succeeded. *Acta* provided a medium in which mathematicians could publish unconstrained by national sensitivities and one in which ideas could be rapidly circulated across international boundaries.

Acta also provided Scandinavian mathematics with a much-needed international platform. Although there were Scandinavian mathematical journals in existence before the foundation of *Acta*, for example, *Tidsskrift for Mathematik* (1859, formerly *Mathematisk Tidsskrift*) and *Archiv för mathematik og naturvidenskab*, they were national in character and language, and Scandinavian mathematicians seeking an international audience had to publish in a country and a language not their own. For those, unlike Mittag-Leffler, who had had little opportunity for foreign travel or for making foreign contacts, however, it was not always easy to pursue publication abroad. *Acta*, based in Sweden and with an essentially Scandinavian editorial board, gave these mathematicians a much more straightforward option. As attested by the number of Scandinavian papers published in the journal, it was an option they embraced with enthusiasm.

From the publication of its first volume, *Acta* met with critical acclaim, but it was not only for the contents of the journal that Mittag-Leffler deserved congratulation. It was also for the very short time in which he managed to get the whole project off the ground. It took approximately eighteen months from the time of his first conversation with Lie to the time when the first part of the journal was published. The speed of Mittag-Leffler's success was in part due to the fact that, having successfully forestalled the possible antipathy of the German mathematicians, he had met no opposition to the idea.[70] Second, as a result of his network of contacts across Europe, he was able to obtain a good number of high quality articles very quickly, and third, he was an efficient fund-raiser, employing his ability to delegate when the occasion demanded.

As the analysis of the first twenty volumes and the earlier discussion of the foundation of the journal show, *Acta Mathematica* began and remained an international journal for both authors and readers alike. Mittag-Leffler was unstinting in the effort he put in as an editor and succeeded in attracting the best mathematical analysts to the journal, as well as many other first-class mathematicians. They, in their turn, were grateful for the time and effort Mittag-Leffler put in on their behalf, as the following tribute, which he received in acknowledgment of his work as editor on the completion of the twentieth volume in 1897, reveals:

> Sir,
>
> You have rendered to mathematics, by the founding of the *Acta Mathematica*, a service of the highest importance for which

---

[70]This is in contrast to the situation that developed during the late 1890s over the founding of the *Transactions of the American Mathematical Society*. In this case, the American Mathematical Society initially proposed an involvement with the then financially insecure *American Journal of Mathematics*. The involvement was resisted by those at the Johns Hopkins University, who feared for the loss of control of the *American Journal*. The matter was eventually resolved by the establishment of the *Transactions of the AMS* as a separate and distinct journal, and a journal that was not, due to its transactional nature, in direct competition with the *American Journal*. See Raymond C. Archibald, *A Semicentennial History of the American Mathematical Society*, (New York: American Mathematical Society, 1938), pp. 56-65.

> you have earned unanimous recognition. That which was done by Crelle and Liouville for Germany and for France, you have done for Scandinavia. At a time when papers and discoveries were increasing, you undertook the mission—and you have fulfilled it with honor—to contribute to the progress of Mathematics by helping the authors with the publicity of their works.
>
> The journal, to which for thirteen years you have consecrated your devotion and your talents, has had the good fortune of gathering memoirs that will remain of exceptional merit for Analysis.
>
> It has been placed in the highest rank among the current periodical publications; it has given a productive impetus to mathematical studies in the Scandinavian countries that unite with pride the glory of Abel to that of Linné, of Scheele, of Berzelius and of Oersted.
>
> In the name of friends of Analysis, we express the hope that the *Acta Mathematica* will pursue for the good of Science a career begun with brilliance and encouraged by the universal feelings of mathematicians.[71]

It was signed by: Weierstraß, Paul Du Bois-Reymond, Fuchs, Schering, Thomas Craig, Simon Newcomb, Eduard Weyr, Matyas Lerch, Lord Kelvin, Lord Rayleigh, James J. Sylvester, Paul Mansion, Joseph Bertrand, Hermite, Camille Jordan, Darboux, Poincaré, Picard, Appell, Francesco Brioschi, Luigi Cremona, Beltrami, Nikolai Sonin, Andréei Markov, Cyparissos Stephanos, Pieter Schoute, Francisco Gomes Texeira, David Emmanuel, Michel Petrovitch, and Carl Geiser.

Signed by mathematicians from so many countries, it is a glowing testimonial to Mittag-Leffler. He was a mathematician who believed strongly in the internationalization of mathematics and who, on behalf of *Acta*, worked untiringly on an international stage. His dedication to the journal was unparalleled, and his efforts were widely acknowledged. Ganesh Prasad's description of Mittag-Leffler as a man with the instinct of an internationalist who was as much at home in Berlin as in Paris, and as much a *persona grata* in influential circles in London as in Rome, is a fitting one.[72] It would have given Mittag-Leffler great pleasure to know that

---

[71] "Monsieur,

Vous avez rendu aux géomètres, par la fondation des *Acta Mathematica*, un service de la plus haute importance qui vous a mérité la reconnaissance unanime. Ce qu'ont fait Crelle et Liouville pour l'Allemagne et pour la France, vous l'avez fait avec un égal succès pour les pays Scandinaves. A une époque où se multiplient les travaux et les découvertes, vous avez pris la mission, et vous l'avez remplie avec honneur, de concourir au progrès des Mathématiques en facilitant aux auteurs la publicité de leurs oeuvres.

Le journal, auquel depuis treize ans vous avez consacré votre dévouement et votre talent, a eu l'heureuse fortune d'accueillir des mémoires d'un mérite exceptionnel qui resteront à jamais dans l'Analyse.

Il s'est placé au plus haut rang parmi les publications périodiques actuelles; il a donné une impulsion féconde aux études mathématiques dans les pays Scandinaves qui réunissent avec orgueil la gloire d'Abel à celle de Linné, de Scheele, de Berzélius et d'Oersted.

Au nom des amis de l'Analyse, nous vous exprimerons le voeu que les *Acta Mathematica* poursuivent pour le bien de la science, une carrière commencée avec éclat et encouragée par l'universelle sympathie des géomètres."

Niels E. Norlund, "G. Mittag-Leffler," *Acta Mathematica* **50** (1927):i-xxiii on pp. iii-iv.

[72] Prasad, p. 224.

in 1928, the year after his death, G. H. Hardy declared *Acta* to be "the most completely international of all mathematical journals."[73]

References

Archival Sources

Mittag-Leffler, Gösta. Correspondence. Institut Mittag-Leffler, Djursholm, Sweden.

Printed Sources

Acta Mathematica **1** (1882).

Acta Mathematica **39** (1923).

Archibald, Raymond C. *A Semicentennial History of the American Mathematical Society.* New York: American Mathematical Society, 1938.

Barrow-Green, June E. *Poincaré and the Three Body Problem.* HMATH. Vol. 11. Providence: American Mathematical Society and London: London Mathematical Society, 1997.

Birkeland, Bent. "Ludvig Sylow's Lectures on Algebraic Equations and Substitutions, Christiana (Oslo), 1862: An Introduction and a Summary." *Historia Mathematica* **23** (1996):182-199.

Bottazzini, Umberto. *The Higher Calculus: A History of Real and Complex Analysis from Euler to Weierstraß.* New York: Springer-Verlag, 1986.

Brock, William H. and McLeod, Roy M., Ed. *Natural Knowledge in a Scientific Context: The Journals of Thomas Archer Hirst, F.R.S.* London: Mansell, 1980.

Cahiers de Séminaire d'histoire des mathématiques **7** (1986):103-105.

Cantor, Georg. "De la puissance des ensembles parfaits de points." *Acta Mathematica* **4** (1884):381-392.

———. "Extension d'un théorème de la théorie des séries trigonométriques." *Acta Mathematica* **2** (1883):336-348.

———. "Fondements d'une théorie générale des ensembles." *Acta Mathematica* **2** (1883):381-408.

———. "Sur divers théorèmes de la théorie des ensembles des points situés dans un espace continu à N dimensions." *Acta Mathematica* **2** (1883):409-414.

———. "Sur les ensembles infinis et linéaires de points: I-IV." *Acta Mathematica* **2** (1883):349-380.

———. "Sur les séries trigonométriques." *Acta Mathematica* **2** (1883):329-335.

———. "Sur une propriété du système de tous les nombres algébriques réels." *Acta Mathematica* **2** (1883):305-310.

———. "Über verschiedene Theoreme aus der Theorie der Punctmengen in einem n-fach ausgedehnten stetigen Raume G_n." *Acta Mathematica* **7** (1885):105-124.

———. "Une contribution à la théorie des ensembles." *Acta Mathematica* **2** (1883):311-328.

Cooke, Roger. *The Mathematics of Sonya Kovalevskaya.* New York: Springer-Verlag, 1984.

Crawford, Elisabeth. *The Beginnings of the Nobel Institution.* Cambridge: University Press, 1984.

[73]Hardy, p. v.

Dauben, Joseph. *Georg Cantor: His Mathematics and Philosophy of the Infinite.* Cambridge: Harvard University Press, 1979.

Demidov, Serguei S. "La Revue *Matematicheskii Sbornik* dans les années 1866–1935." In *Messengers of Mathematics: European Mathematical Journals (1800–1946).* Ed. Mariano Hormigón and Elena Ausejo. Zaragoza: Siglo XXI de España Editores, 1993, pp. 235-256.

Dhombres, Jean and Otero, Mario. "Les *Annales de mathématiques pures et appliquées*: Le journal d'un homme seul au profit d'une communauté enseignante." In *Messengers of Mathematics: European Mathematical Journals (1800–1946).* Ed. Mariano Hormigón and Elena Ausejo. Zaragoza: Siglo XXI de España Editores, 1993, pp. 3-70.

Domar, Yngve. "On the Foundation of *Acta Mathematica.*" *Acta Mathematica* **148** (1982):3-8.

Eneström, Gustav. "Table des matières des tomes 1-10 composée par G. Eneström." *Acta Mathematica* **10** (1887):349-397.

Gårding, Lars. *Mathematics and Mathematicians: Mathematics in Sweden before 1950.* HMATH. Vol. 13. Providence: American Mathematical Society and London: London Mathematical Society, 1998.

Gispert, Hélène. "Une comparison des journaux français et italiens dans les années 1860–1875." In *L'Europe mathématique–Mathematical Europe.* Ed. Catherine Goldstein, Jeremy J. Gray, and Jim Ritter. Paris: Éditions de la Maison des sciences de l'homme, 1996, pp. 389-406.

Grattan-Guinness, Ivor. "An Unpublished Paper by Georg Cantor." *Acta Mathematica* **124** (1970):65-107.

Hardy, Godfrey H. "Gösta Mittag-Leffler, 1846-1927." *Proceedings of the Royal Society of London* A **119** (1928):v-viii.

Hill, George W. "On the Part of the Motion of the Lunar Perigee Which Is a Function of the Mean Motions of the Sun and Moon." *Acta Mathematica* **8** (1886):1-36.

Koblitz, Ann Hibner. *A Convergence of Lives: Sofia Kovalevskaia: Scientist, Writer, Revolutionary.* Boston: Birkhäuser Verlag, 1983.

Kovalevskaya, Sonya. "Sur le problème de la rotation d'un corps solide autour d'un point fixe," *Acta Mathematica* **12** (1889):177-232.

———. "Sur un théorème de M. Bruns." *Acta Mathematica* **15** (1891):45-52.

———. "Sur une propriété du système d'équations différentielles qui définit la rotation d'un corps solide autour d'un point fixe." *Acta Mathematica* **14** (1890): 81-94.

———. "Über die Brechung des Lichtes in cristallinischen Mitteln," *Acta Mathematica* **6** (1885):249-304.

———. "Über die Reduction einer bestimmten Classe Abel'scher Integrale 3ten Ranges auf elliptische Integrale." *Acta Mathematica* **4** (1884):393-414.

Lehto, Olli. *Mathematics Without Borders: A History of the International Mathematical Union.* New York: Springer-Verlag, 1998.

Liapunov, Alexander. *On the Stability of Ellipsoidal Forms of Equilibrium of a Rotating Liquid.* St. Petersburg: N.p., 1884: Trans. "Sur la stabilité des figures ellipsoidales d'équilibre d'un liquide animé d'un mouvement de rotation." *Annales de la Faculté des sciences de l'Université de Toulouse* (2) **6** (1904):5-116.

Lützen, Jesper. *Joseph Liouville 1809-1882: Master of Pure and Applied Mathematics.* New York: Springer-Verlag, 1990.

Mittag-Leffler, Gösta. *Acta Mathematica* **39** (1923):iii.

———. "Procès-verbaux." *Compte rendu du Congrès des mathematicians tenu à Stockholm, 22-25 septembre 1909.* Leipzig: N.p., 1910.

———. "Sur la répresentation analytique des fonctions monogènes uniformes d'une variable indépendante." *Acta Mathematica* **4** (1884):1-79.

———. "Sophie Kovalevsky, Notice Biographique." *Acta Mathematica* **16** (1892): 385-390.

———. "Sophus Lie." *Acta Mathematica* **22** (1899):i-ii.

———. "Weierstraß and Sonja Kowalewsky." *Acta Mathematica* **39** (1923):133-198.

Nabonnand, Philippe, Ed. *La correspondance entre Henri Poincaré et Gösta Mittag-Leffler.* Basel: Birkhäuser Verlag, 1999.

———. "The Poincaré-Mittag-Leffler Relationship." *The Mathematical Intelligencer* **21** (2) (1999):58-64.

Nature **30** (12 June, 1884):153.

Norlund, Niels E. "G. Mittag-Leffler." *Acta Mathematica* **50** (1927):i-xxiii.

Parshall, Karen Hunger and Rowe, David E. *The Emergence of the American Mathematical Research Community, 1876-1900: J. J. Sylvester, Felix Klein, and E. H. Moore.* HMATH. Vol. 8. Providence: American Mathematical Society and London: London Mathematical Society, 1994.

Poincaré, Henri. *Oeuvres de Henri Poincaré.* 11 Vols. Paris: Gauthier-Villars, 1916–1956.

———. "Remarques sur les intégrales irrégulières des équations linéaires. Réponse à M. Thomé," *Acta Mathematica* **10** (1887):310-312. In *Oeuvres de HP*, 1:333-335.

———. "Sur l'équilibre d'une masse fluide animée d'un mouvement de rotation." *Acta Mathematica* **7** (1885):259-380. In *Oeuvres de HP*, 7:40-140.

———. "Sur le problème des trois corps et les équations de la dynamique." *Acta Mathematica* **13** (1890):1-270. In *Oeuvres de HP*, 7:262-490.

———. "Sur les intégrales irrégulières des équations linéaires." *Acta Mathematica* **8** (1886):295-344. In *Oeuvres de HP*, 1:290-332.

Prasad, Ganesh. *Some Great Mathematicians of the Nineteenth Century.* Benares: Benares Mathematical Society, 1934.

Quarterly Journal of Pure and Applied Mathematics **19** (1881–1882):189.

Rowe, David E. "Three Letters from Sophus Lie to Felix Klein on Parisian Mathematics during the Early 1880's." *The Mathematical Intelligencer* **7** (3) (1985):74-77.

Schoenflies, Arthur M. "Die Krisis in Cantor's mathematisichem Schaffen." *Acta Mathematica* **50** (1927):1-23.

Sylvester, James J. "On a Funicular Solution of Buffon's 'Problem of the Needle' in Its Most General Form." *Acta Mathematica* **14** (1890–1891):185-206.

Tannery, Jules. "Revue." *Bulletin des sciences mathématiques et astronomiques* **8** (2) (1884):136-171.

Thomé, Ludwig, W. "Bemerkung zur Theorie der linearen Differentialgleichen." *Journal für die reine und angewandte Mathematik* **101** (1887):203-208.

Thomson, Willliam. "On the Division of Space with Minimum Partition Area." *Acta Mathematica* **11** (1887–1888):121-134.

Volterra, Vito. "Sur les vibrations lumineuses dans les milieux biréfringents," *Acta Mathematica* **16** (1892):153-206.

Weil, André. "Mittag-Leffler As I Remember Him." *Acta Mathematica* **148** (1982):9-13.

CHAPTER 9

An Episode in the Evolution of a Mathematical Community: The Case of Cesare Arzelà at Bologna

Laura Martini*
University of Virginia (United States)

The Bolognese Context

After the unification of Italy and its associated modifications of the political regime, the regulations governing the Italian universities and their teaching staffs changed. At last unified as a nation, Italy strove to revive the intellectual power of the state by improving its university studies. In 1860, chairs of higher mathematics were founded in the country's principal universities: Enrico Betti (1823–1892) and Francesco Brioschi (1824–1897) obtained the chairs of higher analysis in Pisa and Pavia, respectively, while Giuseppe Battaglini (1826–1894) and Luigi Cremona (1830–1903) inaugurated their courses in higher geometry in Naples and Bologna.[1] Following the unification, in fact, the Faculty of Mathematics at Bologna gained three new professors. In addition to Cremona, Quirico Filopanti (1812–1894) joined the faculty as Professor of Applied Mathematics and Eugenio Beltrami (1835–1900) became Extraordinary Professor of Complementary Algebra. Together with Domenico Chelini (1802–1878), who had served as Professor of Mechanics and Hydraulics since 1851, these men constituted the new Faculty of Mathematics at the University of Bologna.[2] Under their guidance, the first ten years of unification witnessed intense scientific activity at Bologna.

First, the University had three of Italy's best mathematicians on its faculty: Cremona, Beltrami, and Chelini. Such a concentration of talent had been unknown for more than a century. Unfortunately, from an institutional standpoint, things did not markedly improve. Despite Cremona's attempts to bring Bologna's mathematics teaching up to international standards, the University was still unable to offer a full baccalaureat course of study. This only began to change in the early 1880s when Cesare Arzelà (1847–1912) was named Professor of Higher Analysis in 1880–1881 and Salvatore Pincherle (1853–1936) followed as Professor of Algebra and Analytical Geometry a year later. In 1880, Luigi Donati (1846–1932), who had

---

*The present paper follows—but has a differently oriented argument from—Laura Martini, "The First Lectures in Italy on Galois Theory: Bologna, 1886–1887," *Historia Mathematica* **26** (1999):201-223.

[1] Ettore Bortolotti, *Storia della matematica nella Università di Bologna* (Bologna: Nicola Zanichelli Editore, 1947), p. 211, and Luigi Pepe, "I matematici bolognesi fra ricerca avanzata e impegno civile 1880–1920," *Archimede* **46** (1994):19-27 on p. 20.

[2] Pepe, pp. 21-22.

been teaching at the Engineering School of Bologna for three years, was named Professor of Mathematical Physics in the Faculty of Science.[3] A new age in Bolognese mathematics finally began with their arrival. It was in this particular atmosphere of renewal and intellectual ferment and growth that Arzelà gave a course on Galois theory in 1886–1887 that represented the first known public course on the subject in Italy.

The audience that year included a student who later became famous as a mathematician and historian of mathematics, Ettore Bortolotti (1866–1947). Bortolotti compiled the text of the lectures; the set of notes forms a substantial volume, entitled *Teoria delle sostituzioni*.[4] As Professor of Higher Analysis, Arzelà ostensibly lectured on that topic, but the 1886–1887 course was actually on the theory of substitutions and Galois theory.

Galois Theory in the European Curriculum

Contrary to what is commonly believed,[5] Enrico Betti was not the first in Italy to offer a public cycle of lectures on Galois theory. Although he was the first Italian mathematician to devote himself to the study of Galois theory, he never taught it as part of the curriculum at the University of Pisa. As Betti wrote in a letter to Placido Tardy (1816–1914) in 1859,[6] he did give lectures on the most important parts of algebra, but only to a few talented students in his home and not in the public forum of the university.[7] Others had assumed that Luigi Bianchi's (1856–1928) course on Galois theory at the *Scuola normale superiore* of Pisa in the academic year 1896–1897 was the first such presentation because it was the first actually to be published.[8] Arzelà gave his course, however, ten years before Bianchi.

It is indisputable that the first Italian mathematician to study questions related to the solvability of algebraic equations by radicals and Galois theory was Enrico Betti. His first note on the subject, "Sopra la risolubilità per radicali delle equazioni irriduttibili di grado primo," appeared in 1851.[9] A year later, he published "Sulla

---

[3]Pepe, pp. 22-23.

[4]Cesare Arzelà, *Teoria delle sostituzioni*, manuscript lecture notes taken by Ettore Bortolotti in the academic year 1886–1887, Bortolotti Library, University of Bologna.

[5]See, for example, Bent Birkeland, "Ludvig Sylow's Lectures on Algebraic Equations and Substitutions, Christiania (Oslo), 1862: An Introduction and a Summary," *Historia Mathematica* **23** (1996):182-199 on p. 184.

[6]The letter has never been published but is held in the "Fondo Betti" at the *Scuola normale superiore* in Pisa.

[7]As Betti explained to Tardy, he gave the twice-weekly lectures to four students, "per esporre loro le parti più elevate dell'algebra che non posso esporre nel corso che fo all'Università. Per ora ho esposto la teorica delle equazioni abeliane all'applicazione [sic] alla teorica della divisione del circolo. Passerò presto ad esporre la teorica della risoluzione algebraica in tutta la sua generalità [in order to convey to them the highest parts of algebra which I am not able to include in my university course. Up to now, I have spoken on the theory of abelian equations as applied to [?] the theory of the division of the circle. I will soon move to the theory of algebraic resolution in all of its generality]."

[8]Luigi Bianchi, *Lezioni sulla teoria dei gruppi di sostituzioni e delle equazioni algebriche secondo Galois* (Pisa: E. Spoerri, 1899).

[9]Enrico Betti, "Sopra la risolubilità per radicali delle equazioni algebriche irriduttibili di grado primo," *Annali di scienze matematiche e fisiche* **2** (1851):5-19, in Enrico Betti, *Opere matematiche di Enrico Betti*, 2 vols. (Milan: U. Hoepli, 1903–1913), 1:17-27.

risoluzione delle equazioni algebriche."[10] Given that Betti assumed the chair of algebra at the University of Pisa in 1857, Italy could have become an early European leader in Galois theory and group theory, but Betti taught only the traditional algebraic topics, never including Galois theory in his university courses.[11] Moreover, he never wrote a textbook on the subjects to which he had dedicated himself during the first years of his research activity.[12]

In 1859, Betti moved into the chair of higher analysis at Pisa, succeeded in the chair of algebra by Giovanni Novi (1827–1866). Novi planned to write a three-volume treatise on higher algebra, but only one volume of the *Trattato di algebra superiore* appeared in 1863.[13] In the preface, he explained that he had followed Betti's lecture notes in compiling the treatise; thus, once again, Galois theory failed to reach a broader Italian audience.[14] In Germany and elsewhere, the situation was quite different.

The first university course on Galois theory was given by the German mathematician, Richard Dedekind, at the University of Göttingen in the winter semester of the academic year 1856–1857.[15] So interested was Dedekind in the topic that he gave a second course on it the following winter semester. His written text of the lectures provided not only the first organic exposition of a large part of Galois theory (at that time the solvability conditions by radicals were not completely clear) but also a basic contribution to group theory of which he, together with Galois, is considered a founder.

Five years after Dedekind gave his second course on Galois theory in Germany, Ludvig Sylow lectured on the subject at the University of Oslo (at that time called Christiania) in Norway. In his presentation, Sylow gave the criterion of solvability for irreducible equations of prime degree, but his exposition of the condition for the general equation of degree higher than four was not clear. Among the students who heard his explanations, however, was the twenty-year-old Sophus Lie.[16]

In France, the first university text to include a chapter on Galois theory was the 1866 edition of the *Cours d'algèbre supérieure* by Joseph Alfred Serret. Serret's treatise was widely used as a textbook. As early as 1867, it had been adopted in the

---

[10] Enrico Betti, "Sulla risoluzione delle equazioni algebriche," *Annali di scienze matematiche e fisiche* **3** (1852):49-119, in Enrico Betti, *Opere matematiche di Enrico Betti*, 2 vols. (Milan: U. Hoepli, 1903–1913), 1:31-80.

[11] Betti treated the following subjects in his algebra course: infinite series, the theory of derivatives (Taylor series), the theory of homogeneous functions, invariant theory, general principles of equations of any degree, symmetric functions of the roots of an equation, equations with more than one unknown, limits of roots, Descartes's theorem, separation of roots, irreducible equations, Newton's method as improved by Fourier, numerical resolution of equations by continuous fractions, manipulation of equations, binomial equations, algebraic resolution of third degree equations, and algebraic resolution of fourth degree equations. Umberto Bottazzini, "Enrico Betti e la formazione della scuola matematica pisana," in *Atti del convegno "La storia delle matematiche in Italia," Cagliari 1982* (Bologna: Monograf, 1984), pp. 229-276 on p. 244.

[12] For a detailed account of the contribution of Enrico Betti to Galois theory, see Pasquale Mammone, "Sur l'apport d'Enrico Betti en théorie de Galois," *Bollettino di storia delle scienze matematiche* **9** (2) (1989):143-169.

[13] Giovanni Novi, *Trattato di algebra superiore* (Firenze: Le Monnier, 1863).

[14] Laura Toti Rigatelli, *La mente algebrica: Storia dello sviluppo della teoria di Galois nel XIX secolo* (Busto Arsizio: Bramante, 1989), p. 66.

[15] Richard Dedekind, *Lezioni sulla teoria di Galois*, ed. Laura Toti Rigatelli (Firenze: Sansoni, 1990).

[16] See Birkeland.

United States, and a German translation appeared a year later in 1868.[17] Serret's *Cours*, in a seventh and final edition in 1928, had a great impact on students of algebra well into the twentieth century. As detailed below, Cesare Arzelà based part of his university lectures on Galois theory in Italy on the fifth edition (published in 1885) of this influential book.[18]

Serret's text was soon followed by Camille Jordan's ground-breaking *Traité des substitutions et des équations algébriques* of 1870.[19] There, Jordan gave the first cogent explanation of the conditions for solvability by radicals. Moreover, he recognized that the concept of a group could be fruitfully applied outside the theory of algebraic equations.

Like Serret's work, Jordan's *Traité* proved extremely influential to a generation of mathematicians in France, Europe, and abroad. Beginning in the mid 1870s, for example, Julius Petersen gave courses on the theory of algebraic equations at the Polytechnic School of Copenhagen in Denmark. Based on his lectures, Petersen wrote a two-volume book, which was published in Copenhagen in 1878 and which treated the theory of algebraic equations, the theory of substitutions, and Galois theory.[20]

In Germany, studies on Galois theory and group theory also proliferated. In 1881, Paul Bachmann, who had been a student in that first course by Dedekind in 1856–1857, published the article "Ueber Galois' Theorie der algebraischen Gleichungen" in the *Mathematische Annalen.*[21] There, he based his analysis of Galois theory not on the concept of group, but on the new concept of a division ring. Almost immediately, Eugen Netto, a former student of Leopold Kronecker in Berlin, published his textbook, *Substitutionentheorie und ihre Anwendungen auf die Algebra*, in 1882.[22] In 1881, Kronecker had written a very long memoir entitled "Grundzüge einer arithmetischen Theorie der algebraischen Grössen," which had appeared in Crelle's *Journal* the following year.[23] In this important work, after introducing the concept of field of rationality,[24] Kronecker defined the notion of a family as an enlargement of the field of rationality. Netto's text advocated Kronecker's notion of a family over Bachmann's concept of a division ring[25] and emphasized that in the 1880s Galois theory was developing in a number of different ways. Moreover, the brief discussion of Galois theory in the European curriculum thus far shows that by the 1880s the subject was well-entrenched in France and Germany and even in Norway and Denmark.

---

[17]Toti Rigatelli, p. 87.

[18]Joseph Alfred Serret, *Cours d'algèbre supérieure*, 2 vols. (Paris: Gauthier-Villars, 1885).

[19]Camille Jordan, *Traité des substitutions et des équations algébriques* (Paris: Gauthier-Villars, 1870).

[20]Julius Petersen, *Theorie der algebraischen Gleichungen* (Kopenhagen: A. F. Host & Sohn, 1878).

[21]Paul Bachmann, "Ueber Galois' Theorie der algebraischen Gleichungen," *Mathematische Annalen* **18** (1881):449-468.

[22]Eugen Netto, *Substitutionentheorie und ihre Anwendungen auf die Algebra* (Leipzig: B. G. Teubner Verlag, 1882).

[23]Leopold Kronecker, "Grundzüge einer arithmetischen Theorie der algebraischen Grössen," *Journal für die reine und angewandte Mathematik* **92** (1882):1-122, in Leopold Kronecker, *Werke*, ed. Kurt Hensel, 5 vols. (Leipzig: B. G. Teubner Verlag, 1895-1930), 2:237-388.

[24]For Kronecker, a "field of rationality" or "*Rationalitäts-Bereich*" of magnitudes R', R'', R''', ... was the collection of all rational functions of R', R'', R''', ...

[25]Toti Rigatelli, p. 117.

In Italy, however, Betti's Galois-theoretic work of the 1850s represented only an isolated case. Group theory did not enter the Italian research arena until the mid-1870s when Alfredo Capelli (1855–1910) published over a dozen memoirs on groups of substitutions and on the theory of algebraic equations. In particular, Capelli wrote five articles on group theory, mainly during the first years of his scientific activity.[26] At the end of the introduction of his work "Sopra l'isomorfismo dei gruppi di sostituzioni," Capelli stated that he had attended a course on group theory given by Giuseppe Battaglini in Rome during the academic year 1875–1876.[27] From Capelli's remarks it is not clear whether the content of the course included Galois theory.[28] By 1885, Giovanni Frattini (1852–1925) had also added his algebraic works, notably "Intorno alla generazione di gruppi di operazioni," in which he characterized that subgroup of a group that has been named after him.[29] Also in 1885, the Italian translation by Giuseppe Battaglini of Netto's *Substitutionentheorie* appeared.[30] It was followed in 1891 by an Italian version of Petersen's treatise as well.[31] In this atmosphere of renewed interest in group theory, and almost fifty years after Betti's studies, Italy finally saw the publication of its first lectures on Galois theory. During the 1896–1897 academic year, Luigi Bianchi gave a course on this topic at the *Scuola normale superiore* in Pisa; the text of his lectures appeared in print in 1899.[32]

Arzelà's Sources for the Lectures

While Bianchi's course in Pisa may have been the first to be published in Italy, it was not, as noted above, the first to be given. That distinction belongs to Cesare Arzelà a decade earlier at the University of Bologna. Reflective of the active interest in Galois theory outside of Italy, Arzelà consciously drew from the available European texts in introducing his students to this subject. In order to present a clear and, at the same time, substantial course on the theory of substitutions and Galois theory, Arzelà referred to the most significant published texts on the subject in Europe. In selecting the specific material for his presentation, he chose the best works the international community had to offer: Netto's *Substitutionentheorie*, Peter Lejeune Dirichlet's *Vorlensungen über Zahlentheorie*,[33] Serret's *Cours d'algèbre*

---

[26]Alfredo Capelli, "Dimostrazione di due proprietà numeriche offerte dalla teoria delle sostituzioni permutabili con una sostituzione data," *Giornale di matematiche* **14** (1876):66-74; "Intorno ai valori di una funzione lineare di più variabili," *Giornale di matematiche* **14** (1876):141-145; "Sopra l'isomorfismo dei gruppi di sostituzioni," *Giornale di matematiche* **16** (1878):32-87; "Sopra la composizione dei gruppi di sostituzioni," *Memorie della reale Accademia dei Lincei* **19** (1884):262-272; and "Sulle generatrici del gruppo simmetrico delle sostituzioni di n elementi," *Giornale di matematiche* **35** (1897):354-355.

[27]Capelli, "Sopra l'isomorfismo dei gruppi di sostituzioni," p. 32.

[28]For an interesting account of the contributions of Alfredo Capelli to group theory, see Giuseppina Casadio and Guido Zappa, "I contributi di Alfredo Capelli alla teoria dei gruppi," *Bollettino di storia delle scienze matematiche* **11** (2) (1991):25-54.

[29]Giovanni Frattini, "Intorno alla generazione dei gruppi di operazioni," *Rendiconti dell'Accademia dei Lincei* **4** (1885):281-285 and 455-457.

[30]Eugenio Netto, *Teoria delle sostituzioni e sue applicazioni all'algebra* (Torino: Loescher, 1885).

[31]Julius Petersen, *Teoria delle equazioni algebriche*, trans. Gerolamo Rizzolino and Giuseppe Sforza (Napoli: Libreria B. Pellerano, 1891).

[32]Bianchi.

[33]Peter G. Lejeune Dirichlet, *Vorlesungen über Zahlentheorie* (Braunschweig: Vieweg Verlag, 1879). It should be noted that Dirichlet's lectures on number theory were translated into Italian in

supérieure, and Jordan's *Traité des substitutions*. Arzelà used Netto's work as a principal reference for his course, and this emphasis may be reflected in the title Bortolotti chose for his notebook, namely, *Teoria delle sostituzioni*.

The great impact the reading of Netto's treatise had on Arzelà is evident in a large part of the lectures. In the first part of the course, Arzelà drew his presentation of the theory of symmetric and multiple-valued functions and that of the theory of groups of substitutions in all its generality from Netto's treatise. In assessing the various presentations, Arzelà chose what he viewed as the best of the best. Again in the first part of the course, he followed Dirichlet's *Zahlentheorie* for his discussion of number theory, while in the course's second part, he used Serret's *Cours* for the explanation of algebraic functions and abelian equations, and Jordan's *Traité* for the Galois theory and the exposition of the solvability conditions by radicals.

The Ruffini-Abel Theorem

Of particular interest here, Arzelà devoted the second part of his course to the question of the resolution of algebraic equations. Once he had concluded the long and detailed discussion of the theory of algebraic functions, he stated and proved the following theorem, "Equations of degree higher than the fourth cannot be solvable algebraically,"[34] which has been known, since the end of the nineteenth century, as the Ruffini-Abel theorem. Despite the fact that Arzelà intended to lecture about Galois theory and, therefore, to state and prove the solvability conditions by radicals,[35] he gave the proof of the Ruffini-Abel theorem. In this way, he gave a somewhat more historical presentation of the material, moving from the result of Ruffini and Abel to the work of Galois.

As is well-known, Ruffini published the result of his first studies on the solvability of algebraic equations in 1799 in a two-volume work, entitled *Teoria generale delle equazioni in cui si dimostra impossibile la soluzione algebraica delle equazioni generali di grado superiore al quarto*.[36] Following the debate generated by the publication of his work, Ruffini presented four new proofs of the theorem. The last, published in 1813 as part of the memoir, *Riflessioni intorno alla soluzione delle equazioni algebraiche generali*,[37] is the fifth, the simplest, and the clearest of Ruffini's proofs, and it essentially coincides with what would later be called the modification of Abel's proof, published in 1845 by the French mathematician, Pierre Laurent Wantzel.[38] Ruffini's work, however, was difficult to understand.

Despite his attempts to explain the validity of his work to his colleagues, his proofs were not completely accepted by the European mathematical community.

1881 as Peter G. Lejeune Dirichlet, *Lezioni sulla teoria dei numeri* (Venice: Tipografia Emiliana, 1881.)

[34] "Le equazioni di grado superiore al quarto non si possono risolvere algebricamente." See Arzelà, Section 150.

[35] Arzelà, Section 197.

[36] Paolo Ruffini, *Teoria generale delle equazioni, in cui si dimostra impossibile la soluzione algebraica delle equazioni generali di grado superiore al quarto* (Bologna: Stamperia di S. Tommaso d'Aquino, 1799) in Paolo Ruffini, *Opere matematiche di Paolo Ruffini*, ed. Ettore Bortolotti, vol. 1 (Palermo: Tipografia matematica di Palermo, 1915); vols. 2-3 (Rome: Edizioni Cremonese, 1953-1954), 1:1-324.

[37] Paolo Ruffini, *Riflessioni intorno alla soluzione delle equazioni algebraiche generali* (Modena: Società tipografica, 1813), in Ruffini, *Opere matematiche*, 2:157-268.

[38] Pierre Laurent Wantzel, "De l'impossibilité de résoudre toutes les équations algébriques avec des radicaux," *Annales de mathématiques pures et appliquées* **4** (1845):57-65.

Abel, for example, still believed, in 1821, that he had found a solution by radicals of the general quintic equation.[39] The young Norwegian mathematician soon discovered his own error and, in 1824, proved that the general quintic equation is not algebraically solvable. By 1826, he had a new proof of the impossibility of solving algebraic equations by radicals, independent of Ruffini's work.

As Heinrich Burckhardt noted in an article in 1892,[40] Ruffini's work seemed to be forgotten. Several nineteenth-century mathematicians—William Rowan Hamilton in Ireland,[41] Serret in France,[42] Petersen in Denmark,[43] Joseph Antoine Carnoy in Belgium,[44] and Leopold Kronecker in Germany[45]—worked on the question of the solvability of equations, but they all referred mainly to Abel's research. Therefore, at the time in which Arzelà gave his lectures, he practically had at his disposal no studies on Ruffini's works and no papers in which Ruffini's proof was explicitly mentioned, except for Wantzel's article.[46] However, at the end of his proof, Arzelà stated categorically that "[t]his proof was given for the first time by Ruffini."[47] It is possible that Arzelà had read Ruffini's first proof which, after all, had been published privately in Bologna and which was well-known to several of Ruffini's Italian contemporaries in mathematics.

For the presentation of the classification of algebraic functions according to order and degree, for the construction of the most general expression to represent an algebraic function of order μ and degree m, and for the study of the algebraic functions that satisfy a given equation, Arzelà followed Serret's exposition.[48] Serret's next step into the theory of algebraic equations was the presentation of the proof that general equations of degree higher than four are not solvable algebraically. In the introduction that precedes the proof, Serret noted that "[t]his theorem was

[39]Ludvig Sylow, "Notes aux mémoires du tome I," in *Oeuvres complètes de Niels Henrik Abel*, ed. Ludvig Sylow and Sophus Lie, 2 vols. (Christiania: Imprimerie de Grøndal & Søn, 1881), 2:290-323 on pp. 290-291.

[40]Heinrich Burckhardt, "Die Anfänge der Gruppentheorie und Paolo Ruffini," *Abhandlungen zur Geschichte der Mathematik* **6** (1892):119-159.

[41]William Rowan Hamilton, "On the Argument of Abel, Respecting the Impossibility of Expressing a Root of Any General Equation above the Fourth Degree, by Any Finite Combination of Radicals and Rational Functions," *Transaction of the Royal Irish Academy* **18** (1839):171-259, in *The Mathematical Papers of Sir William Rowan Hamilton*, vol. 1, ed. A. W. Conway and J. L. Synge, vol. 2, ed. A. W. Conway and A. J. McConnell, vol. 3, ed. Heini Halberstam and R. E. Ingram (Cambridge: University Press, 1931–1967), 3:517-569.

[42]Serret included the proof of the theorem of impossibility in the third edition of his *Cours d'algèbre supérieure*, which was published in 1866.

[43]Petersen.

[44]Joseph Antoine Carnoy, *Cours d'algèbre supérieure* (Louvain: A. Uystpruyst, 1892).

[45]Leopold Kronecker, "Einige Entwicklungen aus der Theorie der algebraischen Gleichungen," *Monatsberichte der Berliner Akademie* (1879):205-229, in Leopold Kronecker, *Werke*, ed. Kurt Hensel, 5 vols. (Leipzig: B. G. Teubner Verlag, 1895-1930), 4:75-96.

[46]Wantzel wrote about Ruffini's work on the solvability of algebraic equations in these terms: "Plusieurs années auparavant, [with respect to Abel's work] Ruffini, géomètre italien, avait traité la même question d'une manière beaucoup plus vague encore, et avec des développements insuffisants, quoiqu'il soit revenu plusieurs fois sur le même sujet [Several years before, the Italian geometer Ruffini treated the same question in an even more vague manner and with insufficient proofs, although he returned several times to the same subject]." See Wantzel, p. 57. However, before he presented his proof, Wantzel stated his intention to face the problem from the same point of view "envisagé dans les mémoires d'Abel et de Ruffini [envisioned in the memoirs of Abel and Ruffini]." See Wantzel, p. 58.

[47]"Questa dimostrazione fu data la prima volta da Ruffini." See Arzelà, Section 150.

[48]Serret, 2:497-512.

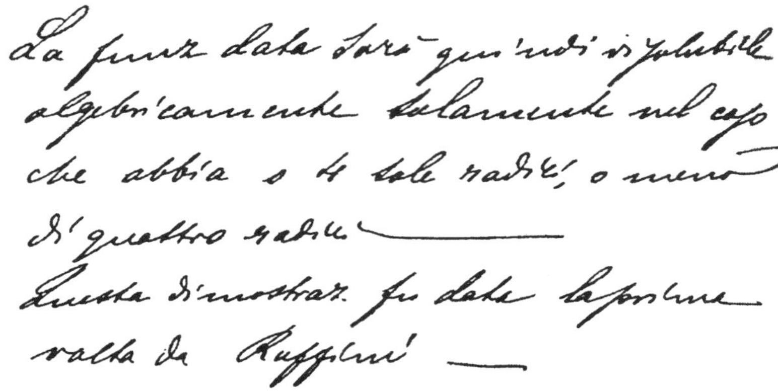
La funz. data sarà quindi risolubile
algebricamente solamente nel caso
che abbia 3 o 4 sole radici; o meno
di quattro radici —
Questa dimostraz. fu data la prima
volta da Ruffini —

FIGURE 1. Final paragraph of Cesare Arzelá's proof of the Ruffini-Abel Theorem from the handwritten lecture notes of his course by Ettore Bortolotti, 1886–1887

proved for the first time in a rigorous way by Abel; here I give the simplest proof due to Wantzel."[49] Thus, Serret presented the proof known as Wantzel's modification of Abel's proof, without mentioning Ruffini's works. Since Arzelà was utilizing Serret's text, he most likely read the proof, but he decided not to present it to his students. As the good professor and teacher that he was, Arzelà sought the clearest and simplest proof of that theorem of fundamental significance in the theory of algebraic equations. Once more, he cast about in the literature—a literature that had been written in France, Germany, Norway, and Denmark, but not in Italy—for the best proof for his students, and he found such a demonstration in Netto's treatise.[50]

The structure of the proof is clearer and simpler than those of Serret and Ruffini, and probably this is the reason for Arzelà's choice. Moreover, having utilized Netto's text for his lectures on the theory of symmetric and multiple-valued functions, Arzelà had the right background and the tools to make the proof understandable to his students. Arzelà did not restrict himself to Netto's exposition; rather, he filled in the details that Netto took for granted for the benefit of his students.

Compare, for example, the same crucial step in Netto's proof and in the proof Arzelà presented in his lecture. Arzelà presented the proof in this way:

> [Theorem:] Equations of degree higher than the fourth cannot be algebraically solvable.
>
> Let
> $$x^m + a_1 x^{m-1} + \cdots = 0$$
> be an algebraic equation.
>
> To find the roots, that is, to find an algebraic expression of the coefficients which satisfies the given equation, we will start combining the coefficients rationally.

[49] "Ce théorème a été démontré, pour la première fois, d'une manière rigoureuse par Abel; je présenterai ici la démonstration plus simple que l'on doit à Wantzel." See Serret, 2:512.

[50] Eugenio Netto, p. 245.

But we know that it is possible to express only the roots of functions of the first degree by rational expressions of the coefficients.

Thus, it is necessary to apply some radicals to the combination of algebraic operations

$$\phi(a_1 a_2 \dots a_m)$$

that we found.

I can always suppose that the first radical I apply is of prime order m_1, I claim that it must be $m_1 = 2$.

And, in fact, by the previous result,[51] it must be

$$\sqrt[m_1]{\phi(a_1 a_2 \dots a_n)} = \psi(x_1 x_2 \dots x_n)$$

which means, a rational function of the roots.

Namely, ψ has to be an m_1-valued function whose m_1^{th} power

$$\psi^{m_1}(x_1 x_2 \dots x_n) = \phi(a_1 a_2 \dots a_n)$$

is a single-valued function.

Such functions exist only if $m_1 = 2$.[52]

For the same reason, we could not continue to apply to ϕ a radical of index higher than the second; besides, applying quadratic radicals, we can solve equations only of the second degree because the corresponding functions of x are two-valued functions.

Thus, to solve equations of degree higher than the second, we must apply to the radical just found new radicals, for instance $\sqrt[p_1]{\ }$

$$\sqrt[p_1]{\sqrt{\phi(a_1 a_2 \dots)}}$$

But even this algebraic expression of the coefficients has to be a rational expression of the roots, that is:

$$\sqrt[p_1]{\sqrt{\phi(a_1 a_2 \dots)}} = \psi(x_1 x_2 \dots)$$

which implies

$$\sqrt{\phi(a_1 a_2 \dots)} = (\psi(x_1 x_2 \dots))^{p_1}$$

ψ has to be a $2p_1$-valued function, such that its p_1^{th} power is a two-valued function.

Now, such functions do not exist if the number of the elements on which they operate is greater than four.[53]

The given function will be algebraically solvable only in the case in which we have either four roots or less than four roots.

[51] In the previous paragraph, Arzelà proved that all the algebraic functions of the coefficients involved in the resolution of an algebraic equation of degree m are rational functions of the roots of the equation. Arzelà, Section 149.

[52] Arzelà had previously proved that the only functions which, raised to a certain power can become symmetric, are alternating functions. Arzelà, Section 51.

[53] Previously, Arzelà had stated and proved the following theorem: it is not possible to find a function of more than four elements which, when raised to a prime power, can become a two-valued function. Arzelà, Section 53.

> This proof was given, for the first time, by Ruffini.[54]

Netto's proof ran this way:

> Theorem III. The general equations of degree higher than the fourth are not algebraically solvable.
>
> For if the n quantities $x_1, x_2, \ldots x_n$, which in the case of the general equation are independent of one another, could be algebraically expressed in terms of $R', R'', \ldots$, then the first introduced irrational function of the coefficients, V_ν, would be the p_ν^{th} root of a rational function of $R', R'', \ldots$. Since, from Theorem II,[55] V_ν is a rational function of the roots, it appears that V_ν, as a p_ν-valued function of $x_1, x_2, \ldots, x_n$, the p_ν^{th} power of which is symmetric, is either the square root of the discriminant, or differs from the latter only by a symmetric factor. Consequently, we must have $p_\nu = 2$ (§57). If we adjoin the function $V_\nu = S_1\sqrt{\Delta}$ to the rational domain, the latter then includes all the one-valued and two-valued functions of the roots. If we are to proceed further with the solution, as is necessary if $n > 2$, there must be a rational function $V_{\nu-1}$ of the roots, which is $(2p_{\nu-1})$-valued, and of which the $(p_{\nu-1})^{th}$ power is two-valued. But such a function does not exist if $n > 4$ (§59). Consequently, the process, which should have led to the roots, cannot be continued further. The general equation of a degree above the fourth therefore cannot be algebraically solvable.[56]

As the citation above makes clear, Arzelà began his proof by considering an algebraic equation of degree m and stated that, to find its roots, he needed to start combining the coefficients rationally. Then he proved that the first radical to be applied to the rational combination of the coefficients of the equation has to be of order two. He next proceeded with the proof, filling in the details and explaining clearly the passages to his students. At this point, Netto says merely: "If we are to proceed further with the solution, as is necessary if $n > 2$, there must be a rational function $V_{\nu-1}$ of the roots, which is $(2p_{\nu-1})$-valued, and of which the $(p_{\nu-1})^{th}$ power is two-valued. But such a function does not exist if $n > 4$." Arzelà, on the other hand, provides the construction in complete detail. The result is an easily understandable proof, impeccable from a pedagogical point of view.

---

[54] Arzelà, Section 150. For the original Italian, see the Appendix.

[55] Theorem II. The explicit algebraic function x_0, which satisfies a solvable equation $f(x) = 0$, can be expressed as a rational integral function of quantities $V_1, V_2, V_3, \ldots, V_\nu$, with coefficients which are rational functions of the quantities $R', R'', \ldots$. The quantities V_λ are on the one hand rational integral functions of the roots of the equation $f(x) = 0$ and of primitive roots of unity, and on the other hand they are determined by a series of equations $V_a^{p_a} = F(V_{a-1}, V_{a-2}, \ldots, V_\nu; R', R'', \ldots)$. In these equations the $p_1, p_2, p_3, \ldots, p_\nu$ are prime numbers, and $F_1, F_2, F_3, \ldots, F_\nu$ are rational integral functions of their elements V and rational functions of the quantities $R', R'', \ldots$, which determine the rational domain. Eugenio Netto, p. 245.

[56] Eugenio Netto, p. 245. The English translation is from Eugen Netto, *The Theory of Substitutions and Its Applications to Algebra*, trans. Frank N. Cole (Bronx, NY: Chelsea Publishing Co., 1964), pp. 250-251. We do not supply the original German or the Italian translation of the German.

Conclusions

The case of Arzelà's course on Galois theory at the University of Bologna in 1886–1887 brings a number of issues about the evolution of mathematics in late-nineteenth-century Italy into relief. In the years following the unification, the mathematical community of Bologna worked diligently to bring its mathematical teaching up to international standards. The 1880s saw the result of these efforts: the improvement of the teaching faculty and the deepening of the mathematical offerings. The course Arzelà gave on Galois theory can be considered a significant example of this process. Arzelà, drawing from the best books the international publication community had to offer, brilliantly succeeded in organizing a cogent and pedagogically sound course of lectures. His course exemplifies the international exchange of mathematical ideas and underscores the fruitfulness of such exchange for developing mathematical communities like that of Italy in the 1880s and 1890s.

References

Abel, Niels H. "Démonstration de l'impossibilité de la résolution algébrique des équations générales qui passent le quatrième degré." In *Oeuvres*, 1:66-87.

———. "Notes aux mémoires du tome I." In *Oeuvres*, 2:290-323.

———. *Oeuvres complètes de Niels Henrik Abel.* Ed. Ludvig Sylow and Sophus Lie. 2 Vols. Christiania: Imprimerie de Grøndal & Søn, 1881.

Arzelà, Cesare. *Teoria delle sostituzioni.* Manuscript lecture notes taken by Ettore Bortolotti in the academic year 1886-1887. Bortolotti Library. University of Bologna.

Bachmann, Paul. "Ueber Galois' Theorie der algebraischen Gleichungen." *Mathematische Annalen* **18** (1881):449-468.

Betti, Enrico. *Opere matematiche di Enrico Betti.* 2 Vols. Milan: U. Hoepli, 1903–1913.

———. "Sopra la risolubilità per radicali delle equazioni algebriche irriduttibili di grado primo." *Annali di scienze matematiche e fisiche* **2** (1851):5-19. In *Opere*, 1:17-27.

———. "Sulla risoluzione delle equazioni algebriche." *Annali di scienze matematiche e fisiche* **3** (1852):49-119. In *Opere*, 1:31-80.

Bianchi, Luigi. *Lezioni sulla teoria dei gruppi di sostituzioni e delle equazioni algebriche secondo Galois.* Pisa: E. Spoerri, 1899.

Birkeland, Bent. "Ludvig Sylow's Lectures on Algebraic Equations and Substitutions, Christiania (Oslo), 1862: An Introduction and a Summary." *Historia Mathematica* **23** (1996):182-199.

Bortolotti, Ettore. *Storia della matematica nella Università di Bologna.* Bologna: Nicola Zanichelli Editore, 1947.

Bottazzini, Umberto. "Enrico Betti e la formazione della scuola matematica pisana." In *Atti del convegno "La storia delle matematiche in Italia," Cagliari 1982.* Bologna: Monograf, 1984, pp. 229-276.

Burckhardt, Heinrich. "Die Anfänge der Gruppentheorie und Paolo Ruffini." *Abhandlungen zur Geschichte der Mathematik* **6** (1892):119-159. Italian translation: Pascal, Ernesto. "Paolo Ruffini e i primordi della teoria dei gruppi." *Annali di matematica* **22** (1894):175-212.

Capelli, Alfredo. "Dimostrazione di due proprietà numeriche offerte dalla teoria delle sostituzioni permutabili con una sostituzione data." *Giornale di matematiche* **14** (1876):66-74.

______. "Intorno ai valori di una funzione lineare di più variabili." *Giornale di matematiche* **14** (1876):141-145.

______. "Sopra l'isomorfismo dei gruppi di sostituzioni." *Giornale di matematiche* **16** (1878):32-87.

______. "Sopra la composizione dei gruppi di sostituzioni." *Memorie della reale Accademia dei Lincei* **19** (1884):262-272.

______. "Sulle generatrici del gruppo simmetrico delle sostituzioni di n elementi." *Giornale di matematiche* **35** (1897):354-355.

Carnoy, Joseph A. *Cours d'algèbre supérieure.* Louvain: A. Uystpruyst, 1892.

Casadio, Giuseppina and Zappa, Guido. "I contributi di Alfredo Capelli alla teoria dei gruppi." *Bollettino di storia delle scienze matematiche* **11** (2) (1991):25-54.

Dedekind, Richard. *Lezioni sulla teoria di Galois*, Ed. Laura Toti Rigatelli. Firenze: Sansoni, 1990.

Frattini, Giovanni. "Intorno alla generazione dei gruppi di operazioni." *Rendiconti della reale Accademia dei Lincei* **4** (1885):281-285 and 455-457.

Hamilton, William R. "On the Argument of Abel, Respecting the Impossibility of Expressing a Root of Any General Equation above the Fourth Degree, by Any Finite Combination of Radicals and Rational Functions." *Transactions of the Royal Irish Academy* **18** (1839):171-259. In *The Mathematical Papers of Sir William Rowan Hamilton.* Vol. 1. Ed. A. W. Conway and J. L. Synge. Vol. 2. Ed. A. W. Conway and A. J. McConnell. Vol. 3. Ed. Heini Halberstam and R. E. Ingram. Cambridge: University Press, 1931–1967, 3:517-569.

Jordan, Camille. *Traité des substitutions et des équations algébriques.* Paris: Gauthier-Villars, 1870.

Kronecker, Leopold. "Einige Entwicklungen aus der Theorie der algebraischen Gleichungen." *Monatsberichte der Berliner Akademie* (1879):205-229. In *Werke*, 4:75-96.

______. "Grundzüge einer arithmetischen Theorie der algebraischen Grössen." *Journal für die reine und angewandte Mathematik* **92** (1882):1-122. In *Werke*, 2:237-388.

______. *Werke.* Ed. Kurt Hensel. 5 Vols. Leipzig: B. G. Teubner Verlag, 1895–1930.

Lejeune Dirichlet, Peter G. *Vorlesungen über Zahlentheorie.* Braunschweig: Vieweg Verlag, 1879. Italian translation: *Lezioni sulla teoria dei numeri.* Venice: Tipografia Emiliana, 1881.

Mammone, Pasquale. "Sur l'apport d'Enrico Betti en théorie de Galois." *Bollettino di storia delle scienze matematiche* **9** (2) (1989):143-169.

Martini, Laura. "The First Lectures in Italy on Galois Theory: Bologna, 1886-1887." *Historia Mathematica* **26** (1999):201-223.

Netto, Eugen. *Substitutionentheorie und ihre Anwendung auf die Algebra.* Leipzig: B. G. Teubner Verlag, 1882.

______. *The Theory of Substitutions and Its Applications to Algebra.* Trans. Frank N. Cole. Bronx, NY: Chelsea Publishing Co., 1964.

Netto, Eugenio. *Teoria delle sostituzioni e sue applicazioni all'algebra.* Torino: Loescher, 1885.

Novi, Giovanni. *Trattato di algebra superiore.* Firenze: Le Monnier, 1863.

Pepe, Luigi. "I matematici bolognesi fra ricerca avanzata e impegno civile 1880-1920." *Archimede* **46** (1994):19-27.

Petersen, Julius. *Theorie der algebraischen Gleichungen.* Kopenhagen: A. F. Host & Sohn, 1878. Italian translation: *Teoria delle equazioni algebriche.* Trans. Gerolamo Rizzolino and Giuseppe Sforza. Napoli: Libreria B. Pellerano, 1891.

Ruffini, Paolo. *Opere matematiche di Paolo Ruffini.* Ed. Ettore Bortolotti. Vol. 1. Palermo: Tipografia matematica di Palermo, 1915; Vols. 2-3. Rome: Edizioni Cremonese, 1953–1954.

———. *Riflessioni intorno alla soluzione delle equazioni algebraiche generali.* Modena: Società tipografica, 1813. In *Opere*, 2:157-268.

———. *Teoria generale delle equazioni, in cui si dimostra impossibile la soluzione algebraica delle equazioni generali di grado superiore al quarto.* Bologna: Stamperia di S. Tommaso d'Aquino, 1799. In *Opere*, 1:1-324.

Serret, Joseph A. *Cours d'algèbre supérieure.* 2 Vols. Paris: Gauthier-Villars, 1885.

Toti Rigatelli, Laura. *La mente algebrica: Storia dello sviluppo della teoria di Galois nel XIX secolo.* Busto Arsizio: Bramante, 1989.

Wantzel, Pierre L. "De l'impossibilité de résoundre toutes les équations algébriques avec des radicaux." *Annales de mathématiques pures et appliquées* **4** (1845): 57-65.

Appendix

The following is the original Italian text of Arzelà's presentation of the Ruffini-Abel theorem [Arzelà, Section 150].

[Teorema:] Le equazioni di grado superiore al quarto, non si possono risolvere algebricamente.

Si abbia una equaz algebrica

$$x^m + a_1 x^{m-1} + \cdots = 0$$

Per trovarne le radici, per trovare cioè un'espressione algebrica dei coeff che sostituita per x la renda identica si incomincierà col combinare raz fra loro i coefficienti

Ma noi sappiamo che mediante espressioni raz dei coeff non si possono esprimere che le radici di funzioni del primo grado

Quindi al complesso di operaz algebriche

$$\phi(a_1 a_2 \dots a_m)$$

trovato bisognerà applicare dei radicali.

Posso sempre supporre che il primo radicale che si impiega sia di ordine primo m_1, dico che deve essere $m_1 = 2$.

Ed infatti. Per quanto abbiamo trovato deve essere

$$\sqrt[m_1]{\phi(a_1 a_2 \dots a_n)} = \psi(x_1 x_2 \dots x_n)$$

cioè funz raz delle radici.

Ossia la ψ deve essere una funz ad m_1 valori la cui potenza m_1^{esima}

$$\psi^{m_1}(x_1 x_2 \dots x_n) = \phi(a_1 a_2 \dots a_n)$$

è ad un sol valore.

Tali funzioni non esistono se non nel caso di $m_1 = 2$.

Per la stessa ragione non si potrà continuare coll'applicare alla ϕ un radicale di indice sup. al secondo, e siccome d'altra parte con radicali quadrati non si possono risolvere che equaz di secondo ordine perchè le funz di x corrisp. sono a due soli valori.

Dunque per risolvere le radici di equ. di ordine sup. al secondo sarà giocoforza applicare nuovi radicali es $\sqrt[p_1]{\ }$ al radicale trovato

$$\sqrt[p_1]{\sqrt{\phi(a_1 a_2 \dots)}}$$

Ma anche questa espressione algeb. dei coeff dovrà essere raz nelle radici, sarà cioè:

$$\sqrt[p_1]{\sqrt{\phi(a_1 a_2 \dots)}} = \psi(x_1 x_2 \dots)$$

da cui:

$$\sqrt{\phi(a_1 a_2 \dots)} = (\psi(x_1 x_2 \dots))^{p_1}$$

La ψ dovrà essere una funzione a $2p_1$ valori e tale che la sua potenza p_1^{esima} sia una funzione a due soli valori.

Ora tali funzioni non esistono se il numero degli elementi su cui operano è maggiore di quattro.

La funz data sarà quindi risolubile algebricamente solamente nel caso che abbia o 4 sole radici, o meno di quattro radici.

Questa dimostraz. fu data la prima volta da Ruffini.

CHAPTER 10

The First International Mathematical Community: The *Circolo matematico di Palermo*

Aldo Brigaglia
Università di Palermo (Italy)

Introduction

The rise in Palermo of the *Circolo matematico* as a major international association was highly unusual. As a city on the island of Sicily, Palermo is located geographically at what has historically been a crossroads of cultures. As a city within the Italian nation, it fostered an organization for the promotion of mathematics that flourished within the broader context of Italian mathematics. Transcending both the Sicilian and the national settings, that organization was, at least in the opening two decades of the twentieth century, regarded as *the* international association of mathematicians. How and why did such a mathematical society develop in a city so far from the forefront of advanced mathematics? An answer to this question emerges by viewing the context of the *Circolo*'s evolution from the Sicilian, the Italian, and the international points of view, respectively.

The *Circolo matematico di Palermo* in the Sicilian Cultural Milieu

At the end of the nineteenth century, the Sicilian bourgeoisie experienced an extraordinary, if ephemeral, development in both industry and culture that resulted in a thriving intellectual environment. Among the main industrial activities were shipbuilding, navigation, furniture manufacture, Marsala wine production, sulfur mining, and banking. On the cultural side, architecture flourished in the liberty style, the main trend in *fin de siècle* European architecture that was called "*art nouveau*" in France and "*Jugendstyle*" in Germany. The Sicilian architect, Ernesto Basile, is now considered one of the main interpreters of this great international artistic style. Similarly, the novelist and Nobel laureate, Luigi Pirandello, exemplified the lively cultural atmosphere of which the *Circolo* was one very important, but not exclusive, manifestation.

The *Circolo* was founded in Palermo on 2 March, 1884, one in a string of cultural events in that city, which included the great international exposition in 1893 and the opening four years later of the *Teatro massimo*, the second largest theater in Europe.[1] The cultural growth of which these developments seemed indicative was, however, not deeply rooted. Growing problems within Sicilian society hampered further progress. For example, in 1893, the former mayor of Palermo, Emanuele Notarbartolo, was murdered by the Mafia; in 1893-1894, the "*fasci siciliani*," the largest farmers' rebellion to date, was put down amidst violence and bloodshed;

[1] Only the *Opéra* in Paris is larger.

and in 1909, Joseph Petrosino, a detective from New York, was assassinated. These events were by no means independent. Sicily's rapid industrial development was largely based on specific economic circumstances: favorable conditions for the export of some of the typical Sicilian products (wine, oranges, sulfur), low salaries, and a plentiful supply of cheap labor from rural areas. These conditions soon changed, however. The shift internationally from free-market to protectionist policies, the end of the Sicilian monopoly in the production of strategically important sulfur, the beginning of massive emigration to the United States, the sharp rise in the organization of trade unions, and the emergence of a strong socialist party all influenced these changes.

Under these new conditions, the Sicilian middle class was unable to strengthen industrial and cultural growth, to diffuse social well-being, and to create a widespread network of entrepreneurs. Instead, it relied on the Mafia to bring union demands under control, thus revealing itself as outwardly strong but ultimately feeble.

To take this analysis further would require a close examination of the links between a variety of intricate social issues. It is, however, crucial to stress that in some way the history of the *Circolo*'s first three decades accurately reflected Palermian society at the turn of the century. The society grew in a fertile, but ultimately fragile, international environment. Indeed with hindsight, the *Circolo*'s decline following the death in 1914 of its founder, Giovan Battista Guccia (1855–1914), and the change in international relations due to the outbreak of the First World War can be seen as almost inevitable.

The *Circolo matematico di Palermo* in the Italian Mathematical Milieu

As is well-known,[2] the flourishing of mathematics in nineteenth-century Italy is strongly linked to the unification of the Italian states in 1860 and to the attendant atmosphere of cultural internationalization. It is thus no coincidence that Vito Volterra, speaking about the evolution of mathematics in post-unification Italy, stressed the role played by the strong international links of its founders:

> During the autumn of the year 1858, three young Italian mathematicians left together for a scientific tour of foreign universities, to get in touch with the most celebrated foreign scholars, to get to know their ideas, and to communicate their own scientific contributions. This trip by Betti, Brioschi, and Casorati marks a significant date: Italy was on the verge of achieving unity and was ready to make its own contribution to the international scientific community.[3]

---

[2]See, for instance, Umberto Bottazzini, *Va' Pensiero: Immagini della matematica nell'Italia dell'Ottocento* (Bologna: Il Mulino, 1994).

[3]"Nell'Autunno del 1858 tre giovani matematici italiani partivano insieme per un viaggio scientifico allo scopo di visitare le Università straniere e mettersi in rapporto con i più celebri scienziati esteri in modo da conoscere le loro idee, e da rendere noti al tempo stesso i propri lavori scientifici. Questo viaggio di Betti, Brioschi e Casorati segna una data meritevole di ricordo: l'Italia stava per costituire la propria unità e prender parte ai lavori scientifici internazionali apportandovi il proprio contributo." Vito Volterra, "Betti, Brioschi, Casorati: Tre analisti e tre modi di considerare le questioni di analisi," *Compte rendu du deuxième Congrès international des mathématiciens* (Paris: Gauthier-Villars, 1902), pp. 43-57; also in Vito Volterra, *Saggi scientifici* (Bologna: Zanichelli, 1990).

Political involvement and strong international connections were always considered two main characteristics of the growth of Italian mathematics. The principal Italian mathematicians in the second half of the nineteenth century—Brioschi, Betti, Cremona, Casorati, and Beltrami—all looked abroad chiefly as a means of modernization. In that spirit, they made rapid advances in the latest areas of mathematical research.

Francesco Brioschi[4] was one of the first to engage in the theory of determinants, and he published one of the first textbooks devoted to it. His main mathematical achievements, however, were the solutions, by means of elliptic and hyperelliptic functions, of algebraic equations of fifth (in which he found formulas different from those of Charles Hermite) and sixth degree. Beyond his mathematical researches, Brioschi was an engineer, who founded Milan's *Istituto politecnico* in 1863; the editor-in-chief of the influential research journal, *Annali di matematica pura ed applicata*; and, for a time, the Secretary of the Italian Ministry of Education. Brioschi's friend, Enrico Betti, [5] was highly significant in the diffusion into Italy of Galois theory[6] and of Riemann's work in mathematical physics. Betti was also the director of the famous Italian *Scuola normale superiore* in Pisa, a member of the *Annali*'s editorial board, and a Senator in the Italian Parliament.

Luigi Cremona,[7] remembered as the founder of the prestigious Italian school of algebraic geometry, was among the principal propagators of the work in projective geometry of Christian von Staudt and Michel Chasles as well as of Carl Culmann's graphical statics. He likewise admired Sophus Lie's work on transformations. Cremona's geometrical books had a wide circulation and were translated into French, English, German, and other European languages. Like Brioschi and Betti, he was also deeply involved both in editing the *Annali* and in politics, serving as the Vice President of Italian Senate and Minister of Public Instruction.

Finally, Felice Casorati[8] developed Karl Weierstraß's works in complex analysis and held a chair at Bologna, while Eugenio Beltrami[9] was one of the principal figures in the acceptance among European mathematicians of both non-Euclidean geometry and Riemann's geometrical ideas. His extensive correspondence with Gaston Darboux and Jules Hoüel shows the high regard in which he was held by the French mathematical community.

As the careers of these five suggest, Italian mathematics acquired a widespread reputation among young European mathematicians not only for its achievements

---

[4]Besides the above-mentioned book by Bottazzini, a mathematical biography of Brioschi is given in the obituary by Max Noether in *Mathematische Annalen* **10** (1898):477-491.

[5]See his obituary by Francesco Brioschi in *Annali di matematica pura ed applicata* **20** (1892):256. The very interesting correspondence between Betti and Cremona has been published by Romano Gatto in Marta Menghini, ed., *La corrispondenza di Luigi Cremona*, vol. 3 (Milano: Quaderni PRISTEM, 1996).

[6]Compare Laura Martini's chapter above.

[7]An important obituary of Cremona was published by Max Noether in *Mathematische Annalen* **59** (1904):1-19. His correspondence is now being published under the supervision of Giorgio Israel.

[8]See the obituary by Eugenio Bertini in *Rendiconti dell'Istituto Lombardo* (2) **25** (1892):1206-1236.

[9]See his obituary by Luigi Cremona in the *Rendiconti del Circolo matematico di Palermo* **14** (1900):275-289. Also of interest is the correspondence between Beltrami and Jules Hoüel, contained in Luciano Boi, Livia Giacardi, and Rossana Tazzioli, ed., *La découverte de la géométrie non euclidienne sur la pseudosphere* (Paris: Blanchard, 1993).

but also for its open-minded attitude toward new and modern trends in mathematical research. This contrasted sharply with the prevailing attitudes in old-fashioned academic circles.

In this context, an extract from a letter from Darboux to Hoüel is highly significant, revealing that young French mathematicians were deeply impressed by the progressive nature of Italian mathematicians and by their enthusiastic endorsement of new international trends. "I believe," Darboux declared, "that if this kind of thing continues, the Italians will surpass us before long. So let us try, with our *Bulletin* [*des sciences mathématiques et astronomiques*], to wake up the holy fire and help the French to understand that things in the world work differently and that, even if we are still the G-r-r-eat nation, nobody abroad may notice this."[10]

A very important feature of Italian mathematics was the belief in the need for great scholars to devote themselves not only to strictly scientific issues but also to organizational and political matters (as reflected in the short biographies above). As Giuseppe Veronese remarked in his obituary of Cremona:

> Foreigners may perhaps observe that our scholars adhere to political life more often than in their countries. In this way they are obliged to sway from science, which, like an intrusive lover, wants to receive all their embraces. But this may be explained by the fact that Italy has not yet gained a definitive arrangement in its public life, which could be adequate for its traditions and its needs.[11]

This circumstance also prompted Italian mathematicians to engage in organizational and educational tasks. The (re)establishment of a leading mathematical journal, the *Annali di matematica pura ed applicata* in 1858 by Betti and Brioschi, the founding of the *Istituto politecnico* in Milan, and the establishment of new chairs (above all in the so-called "*geometria superiore*" and in applied mathematics such as graphical statics and mathematical physics) showed that Italian mathematicians were well-aware that organizational structures were vital in building a sound scientific community. Educational issues also interested them deeply. For example, Cremona had been engaged (with Betti and Brioschi) in reforming the high school geometry curriculum, introducing the study of Euclid's *Elements* at the secondary level and projective geometry into technical high schools. All of these initiatives were highly influential in laying the groundwork for the founding of the *Circolo matematico di Palermo*.

---

[10] "Je crois que si cela continue les Italiens nous dépasseront avant peu. Aussi tâchons avec notre *Bulletin* de réveiller ce feu sacré et faire comprendre aux français qu'il y a un tas de choses dans le monde dont ils ne se doutent pas, et que si nous sommes toujours la Grrrande nation, on ne s'en aperçoit guère l'étranger." Letter from Gaston Darboux to Jules Hoüel, undated, in Hélène Gispert, "Lettres de G. Darboux à J. Hoüel," *Cahiers du Séminaire d'histoire des mathématiques* **8** (1987):67-202.

[11] "Forse dagli stranieri si osserva che i nostri scienziati più di frequente che altrove partecipino alla vita politica, tanto da distoglierli dalla scienza che, indiscreta amante vuole tutti gli amplessi per sé. Ma il fatto è più spiegabile perché l'Italia non ha ancora raggiunto nei pubblici ordinamenti un assetto definitivo che corrisponda alle sue tradizioni e ai suoi bisogni." Giuseppe Veronese, "Commemorazione del Socio Luigi Cremona," *Atti della reale Accademia dei Lincei* (5) **12** (1903):664-678.

The Role of Giovan Battista Guccia in the *Circolo*

Unlike many mathematically advanced countries, Italy had no national mathematical association throughout the nineteenth century; in fact, the first such body, the *Unione matematica italiana*, did not come into being until 1922. Still, on 2 March, 1884, a local mathematical association was founded in Sicily under the name of the *Circolo matematico di Palermo*.[12] Among its twenty-seven original members were nine engineers, seven secondary school teachers, and nine professional mathematicians, one of whom, a young algebraic geometer, Giovan Battista Guccia, played a crucial role in the *Circolo*'s formative years. Indeed, there is no doubt that in the history of the *Circolo*, this man's extraordinary personality was of fundamental importance and influence.

Belonging to a wealthy family (the marquises of Ganzaria), Guccia used much of his money—and part of his palace—to finance the *Circolo*. He was a very good, although not first-rate, mathematician[13] and a strict adherent to the purist geometrical style of his master, Cremona. He published about forty mathematical papers, mainly on Cremona transformations.

In the summer of 1880, some months before his graduation, Guccia was sent by Cremona, with many letters of recommendation, to the meeting of the *Association française pour l'avancement des sciences* (AFAS) in Reims. There, Guccia found a lively atmosphere, with discussions about the needs of modern mathematics and, above all, about the need to overcome national barriers to allow a steady flow of ideas.

The Reims gathering was much more than a national affair. The participants included some outstanding scholars from abroad as well as founders and presidents of recently founded national mathematical associations, such as Eugène Catalan, Arthur Cayley, Gaston Darboux, Edmond Laguerre, James Joseph Sylvester, Felice Casorati, Charles Laisant, and Thomas Archer Hirst. It was probably on this occasion that Guccia developed the idea of creating the *Circolo*.[14]

At this time, there was a sharp contrast between two different viewpoints. Within the mathematical (and the broader scientific) community, there was an ever-growing need to develop mutual international exchanges between different communities to keep up with the growth of new ideas and methods. The political atmosphere, however, was very hostile towards internationalism: the wounds of the Franco-Prussian War of 1870 were still fresh, and the threat of new clashes between the great powers was growing stronger.

It should be remembered that in 1880 the first national mathematical organizations were still very recent creations: the Moscow Mathematical Society dated from 1864; the London Mathematical Society (LMS) from 1865; the *Société mathématique de France* from 1872; and the Tokyo Mathematical Society from 1877. Thus, to establish and to develop such societies was still an important *national* task in the late 1800s. These societies were quickly followed by the *Circolo matematico*

---

[12]For further references, see Aldo Brigaglia and Guido Masotto, *Il Circolo matematico di Palermo* (Bari: Dedalo, 1982) and the other papers by the author cited in the bibliography.

[13]On Guccia's mathematical work, see Michele De Franchis, "Giovan Battista Guccia," *Rendiconti del Circolo matematico di Palermo* **39** (1914):1-10.

[14]References to this meeting, with detailed programs, list of conferences, and comments by the participants may be found in the *Circolo*'s archives. On the role of the *Association française pour l'avancement des sciences* in French mathematical life, see the chapter by Hélène Gispert above.

di Palermo in 1884, the New York (later American) Mathematical Society in 1888, and the *Deutsche Mathematiker-Vereinigung* in 1890.

The foundation of these national societies was an important scientific task even from an international perspective. They were needed not only to bring mathematicians together nationally but also to inform national communities of scientific developments abroad. Sir Edward Collingwood alluded especially to the latter function in his history of the London Mathematical Society. "It was quite exceptional in the first half of the 19th century," he wrote retrospectively, "for a British mathematician to study abroad or to visit professional colleagues in foreign centres. The difficulty of travel ... was no doubt an obstacle, but not the only one. Tradition and prejudice played a part as well .... The absence of analysis reflected the isolation of British pure mathematics from the continent ... [but] by the turn of the century the isolation had been broken."[15] Michel Chasles made a similar point, but with dramatic nationalistic overtones, in 1870:

> Mathematics is making considerable advances abroad. The variety and level of the problems that they are now tackling in a great number of periodical journals ... prove this without any doubt. A simple fact would suffice to show everyone how much we should fear being left behind in this part of the sciences. We have, in our *Société philomatique* a mathematical section, with a small number of members, whose communications only appear after long intervals [and then] together with papers on other topics in a very limited, trimestrial *Bulletin*; however, a mathematical society was founded in London in 1865 with a membership of one hundred, and this number is increasing; a Society whose *Proceedings*, like those of the Royal Society of London, ... publishes abstracts, more or less extended, of many papers. Is not this initiative, which we applaud, an ingredient of future superiority in mathematical culture that should worry us?[16]

Guccia was very impressed by these words, which he called "*le cri d'alarme de Chasles*" and which he very often cited in his correspondence. He took Chasles's concerns to heart in his own association with the *Circolo* and its publications.

Despite his own rather derivative research interests, Guccia was well-aware of the most recent trends in mathematical research and very well-versed in them.

---

[15] Edward F. Collingwood, "A Century of the London Mathematical Society," *Journal of the London Mathematical Society* **41** (1966):577-594 on pp. 578 and 590.

[16] "Les Mathématiques prennent, à l'étranger des développements considérables. La variété et l'élévation des matières qui s'y traitent dans de nombreux recueils périodiques ... le prouvent incontestablement. Mais un simple fait suffirait pour montrer aux yeux de tous combien nous devons craindre de nous laisser arriérer dans cette partie des sciences. Nous possédons dans notre Société philomathique une section des Mathématiques, d'un nombre de membres limité, dont les communications ne paraissent que de loin en loin avec d'autres matières dans un bulletin trimestriel fort restreint; or il s'est formé à Londres, en 1865, une Société mathématique d'une centaine de membres, et le nombre s'en accroît encore; société dont les *Proceedings*, à l'instar de la Société royale de Londres ... , font connaître les travaux par des analyses plus ou moins étendues. Ce fait, auquel nous applaudissons, n'est-il pas, dans la culture des Mathématiques, un élément de supériorité future qui doit nous préoccuper?" Michel Chasles, *Rapport sur les progrès de la géométrie* (Paris: Imprimerie national, 1870), pp. 378-379. Compare Hélène Gispert's contribution to the present volume for a fuller sense of the context of Chasles's remarks.

Under his direction, the *Circolo*'s publications were always open to contributions related to anything new in the mathematical research arena. Guccia also helped assure that that work came out in a timely and polished form.[17] The production of the *Circolo* under his supervision was very highly regarded, since it drastically reduced the time required to publish a paper. Edmund Landau marveled that "[t]he trip from my city [Göttingen] to Palermo lasts three days, but sometimes I got the proofs of my papers eight days after sending the manuscript. And thanks to the perfection of the typing I didn't need to correct much."[18]

Guccia not only personally oversaw all of the production details of the *Circolo*'s publications, but he also kept in close touch with the main European mathematical publishers (above all Gauthier-Villars in Paris and Teubner in Leipzig), with the editors of the key mathematical journals, and with the leading members of other mathematical societies and associations. That he felt in intense, but friendly, competition with them came through clearly in his correspondence. In 1905, for example, he wrote to Valentino Cerruti that "I need more time ... to defeat our four sisters of London, Paris, New York, Germany! But I will succeed ... ! I will succeed because we have at our disposal means, methods, and organizations that they do not."[19]

Guccia also had a remarkably open-minded stance towards emerging young mathematicians, which worked to the *Circolo*'s advantage. Landau provided confirmation of this policy when he recounted that "[t]wo of us had already been invited to join his [Guccia's] editorial board, ... without even holding any official position as professors in our country. ... However, one of the most valuable of Guccia's merits was that he evaluated mathematicians by looking only at their works, without any regard to their age or official position, and he helped many a beginner—as I myself was a dozen years ago—to publish their papers in his important journal."[20] This youth-oriented editorial policy had a great international impact, for the *Rendiconti* published the first important memoirs of such mathematicians as Maurice Fréchet, Hermann Weyl, and Émile Borel.

[17] On this aspect, see Domenico De Masi, *L'emozione e la regola: I gruppi creativi in Europa, dal 1850 al 1990* (Bari: Laterza, 1990). As an editor, Guccia shared many of the attributes of Mittag-Leffler. Compare June Barrow-Green's chapter above and see below.

[18] "Il faut trois jours pour venir de ma ville à Palerme, et il m'est arrivé d'avoir les épreuves huit jours après l'expédition du manuscrit; et grâce à la perfection de l'imprimerie, il n'y avait pas grande chose à corriger." Edmund Landau, "Discorso del prof. Dott. Edmund Landau della reale Università di Gottinga," *Supplemento ai Rendiconti del Circolo matematico di Palermo* **9** (1914):12-13; now also in Pietro Nastasi, ed., "Documenti della vita del Circolo matematico di Palermo," *Supplemento ai Rendiconti del Circolo matematico di Palermo* (2) **20** (1988).

[19] "Qualche altro anno di tempo mi è necessario ... prima di battere vittoriosamente tutte e quattro le nostre consorelle di Londra, Parigi, New York e Germania! Ma vi riuscirò ... ! Vi riuscirò perché noi disponiamo di mezzi, metodi e organizzazioni che esse non hanno." Letter from Guccia to Valentino Cerruti, 15 May, 1905, Archives of the Circolo matematico di Palermo (hereinafter ACMP).

[20] "Deux entre nous avaient été invités par lui à faire partie de son Comité de rédaction, ... avant même que nous n'ayons aucune position officielle dans l'enseignement de notre pays. ... En tous cas—et c'est parmi les grands et immortel mérites de M. Guccia un des plus remarquables—il a toujours jugé les mathématiciens uniquement d'après leurs travaux, sans s'inquiéter de leur âge ni de leur rang officiel, et il a secouru bien des commençants—comme j'en étais un, il y a un dizain d'années—à publier leurs recherches dans son important journal." Landau, "Discorso," pp. 12-13.

The *Circolo* and the Internationalization of Mathematical Research

During the years from 1886 to 1888, the *Circolo* emerged, unofficially at least, as *the* Italian mathematical society. Its members included almost every significant Italian mathematician: Giuseppe Battaglini (joined 1886); Ernesto Cesàro, Corrado Segre, Francesco Gerbaldi, Brioschi, Cremona, Enrico D'Ovidio, Giuseppe Peano, Vito Volterra, and Betti (1887); Beltrami, Casorati, Eugenio Bertini, Cesare Arzelà, Salvatore Pincherle, and Giuseppe Veronese (1888). Guido Castelnuovo, Federigo Enriques, Tullio Levi-Civita, and Francesco Severi followed.

In 1887, the first volume of the *Rendiconti del Circolo matematico di Palermo* was published, containing papers by Segre, Eugène Catalan, Thomas Archer Hirst, and Pieter Heinrich Schoute. One year later, the second volume was equally impressive with contributions by Betti, Georges Halphen, Émile de Jonquières, Camille Jordan, Peano, Segre, Alexis Starkov, and Volterra.

Also in 1888, the *Circolo* adopted a new statute, concerning the editorial board of the *Rendiconti* (which also acted as a "*direttivo*," namely, a council concerned not only with editing the *Rendiconti* but also with the *Circolo*'s cultural activities). This was chosen by the members of the *Circolo* through a general election. The new editorial board was largely representative of the principal universities and mathematical schools throughout Italy, comprising five members from Palermo (Giuseppe and Francesco Albeggiani, Francesco Caldarera, Michele Gebbia, and Guccia); three from Pavia (Beltrami, Bertini, and Casorati); three from Pisa (Betti, Riccardo De Paolis, and Volterra); two from Naples (Battaglini and Pasquale Del Pezzo); two from Milan (Brioschi and Giuseppe Jung); two from Rome (Cerruti and Cremona); two from Turin (D'Ovidio and Segre); and one from Bologna (Pincherle). In 1891, Henri Poincaré joined the board as its first foreign member.[21] In 1894, Gösta Mittag-Leffler, editor-in-chief of *Acta Mathematica*, became the second non-Italian member of the *direttivo*; his journal had an inter-Scandinavian, if not truly international, editorial board.

Owing to contemporary political circumstances, it was inconceivable that the three major mathematical communities (i.e., those in Germany, France, and Britain) would have actively worked to initiate a truly international—as opposed to a competitively national—mathematical community. From this point of view, it was only natural that the first steps in this direction were taken by countries then considered neutral: Sweden (*Acta Mathematica*, 1882), Italy (*Rendiconti*, 1887), and Switzerland (*L'Enseignement mathématique*, 1899). As is well-known, *Acta Mathematica* may be considered the first international journal: this is true not only from a linguistic point of view but also from a scientific one, since it contributed significantly to the diffusion of Cantor's works in France and of the ideas of Poincaré in Germany.[22] Guccia was in close touch with Mittag-Leffler[23] and was very impressed by *Acta* and the outward focus of its editorial policies. There is no doubt that Mittag-Leffler's journal constituted the prime model for the *Rendiconti*.

---

[21] In 1855, the *Quarterly Journal of Pure and Applied Mathematics* was founded in England with Charles Hermite as one of its editors. (See the chapter by Sloan Despeaux above.) The editors of this journal, like those of the *Rendiconti*, recognized the potential importance of a more widely cast, internationalized editorial board both for attracting contributions from foreign countries and for bringing new mathematical ideas before the readership.

[22] See the chapter by June Barrow-Green in this volume.

[23] Mittag-Leffler had many Italian links, owing to the fact that his wife was the sister of the Neapolitan mathematician, Pasquale Del Pezzo.

There is also much evidence to suggest that, from the very beginning, Guccia intended the *Circolo* to be a truly international organization. Nevertheless, as noted, it initially operated purely at the national level. Its international evolution was gradual. Guccia followed the development of the different initiatives very attentively, aiming at the establishment of sound international links among the national communities. The *Circolo*'s archives still contain some of the documents that were influential in shaping his ideas. Chief among them is a "Project" by Charles Laisant in which he expressed a clear will to establish a "Society ... whose aim is to establish a moral and scientific bond among all people working in pure and applied mathematics today."[24] A more precise project was approved in 1895 during the meeting of the AFAS in Caen. One of its components was the proposal for a gathering that eventually evolved into the International Congress of Mathematicians (ICM) in Zürich in 1897.[25]

By 1900, the *Circolo* was well-established as the (unofficial) Italian national mathematical society with a strong international reputation. Its international flavor was further reinforced by the fact that its *Rendiconti* permitted seven languages—Italian, French, German, Spanish, English, Esperanto, and Peano's Latino sine flexione—for publication.[26] With the society and its publications firmly grounded, Guccia was in a position to take the first steps toward the realization of his great dream of transforming the *Circolo* into a great international society.

A Decade of Great Development, 1904–1914

In the early 1900s, the *Circolo* was basically a national organization with many international links. For instance, there were only two foreigners on the editorial board (Poincaré and Mittag-Leffler) out of twenty members. Among the 200 members in 1904, forty (20%) were foreigners, but only four of these were Germans.[27]

The project to transform the *Circolo* into a fully international association began to take shape in 1904, when it was charged (together with the mathematical section of the prestigious *Accademia dei Lincei*) with the organization of the third ICM, to be held in Rome in 1908. Guccia's intention was that, by the time of this Congress, the *Circolo* would have been transformed into an international association (or, to put it more accurately, into *the* international association) free, even formally, from Italian dominance. To realize this ambition, the *Circolo*'s members should be from all over the world, and its *direttivo* should be representative of the mathematicians from the most advanced communities.

This project required a great deal of organizational effort. Guccia stated his ideas in many letters to his friends during the hectic years from 1904 to 1909. In one from November 1905, he expressed the aims of the association to Mittag-Leffler: "To internationalize, to diffuse, and to expand the mathematical production of the whole world, making full use of the progress made by modern civilization

[24] "Associazione ... che ha per obiettivo di stabilire un legame morale e scientifico tra tutte le persone che coltivano oggi nel mondo le scienze matematiche pure ed applicate." Letter from Charles Laisant to Guccia, 6 November, 1903, ACMP. The project to which Laisant referred was from 1891.

[25] The information about this Congress is drawn from its final resolution, a copy of which may be found in the *Circolo*'s archives. Compare the chapter by Olli Lehto in this volume.

[26] For more on this issue of the language of mathematics, see Jeremy Gray's chapter below.

[27] Given that Germany was viewed as the mathematical standard-bearer, the scarcity of German members at this time reflected a certain lack of regard for the *Circolo* as a mathematical force.

in international relations."[28] He wrote more extensively in June of the following year to Volterra (who at that time was the most active of his collaborators): "You will understand that we are now engaged with so many commitments towards our mathematical audience (1. Guccia's medal, 2. Congress of Rome, 3. 1,000 pages of the *Rendiconti*, *Supplemento*, and *Annuario* to be sent each year to our members, 4. *Réclame* to get 500 members!) that we cannot divert a single part of our strength to any other enterprise!"[29] This was, in fact, an outline of Guccia's project for the *Circolo*'s transformation, although he succeeded in only some of his objectives.

Membership was a notable success. In 1908, the *Circolo* numbered 605 members, a figure that amounts to more than three times the figure of 1904 and 20% more than Guccia's hopes in 1906; in the same year, the number of German members had grown to seventy-eight. By the time of Guccia's death in 1914, the *Circolo* had almost one thousand members and was by far the biggest mathematical association in the world. This is not only a matter of quantity but also—and most importantly—of quality. (See Table 1.)

Another interesting aspect is the composition of the associates. In 1914, the foreign members numbered 618 out of a total of 924, jumping from 20% in 1904 to 66.6% in 1914! Among the associates were 140 Germans, 140 from the United States, seventy-seven Austrians, sixty-seven French, forty-four Russians, and twenty-nine British. Guccia's pride was justified when he claimed that the yearbook of the *Circolo* was a kind of *Almanach de Gotha* of the mathematical sciences. The effect of Guccia's *réclame* may be seen at once by looking at Figure 1, which contains the number of *Circolo* members in the thirty years from its foundation in 1884 to 1914. The sharp increase after 1904 is instantly noticeable.

It may also be useful to compare the number of members of the *Circolo* with those of comparable mathematical societies. Taking the figures from 1914, the *Circolo*, as noted, had 924 members compared to 769 in the *Deutsche Mathematiker-Vereinigung*, 703 in the American Mathematical Society, 320 in the London Mathematical Society, and 298 in the *Société mathématique de France*.[30]

These impressive results were not achieved without the exploitation of the many links Guccia had established with the most important mathematicians and mathematical associations of Europe. For example, on 3 November, 1907, Guccia gave a dinner in Paris in the restaurant of the Hotel Continental, inviting Darboux, Jordan, Poincaré, Appell, Painlevé, Humbert, Hadamard, Borel, Desiré André, Laisant, Georges Fouret, Jules Drach, Louis Olivier, and Pierre Boutroux. These were not names chosen by chance. To give the meeting an official flavor, Volterra sent a telegram to underline his approval, and *L'Enseignement mathématique* published

[28] "Bien internationaliser, diffondre et répandre la production mathématique du monde entier en profitant des progrès accompli par la civilisation moderne dans les rapports internationaux." Letter from Guccia to Gösta Mittag Leffler, 11 September, 1905, Archive of the Institut Mittag Leffler, Djursholm, Stockholm. See also Antonio Pillitteri, "I rapporti tra Giovan Battista Guccia e Gösta Mittag-Leffler," (University of Palermo, unpublished dissertation, 1991), p. 68.

[29] "Ella saprà ben comprendere che noi, in questo momento, abbiamo assunto tali e tanti impegni verso il pubblico matematico (1° Medaglia Guccia; 2° Congresso di Roma, per la parte che ci riguarda; 3° 1.000 pagine all'anno da mandare ai soci, fra *Rendiconti*, *Supplemento*, ed *Annuario*; 4° Réclame per arrivare al numero di 500 soci!) che non abbiamo il coraggio di distrarre parte delle nostre forze in altre imprese!" Letter from Guccia to Vito Volterra, 28 June, 1906, ACMP.

[30] All figures are drawn from the *Circolo*'s yearbook for 1914.

an account of the dinner. The write-up emphasized that the main French mathematical associations—the *Académie des sciences* (Poincaré and Hadamard) and the *Société mathématique de France* (Darboux and Jordan)—were both represented as were the main French-language journals: the *Journal de mathématiques pures et appliquées*, the *Bulletin des sciences mathématiques* (Darboux), the *Nouvelles annales des mathématiques* (Laisant), the *Intermédiaire des mathématiciens* (Laisant), the *Revue du Mois* (Borel), *L'Enseignement mathématique* (Laisant again), and the *Revue générale des sciences* (Olivier).[31] Significantly, the account also quoted Guccia's claim that his project aimed to "create an international mathematical association."[32]

TABLE 1: FOREIGN MEMBERS OF THE *Circolo* BY YEAR

1886:	E. Catalan
1887:	T. A. Hirst, G. Humbert
1888:	G. Mittag-Leffler
1889:	C. Jordan
1890:	H. Poincaré, É. Picard
1891:	P. Appell
1894:	G. B. Halsted
1895:	P. Painlevé
1904:	M. Noether
1905:	F. Klein, H. Zeuthen, W. F. Osgood, G. Cantor, J. Lüroth, O. Veblen, G. Darboux, E. Landau, E. H. Moore
1906:	I. Fredholm, É. Borel, M. Fréchet, D. Hilbert, J. Hadamard, J. H. M. Wedderburn
1907:	L. Sylow, P. Duhem, K. Hensel, H. Lesbegue
1908:	M. Riesz, M. Dehn, E. Zermelo, H. Weyl, E. Noether
1909:	A. Hurwitz, H. Bohr, W. Sierpinski
1910:	L. Bieberbach, R. Courant, F. Hausdorff, G. H. Hardy, W. Burnside
1911:	J. L. Coolidge, F. Noether
1912:	B. Russell, G. Polya
1913:	H. Steinhaus, G. D. Birkhoff
1914:	S. Lefschetz, A. Fraenkel

Guccia's tactics—merging personal contacts with official relations in an effort to build an association fully representative of the different mathematical communities and organizations—are illustrated well here. It is highly likely that he made a

[31] "Un diner mathématique," *L'Enseignement mathématique* **9** (1907):491-492.
[32] *Ibid.*

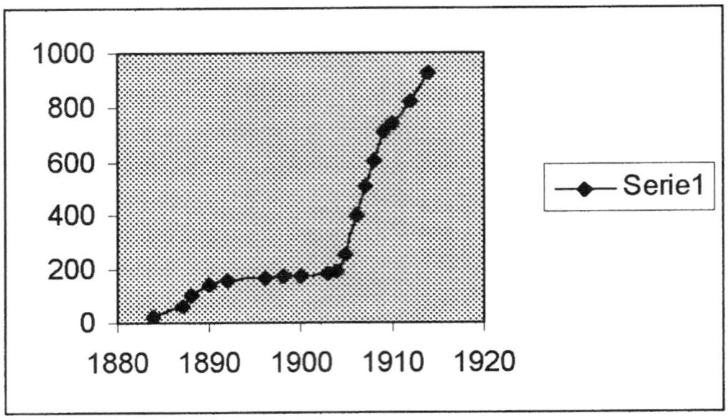

FIGURE 1. Membership profile of the *Circolo Matematico di Palermo*, 1884–1914

similar effort to secure the participation of German mathematicians, since, as noted above, by 1906 the number of German members had increased dramatically, but no corroborative documentary evidence has been found to date. In the summers of 1905 and 1906, Guccia was in Göttingen, and it is probable that it was there that he had his first meeting with Edmund Landau, who, along with Max Noether, was one of the German mathematicians most actively involved with the *Circolo*.

In 1908, the *Circolo* adopted another new statute specially prepared by Guccia. It stated that thenceforth the majority of members of the editorial board were to be mathematicians from abroad. The new editorial board, elected in 1909, was thus composed of fifteen Italians and twenty-five foreign mathematicians. (See Table 2.) It appears that Guccia had intended the Rome ICM of 1908 to act as some sort of official inauguration of the *Circolo*'s new status as the international organization of mathematicians. However, if this was his objective, he failed entirely.

In order officially to become the international organization of mathematicians, the *Circolo* first needed the complete support of the leading *Italian* mathematicians. But this support was not forthcoming during the Rome Congress. From the very opening day, the *Accademia dei Lincei*, which, together with the *Circolo*, had been entrusted with the organization of the meeting, completely assumed scientific and organizational control, failing to acknowledge the *Circolo*'s role. Guccia chronicled the defeat in a letter to Mittag-Leffler:

> Regarding the Congress of Rome, as you may have noticed, our Society has been troubled because, during the inaugural session, the President of the *Accademia dei Lincei* announced that it had undertaken the organization of the Congress. This fact, unexpected and strange, contradicted the previous circular letters and amazed the audience. *It was necessary, without any fuss, to safeguard the dignity of our great international association ... which is, from every point of view, greatly autonomous*, and which has among its members the most famous and illustrious mathematicians in the world. I don't think that I have done anything wrong, in my capacity as representative of the

> *Circolo* at the Congress, in rescinding, advancing a likely excuse, ... our agreements with the organizational Committee .... The only thing that I find really painful in this affair is that among the members of the organizational Committee ... there were three members of the *Circolo*'s editorial board, who are my good friends, but who on this occasion completely forgot our Society! Between the *Accademia dei Lincei* (renowned and famous) which is interested in everything, even Gorgonzola cheese, and our modest Society which is interested only in mathematics, they had no doubt: they sided with the *Accademia*, to advance its reputation on the occasion of the Congress![33]

As far as is known, Guccia's correspondence does not name those responsible for this move against the *Circolo*, but the three members of the *Rendiconti*'s editorial board who were also on the *Lincei*'s organizational committee were Cerruti, Alberto Tonelli, and Volterra. Perhaps it is no coincidence that none of these three was elected to the prestigious editorial board in 1909. Cerruti and Tonelli, although not of the highest rank scientifically, were powerful in academic circles, having both been rectors of the University of Rome. But in 1908, they were not on good terms with Guccia. The omission of Volterra is particularly surprising, since he had been one of the principal supporters of Guccia's efforts and was widely viewed as Italy's most important mathematician at this time. However, it seems that by 1910 he and Guccia were once again collaborating.

The letter to Mittag-Leffler is highly indicative of the many clashes between Guccia and the Italian academic establishment. Guccia's idea of building an international association at once *official* and *independent* was wholly utopic. The established academic circles could not allow the growth of a society completely free from their influence.[34] As a result, after 1908, the *Circolo* did indeed grow very fast, but as a completely *private* association of mathematicians.

---

[33]"Quant au Congrès de Rome, ainsi que vous l'avez pu remarquer sur la place, notre Société c'est effacé dès que dans la Séance d'inauguration l'Académie des Lincei a fait annoncer par son Président qu'elle prenait la direction du Congrès. Cette circonstance, aussi imprévue que bizarre, étant en désaccord avec les circulaires précédents, a un peu étonné le public. *Il fallait donc, sans faire du bruit, sauvegarder la dignité de notre grande association internationale ... qui jouit, à tous les points de vue, d'une grande indépendance*, et à laquelle appartiennent les plus illustres et célèbres mathématiciens du monde entier. Je ne crois pas donc avoir mal agi, dans ma qualité de délégué du Circolo au Congrès, en résiliant, sous un prétexte avouable et vraisemblable ... nos accords avec le Comité d'organisation ... La seule chose qui m'est vraiment pénible dans cette affaire, c'est que, parmi les membres du Comité "organisateur" (le seul responsable vis-à-vis du public) il y avait trois membres du Conseil de Direction du *Circolo*, qui sont de mes bons et excellents amis, mais qui, dans la circonstance, ont oublié tout à fait notre Société! Entre l'Accademia dei Lincei, illustre et célèbre qui s'occupe de tout, même du fromage Gorgonzola et notre modeste société qui ne s'occupe d'autre que des mathématiques le choix n'était pas douteux pour eux: ils se sont tournés vers l'Académie, pour accroître sa gloire à l'occasion du Congrès." Letter from Guccia to Gösta Mittag Leffler, 12 July, 1908, ACMP; also in Pillitteri, p. 78. Guccia's emphasis.

[34]In the same period (starting in 1907), Volterra tried to build an independent scientific association, the *Società italiana per l'avanzamento delle scienze*, but his efforts were ultimately unsuccessful. For information about the efforts of Italian mathematicians to build a new scientific society, see Aldo Brigaglia, "La matematica italiana dell'inizio del secolo e le sue proiezioni all'esterno," in *Accademia dei XL: Scritti di storia della scienza in onore di Giovanni Battista Marini-Bettolo* (1990), pp. 393-407.

TABLE 2: THE EDITORIAL BOARD OF THE *Circolo* 1909

Italy:	G. Bagnera (Palermo); M. De Franchis (Catania); C. Segre and C. Somigliana (Turin); G. Loria (Genoa); G. Vivanti (Pavia); T. Levi-Civita and F. Severi (Padua); F. Enriques and S. Pincherle (Bologna); E. Bertini, L. Bianchi, and U. Dini (Pisa); R. Marcolongo and E. Pascal (Naples)
France:	É. Borel, J. Hadamard, G. Humbert, É. Picard, and H. Poincaré (all from Paris)
Germany:	D. Hilbert, E. Landau, and F. Klein (Göttingen); C. Carathéodory (Hannover); M. Noether (Erlangen); P. Staeckel (Karlsruhe)
England:	A. R. Forsyth (Cambridge) and A. Love (Oxford)
Austria-Hungary:	L. Fejer (Koloszovar); F. Mertens and W. Wirtinger (Vienna)
United States:	E. H. Moore (Chicago) and W. F. Osgood (Cambridge)
Russia:	A. Liapunov and A. Steklov (both from St. Petersburg)
Sweden:	E. I. Fredholm and G. Mittag-Leffler (both from Stockholm)
Greece:	C. Stephanos (Athens)
Belgium:	C. J. de la Vallée Poussin (Louvain)
Denmark:	H. G. Zeuthen (Copenhagen)

The third of Guccia's objectives aimed at the development of the *Rendiconti*. This was another indubitable success. The journal grew both in quantity and in quality. With a circulation of 2,000 copies of each of the two volumes (more than 1,200 pages) edited during the year, it became the most widespread mathematical journal in the world. Among the many fundamental papers published there during the years 1904–1914, nine were by Poincaré (one each year till his death in 1912). Among them were such fundamental contributions as: "5e complément à l'analysis situs" (the first paper in this series had appeared in the *Rendiconti* in 1899); "Nouvelles remarques sur les groupes continus" (the first of his papers on this subject had appeared in 1901); "Sur la dynamique des électrons;" and his last, famous paper, "Sur un théorème de géométrie."[35] The latter paper is also a testimonial to the close links between Poincaré and Guccia, and its introduction (in which Poincaré asked Guccia to publish the paper even if incomplete, as he did not know if he would ever be able to complete it) amounts to a real scientific testimonial.[36] The paper was published a few days before Poincaré's death and, as is well-known, a young American mathematician, George David Birkhoff, solved the problem the following year.[37] As we have already noted, Guccia gave particular emphasis to the publication of work by promising young mathematicians, and this period witnessed innovative papers by Fréchet, Weyl, Borel, Landau, Vitali,[38] and many others.

During this period, Guccia also conceived many projects aimed at the extension of the scope of the *Rendiconti* and the other publications of the *Circolo*. These projects had little or no success, but a sketch of them provides a more complete picture of Guccia's cultural politics. The first and most ambitious project had as its objective the establishment of a new journal (to accompany the *Rendiconti*) in applied mathematics. Guccia discussed this project at length in a letter to Valentino Cerruti, expressing some interesting ideas about the meaning of applied mathematics: "I would like to be very clear with the audience about the word 'applied' ... if not we are lost! We want the applications to stem from the higher and more modern theories and findings of pure mathematics and not at all from elementary or trivial mathematics because, in that case, we would go downhill immediately ... and everything would be included!"[39] Perhaps because Guccia failed to find anyone who could give him the necessary aid in this new enterprise,

---

[35]See Henri Poincaré, "5e complément à l'analysis situs," *Rendiconti del Circolo matematico di Palermo* **18** (1904):45-110; "Complément à l'analysis situs," *op. cit.*, **13** (1899):285-343; "Nouvelles remarques sur les groupes continus," *op. cit.*, **25** (1908):81-130; "Sur la dynamique des électrons," *op. cit.*, **21** (1906):129-176; and "Sur un théorème de géométrie," *op. cit.*, **33** (1912):375-407, respectively.

[36]In a letter published in the *Supplemento ai Rendiconti del Circolo matematico di Palermo* **8** (1913):28-29, and partially printed in the preface of his paper (p. 375), Poincaré wrote: "My dear friend, I already spoke to you, during your last visit, of a paper which has interested me for more than two years. I have not made any progress, so I have decided to abandon it for a while. This could be wise if I was sure to be able to work on it again some day. But my age cannot assure me of it. In this situation, do you think that it would be convenient to publish a paper not yet complete, where I could expose my aims .... Please tell me what you think about this matter and what you may suggest. [Mon cher ami, Je vous ai parlé, lors de votre dernière visite, d'un travail qui me retient depuis deux ans. Je ne suis pas plus avancé et je me décide à l'abandoner provisoirement pour lui donner le temps de mûrir. Cela serait bien si j'étais sûr de pouvoir le reprendre; à mon âge je ne puis en répondre ... Ditez moi, je vous prie, ce que vous pensez de cette question et ce que vous me conseillez.]"

[37]George David Birkhoff, "Proof of Poincaré's Geometric Theorem," *Transactions of the American Mathematical Society* **14** (1913):14-22.

[38]Maurice Fréchet, "Sur quelques points du calcul fonctionnel," *Rendiconti del Circolo matematico di Palermo* **22** (1906):1-74; Hermann Weyl, "Über beschränkte quadratische Formen, deren Differenz, vollstetig ist," *op. cit.*, **27** (1909):373-392; Émile Borel, "Les probabilités dénombrables et leurs applications arithmétiques," *op. cit.*, **27** (1909):247-271; Edmund Landau, "Über die Multiplikation Dirichlet'schen Reihen," *op. cit.*, **24** (1907):81-160; and Giuseppe Vitali, "Sulle funzioni a integrale nullo," *op. cit.*, **20** (1905):136-141.

[39]"Vorrei che nella parola 'applicate' ci intendessimo bene col pubblico ... se no siamo perduti! Vogliamo che le applicazioni si traggano dalle più alte e moderne dottrine e scoverte delle matematiche pure, non mai dalla matematica elementare o media, perché in tal caso si andrebbe subito giù ... si comprenderebbe ogni cosa!" Letter from Guccia to Valentino Cerruti, 11 June, 1905, ACMP.

it never came to fruition. However, of the new papers appearing in the *Rendiconti*, many were important contributions to applied mathematics, particularly those by Poincaré (besides the above-mentioned "Sur la dynamique de l'électron," "Sur la diffraction des ondes hertiennes" was notable[40]) and by Levi-Civita.[41] Levi-Civita also published the highly influential, "Nozione di parallelismo su di una varietà qualunque e conseguente specificazione geometrica della curvatura riemanniana," which was widely used by Einstein in the mathematical formulation of his theory of general relativity.[42]

Guccia clearly expressed his stand on the subject of applied mathematics in the *Rendiconti* in a letter to the young Italian engineer, Gustavo Colonnetti:

> I hope to extend the contributions of the *Rendiconti* in the field of modern higher applied mathematics, following a continually growing trend. The paper by Levi-Civita was a first step taken in this order of ideas; as well as another very important paper [by Poincaré, "Sur la dynamique de l'électron"], which adds a great contribution to the works of Lorentz and others on the question and which will be published in the next few days.[43]

In this same vein, Guccia invited a great Sicilian bank, the *Cassa di risparmio*, to sponsor an international prize competition in applied mathematics with a jury to be headed by Poincaré, but this idea was never realized.

Another field of editorial activity for the *Circolo* was the history of mathematics, in which Guccia conceived a project of publishing classical mathematical works. In this case, the plan was partially successful, and, indeed, the *Circolo* commissioned the eminent Italian historian, Ettore Bortolotti, to prepare an edition of the mathematical works and correspondence of Paolo Ruffini. The first volume of a planned set of three was published in 1914; the second, ready for press in 1914, was shelved after the First World War and eventually published by the *Unione matematica italiana* in 1943; the third was not published until the 1950s.

Guccia also had interesting ideas in the field of the popularization of mathematics. He wrote to Poincaré and Mittag-Leffler on this subject:

> I have not given up my plans ... to publish every year in the *Annuario* ... a paper by a master in science on a subject which may be interesting even to people with a sparse scientific background. This could give the vast public an idea of the great recent achievements of science and increase the number of friends

[40]Henri Poincaré, "Sur la diffraction des ondes hertiennes," *Rendiconti del Circolo matematico di Palermo* **29** (1910):169-260.

[41]Tullio Levi-Civita, "Sopra un problema di elettrostatica che si è presentato nella costruzione dei cavi," *Rendiconti del Circolo matematico di Palermo* **20** (1905):173-228, and "Scie e leggi di resistenza," *op. cit.*, **23** (1907):1-37.

[42]Tullio Levi-Civita, "Nozione di parallelismo su di una varietà qualunque e conseguente specificazione geometrica della curvatura riemanniana," *Rendiconti del Circolo matemaitco di Palermo* **42** (1917):173-205.

[43]"Io spero di dare ai *Rendiconti* maggiore estensione nel campo delle applicazioni delle matematiche superiori moderne, seguendo in ciò un movimento che va sempre più accentuandosi. La memoria di Levi-Civita fu il primo passo verso questo ordine di idee; come un altro è un'importantissima memoria di che comparirà fra pochi giorni e che porta un notevole contributo ai lavori di Lorentz ed altri sulla questione." Letter from Guccia to Gustavo Colonnetti, 24 February, 1906, ACMP.

> of mathematics. The general public, and sometimes even scholars, do not have a clear idea of the links between pure and applied science. Therefore, to show clearly (speaking to the general public) everything that is derived from higher mathematics in great modern practical discoveries, could be useful.[44]

Guccia already had in mind the first two papers to be published: the first was to be by Poincaré on the contributions of higher mathematics to the development of electricity; the second was to be a documented biography of Weierstraß by Mittag-Leffler, which he hoped would give an insight into the development of classical analysis. Sadly, even in this case, Guccia's ideas, which were indeed very progressive, never came to fruition.

The fourth of Guccia's projects, the *Medaglia Guccia*, was to be an international mathematical prize conferred during each International Congress to a prominent mathematician. This was in some way similar to the idea behind the Fields Medal later, but Guccia chose to follow the old-fashioned style of the Paris Academy, where participants were required to send the jury a paper on a predetermined subject. In 1908, Guccia chose the topic of algebraic curves in space, but in the absence of any significant entries, the jury (Poincaré, Max Noether, and Segre) gave the medal to Francesco Severi for his overall contribution to geometry. This was the only time the medal was awarded, and during the next ICM in Cambridge in 1912, Guccia did not attempt to take the idea any further.

The Decline

Guccia died on 28 October 1914, shortly after the outbreak of World War I. Although, at the time of his death, his dream of a great international mathematical association centered in Sicily seemed to have been successfully fulfilled, in reality the favorable conditions that had made it possible were coming to an end. It was not only a matter of the physical death of the founder. A whole world-order was dying.

The thriving Palermian middle class was rapidly declining. The new generations quickly dissipated the huge patrimony accumulated by their grandfathers. Furthermore, Italian mathematics was losing its leadership in many fields. In algebraic geometry, Severi and Enriques had reigned supreme for some time, but by the end of the 1920s, they and their pupils could not follow the new languages and methods elaborated by Solomon Lefschetz, Oscar Zariski, and Baertel van der Waerden. In mathematical logic, Peano's school was completely surpassed by those of the British, the Germans, and, above all, the Polish. In functional analysis, the barycenter had moved from the school of Volterra to those of Hilbert, Stefan Banach, and Erik Fredholm. The list goes on, but it is clear that the image of a daring, open-minded, and pioneering Italian school of mathematics was swiftly changing into an academic and provincial one.

---

[44] "Non ho abbandonato l'idea ... di pubblicare ogni anno nell'Annuario ... un articolo di un maestro della scienza su di un argomento che possa interessare anche coloro che non hanno che una modesta cultura scientifica. Ciò dovrebbe ... dare un'idea al grande pubblico delle grandi conquiste della scienza in questi ultimi tempi, ed aumentare il numero degli amici della matematica. Il legame tra scienza pura e scienza applicata sfugge al grande pubblico e talvolta anche agli studiosi. Sarebbe quindi utile ... fare bene risaltare (indirizzandosi al grande pubblico) tutto ciò che proviene a pieno diritto dalla matematica superiore, nelle grandi scoperte moderne di ordine pratico." Letter from Guccia to Henri Poincaré, 19 December, 1905, ACMP.

The last, and perhaps heaviest, blow to the vitality of the *Circolo* derived from the decline in the international atmosphere. After the war, there was no room left for an *independent* international mathematical association. Nevertheless, Guccia's successor, Michele De Franchis, made great and partially successful efforts to preserve the noble international traditions of the *Circolo*. At a time when all international scientific organizations (including the 1920 ICM in Strasbourg) had expelled their German associates, De Franchis succeeded in preserving the composition of the editorial board of the *Rendiconti*. A Sicilian may be justifiably proud to observe that during the years immediately following the First World War, the *Circolo* was the only scientific organization in the world where one could see such names as Hilbert, Klein, Noether, Landau, Picard, de la Vallée Poussin, Hadamard, Borel, and Lebesgue side by side.

De Franchis firmly opposed the efforts of Picard and de la Vallée Poussin to obtain the expulsion of German members from the *Circolo*.[45] In 1919, it seemed probable that if he had not accepted Picard's dictate, the *Circolo* would have been forced to dissolve. His fierce answer reveals the fervor with which he guarded the internationalism that he and Guccia had worked so hard to foster:

> I think that, just as I would not be held responsible for any detestable undertaking of a minority of factious people influenced by false mirages, so I cannot attribute to an entire people, and above all to a distinguished group of scholars, all the weight of the hideous crimes performed by German imperialism (and which any imperialism, driven to this extreme, by a group of interested people could accomplish as well). In my opinion, the main enemy of humanity is imperialism, which these days is completely anachronistic. ... I cannot agree ... that it is right to expel Hilbert ... just because Hilbert was born in Germany. ... There is no doubt that the divisions arising among scientists according to their nationality will remove any international character from science for a long time. ... In a few years scientific collaboration will surely be renewed, but in the meantime our society will die.[46]

He continued these thoughts a few months later. "The *Circolo* does not make and will never make any racial or national distinction between mathematicians," De

[45]For more on Picard's position, see the chapters by Sanford Segal and Olli Lehto below.

[46]"Io ritengo che, come non vorrei essere tenuto responsabile di un'impresa abominevole eventualmente condotta da una minoranza di faziosi col concorso del popolo suggestionato da falsi miraggi, così non posso addossare su tutto un popolo e tanto meno su una eletta schiera di studiosi il peso degli orribili misfatti dei quali si è macchiato l'imperialismo germanico e dei quali, tra parentesi, sarebbe capace di macchiarsi qualsiasi altro imperialismo spinto alle ultime conseguenze da un manipolo di persone interessate. Per me, il principale nemico dell'umanità è l'imperialismo il quale costituisce ai nostri tempi un anacronismo. ... Non posso convenire ... che sia giusto radiare un Hilbert ... solo perché Hilbert è nato in terra germanica. ... È fuori di dubbio che le distinzioni che si vogliono fare tra gli scienziati a seconda del paese di origine toglieranno per lungo tempo alla scienza il carattere internazionale ... E badi che ... tra qualche anno la collaborazione scientifica è fatale che si riattivi, ma intanto la nostra società sarà morta." Letter from Michele De Franchis to Luigi Bianchi, 1 March, 1919, ACMP.

Franchis declared in a letter to Segre.[47] This staunch position made a deep impression on many mathematicians, above all the Germans. Landau, Weyl, and Rothe[48] but also Hadamard, Lefschetz, and Birkhoff, among many others, expressed their deep feelings of pride in the *Circolo*'s moral conviction. This allowed the *Circolo* to regain and retain a prominent position among the mathematical associations of the world for many years. Even though its journal had lost its premier scientific position, the *Circolo* remained a great symbol of the internationality of science and was highly valued by many mathematicians.

By 1934, the *Circolo*'s editorial board was again representative of the most prominent mathematical communities in the world.[49] It is worth noting that, by this time, Volterra had already refused the prescribed fidelity oath to the fascist regime and had consequently been excluded from university teaching and from the *Accademia dei Lincei*. In the meantime, Landau and Courant (who, like Volterra, were Jewish) were about to lose their respective positions at Göttingen. De Franchis had preserved well the *Circolo*'s tradition of total independence.

Obviously, this kind of independence was totally unacceptable to Mussolini's fascist regime. During the thirties, and above all after the invasion of Ethiopia in 1934, the dictatorship could not accept any kind of independent association. Internationalism was synonymous with anti-patriotic activity. Indeed, even the prestigious *Accademia dei Lincei* was viewed with suspicion by the authorities and the parallel *Accademia d'Italia* was founded. The major blow to the *Circolo* came directly from the government with the imposition (by "*regio decreto*") in 1934 of a new statute, rescinding its international status. According to the old statutes, every modification had to be approved by a two-thirds majority of all members (Italian or otherwise). Indeed, the *Circolo* had been proud of its two-thirds foreign membership. The new statute imposed the restriction that "the number of foreign members of the *Circolo* cannot be greater than half the number of Italian

---

[47] "Il Circolo non fa e non farà mai distinzioni di nazionalità o di razze tra i matematici." Letter from Michele De Franchis to Corrado Segre, 7 October, 1919, ACMP.

[48] Rudolf Rothe wrote to De Franchis that "I must confess sincerely that I was hesitant for a long time, because in the *Circolo Matematico* there are still members like E. Picard, Ch. De La Vallée Poussin and others, who even now are showing a deep hatred for anything German .... But now I want to express my greetings for your kindness and my hope that the *Circolo Matematico* may be successful in its efforts to rebuild the former peaceful relations among mathematicians. [Le confesso francamente che sono stato per molto tempo in dubbio al riguardo perché del *Circolo Matematico* sono ancora soci persone come E. Picard, Ch. De La Vallée Poussin ed altri, che fino al momento attuale hanno manifestato il loro odio contro tutto ciò che è tedesco .... Però voglio manifestarle i miei più sentiti ringraziamenti per la sua gentilezza ed esprimere la speranza che il *Circolo Matematico* possa riuscire mediante i suoi sforzi a ripristinare le antiche pacifiche relazioni tra matematici.]" Letter from Rudolf Rothe to De Franchis, 1921, (day and month illegible), ACMP.

[49] The list is highly prestigious: Bertini from Pisa; Francesco Paolo Cantelli, Castelnuovo, Enriques, Levi-Civita, Volterra, and Severi from Rome; Guido Fubini and Carlo Somigliana from Turin; Roberto Marcolongo, Ernesto Pascal, Gaetano Scorza, and Mauro Picone from Naples; Pincherle and Leonida Tonelli from Bologna; Giuseppe Vivanti from Milan; Borel, Hadamard, and Picard from Paris; G. Valiron from Strasbourg; Hilbert, Richard Courant, Weyl, and Landau from Göttingen; Constantine Carathéodory from Münich; Leon Lichtenstein from Leipzig; Hardy and Augustus Love from Oxford; John Littlewood from Cambridge; Harald Bohr and Niels Nordlund from Copenhagen; José Maria Plana from Madrid; Charles de la Vallée Poussin from Louvain; Leopold Fejer from Budapest; Wilhelm Wirtinger from Vienna; George David Birkhoff from Cambridge, MA; Eliakim Hastings Moore and Leonard Eugene Dickson from Chicago; and Lefschetz from Princeton.

members."[50] The association was no longer free; instead, members were chosen by the presidency. The editor and the editorial board were no longer elected by the membership but were nominated by the government. In short, the *Circolo*, as envisaged by Guccia, was no more.

The racial laws of 1938 dealt the final blow to the *Circolo*, deeply affecting Italian mathematics. Eminent mathematicians like Volterra, Castelnuovo, Enriques, and Levi-Civita were completely excluded from Italian scientific life and consequently from the editorial board of the *Circolo*. The end finally came during the Second World War, when the society was formally dissolved.

After the War, around 1951–1952 and thanks to the efforts of many eminent mathematicians like Garrett Birkhoff, Hadamard, Lefschetz, John von Neumann, and many others, the *Circolo* was re-founded under the guidance of Edoardo Gugino. It still exists as an international association, publishing a widely circulated journal, the *Rendiconti*, but it has never regained its former position in the international arena.

Conclusions

A complete evaluation of the *Circolo*'s historical role remains elusive. It initially emerged as a national mathematical society of the same kind as the London Mathematical Society. But it must be remembered that, in great contrast to the LMS, it developed in an environment far from the main Italian mathematical centers, and, indeed, it was never completely recognized by the Italian mathematical community as its national mathematical society.

Guccia had deeply held ideological beliefs and was imbued with the idea that the community of scholars should in no way be limited to distinct national venues. From the beginning, he conceived of the *Circolo* as a great international society; the main attributes he directed toward this goal were his many personal links with eminent mathematicians (above all Poincaré) and his indubitable organizational capacity. Another factor in his success, and in that of the *Circolo*, was the unlikelihood (for political reasons) of establishing an international center in France or Germany. In this sense, the role played by the *Circolo* and by its *Rendiconti* is not very different from that of journals like *L'Enseignement mathématique* and *Acta Mathematica*, which, as remarked above, came from the neutral nations of Switzerland and Sweden, respectively.

Guccia was successful in building a great international society and, above all, a very important international journal. Nevertheless, the feeble links with the Italian mathematical community, the peripheral position of Palermo, the decline of its mathematical school (many young Sicilian mathematicians left Palermo during the 1920s and 1930s), the new political situation after World War I, and the growth of Italian nationalism were all conditions adverse to its consolidation. In the difficult years immediately following the Second World War, the *Circolo* played no decisive role in the international mathematical community, although in the early 1950s, it briefly symbolized internationality in mathematics. In some sense, the building of the *Circolo* into a truly international society for mathematicians was never fully accomplished.

---

[50] *Rendiconti del Circolo matematico di Palermo* **59** (1935):1-2.

References

Boi, Luciano; Giacardi, Livia; and Tazzioli, Rossana, Ed. *La découverte de la géométrie non euclidienne sur la pseudosphere*. Paris: Blanchard, 1993.

Borel, Émile. "Les probabilités dénombrables et leurs applications arithmétiques." *Rendiconti del Circolo matematico di Palermo* **27** (1909):247-271.

Bottazzini, Umberto. *Va' Pensiero: Immagini della matematica nell'Italia dell'Ottocento*. Bologna: Il Mulino, 1994.

Brigaglia, Aldo. "Il Circolo matematico di Palermo." *Symposia Mathematica* **27** (1986):265-285.

———. "The Circolo matematico di Palermo and its *Rendiconti*: The Contribution of the Italian Mathematical Community to the Diffusion of International Mathematical Journals 1884–1914." In *Messengers of Mathematics: European Mathematical Journals (1800–1946)*. Ed. Mariano Hormigón and Elena Ausejo. Zaragoza: Siglo XXI de España Editores, 1993, pp. 71-93.

———. "La matematica italiana dell'inizio del secolo e le sue proiezioni all'esterno." In *Accademia dei XL: Scritti di storia della scienza in onore di Giovanni Battista Marini-Bettolo*. 1990, pp. 393-407.

Brigaglia, Aldo and Masotto, Guido. *Il Circolo matematico di Palermo*. Bari: Dedalo, 1982.

Chasles, Michel. *Rapport sur les progrès de la géométrie*. Paris: Imprimerie national, 1870.

Collingwood, Edward F. "A Century of the London Mathematical Society." *Journal of the London Mathematical Society* **41** (1966):577-594.

De Franchis, Michele. "Giovan Battista Guccia." *Rendiconti del Circolo matematico di Palermo* **39** (1914):1-10.

De Masi, Domenico. *L'emozione e la regola: I gruppi creativi in Europa, dal 1850 al 1990*. Bari: Laterza, 1990.

Fréchet, Maurice. "Sur quelques points du calcul fonctionnel." *Rendiconti del Circolo matematico di Palermo* **22** (1906):1-74.

Gatto, Romano, Ed. "Lettere di Luigi Cremona a Enrico Betti (1860–1890)." In *La corrispondenza di Luigi Cremona (1830–1903)*. Ed. Marta Menghini. Palermo: Quaderni PRISTEM, 1996, pp. 7-90.

Gispert, Hélène. "Lettres de G. Darboux à J. Hoüel." *Cahiers du Séminaire d'histoire des mathématiques* **8** (1987):67-202.

Landau, Edmund. "Discorso del prof. Dott. Edmund Landau della reale Università di Gottinga." *Supplemento ai Rendiconti del Circolo matematico di Palermo* **9** (1914):12-13.

———. "Über die Multiplikation Dirichlet'schen Reihen." *Rendiconti del Circolo matematico di Palermo* **24** (1907):81-160.

Levi-Civita, Tullio. "Nozione di parallelismo su di una varietà qualunque e conseguente specificazione geometrica della curvatura riemanniana." *Rendiconti del Circolo matematico di Palermo* **42** (1917):173-205.

———. "Scie e leggi di resistenza." *Rendiconti del Circolo matematico di Palermo* **23** (1907):1-37.

———. "Sopra un problema di elettrostatica che si è presentato nella costruzione dei cavi." *Rendiconti del Circolo matematico di Palermo* **20** (1905):173-228.

Nastasi, Pietro, Ed. "Documenti della vita del Circolo matematico di Palermo." *Supplemento ai Rendiconti del Circolo matematico di Palermo* (2) **20** (1988).

Pillitteri, Antonio. "I rapporti tra Giovan Battista Guccia e Gösta Mittag-Leffler." University of Palermo. Unpublished dissertation, 1991.

Poincaré, Henri. "5e complément à l'analysis situs." *Rendiconti del Circolo matematico di Palermo* **18** (1904):45-110.

———. "Complément à l'analysis situs." *Rendiconti del Circolo matematico di Palermo* **13** (1899):285-343.

———. "Nouvelles remarques sur les groupes continus." *Rendiconti del Circolo matematico di Palermo* **25** (1908):81-130.

———. "Sur la diffraction des ondes hertiennes." *Rendiconti del Circolo matematico di Palermo* **29** (1910):169-260.

———. "Sur la dynamique des électrons." *Rendiconti del Circolo matematico di Palermo* **21** (1906):129-176.

———. "Sur un théorème de géométrie." *Rendiconti del Circolo matematico di Palermo* **33** (1912):375-407.

Veronese, Giuseppe. "Commemorazione del Socio Luigi Cremona." *Atti della reale Accademia dei Lincei* (5) **12** (1903):664-678.

Vitali, Giuseppe. "Sulle funzioni a integrale nullo." *Rendiconti del Circolo matematico di Palermo* **20** (1905):136-141.

Volterra, Vito. "Betti, Brioschi, Casorati: Tre analisti e tre modi di considerare le questioni di analisi." *Compte rendu du deuxième Congrès international des mathématiciens.* Paris: Gauthier-Villars, 1902, pp. 43-57. In Vito Volterra. *Saggi scientifici.* Bologna: Zanichelli, 1990, pp. 35-54.

Weyl, Hermann. "Über beschränkte quadratische Formen, deren Differenz, vollstetig ist." *Rendiconti del Circolo matematico di Palermo* **27** (1909):373-392.

CHAPTER 11

Languages for Mathematics and the Language of Mathematics in a World of Nations

Jeremy J. Gray
Open University (United Kingdom)

Introduction

This chapter focuses on some of the ways in which mathematics was considered a language around 1900. It first explores aspects of the creation of an international language for mathematics within the context of the wider political movements for international languages in general in order to highlight the *multi*national, in contradistinction to the *inter*national, aspects of the mathematical community.[1] Then, a more properly linguistic perspective provides a new view on early-twentieth-century debates about the foundations of mathematics. Positions on the correct relationship between thought, language, logic, mathematics, and the world ranged widely and were sometimes held with an intensity that can only be explained on historical and political grounds. The simultaneous searches for a truly international language and for an appropriate language and foundation for mathematics not only involved several of the same people but also raised difficult questions of syntax, semantics, and the relation between them. Finally, reflections on Hilbert's philosophy of mathematics suggest a number of reasons why it is possible actually to speak of a distrust of language at the time. Shifting the emphasis from logic to language underscores confusions about difficult linguistic issues and deepens our understanding of the old debate about the relationship of logic to mathematics. By situating that debate in its fuller historical context of debates about the nature of language—natural, artificial, and mathematical—it becomes evident that much more was at stake than the nature of proof and of axiomatic reasoning. Intellectually, and often politically, the issue was one of communication and, therefore, of community in mathematics.

A World of Nations

It is commonly, and rightly, said that the years around 1900 saw the emergence of an international mathematical community, marked by the first International Congresses of Mathematicians (ICM)[2] and the emergence of internationally minded mathematical research communities in many countries, for example, the

[1]On the distinctions between "international" and "multinational," recall the discussion in Parshall and Rice's chapter above.

[2]June Barrow-Green, "International Congresses of Mathematicians from Zürich 1897 to Cambridge 1912," *The Mathematical Intelligencer* **16** (2) (1994):38-41.

United States as carefully and well studied by Karen Parshall and David Rowe.[3] Journals too acquired a transnational flavor; some, like *Acta Mathematica*, were set up to reach at least two different countries,[4] while others deliberately set out to bring foreign work to a native audience. However, as has often been noted,[5] it is more accurate to say that around 1900, there were several national communities aware that they could, and even should, compete on an international stage. Truly international cooperation was rare. The German delegation to the World's Fair in Chicago in 1893 was designed to show how Germany led the way. The first International Congress of Mathematicians was held in Zürich in 1897 because Switzerland was neutral territory, equally acceptable to French and Germans who might not so easily have allowed the other country to host the first such event—the Franco-Prussian War of 1870–1871, with the loss of Alsace and Lorraine to Germany, was a vivid memory at the time.

As these examples suggest, it was a world of nations, not an international world. Moreover, the number of nations was small and the national idea was often fragile. Germany, like Italy, had only been unified in the 1860s. The Union of the United States, "one nation under God," had also been re-forged in battle in the same decade,[6] and the extension of the country to the West, "from sea to shining sea," was still being consolidated.[7] The tensions arising from the creation of the nation-state in Europe—and their implications for the idea of "Europe"—were visible, for example, in many a national mathematical society. It took the Germans until 1890 to found their *Deutsche Mathematiker-Vereinigung*,[8] and rivalries between northern and southern Germany,[9] and between Göttingen and Berlin,[10] dominated academic politics there for many years. In America, the passage from the New York Mathematical Society to the American Mathematical Society, although more amicable, highlighted similar problems of regionalism.[11]

---

[3]Karen Hunger Parshall and David E. Rowe, *The Emergence of the American Mathematical Research Community, 1876-1900: J. J. Sylvester, Felix Klein, and E. H. Moore*, HMATH, vol. 8 (Providence: American Mathematical Society and London: London Mathematical Society, 1994).

[4]Yngve Domar, "On the Foundation of *Acta Mathematica*," *Acta Mathematica* **148** (1982):3-8; David E. Rowe, "Klein, Mittag-Leffler, and the Klein-Poincaré Correspondence of 1881-1882," in *Amphora: Festschrift für Hans Wussing zu seinem 65. Geburtstag*, ed. Menso Folkerts *et al.* (Boston-Basel-Berlin: Birkhäuser Verlag, 1992), pp. 597-618; Philippe Nabonnand, "The Poincaré–Mittag-Leffler Relationship," *The Mathematical Intelligencer* **21** (2) (1999):58-63; and June Barrow-Green's chapter in the present volume. The original idea of a journal to showcase Scandinavian mathematics was due to Lie, but Mittag-Leffler took it over and worked hard to ensure the journal's success in France and Germany.

[5]See, for example, Aldo Brigaglia's chapter in the present volume.

[6]This phrase from the United States's oath of allegiance echoes a phrase from Abraham Lincoln's Gettysburg Address.

[7]This phrase comes from Katherine Lee Bates's lyrics to the song, "America the Beautiful."

[8]See Hélène Gispert and Renata Tobies, "A Comparative Study of the French and German Mathematical Societies before 1914," in *L'Europe mathématique–Mathematical Europe*, ed. Catherine Goldstein, Jeremy J. Gray, and Jim Ritter (Paris: Éditions de la Maison des sciences de l'homme, 1996), pp. 407-430 and the references cited there.

[9]Gert Schubring, "Changing Cultural and Epistemological Views on Mathematics and Different Institutional Contexts in Nineteenth-century Europe," in *op. cit.*, pp. 361-388, especially on pp. 372-376.

[10]David E. Rowe, "Episodes in the Berlin-Göttingen Rivalry," *The Mathematical Intelligencer* **22** (1) (2000):60-69.

[11]Parshall and Rowe, pp. 401-411.

Despite these internal political tensions, individual mathematicians traveled increasingly as their world expanded. Some even emigrated for years to study. First Paris, then Berlin, and then Göttingen attracted foreigners from several nations, and depending on the vagaries of their home economies, some remained abroad.[12]

National Languages

In this world of nations, the novel twist was that by 1900 there was an international arena for conferences and societies and advanced education in mathematics. Many different languages were spoken in this arena, so it is interesting to ask: in what language or languages did mathematicians seek to speak to each other?

Plainly, the mathematician who went to Germany had to learn German. He, and in the German case sometimes she, conversed in it, heard lectures in it, read it, and in due course very often wrote in it. Such was the high, indeed dominant, status of German mathematics that it probably brought one a larger audience to publish in German than English, and Americans such as Mellen Haskell, Edward Burr Van Vleck, and William Fogg Osgood did.[13] Osgood was such a Germanophile that, even though he taught at Harvard, he published his treatise on complex function theory in German, something that had not been the practice of others, like James Harkness and Frank Morley.[14] That the French wrote in French goes without saying. People from nations where it might be conceded that the national tongue was not well-known abroad had a difficult problem. Growing national pride might require that one write in the language of one's birth, and then perhaps translate one's work into French or German. This was the case for Scandinavians, such as Lie and Mittag-Leffler, and for Russians.[15] Somewhat later, in the Czech region or Poland, the creation of journals publishing in the native language, and the decision to publish in it, were political acts.[16]

Between 1850 and 1914, French and German mathematical journals dominated the field, with *Acta Mathematica* and the best of the Italian, British, and American journals following along behind. Faced with this situation, mathematicians either tried to write in French for French journals and in German for German ones, or they hoped that the journals would accommodate English or Italian. French or German mathematicians, like English and American ones today, did not have these same worries. In the former two nations, the bulk of mathematicians were monoglots who

---

[12]Thomas Archibald and June Barrow-Green have considered this process for the British Empire. See Thomas Archibald and Louis Charbonneau, "Mathematics in Canada before 1945: A Preliminary Survey," in *Canadian Mathematical Society/Société mathématique du Canada 1945-1995: Mathematics in Canada/ Les mathématiques au Canada*, ed. Peter Fillmore, 2 vols. (Ottawa: Canadian Mathematical Society, 1995), 1:1-44, and June Barrow-Green (private communication). Parshall and Rowe have looked at the German-American case in rich detail.

[13]Mellen W. Haskell, "Über die zu der Curve $\lambda^3\mu + \mu^3\nu + \nu^3\lambda = 0$ im projektiven Sinne gehörende mehrfache Überdeckung der Ebene," *American Journal of Mathematics* **13** (1890):1-52; Edward Burr Van Vleck, "Zur Kettenbruchentwicklung hyperelliptischer und ähnlicher Integrale," *American Journal of Mathematics* **16** (1894):1-91; and William F. Osgood, *Lehrbuch der Funktionentheorie*, 2 vols. (Leipzig: B. G. Teubner Verlag, 1907–1924; reprint ed., New York: Chelsea Publishing Co., 1965).

[14]James Harkness and Frank Morley, *A Treatise on the Theory of Functions*, (New York: Stechert, 1893).

[15]Compare June Barrow-Green's chapter in the present volume on Sonya Kovalevskaya's translation efforts on behalf of some of her Russian colleagues.

[16]Roman Duda, "*Fundamenta Mathematica* and the Warsaw School of Mathematics," in *L'Europe mathématiques–Mathematical Europe*, pp. 479-498.

seldom wrote outside their own linguistic boundary. Thus, when Hilbert published in the *Proceedings of the London Mathematical Society*, he did so in German.[17] Interestingly, Felix Klein, the most ardent advocate of Germany's leading role in the mathematical world, chose to publish in English when addressing an American or English audience.[18]

The importance of translations is equally clear. In France, this torch was carried by Joseph Liouville and later by Guillaume Jules Hoüel.[19] It is quite striking how much farther the American mathematical community went than the British in this direction. It is well-known that Hilbert's twenty-three problems, presented in 1900 to the Paris ICM, were rapidly published in the *Bulletin of the American Mathematical Society* (and also in a French translation in the *Comptes rendus* of the Congress),[20] and the English translations of Poincaré's famous books of essays are largely, if not entirely, American in origin. The American Mathematical Society was not the only body active in this way; Paul Carus's journal, *The Monist*, was also an important source. It was not just a matter of translations; summaries, reports, digests, and commentaries on European work fill the pages of the *Bulletin of the American Mathematical Society* until at least 1914.

All this is more or less predictable, although some of the disparities and lack of uniformity are interesting. There were four languages for international use in mathematics around 1900. French and German because of the indigenous strength of the mathematics profession in those countries, English because of the strength of the British Empire and the growing power of the United States, and perhaps Italian, again, because there were so many good Italian mathematicians. The use of any other language raises questions about the intended communication of the paper or book and about the situation of the corresponding indigenous mathematical community. Yet, even the use of one of these four accepted languages had the potential to highlight interesting extra-mathematical issues. According to Ernst Schröder[21] in his address to the first ICM in Zürich in 1897, for example, there had been talk of making English the official language of the Congress because it was neutral ground between French and German.[22] This incident underscored the fact that there was no recognized *lingua franca* in which mathematicians communicated.

---

[17]David Hilbert, "Über den Satz von der Gleichheit der Basiswinkel im gleichschenkligen Dreieck," *Proceedings of the London Mathematical Society* **35** (1902):50-68.

[18]Felix Klein, *Klein's Evanston Colloquium Lectures and Other Works*, ed. David E. Rowe and Jeremy J. Gray (New York: Springer-Verlag), to appear.

[19]On Liouville's efforts to generate translations into French, compare Jesper Lützen's chapter in the present volume. In his chapter, Ivor Grattan-Guinness also considered the role of translations, but from French into other languages.

[20]David Hilbert, "Sur les problèmes futurs des mathématiques," trans. L. Laugel, *Compte rendu du deuxième Congrès international des mathématiciens* (Paris: Gauthier-Villars, 1902), pp. 58-114, and David Hilbert, "Mathematical Problems," *Bulletin of the American Mathematical Society* **8** (1902):437-479; reprint ed., "Mathematical Problems," in *Mathematical Developments Arising from Hilbert Problems*, Proceedings of Symposia in Pure Mathematics, 2 vols. (Providence: American Mathematical Society, 1976), 1:1-34.

[21]Ernst Schröder, "Über Pasigraphie, ihren gegenwärtigen Stand und die pasigraphische Bewegung in Italie," in *Verhandlungen des ersten Internationalen Mathematiker-Kongresses vom 9. bis 11. August, 1897*, (Leipzig: B. G. Teubner Verlag, 1898), and Ernst Schröder, "On Pasigraphy: Its Present State and the Pasigraphic Movement in Italy," *The Monist* **9** (1899):44-62 (Corrigenda, p. 320) on p. 44.

[22] In the event, only ten English speakers attended, and many more British and American mathematicians instead went to the meeting of the British Association for the Advancement of Science held concurrently in Toronto.

International Languages in General

A much less predictable development was the creation of one, and indeed several rival, *linguæ francæ*. Moreover, there were vigorous groups that aimed to create or adopt one or another international language for mathematical, or even all scientific, work. The opposition to such schemes came in various forms. There is little to say about speakers of any hegemonic language who wished their language to prevail. Many such people thought that the sensible thing to do was to keep quiet and let things carry on as they were. In 1900, the majority position was that there were a number of mainstream languages, and one had to learn one, or two, or three. Since there were other cultural advantages as well as a number of educational opportunities in doing so, there was no reason to challenge the *status quo*. The idea that the entire world should learn either French or English was explicitly advocated by some, including the linguist, Michel Bréal, one of the inventors of semantics.[23]

Of greater interest is the minority position, namely, that there should be a new language created specifically for the purpose of enhancing international communication. People who advocated this sometimes did so for mundane reasons, sometimes animated by a loftier goal. Once, there had been Latin. Scholars and clerics from all over Europe could read each other's work and could converse when they met. The part of Paris around the university was and is called the Latin Quarter (precisely, the *Quartier Latin*) for that very reason. But first the French and then the British broke with that. By the mid-nineteenth century, some wanted to return to the idea of a universal means of communication, motivated by the profusion of tongues spoken on the international stage. The loftier goal was that an international language would help bring about a truly international world, one in which national rivalries and the dangers of war were diminished. Two appalling World Wars later, one should not be blind to the momentum such efforts acquired in their day.

Books have been written on the history of these artificial languages,[24] and it is very easy to find the whole enterprise rather comic, as most authors do. There is something naïve about believing that immense benefits would follow if all people spoke the same language. When this is coupled with the immense egotism of believing that it is one's own language that will bring this about, the net effect is risible. There is also something rather odd about some of the languages themselves, as will soon become clear.

The language issue is worth considering in its own right. English abolished gender centuries ago, and it dispensed almost completely with declensions in nouns. Its verbs are largely German (English is a Germanic language) but simplified. So far, so good, but, as every non-native speaker and many an English or American child complains, the gap between spelling and pronunciation in English is unpleasantly large, even without entering the realm of dialects. French, on the other hand, is laden with grammatical issues English is not but manages to be more nearly phonetic. Latin, itself a much more inflected language, has the virtue of allowing a very flexible word order, but this makes it very difficult to learn. In the common opinion of many in the nineteenth century, mastering four conjugations and five

[23]Michel Bréal, "Le choix d'une langue internationale," *Revue de Paris* (4) **8** (1901):229-246.

[24]See, for example, Albert L. Guérard, *A Short History of the International Language Movement* (London: Fisher Unwin, 1922), and Umberto Eco, *The Search for a Perfect Language*, trans. James Fentress (London: Fontana, 1995), together with their associated bibliographies.

FIGURE 1. The Rue des Nations at the Paris Exposition of 1900

declensions, with a few irregular cases, was harder than learning to spell word by word with very few rules to guide.

This profusion of possibilities arose for good reasons that have to do more with the nature of language and less with the concerns of school grammarians. For one, speech patterns change over time (to give one example, there is something called the great vowel shift in English, still audible in popular songs, when "die" no longer rhymes with "be"). Also, one language may draw an extensive part of its vocabulary from other languages. English, for example, borrowed vocabulary from French and Latin and many verbs from German; such words have spellings that reflect their origins. Attempts to override these linguistic facts artificially confront, or perhaps fail to confront, more complicated problems that are apparent at first sight.

The issue of pronunciation is also important. One knows that Church Latin is and was spoken with large regional variations. One of the aims of the proponents of the new international languages was that scientific papers could be read aloud, understood, and the speaker questioned—all in the new language. This aim was more difficult to realize than the production of an international written language. Thus, a universally intelligible spoken language—a pasilalia—was much harder to attain than a universal written language or pasigraphy.

International Languages in Particular

Moves to bring about an international language for science were under way by 1900. The International Congress of Philosophy met in Paris 2-7 August, 1900. Louis Couturat was one of the organizers, and he invited Gottlob Frege to join the Organizing Committee. Frege agreed, and a correspondence ensued in which

Couturat broached the topic of an international auxiliary language. Frege seems to have been receptive.[25] To judge by Couturat's reply, Frege could support a scientific, commercial, and utilitarian language, but worried about the creation of neologisms.

The International Congress of Mathematicians, which met immediately after the Congress of Philosophy, heard a paper by Charles Méray of France on the merits of Esperanto.[26] He observed that a previous international language, Volapük, had died but observed that the real difficulties even highly educated professors had in understanding one another at the Congress surely suggested that Zamenhof's simpler creation, Esperanto, had some claim on everyone's attention. The paper led to a discussion of a resolution proposed by Léopold Leau to adopt a universal scientific and commercial language to be realized by the international academies. Couturat, Leau, Charles Laisant, and Alessandro Padoa were among those in favor, Schröder and the Russian Aleksandr Vasil'ev[27] against. In the end, it was Vasil'ev's counterproposal that "the academies and learned societies of all countries study the proper means for remedying the evils that arise from the increasing variety of languages used in the scientific literature" that was carried.[28]

The case of Volapük illustrates many of the issues under examination here. It was the creation of a German Roman Catholic priest, Father Schleyer, who presented it to the world in 1880. It was what Couturat and Leau, in their major book (of over 600 pages) on international languages,[29] classified as an *a posteriori* language because it was modeled on existing languages. Despite attempts to choose the vocabulary so that Europeans would find much of it familiar, it succeeded in looking strange because the grammatical forms gave it a markedly Slavonic feel, and its vowel sounds were nothing like English. (Who would guess that its very name means "world-speak"?) But its grammar was at least regular, its pronunciation standardized. By 1889, its adherents claimed it was spoken by upwards of 200,000 people, yet in 1890, it was dying, torn apart by the failure of its creator to let its speakers modify it.

Esperanto was the creation of a Russian Jew from Bialystock, Louis Lazarus Zamenhof, who hoped his language would bring about the reconciliation of races and nations only too obviously hostile to each other in his native region. He presented it to the world in 1887 as "*La Lingvo Internacia de la Doktoro Esperanto* [The International Language of Doctor Hopeful]," and his pseudonym happily became

---

[25]His replies are apparently lost. See Gottlob Frege, *Philosophical and Mathematical Correspondence* (Chicago: University Press, 1980).

[26]Charles Méray, "Sur la langue internationale auxiliaire de M. de Zamenhof, connue sur le nom d'Esperanto," in *Compte rendu du deuxième Congrès international des mathématiciens* (Paris: Gauthier-Villars, 1902), pp. 429-432. The paper was actually read by Léopold Leau.

[27]Vasil'ev, a Russian mathematician at Kasan, had been influential in establishing the Lobachevskii prize, first awarded in 1897 to Sophus Lie. He went on to write a book on relativity theory that was translated into English and published with a preface by Bertrand Russell. See Aleksandr Vasil'ev, *Space Time Motion: An Historical Introduction to the General Theory of Relativity* (London: Chatto and Windus, 1924). His philosophy of mathematics is described in Alexander Vucinich, "Mathematics and Dialectics in the Soviet Union: The Pre-Stalin Period," *Historia Mathematica* **26** (1999):107-124.

[28]See Hubert C. Kennedy, *Peano: Life and Works of Giuseppe Peano* (Dordrecht and Boston: D. Reidel Publishing Co., 1980), p. 98.

[29]Louis Couturat and Léopold Leau, *Histoire de la langue universelle* (Paris: Hachette, 1903). The information on Volapük and Esperanto presented here is drawn largely from this source.

attached to the language. Unlike Volapük, its written form is more recognizable and more euphonious, being close to French and Spanish, but like its predecessor, it has standardized declensions and conjugations (Zamenhof was inspired in this respect by English but did not go all the way). Esperanto began to prosper in 1898 when the French took it up.

Six months of international conferences in Paris in 1900 passed without international languages being a main agenda item for any (there were congresses of Volapük, and international languages were mentioned in various congresses), but in 1901, the Delegation for the Adoption of an International Language was organized, the Touring Club of France (membership over 100,000) gave Esperanto their support, and Hachette began to publish Esperantist literature. Large international conferences followed, and it seems that converts found it easy enough to learn, although Esperanto was easier to write and read than to speak. A large indigenous literature was created, to which were added translations of many great works of literature.

The many adherents to the cause were a blessing and a difficulty. The difficulty was fueled by the great moral claims made for it and provoked by the awkwardnesses of the original creation. Of Zamenhof's original choices, the use of "-*j*" for plurals and "-*n*" for the accusative are perhaps the most unfortunate. The language was also not sufficiently *a posteriori*, and so it left room for alternatives. Of these, the one that seemed most obvious was simplified Latin. A strong early contender was one called Idiom Neutral, first presented in 1903 and significantly revised in 1907. Another was Latino sine Flexione, the creation of Giuseppe Peano.[30] He based it on a rigorous attempt to identify the basic vocabulary of all European languages and gave it a simplified grammar that resembled Latin but sounded like Italian. He promoted it in his mathematical journals as well as in others of a more general character and, after 1908, through the formerly Volapükist *Akademi de lingu universal*, renamed the *Academia pro interlingua*. His was a well-thought-out creation. Romance speakers could be expected to acquire fluency, and the rest of the international professoriate had almost certainly studied Latin for some years at school. They might reasonably be allowed to forget most of the word endings and even be happy to use the word order of English or French rather than chase around finding out what agreed with what. Spelling was reasonably phonetic, vocabulary often English. Calling the resulting hybrid "Latin without inflections" avoided the problems of nationalism that would have arisen if, instead, Peano had called it "standard Italian."[31]

Whatever its intrinsic merits, Latino sine Flexione or Interlingua (as it also became called) could not have competed successfully against Esperanto had that popular tongue not fallen into schism. In circumstances too complicated to relate here, a special committee was established in 1907 to decide which international language should henceforth be actively promoted.[32] It drew in a number of prominent

---

[30] Giuseppe Peano, "De Latino sine flexione—lingua auxiliare internationale," *Rivista di matematica* **8** (1903):74-83, or Giuseppe Peano, *Opere scelte*, ed. Ugo Cassina, 3 vols. (Rome: Edizione Cremonese, 1958), 2:439-447. See also the thorough account in Kennedy.

[31] C. K. Ogden's Standard English, a systematic attempt to define a basic vocabulary and a basic set of grammatical forms, only succeeded within the bounds of the British Empire. These days, the job of promoting simplified English is done by Hollywood, quite possibly with the same self-imposed linguistic restrictions.

[32] See Guérard, pp. 145-160.

academic figures, among them, Peano as well as the distinguished chemist, Wilhelm Ostwald, and the professional linguist, Otto Jespersen. Zamenhof sent the Marquis de Beaufront to speak for Esperanto. For the occasion of the committee meetings, which took place at the *Collège de France* 15-24 October, 1907, Couturat and Leau produced a new book, *Les nouvelles langues internationales.*[33]

The decision of the committee was ambiguous. Some thought that it endorsed Esperanto, or at least a simplification of it proposed by the Marquis de Beaufront and called Ido. Others thought that it found no international language acceptable and proposed to start again with Ido. Esperantists, Zamenhof among them, could not agree because there deliberately was no official body capable of deciding changes in Esperanto. Peano was among those who denied that the committee could legislate as it wished. The new language, Ido, was surely an improvement on Esperanto, particularly in its plurals and in its limited use of the "-*n*" ending, but whether, as further polished by Couturat, it was significantly better was an issue thrown forever into the shade of the acrimonious dispute between the Esperantists and the Idoists. An international language can surely only succeed when it is unique in its claim on people's attention, a pacifist cause only when its leaders are not themselves at war. The quality of the arguments of Couturat, Jespersen, Ostwald, and others for Ido and against Esperanto pale by comparison with the disagreeable sectarian nature of their rhetoric.[34]

As this account makes clear, mathematicians, in common with many people at the time, were drawn into debates about the merits of an international language for their subject. Some hoped for a rich language one could speak, but their efforts foundered in repeated splits. Despite widespread recognition that it was difficult, and not always possible, to understand papers at conferences, and despite the feeling that as mathematicians were drawn from more and more countries this problem could only get worse, the mathematical community could not agree on what should be done. Mathematics might be crossing frontiers, and a sense of an international community might be spreading, but the fundamental national divisions occasioned by language remained.

A more modest enterprise did get under way based on the special character of mathematics; mathematics itself has been viewed as a language. The question was could, indeed should, there be a purified mathematical language that would not only obviate error but speak to all mathematicians equally? It is first to efforts to address this issue and then to the overlapping but distinct perspectives that logic and language afford of mathematics that we now turn.

Language, Meaning, and Mathematics: Calculation and Pasigraphy

In his address to the ICM in Zürich in 1897, Ernst Schröder came out strongly in favor of a pasigraphy, one almost entirely of his own devising.[35] The project, as he saw it, was the creation of an international scientific language. This involved first isolating the minimum number of primitive notions or categories that express

---

[33]Louis Couturat and Léopold Leau, *Les nouvelles langues internationales* (Paris: Hachette, 1907).

[34]I am indebted to one of the referees for the information that Christine Ladd-Franklin favored Ido and Rudolph Carnap Esperanto.

[35]As noted above, this address was published in English with minor modifications in 1899 in *The Monist.* On Schröder and pasigraphy, see Volker Peckhaus, "Ernst Schröder und die 'pasigraphische Systeme' von Peano und Peirce," *Modern Logic* **1** (1991):174-205.

all necessary concepts and then combining them using purely logical operations that would, in their turn, be handled using only the laws of ordinary logic. All this was to be carried out with absolute consistency and rigor in terms of easy signs and simple unambiguous symbols. In his view, such an approach would put the whole enterprise on a much higher plane than the merely linguistic aims of the Volapükists. His reasons are not clear, but he seems to have thought that the Volapükists aimed at supplementing natural languages, whereas his pasigraphy would supplant them for scientific purposes.

Schröder acknowledged that his pasigraphy could never be spoken, but he persisted in believing in its ultimate utility. He regarded mathematics as the branch of logic that deals with number. This being the case, he argued, Charles S. Peirce's logic of relatives had shown that five (and perhaps only four) notions were necessary: equality, set-theoretic intersection, negation, an operation he called conversion that inverts a relation (turning "parent of" into "child of" in his example), and relation in general. Schröder then gave his reasons for preferring Peirce's system (with his own slight modifications) to that of Peano and his school; despite the many laudable achievements of the Peanists, their system lacked what he viewed as the crucial category of relation in general. Schröder then demonstrated some of the things he believed his pasigraphy could establish, before indicating that the main problem facing the project was the development of a flexible calculus.

In passing, Schröder caustically dismissed Frege for taking immense pains to perform what had already been done much better and, in so doing, delivering a still-born child. Although Schröder did not say so, Frege's *Begriffschrift* was presumably meant here; Schröder had criticized it in a lengthy review for its neglect of George Boole and for its cumbersome notation.[36] As Schröder was by this point well-aware, Frege had already returned the compliment when he criticized the first volume of Schröder's *Algebra der Logik* for not making a sharp distinction between part-whole theory and logic, for not distinguishing (as he observed Peano did) between an element of a set and the subset consisting of just that element, and (typical Fregean themes here) for being mistaken about what definitions are and the relationship between concepts and names.

Schröder died in 1901, leaving behind what eventually became three large volumes of his *Algebra der Logik* and a number of expositors willing to try to put together what he had been saying but at much less length. His inspiration, C. S. Peirce, lived on until 1914, but not in circumstances that were designed to keep his work at the forefront of people's attention.[37] Peirce's and Schröder's ideas were eclipsed in many countries (America is a notable exception) by those of Russell and Hilbert. Recently, there has been much debate about how this happened,[38] but it is generally agreed that algebraic logic entered a fallow period and that Peano's formalism rose instead. Starting in 1903, with the fifth volume of his *Formulario*, Peano wrote mathematics in his pasigraphy and explained it in Interlingua. His

---

[36]Ernst Schröder, "Rezension von Freges Begriffschrift," *Zeitschrift für Mathematik und Physik, Hist.-Lit Abtheilung* **25** (1880):81-94, discussed in Hans Sluga, *Gottlob Frege* (London and New York: Routledge, 1980), pp. 68-76. I have only recently discovered that Sluga's book contains a considerable amount of information concerning Frege that overlaps with the arguments I develop independently here. I hope to return to this matter on another occasion.

[37]See Joseph Brent, *Charles Sanders Peirce: A Life* (Bloomington: Indiana University Press, 1993).

[38]See Irving Anellis, "Peirce Rustled, Russell Pierced: How Charles Peirce and Bertrand Russell Viewed Each Other's Work in Logic, etc.," *Modern Logic* **5** (1995):270-328.

pasigraphic language, rudely dismissed by critics, was a purely written language with a stunted vocabulary and grammar, in which all of the mathematical relations were expressed by symbols. This written script, which filled up the pages of several Italian journals, notably Peano's *Rivista di matematica*, was intended to be free of all the obscurities of a natural language.[39] It is not clear whether one was meant to think in it, or to translate one's papers into it, or simply to ensure that one's prose was translatable into it. Significantly, papers published in this language were often accompanied by translation into plain Italian. On any interpretation, it is clear that Peano and his followers distrusted natural languages and sought to replace them.

Is Mathematics a Language?

Attempts to provide a pasigraphy for mathematics, and with it an internationally comprehensible script, failed. They represented an early-twentieth-century version of the long-running debate about the quality of mathematical language and the desirability of improving it. They contributed to investigations of the relationship of mathematics to logic, the latter then entering its great revival. It is from the point of view of logic and its resurgence that pasiographic efforts are usually discussed; their linguistic aspect has been remarked upon less.

Mathematics has generally been esteemed for the quality of its reasoning. Many have attested to the apparently compelling quality of mathematical arguments, even when they lead to implausible results. The precision of mathematics has been attributed either to a precision of thought, or to a precision of the language in which that thought is expressed, or to various combinations of the two. Over the centuries, many writers have tried to grapple with the success of mathematics and to explain why there was certain knowledge and progress in mathematics which, by comparison, philosophy lacked. While Immanuel Kant attributed this difference to a special feature of mathematical activity, namely, the constructions by which mathematical objects were made known to the mind, other writers found it in the simplicity of those objects and the consequent limpidity in expressing ideas about them. For Gottfried Leibniz, Johann Lambert, and others, the reliability of mathematical manipulation with its symbols was the key. For the writers of the eighteenth-century French *Encyclopédie*, language and thought advanced together in every sphere, including mathematics. Charles Bossut opened the "Discours préliminaire" to the *Encyclopédie* by saying that mathematics was a chaining together of principles, arguments, and conclusions that was always accompanied by the qualities of being certain and evident. As Thomas Hankins has noted, "Abbé Condillac claimed that algebra was the best language because it had the best symbols. There was no ambiguity in their meaning, and the grammar of this 'language' was such that conclusions followed absolutely rigorously from premises."[40]

The power of mathematical arguments did not arise solely because the objects were particularly clear to the mind. Mathematical arguments had a particular merit. The hope, sometimes stated, sometimes implied, was that other sciences and even branches of philosophy would prove amenable to the same combination, and that the language of mathematics would stand as a paradigm for reasoning in

[39]As Aldo Brigaglia notes in his chapter above, Peano's Latino sine flexione was also one of the languages explicitly accepted by the *Rendiconti del Circolo matematico di Palermo*.

[40]Thomas L. Hankins, *Science and the Enlightenment* (Cambridge: University Press, 1985), p. 109.

other spheres. However, it is hard to find the commonplace view that mathematics is a language spelled out anywhere carefully and at length (whence the quotation marks around the word language in the above passage from Hankins). Certainly, mathematics has many features in common with a natural language: it involves signs and symbols that are manipulated according to certain rules. But in any given paper, book, or lecture, it is not clear where natural language stops and the mathematical language takes over. In particular, and this is a major point for any linguist, it is not clear that mathematics is a spoken language at all, whereas all natural languages are primarily spoken and only secondarily written.

When, in the late nineteenth century, great claims were made for logic, notably that mathematics was but a branch of it, analogous claims were made for logic as a language. It too, suitably enriched by talk of relations that syllogistic logic had lacked, was now seen as a purified language; a properly clarified logical language was seen as the appropriate basis for philosophy. Much of the philosophy of language can be interpreted as debating the boundaries between logic and language. The great question that arose in all those discussions goes today under the name of psychologism: are the laws of logic merely statements about the working of human minds and therefore psychologistic, or are these laws prior?[41]

Contemporary linguistics offers another perspective on these debates in the nineteenth century. Analysts of language in the nineteenth century had particular reasons to consider the relationship of language to thought, especially if language was taken to be a characteristically human activity. As they sought to give internal accounts of language structure and linguistic change, they too negotiated the waters of psychologism and reflected on the relationship between language and the world.

It is hard to imagine anyone in the period around 1900 being expert in the four fields of mathematics, logic, philosophy of language, and linguistics (Wilhelm Wundt was, arguably, the only person even competent to try his hand at all four). The subjects were too large, and the professional divisions too marked. Yet, the partial overlaps were particularly intense, and several interesting contacts and exchanges were established. An analysis of these overlaps, and the ways in which they were perceived, offers a fresh and interesting perspective on debates about the nature of mathematics. Inevitably, the sheer size of the field led to a lack of clarity and awareness of issues, most notably over the question of meaning or semantics in mathematical language, and even Hilbert and Frege were led into confusion. But a consideration of contemporary attitudes about natural and idealized languages, including theories of how they work to convey and advance thought and of how grammar and meaning can be related, situates debates about mathematical language and the truth of mathematics in a productive context.

Universal Language and Calculating Language

It is against this background that protagonists for improving the language of mathematics must be seen. It is helpful to address the question of what an ideal language for mathematics would do and to recall an interesting, even fundamental, position on the nature of language that was introduced into recent discussions of

---

[41]For a thorough account of the many different, changing, and overlapping positions held on this question in the German-speaking world, see Martin Kusch's fascinating account, *Psychologism: A Case Study in the Sociology of Philosophical Knowledge* (London and New York: Routledge, 1995).

logic and language by Jean van Heijenoort, and elaborated by Jaakko and Merrill Hintikka as well as by Martin Kusch.[42] These writers distinguish between language as a universal language and language as a calculus. Prominent upholders of linguistic universalism were Frege and Russell,[43] and of language as an instrument of calculation, Peirce and Husserl. The ideal type of the first view carries the deliberate limitation that no one can escape one's universe of discourse, and the relationship of language to the world is ineffable. Since it is impossible to step outside of language to talk about it, only hints remain; a meta-language for talking about a language borders on the absurd. Typically, the world is all there is. On the second view, there is no fixed universe of discourse. Language is merely properly formulated sets of symbols, and interpretation in terms of possible worlds is allowed. Meta-languages are entirely sensible.

In practice, exponents of the universalist view find semantics difficult, if not impossible, to explain. They cannot escape language and look at the world directly, so they find the relationship between language and the world obscure. Failing to resolve semantic disputes concerning different languages, they incline towards linguistic relativism. Because objects in the world cannot be known "in themselves," the correspondence theory of truth is unattractive to them, while a certain kind of Kantianism becomes attractive. Exponents of the second view take more or less the opposite position on all these issues. When it comes to formalism, universalists and calculators again disagree. Universalists can welcome it because syntax is the part of language that is left to them when semantics becomes inaccessible. They deny, however, that a formal system is open to many interpretations. Calculators urge interpretability and do not wish to be caught in the thickets of mere syntax.

Real existing languages bear a complicated relationship to the abstract concept of "language" (as did real existing socialism to its theoretical accounts). They are also semantically confusing because words have multiple meanings and, beyond that, associations with ideas that may be false. Rhetoric, poetry, even philosophy can seem suspicious to the seeker after truth. From these perceptions, some seventeenth-century writers sought to distill or to create from scratch a language in which each word had a single meaning, and reasoning would be automatic (an *a priori* language in the terminology of Couturat and Leau). This forced them to decide what the primitive notions were and to arrange them in classificatory schemes. Words (either written or spoken) were then constructed to reflect the formerly latent, now explicit order of nature. While not the first of such enthusiasts, Leibniz pushed the project with remarkable energy.[44] As Umberto Eco characterized him, Leibniz was pluralist and ecumenical in his views and hoped that a universal language would bring about world peace. He was not, however, an Esperantist *avant la lettre* because he placed more hope in science than in linguistic reform.

Leibniz identified as crucial to his linguistic project a system of deductive rules that could be applied to uninterpreted symbols. Ideally, as he put it, in such a

[42]Jean van Heijenoort, "Logic as Language and Logic as Calculus," *Synthese* **17** (1967):324-330; Jaakko Hintikka and Merrill Hintikka, *Investigating Wittgenstein* (Oxford: Basil Blackwell, 1986); and Martin Kusch, *Language as Calculus vs. Language as Universal Medium* (Dordrecht: Kluwer Academic Publishers, 1989).

[43]"The meaning of the fundamental terms cannot be defined, but only suggested. If the suggestion does not evoke in the reader the right idea, nothing can be done." Bertrand Russell, "On the Axioms of Geometry" (1899), in *The Collected Papers of Bertrand Russell*, ed. Nicholas Griffin and Albert C. Lewis (London: Hyman Unwin, 1990), 2:394-415.

[44]See Eco, pp. 209-268.

system "mental error is exactly equivalent to a mistake in calculation."[45] But the symbols were not always meaningless: they usually had interpretations built in to prevent, for example, a valid argument about monkeys being applied to yield incorrect conclusions about man. Leibniz also allowed that the enumeration of primitive notions had to be open-ended, as human knowledge was forever imperfect. Accordingly, his system gave more weight to syntax than semantics.

A century later, the *encyclopédistes* tried to put an end to the construction of *a priori* languages. They argued that thought and language advance together, each influencing the other, and so there was no system of pure thought that could be articulated by the creation of a perfect language. People were trapped in historically changing languages, which could be improved but not escaped. This view became the orthodoxy, and as noted, the exponents of subsequent artificial languages took the *a posteriori* route of modifying existing languages. Improvement was not ruled out, perfection was.

In the second half of the nineteeth century, Leibniz's ideas were revived as a result of Carl Gerhardt's new editions of his writings and of their steady incorporation into contemporary philosophy. The work of Couturat, who edited a number of previously unpublished essays by Leibniz on logical and linguistic issues, should also be mentioned. The influence of this work on others, notably Russell, was profound.[46] These authors stressed Leibniz's ideas on logic, which had impelled him to seek foundations for the axioms of Euclidean geometry and to regard mathematics as that branch of logic dealing with number and quantity (and mass and weight, as it happens). These ideas represent the kernel of Leibniz's hope that even if a truly general universal language for all purposes was beyond reach, a universal language of science was achievable.

Leibniz's ideas now mingled with a number of other developments: the symbolism of George Boole's *Laws of Thought*, Frege's criticisms of contemporary logic, Richard Dedekind's expressed belief that mathematics was the creation of the human mind, Peano's symbolism. Given this diversity of viewpoints, did projects for a universal language take the universalist or the calculating view? Would such a language have a clear syntax but fail to work through semantical issues? Would the semantics be determined by the world and, if so, how? Or would the semantics be pluralistic, even arbitrary? Different projects were at work, and one group that could presume to speak with authority on languages was the growing body of professional linguists.

Nineteenth-century Linguistics

One of the great breakthroughs of nineteenth-century linguistics was the realization that both Sanskrit and Greek belonged to that family of languages that came to be called Indo-European. It was possible to characterize this family, to locate individual languages within it in a systematic way, and to establish that certain languages did not belong. Another development was the profusion of data-oriented approaches that carefully studied many aspects of individual languages, some in danger of extinction.

---

[45]Gottfried W. Leibniz, "De scientia universalis seu calculo philosophico," in *Die philosophischen Schriften von G. W. Leibniz*, ed. Carl Gerhardt, 7 vols. (Berlin: Weidmann, 1875), 7:3-241 on p. 203.

[46]The nineteenth-century part of this story has recently been studied by Volker Peckhaus. See his *Logik, Mathesis universalis und allgemeine Wissenschaft* (Berlin: Akademie Verlag, 1997).

Much of this work has no implications whatsoever for mathematical practice, but the earlier ideas of philosophical grammar do. Wilhelm von Humboldt is a central figure here. He insisted that the great variety of languages did not contradict the idea that there was only one "language," and he held that possession of language was what made us human. For him, language was an "intellectual instinct of reason"[47] that spread and diversified in different human societies. This raises the question of what, if anything, is the difference between logic and language.

For Humboldt, language, and any particular language, are intermediaries between thought and the world; different languages give different views of the world that are not simply inter-translatable. As he showed in the case of Javanese, specific grammatical features of particular languages could be understood as arising from general ways in which, say, verbs operate and are modified. Humboldt took linguistics to be about explaining the regularities of a given language—not merely codifying it (as in school-book grammar) or simply listing it (as in botanical accounts)—and showed that language had a life and rules of its own.

Humboldt distinguished sharply between language and logic and claimed, like Heymann Steinthal in the next generation, that language was prior. This position rejected the tenet of philosophical grammar that all languages share a common underlying structure. On this view, grammatical categories (such as word and sentence) do not map tidily onto logical categories (such as concept and judgment). Consequently, the study of human language ability became part of, or based on, or enabled by, psychology rather than logic. As such, it had an individual and a social aspect (*Völkerpsychologie*).

The next generation of neo-grammarians was different again, and by the end of the century, more linguistic ground was shifting. There was certainly no consensus. There was, at one extreme, the position of Max Müller, a German Sanskrit scholar and from 1868 professor of comparative philology at Oxford, who looked to historical linguistics to solve the problems of philosophy: "If we fully understood the whole growth of every word, philosophy would have and could have no secrets. It would cease to exist."[48] This resembles the view of the Oxford Assyriologist, Arthur Sayce, that Aristotle's philosophy would have been very different if Aristotle had been Mexican, the argument being that Aristotle's logical categories closely resembled those of Greek grammar. Those who saw language in psychologistic terms might not have agreed, but this is not the place to enter into debates about the relative priorities of logic and psychology in the late nineteenth century.[49]

Linguists around 1900 did share some points of agreement, however. For example, they agreed that "snow" is a noun not because it stands for a thing, but because of the places it can appear in sentences and because of its effect on other words in sentences, that is, "because it can appear as the subject of a proposition, can form a plural by adding '-s' . . . ."[50] Likewise the Port-Royal idea that a proposition expresses a judgment by joining a subject and a predicate with a copula ("snow is white": "snow" is the subject, joined by the copula "is" to "white," the predicate),

[47]Quoted in Anna M. Davies, *Nineteenth Century Linguistics*, History of Linguistics, vol. 4, ed. Giulio Lepschy (London and New York: Longmans, Green, and Co., 1998), p. 109. The information presented here on Humboldt is drawn primarily from this source.

[48]In Davies, p. 300.

[49]For these debates, see Kusch.

[50]Henry Sweet, "Words, Logic and Grammar," *Transactions of the Philological Society* (1875–1876):470-503 on p. 487, as quoted in Davies, p. 308.

which Bopp had attempted to revive, collapsed under the weight of sentences that are not of this form. On the other hand, there were functional, grammatical, and psychological analyses of what a sentence is, and the validity of these approaches was much contested (e.g., between Wilhelm Wundt and Anton Marty, a pupil of Franz Brentano).[51]

But if, then as now, linguistics seems more advanced in the study of syntax than semantics, it is important to note that the kind of language mathematics was taken to be could only be seen by a linguist as a syntactically impoverished one. On the other hand, mathematicians concerned to enlarge mathematics by embracing the study of logic, such as Boole, Peirce, and Schröder, were happy to speak of their work as providing a new language for expressing arguments.

Language, Meaning, and Mathematics: Significs

The linguistic approach to semantics is surprisingly recent. Ideas about syntax go back a long way, as far as the ancient Greeks, who used a very similar word. But the word *sémantique* was created in 1883 by Michel Bréal to denote a "*science des significations*" or science of meaning.[52] He sought to emphasize the relationship between form and meaning in language and to make linguistics a human science. He stressed the importance of speech acts, of intention, and of communication.

Another person with a strong interest in the meaning of meaning was Victoria, Lady Welby. She was, among other things, a correspondent of Peirce in his later years and the author of both the book, *What is Meaning?* (1903), and the article on significs—her word for the study of meaning—in the eleventh edition of the *Encyclopaedia Britannica.*[53] She argued against the idea that meanings were universal and for their contingent cultural character. Language, in her view, evolves (in a Darwinian sense) and is used instrumentally. She distinguished between sense, meaning, and significance (reference, intention, and moral aspect). Peirce found her work exciting, but it mattered less in the long run than his own. His, however, exerted little influence in its day, as all Peirce scholars lament, and hers was taken up by the Dutch mathematician and philosopher of mathematics, L. E. J. Brouwer.

The connection was made by another Dutch mathematician, Gerrit Mannoury, the founder of the significs movement in the Netherlands. Mannoury argued that mathematics displayed a close, but not a completely accurate or consistent, relationship between thinking and speaking. It was inevitably inadequate to the task because it attempted to capture infinite, continuous multitudes with a finite use of symbols. Such a usage, Mannoury called a formalization, and he allowed for successive formalizations that improve on their predecessors and allow interpersonal understanding.

Mannoury reviewed Brouwer's thesis on the foundations of mathematics in 1907. In his obituary of Mannoury, Georg van Dantzig quoted the following remarkable passage from Mannoury's response to Brouwer's claim that, even by building

---

[51]See Clemens Knobloch, "Sprache und Denken bei Wundt, Paul und Marty; ein Beitrag zur Problemgeschichte der Sprachpsychologie," *Historiographia Linguistica* **11** (1984):413-484, and Clemens Knobloch, *Geschichte der psychologischen Sprachauffassung in Deutschland von 1850 bis 1920* (Tübingen: Max Niemayer Verlag, 1988).

[52]His influential book, *Essai de sémantique* was published (by Hachette) in 1897.

[53]Lady Victoria Welby, *What is Meaning? Studies in the Development of Significance* (London: Macmillan and Co., 1903), and *Encyclopaedia Britannica*, 11th ed., s.v. "Significs" by Lady Victoria Welby.

language systems, the formalists cannot ensure the reliability of the mathematical properties:

> No, Brouwer, the logicists do not ensure the reliability of the "mathematical properties," but no more will you, with your continuity-intuition ensure it, simply for the reason that it does not exist. Mathematics is human make [sic] and human devise [sic], containing no other truth than in relation to human language, purpose and society. ... Free yourself completely from all conventions and you will come to the conviction: there is no unalterable truth and no unalterable measure for truth, there is no absolute unit, no absolute space and no absolute time, there is no certain knowledge [*Wiskunde* in Dutch, meaning science].[54]

Van Dantzig pointed out that it was only the next year, 1908, that Brouwer first denied the law of excluded middle; the development of his intuitionism dated from 1917. But even in 1905, in his *Levin, Kunst en Mystiek* (*Life, Art, and Mysticism*), Brouwer displayed a remarkable distrust of language: "Living in the intellect carries the impossibility to communicate."[55] It is the source of confusion because it is the slave of the delusion that there is an external reality.[56] Brouwer's distrust extended beyond language and logic to novel ideas of proof. For Brouwer, a proof was an infinite mental construction that could not be described in any language. Writing to his supervisor in 1905, he distinguished sharply between mathematical argumentation and logical argumentation, without, unfortunately, defining either of them precisely.

This distrust of language was characteristic of the significs movement as it developed in the Netherlands, and was initially focused on the tendentious and dishonest language used by participants in the First World War. In a manifesto signed by Brouwer, Mannoury, and others in 1918, they deplored the lack of "a satisfactory store of words of well-considered spiritual value, at least in our western languages" and proposed "to coin words of spiritual value for the languages of western nations" while removing false words.[57] In 1919, the same group called for a new basic vocabulary. In their view, language was always inadequate to represent any part of reality, and meanings were defined exclusively by reference to effects (intended or presumed). From these and later writings, it becomes clear that Brouwer viewed language as built up in levels: words, words in simple relationships, sentences whose meanings depend on the way words are connected, well-regulated language such as scientific language, and finally symbolic language such as mathematical logic and pure mathematics written in pasigraphic form. (Note that this is not a hierarchy of values; if anything, it is the reverse.)

---

[54]Georg van Dantzig, "Gerrit Mannoury's Significance for Mathematics and Its Foundations," *Nieuw Archief voor Wiskunde* (3) **5** (1957):1-18 on p. 14.

[55]Dirk van Dalen, "Brouwer's Dogma of Languageless Mathematics and Its Role in his Writings," in *Significs, Mathematics and Semiotics: The Significs Movement in the Netherlands*, ed. Erik Heijerman and Walter Schmitz (Münster: Nodus Publications, 1991), pp. 33-41 on p. 33.

[56]Christian Thiel, "Brouwer's Philosophical Language Research and the Concept of the Ortho-Language in German Constructivism," in *op. cit.*, pp. 21-32 on p. 32.

[57]*Ibid.*, p. 23.

Language, Meaning, and Mathematics: Hilbert and Husserl

So far, two views have been presented of what an ideal mathematical language could be. One distrusts the usual mathematical language owing to a burdensome concern with meaning and seeks correct definitions (as did Frege) or veridical insight (as did Brouwer). The other distrusts mathematical language and seeks to sharpen mathematical arguments with an effective, reliable syntax. It is natural to ask about Hilbert in this connection because of the significance of his views on the foundations of mathematics.

Hilbert's route into questions about the foundations of mathematics was via geometry. In the nineteenth century, two developments had displaced Euclid's *Elements*. The steady elaboration of projective geometry gave a simpler setting in which Euclidean geometry could be characterized (according to the Kleinian view of geometry[58]) as a special case, and non-Euclidean geometry showed that an alternative geometrical account of physical space was possible. First, Federigo Enriques and then, Ernest Nagel observed that because the principle of duality in plane projective geometry put points and lines on a par, geometers were pushed towards relying on their formalisms (which can cope with duality) and away from intuition (which clearly cannot).[59] On the other hand, non-Euclidean geometry showed only too openly that something about the construction of a geometrical system was problematic and had been flawed. In his book of 1882, Moritz Pasch proposed to distill from experience groups of axioms and to build up geometry from them.[60] The mathematician was to reason with the distilled insights purely axiomatically and formally, and Pasch took the process as far as the elaboration of projective geometry.

Hilbert axiomatized much earlier in the intellectual process than Pasch and with much less regard for reality and experience.[61] He was open to a number of semantical interpretations, that is, different models of his axiom systems. Indeed, he needed different models to establish the independence of various sets of axioms. This is interesting because it distinguishes Hilbert from Frege, on the one hand, and from the Italians active in axiomatizing geometry, on the other. The Hilbert-Frege debate has been much discussed because it interests philosophers of mathematics;[62] less has been said about the Italian axiomatizers, although their efforts were earlier and much more abstract. The significant novelty in Mario Pieri's work, for

---

[58]Felix Klein, *Vergleichende Betrachtungen über neuere geometrische Forschungen* (Erlanger Programm) 1st ed. (Erlangen: Deichert, 1872).

[59]Federigo Enriques, *Problems of Science*, trans. Katherine Royce (Chicago: Open Court Press, 1914) (originally published in 1906 as *Problemi della scienza*), and Ernest Nagel, "The Formation of Modern Concepts of Formal Logic in the Development of Geometry," *Osiris* **7** (1939):142-224.

[60]Moritz Pasch, *Vorlesungen über neuere Geometrie* (Leipzig: B. G. Teubner Verlag, 1882).

[61]See Michael M. Toepell, *Über die Entstehung von David Hilberts "Grundlagen der Geometrie"* (Göttingen: Vandenhoeck & Ruprecht, 1986); Volker Peckhaus, *Hilbertprogramm und kritische Philosophie* (Göttingen: Vandenhoeck & Ruprecht, 1990); and the essay review by David E. Rowe, "Perspective on Hilbert," *Perspectives in Science* **5** (1997):533-570.

[62]Aside from several remarks below, it must be enough here to give a few references to the literature, such as Michael Resnik, *Frege and the Philosophy of Mathematics* (Ithaca and London: Cornell University Press, 1980), and Michael Dummett, *Frege and the Philosophy of Mathematics* (London: Duckworth, 1991). For two recent sightings, see David E. Rowe, "Episodes in the Berlin-Göttingen Rivalry," *The Mathematical Intelligencer* **22** (1) (2000):60-69, and Judson C. Webb, *Mechanism, Mentalism and Meta-mathematics: An Essay on Finitism*, Synthese Library, vol. 137 (London and Dordrecht: D. Reidel Publishing Co., 1980).

example, which distinguished it from Pasch's, was the complete abandonment of any intention to formalize what is given in experience.[63] Instead, as Pieri wrote in 1895, he treated projective geometry "in a purely deductive and abstract manner, . . . , independent of any physical interpretation of the premises." Primitive terms, such as line segments, "can be given any significance whatever, provided they are in harmony with the postulates which will be successively introduced."[64]

This raises the question of an Italian influence—or lack thereof—on Hilbert. The one influence he acknowledged was that of Giuseppe Veronese, but not the more obvious one of Pieri. The usual explanation, offered by Toepell,[65] is that Hilbert only read Italian work in translation. A paper on language and mathematics may be the right place to say this cannot have been from lack of ability. Hilbert studied Latin at school and graduated satisfactorily. He thus had good enough Latin to make reading mathematics in Italian easy, provided he set himself the task. Either he simply did not read this work because he was too busy, or he had his own reasons for choosing not to read it. Those reasons might reflect the evident pleasure he took in his discovery that, by choosing different models, one could analyze axiom systems, and in a much richer way than the Italians were, if indeed he ever saw their work before 1900.[66]

But if Hilbert showed some awareness of the semantic issues involved in axiomatic geometry, his initial insights were far from always sound. In 1900, discussing the second of his twenty-three problems at the ICM in Paris, Hilbert wrote about an axiom system being, as he called it, complete, by which he meant that there was a system of objects obeying the axioms and no larger such system. Hilbert first asserted completeness in this sense for the real numbers, which, he announced, superseded the need for explicit constructions in the manner of Cantor. He then introduced this idea into the second edition of his *Grundlagen der Geometrie* (the first edition had opened with the claim that the axioms were complete without explaining what that meant). For this, he was roundly taken to task by Frege, who shrewdly observed that a completeness axiom of this kind cannot legitimately be invoked when the very existence of objects satisfying the axiom system in question is in dispute.[67] The essential uniqueness of a model for a set of axioms was soon to be better captured by Edward V. Huntington and Oswald Veblen, who introduced the word "categoric" in this context.[68] That usage, and not Hilbert's, will be preserved in what follows.

---

[63]On Pieri, see also Mario Pieri, "I principii della geometria di posizione, composti in sistema logico deduttivo," *Memorie della reale Accademia delle scienze di Torino* (2) **48** (1899):1-62; Mario Pieri, "Della geometria elementare come sistema ipotetico-deduttivo; monografia del punto e del mote," *Memorie della reale Accademia delle scienze di Torino* (2) **49** (1899):173-222; and Elena A. Marchisotto, "Mario Pieri and His Contributions to Geometry and Foundations of Mathematics," *Historia Mathematica* **20** (1993):285-303.

[64]Quoted in Umberto Bottazzini, "Fondamenti dell'aritmetica e della geometria," *Storia della scienza moderna e contemporanea*, ed. Paolo Rossi (Turin: Unione Tipografico Editrice Torinese, 1988), 3:276.

[65]Toepell, p. 56.

[66]Toepell (p. 56) quotes Klein as saying in 1903 that the Italian language posed no essential obstacle to the spread of theoretical work.

[67]Gottlob Frege to David Hilbert, 16 September, 1900, in Frege, pp. 43-48, especially on p. 46.

[68]See John Corcoran, "From Categoricity to Completeness," *History and Philosophy of Logic* **2** (1981):113-119 on p. 117 for the history, and the article as a whole for welcome logical precision on the relationship between completeness and categoricity.

Hilbert was also interested in axiom systems that deliver all of a pre-existing theory. When he spoke in Paris in 1900 of axiomatizing physics, he counseled that "[t]he mathematician will have also to take account not only of those theories coming near to reality, but also, as in geometry, of all logically possible theories. He must be always alert to obtain a complete survey of all conclusions derivable from the system of axioms assumed."[69] It is usually easy to see when the axioms do not deliver all the results: the particular case of what does, and what does not, require or follow from the Euclidean axiom of parallels was well studied by 1900. But how can one be sure that the required consequences are delivered in their entirety? A promising avenue is opened up by agreeing that all and only the theorems in question are delivered by a certain basic set, which, in turn, follow from a system of axioms A'. The task is then obvious: derive the axioms in system A' as theorems from system A. Conversely, to show that the axiom system A does not deliver all that is required, it is enough to find a theorem C with the property that neither C nor its negation contradicts the axioms A. This approach suggests another criterion an axiom system might satisfy. It might be that the axiom system cannot be enlarged: for any statement C one of the axiom systems $A + C$ and $A + \neg C$, where $\neg C$ denotes the negation of C, is consistent, and the other is not. This property of an axiom system is now called completeness. At least informally, and by analogy with the Kleinian view of geometry or with Hilbert's *Grundlagen der Geometrie*, one expects that adding an axiom in a consistent fashion to a system of axioms (enlarging the axioms) will restrict the family of objects obeying the axioms.[70] The risk is, indeed, that adding an axiom (a statement that is not a consequence of the axiom system) will produce an inconsistent axiom system, as it must if the system is complete.[71]

Historically, mathematicians started with objects. In three-dimensional geometry, these might be points, lines, and planes, as described in the thirteen books of Euclid's *Elements*, but suitably reformulated. In this context, the idea that one knows all the theorems of the theory is a comfortable one, and the task is to spell out an appropriate axiom system. In 1882, Pasch approached projective geometry in this way. The axiom system codified the behavior of the objects.

To take an axiom system as the starting point, as various Italian mathematicians did, was much more radical. They approached consistency naïvely, building up axiom systems axiom by axiom in a way that seemingly guaranteed consistency. When the purpose was to clarify, and even define, the nature of the objects, the task (one that Hilbert took on) was even harder. Such an axiom system might be said to create the objects it describes, but when the axiom system is not categoric, the issue of multiple semantical interpretations is important.

Famously, and impossibly, Hilbert claimed in a letter to Frege that "if the arbitrary given axioms do not contradict one another with all their consequences,

---

[69] Hilbert, "Mathematical Problems," p. 15.

[70] Notice in passing that it can be desirable to have axiom systems with unexpectedly isomorphic families of objects, for then results known to be true for one family also apply to the other. But the proofs in one system may be intuitive and those in the other actually unknown.

[71] The relationship between complete and categorical axiom systems requires a degree of logical precision to elucidate. A plausible but false argument suggests that a complete axiom system is categorical and that the reverse implication is also true. With the precision of mathematical logic, one can assert accurate versions of the theorem that categoricity implies completeness, but the claim that completeness implies categoricity is false in the usual formulations of mathematical logic. See Corcoran, p. 118.

then they are true and the things defined by them exist."[72] Existence may be allowed, but truth is a more difficult concept, and one may be reluctant to say of contradictory claims (validly deduced from different and incompatible, but separately consistent, axiom systems) that both are "true." The parallel postulate is a case in point. The best that can be said is that the theorems are true of the objects defined by that axiom system, which is to retreat from "true" to "proved" (the word Hilbert preferred in Paris).

At this early stage in his work on foundational issues, a number of difficult questions were simply invisible to Hilbert. He knew perfectly well that there were consistent but mutually incompatible systems of axioms. But there was no awareness in 1900 that completeness of a theory was a problematic condition to insist upon. Hilbert seems to have thought that to speak of completeness was little different from speaking about truth, relative to the appropriate sphere of objects. In the same way that one may arrive at a categorical system of axioms, it seemed reasonable to Hilbert in the second edition of the *Grundlagen der Geometrie* to believe that an axiom system could define its objects. He was, however, clear neither about these concepts nor about their mutual relationship.[73]

Loose talk about "true" and "proved" distinguished Hilbert (and Husserl) from Frege. Hilbert and Husserl were no longer linguistic universalists. They did not conceive of language as the conveyor of thoughts about the world expressed in as logical a manner as possible. They were on their way to becoming linguistic calculators (of different kinds, in the end) open to many interpretations of structures whose syntax was all they could vouch for. Ultimately, for Frege, ontological questions reduced to determining what was in the world, although the whole thrust of his program was to show that a remarkable amount of that knowledge could be derived logically from very little.[74] For Hilbert, although he cannot have known around 1900 what difficulties he was letting himself in for, multiple semantical interpretations of a formal system was the very nature of the mathematical enterprise, and ontological questions in mathematics were not to be solved by referring to the world.

Language, Meaning, and Mathematics: Hilbert and Schröder

Finally, what implications does this have for a puzzle of some current interest, namely, the eclipse of Ernst Schröder and, behind him, of Peirce? The obvious explanations actually go quite far. Schröder died in 1901, and even his supporters agreed that he had been poor at selling his ideas. Moreover, Russell and Whitehead marginalized both Peirce and Schröder, thus making it seem unattractive after 1910

[72] Hilbert to Frege, 29 December, 1899, in Frege, p. 39.

[73] Husserl's responses to Hilbert's ideas here are somewhat less well-known. They have recently been discussed by Claire Hill and Ullrich Majer in rather different papers with strikingly similar titles: Claire Ortiz Hill, "Husserl and Hilbert on Completeness," in *From Dedekind to Gödel: Essays on the Development of the Foundations of Mathematics*, ed. Jaakko Hintikka, Synthese Library, vol. 251 (Dordrecht: Kluwer Academic Publishing, 1995):143-163, and Ullrich Majer, "Husserl and Hilbert on Completeness," *Synthese* **110** (1997):37-56. Majer shows that Husserl found the idea of completeness of an axiom system quite interesting and that he, too, was attracted to the idea of defining or creating objects by means of a categorical axiom system, although, as Majer describes, his immediate point of interest was defining the integers.

[74] Compare Hallett's comment that for Hilbert the truth of the axioms was reference independent, whereas for Frege, of course, sense determined reference. See Michael Hallett, "Hilbert's Axiomatic Method and the Laws of Thought," in *Mathematics and Mind*, ed. Alexander George (New York and Oxford: Oxford University Press, 1994):158-198 on p. 163.

to go back to their work. This was not the view in the United States, where Peirce and Schröder were better remembered and where the *Principia Mathematica* was less highly esteemed, or in Poland, where their kind of algebraic logic continued to prosper. But the renewed urgency that greeted the foundational questions of mathematics after World War I, especially in Germany, were naturally affected by the fact that Russell and Whitehead were still alive, and neither Peirce nor Schröder was. Hilbert's contribution to this disposition of the honors is consistent with the above linguistic analysis of his position.

Hilbert's project became one of saving mathematics, not as a family of empty languages (pure syntax) and not as a body of truths about the world but as something more complicated. Mathematics had to be rescued from Brouwerian intuition, which, for a time, had caught the sympathies of no less a person than Hermann Weyl, or it would not be able to deliver the richness of analysis and the vast store of applications in science. It had also to deliver familiar truths, in particular about the integers. The linguistic approach opened the perspective that these were, in part, semantic questions, and so they pushed Hilbert away from semantic arbitrariness and towards semantic flexibility or multiplicity.

In unpublished lectures of 1905, Hilbert called "the *a priori* of philosophers" the capability to think things and to denote them through simple signs.[75] Mental ability was reflected in the ability to represent objects and thoughts by signs and to manipulate those signs. In 1918, he introduced his paper, "Axiomatisches Denken," with a painfully topical analogy between the proper relations between nations and those between the sciences and argued that axiomatics helped bring out the fundamental unity of mathematics. By 1922, Hilbert was distinguishing between meaningless and meaningful signs and introducing signs of various kinds (standing, for example, for variables or formulas).[76] Certain formulas would be taken as axioms, and mathematics in the strict sense would be identified with the stock of provable formulas. Statements with content would belong to a new discipline of metamathematics, which would provide proofs of the consistency of sets of axioms. The aim, for Hilbert, was to create mathematics at the level of syntax, and in his paper of 1923, he wrote of a metamathematics that had a semantic aspect: the proofs of the mathematics under investigation. This was Hilbert's famous proof theory, his attempt, as Paul Bernays put it, to transfer the foundations of mathematics from the domain of epistemology to that of an appropriate mathematics.[77]

In the early 1920s, Hilbert advocated the simultaneous strengthening of logic and mathematics.[78] He asserted that "[c]alling on mathematical methods for the investigation of the logical language is not artificial, but fully appropriate and even inevitable .... It is self-evident that, when we exclude the accidental features

---

[75]Quoted in Hallett, p. 179.

[76]David Hilbert, "Neubegründung der Mathematik: Erste Mitteiling," *Abhandlungen aus dem Mathematischen Seminar der Hamburgischen Universität* **1** (1922):157-177, in David Hilbert, *Gesammelte Abhandlungen*, 3 vols. (Berlin: B. G. Teubner Verlag, 1935), 3:157-177, and David Hilbert, "Die logischen Grundlagen der Mathematik," *Mathematische Annalen* **88** (1923):151-165, in *op. cit.*, 3:178-191.

[77]Paul Bernays, "Über Hilberts Gedanken zur Grundlagen der Arithmetik," *Jahresbericht der Deutschen Mathematiker-Vereinigung* **31** (1922):10-19; see p. 19.

[78]Two papers by Hallett show in stimulating detail how attending to semantics and syntax shed light on the progress of Hilbert's thought. See Hallett, "Hilbert's Axiomatic Method," and "Hilbert and Logic," *Québec Studies in the Philosophy of Science* **1** (1995):135-187.

in the derivation of words, then a form of mathematical sign language arises."[79] Moreover, he argued that it would be possible "to frame the rules of grammar in such a surveyable way that logical inference can be carried through automatically by calculation according to simple, determined rules."[80]

How Hilbert attempted to do this, with what success, and to what extent Gödel put an end to the program as originally or even finally conceived is another, much-discussed story. The delicate point in all of it is, as is well-known, Hilbert's idea of a finitary argument. It is enough to note here, however, that the introduction of metamathematics as a language for analyzing mathematics surpasses anything in Schröder. Semantical freedom is required, but not at the expense of dismissing semantic considerations from mathematics altogether. On the other hand, the full-blooded attempt by Russell and Whitehead did at least raise questions about what a set could be and about what relations there could be between mathematics and logic. Once Hilbert began to think that philosophical questions about the nature of mathematics could be answered by re-working them as mathematical questions about mathematical language, the debate naturally moved beyond the simplicities of pasigraphy. The real difficulties, as so often, turned out to be semantic and not syntactic. They have to be confronted not only by thinking how we speak, but, however clumsily, by thinking how we speak about things and what we can speak about.

Conclusion

Questions about the nature of thought and the proper objects of thought are notoriously difficult. They are as much linguistic as logical, they concern the relationship between valid expressions and truths, and they require clarity about syntax and semantics. Clarity was slow in coming, but one important source for it, in the emerging world of nations around 1900, was the strong currents running towards the creation of ideal languages. These currents flowed out of concerns about the nature of science, about communication in science, and about the need for improved communication in general. They spilled over into analyses of the language(s) of mathematics and into the very consideration of mathematics as a language, at levels from the naïve to the technical and from the disinterested to the politically charged. Different attitudes to language, such as linguistic calculationism and linguistic universalism, have implications for syntactic and semantic practice, and help reveal distinctions between Hilbert, Frege, and Russell. When the need for new foundations of mathematics was felt most keenly, in the decades around 1900, questions about the language of mathematics and the existence of mathematical objects were formulated as questions about logic, mathematics, and language in ways that echoed the debates about ideal grammars and vocabularies for novel languages. By reflecting on those debates, the confusions and the strengths of positions taken by Hilbert and those whom he opposed become manifest. Something else comes through as well, and it points to an avenue worthy of exploration by historians of mathematics: discussions one hundred years ago about the nature of mathematics cannot be separated from discussions of language; the activity of mathematics belonged in fundamental ways to rich contemporary debates about

[79]David Hilbert, *Wissen und mathematisches Denken*, ed. William Ackermann (Göttingen: Mathematisches Institut, 1922), p. 130, as quoted in Hallett, "Hilbert and Logic," pp. 180-181.

[80]Hilbert, *ibid.*, p. 79, as quoted in Hallett, *ibid.*, p. 181.

how to communicate in the modern world. This gave a depth and a resonance to vital arguments about mathematics and logic that have hitherto been discussed, however well, from a technical standpoint. Regardless of how mathematicians negotiated the passage from nationalism to internationalism, they could not easily and simply appeal to mathematics as an international language and hope thereby to solve the problems of communication within their growing community.

References

Anellis, Irving. "Peirce Rustled, Russell Pierced: How Charles Peirce and Bertrand Russell Viewed Each Other's Work in Logic, etc." *Modern Logic* **5** (1995): 270-328.

Archibald, Thomas and Charbonneau, Louis. "Mathematics in Canada before 1945: A Preliminary Survey." In *Canadian Mathematical Society/Société mathématique du Canada 1945-1995: Mathematics in Canada/ Les mathématiques au Canada.* Ed. Peter Fillmore. 2 Vols. Ottawa: Canadian Mathematical Society, 1995, 1:1-44.

Barrow-Green, June. "International Congresses of Mathematicians from Zürich 1897 to Cambridge 1912." *The Mathematical Intelligencer* **16** (1994):38-41.

Bernays, Paul. "Über Hilberts Gedanken zur Grundlagen der Arithmetik." *Jahresbericht der Deutschen Mathematiker-Vereinigung* **31** (1922):10-19.

Bottazzini, Umberto. "Fondamenti dell'aritmetica e della geometria." *Storia della scienza moderna e contemporanea.* Ed. Paolo Rossi. Turin: Unione Tipografico Editrice Torinese, 1988, 3:253-288.

Bréal, Michel. *Essai de sémantique.* Paris: Hachette, 1897.

______. "Le choix d'une langue internationale." *Revue de Paris* (4) **8** (1901):229-246.

Brent, Joseph. *Charles Sanders Peirce: A Life.* Bloomington: Indiana University Press, 1993.

Brouwer, Luitzen E. J. *Levin, Kunst en Mystiek* (*Life, Art, and Mysticism*). Delft: Waltman, 1905.

Corcoran, John. "From Categoricity to Completeness." *History and Philosophy of Logic* **2** (1981):113-119.

Couturat, Louis and Leau, Léopold. *Histoire de la langue universelle.* Paris: Hachette, 1903.

______. *Les nouvelles langues internationales.* Paris: Hachette, 1907.

Dalen, Dirk van. "Brouwer's Dogma of Languageless Mathematics and its Role in his Writings." In *Significs, Mathematics and Semiotics: The Significs Movement in the Netherlands.* Ed. Erik Heijerman and Walter Schmitz. Münster: Nodus Publications, 1991, pp. 33-41.

Dantzig, Georg van. "Gerrit Mannoury's Significance for Mathematics and Its Foundations." *Nieuw Archief voor Wiskunde* (3) **5** (1957):1-18.

Davies, Anna M. *Nineteenth Century Linguistics.* History of Linguistics. Vol. 4. Ed. Giulio Lepschy. London and New York: Longmans, Green, and Co., 1998.

Domar, Yngve. "On the Foundation of *Acta Mathematica.*" *Acta Mathematica* **148** (1982):3-8.

Duda, Roman. "*Fundamenta Mathematica* and the Warsaw School of Mathematics." In *L'Europe mathématique–Mathematical Europe.* Ed. Catherine Goldstein, Jeremy J. Gray, and Jim Ritter. Paris: Éditions de la Maison des sciences de l'homme, 1996, pp. 479-498.

Eco, Umberto. *The Search for a Perfect Language.* Trans. James Fentress. London: Fontana, 1995.

Enriques, Federigo. *Problems of Science.* Trans. Katherine Royce. Chicago: Open Court Press, 1914.

Frege, Gottlob. *Philosophical and Mathematical Correspondence.* Chicago: University Press, 1980.

Gispert, Hélène and Tobies, Renata. "A Comparative Study of the French and German Mathematical Societies before 1914." In *L'Europe mathématique–Mathematical Europe.* Ed. Catherine Goldstein, Jeremy J. Gray, and Jim Ritter. Paris: Éditions de la Maison des sciences de l'homme, 1996, pp. 407-430.

Goldstein, Catherine; Gray, Jeremy J.; and Ritter, Jim, Ed. *L'Europe mathématique–Mathematical Europe* Paris: Éditions de la Maison des sciences de l'homme, 1996.

Guérard, Albert L. *A Short History of the International Language Movement.* London: Fisher Unwin, 1922.

Hallett, Michael. "Hilbert's Axiomatic Method and the Laws of Thought." In *Mathematics and Mind.* Ed. Alexander George. New York and Oxford: Oxford University Press, 1994, pp. 158-198.

———. "Hilbert and Logic." *Québec Studies in the Philosophy of Science* **1** (1995):135-187.

Hankins, Thomas L. *Science and the Enlightenment.* Cambridge: University Press, 1985.

Harkness, James and Morley, Frank. *A Treatise on the Theory of Functions.* New York: Stechert, 1893.

Haskell, Mellen W. "Über die zu der Curve $\lambda^3\mu + \mu^3\nu + \nu^3\lambda = 0$ im projektiven Sinne gehörende mehrfache Überdeckung der Ebene." *American Journal of Mathematics* **13** (1890):1-52.

Heijenoort, Jean van. "Logic as Language and Logic as Calculus." *Synthese* **17** (1967):324-330.

Heijerman, Erik and Schmitz, Walter, Ed. *Significs, Mathematics and Semiotics: The Significs Movement in the Netherlands.* Münster: Nodus Publications, 1991.

Hilbert, David. "Axiomatisches Denken." *Mathematische Annalen* **78** (1918):405-418.

———. "Die logischen Grundlagen der Mathematik." *Mathematische Annalen* **88** (1923):151-165.

———. *Foundations of Geometry.* 10th English Ed. Trans. of 2nd German Ed. by L. Unger. Chicago: Open Court Press, 1971.

———. *Gesammelte Abhandlungen.* 3 Vols. Berlin: Springer-Verlag, 1932–1935.

———. *Grundlagen der Geometrie: Festschrift zur Feier der Enthüllung des Gauss-Weber-Denkmals in Göttingen.* Leipzig: B. G. Teubner Verlag, 1899.

———. "Mathematical Problems." *Bulletin of the American Mathematical Society* **8** (1902):437-479; Reprint Ed. "Mathematical Problems." In *Mathematical*

Developments Arising from Hilbert Problems. Proceedings of Symposia in Pure Mathematics. 2 Vols. Providence: American Mathematical Society, 1976, 1:1-34.

———. "Mathematische Probleme." *Archiv für Mathematik und Physik* **1** (1901): 44-63 and 213-237.

———. "Neubegründung der Mathematik: Erste Mitteilung." *Abhandlungen aus dem Mathematischen Seminar der Hamburgischen Universität* **1** (1922):157-177.

———. "Sur les problèmes futurs des mathématiques." Trans. L. Laugel. *Compte rendu du deuxième Congrès international des mathématiciens*. Paris: Gauthier-Villars, 1902, pp. 58-114.

———. "Über den Satz von der Gleichheit der Basiswinkel im gleichschenkligen Dreieck." *Proceedings of the London Mathematical Society* **35** (1902):50-68.

———. *Wissen und mathematisches Denken*. Ed. William Ackermann. Göttingen: Mathematisches Institut, 1922.

Hill, Claire Ortiz. "Husserl and Hilbert on Completeness." In *From Dedekind to Gödel: Essays on the Development of the Foundations of Mathematics*. Ed. Jaakko Hintikka. Synthese Library. Vol. 251. Dordrecht: Kluwer Academic Publishing, 1995, pp. 143-163.

Hintikka, Merrill B. and Hintikka, Jaakko. *Investigating Wittgenstein*. Oxford: Basil Blackwell, 1986.

Husserl, Edmund. *Edmund Husserl: Philosophie der Arithmetik, mit ergrendzenden Texte (1890–1901)*. Ed. Lothar Eley. Husserliana. Vol. 12. Den Haag: Martinus Nijhoff, 1970.

Kennedy, Hubert C. *Peano: Life and Works of Giuseppe Peano*. Dordrecht and Boston: D. Reidel Publishing Co., 1980.

Klein, Felix. *Vergleichende Betrachtungen über neuere geometrische Forschungen* (*Erlanger Programm*). Erlangen: Deichert, 1872.

———. *Klein's Evanston Colloquium Lectures and Other Works*. Ed. David E. Rowe and Jeremy J. Gray. New York: Springer-Verlag, to appear.

Knobloch, Clemens. "Sprache und Denken bei Wundt, Paul und Marty; ein Beitrag zur Problemgeschichte der Sprachpsychologie." *Historiographia Linguistica* **11** (1984):413-484.

———. *Geschichte der psychologischen Sprachauffassung in Deutschland von 1850 bis 1920*. Tübingen: Max Niemayer Verlag, 1988.

Kusch, Martin. *Language as Calculus vs. Language as Universal Medium*. Dordrecht: Kluwer Academic Publishing, 1989.

———. *Psychologism: A Case Study in the Sociology of Philosophical Knowledge*. London and New York: Routledge, 1995.

Leibniz, Gottfried W. "De scientia universalis seu calculo philosophico." In *Die philosophischen Schriften von G. W. Leibniz*. Ed. Carl Gerhardt. 7 Vols. Berlin: Weidmann, 1875, 7:3-241.

Majer, Ullrich. "Husserl and Hilbert on Completeness." *Synthese* **110** (1997):37-56.

Marchisotto, Elena A. "Mario Pieri and His Contributions to Geometry and Foundations of Mathematics." *Historia Mathematica* **20** (1993):285-303.

Méray, Charles. "Sur la langue internationale auxiliaire de M. de Zamenhof, connue sur le nom d'Esperanto." In *Compte rendu du deuxième Congrès international des mathématiciens*. Paris: Gauthier-Villars, 1902, pp. 429-432.

Nabonnand, Philippe. "The Poincaré–Mittag-Leffler Relationship." *The Mathematical Intelligencer* **21** (2) (1999):58-64.

Nagel, Ernest. "The Formation of Modern Concepts of Formal Logic in the Development of Geometry." *Osiris* **7** (1939):142-224.

Osgood, William F. *Lehrbuch der Funktionentheorie.* 2 Vols. Leipzig: B. G. Teubner Verlag, 1907–1924; Reprint Ed. New York: Chelsea Publishing Co., 1965.

Parshall, Karen Hunger and Rowe, David E. *The Emergence of the American Mathematical Research Community, 1876-1900: J. J. Sylvester, Felix Klein, and E. H. Moore.* HMATH. Vol. 8. Providence: American Mathematical Society and London: London Mathematical Society, 1994.

Pasch, Moritz. *Vorlesungen über neuere Geometrie.* Leipzig: B. G. Teubner Verlag, 1882.

Peano, Giuseppe. "De Latino sine flexione—lingua auxiliare internationale." *Rivista di matematica* **8** (1903):439-447.

———. *Opere scelte.* Ed. Ugo Cassina. 3 Vols. Rome: Edizione Cremonese, 1958.

Peckhaus, Volker. *Hilbertprogramm und kritische Philosophie.* Göttingen: Vandenhoeck & Ruprecht, 1990.

———. "Ernst Schröder und die 'pasigraphische Systeme' von Peano und Peirce." *Modern Logic* **1** (1991):174-205.

———. *Logik, Mathesis universalis und allgemeine Wissenschaft.* Berlin: Akademie Verlag, 1997.

Pieri, Mario. "I Principii della geometria di posizione, composti in sistema logico deduttivo." *Memorie della reale Accademia delle scienze di Torino* (2) **48** (1899):1-62.

———. "Della geometria elementare come sistema ipotetico-deduttivo; monografia del punto e del mote." *Memorie della reale Accademia delle scienze di Torino* (2) **49** (1899):173-222.

Rowe, David E. "Klein, Mittag-Leffler, and the Klein-Poincaré Correspondence of 1881-1882." In *Amphora: Festschrift für Hans Wussing zu seinem 65. Geburtstag.* Boston-Basel-Berlin: Birkhäuser Verlag, 1992, pp. 597-618.

———. "Perspective on Hilbert." *Perspectives in Science* **5** (1997):533-570.

———. "Episodes in the Berlin-Göttingen Rivalry." *The Mathematical Intelligencer* **22** (1) (2000):60-69.

Russell, Bertrand. "On the Axioms of Geometry." In *The Collected Papers of Bertrand Russell.* Ed. Nicholas Griffin and Albert C. Lewis. London: Hyman Unwin, 1990, 2:394-415.

Schröder, Ernst. "Rezension von Freges Begriffschrift." *Zeitschrift für Mathematik und Physik. Hist.-Lit Abtheilung* **25** (1880):81-94.

———. *Algebra der Logik.* 3 Vols. 1890–1905. Reprint. Ed. New York: Chelsea Publishing Co., 1966.

———. "Über Pasigraphie, ihren gegenwärtigen Stand und die pasigraphische Bewegung in Italie." In *Verhandlungen des ersten Internationalen Mathematiker-Kongresses vom 9. bis 11. August, 1897.* Leipzig: B. G. Teubner Verlag, 1898.

———. "On Pasigraphy: Its Present State and the Pasigraphic Movement in Italy." *The Monist* **9** (1899):44-62. Corrigenda, p. 320.

Schubring, Gert. "Changing Cultural and Epistemological Views on Mathematics and Different Institutional Contexts in Nineteenth-century Europe." In

L'Europe mathématique–Mathematical Europe. Ed. Catherine Goldstein, Jeremy J. Gray, and Jim Ritter. Paris: Éditions de la Maison des sciences de l'homme, 1996, pp. 361-388.

Sluga, Hans. *Gottlob Frege.* London and New York: Routledge, 1980.

Thiel, Christian. "Brouwer's Philosophical Language Research and the Concept of the Ortho-Language in German Constructivism." In *Significs, Mathematics and Semiotics: The Significs Movement in the Netherlands.* Ed. Erik Heijermann and Walter Schmitz. Münster: Nodus Publications, 1991, pp. 21-32.

Toepell, Michael-M. *Über die Entstehung von David Hilberts "Grundlagen der Geometrie."* Göttingen: Vandenhoeck & Ruprecht, 1986.

Van Vleck, Edward Burr. "Zur Kettenbruchentwicklung hyperelliptischer und ähnlicher Integrale." *American Journal of Mathematics* **16** (1894):1-91.

Vucinich, Alexander. "Mathematics and Dialectics in the Soviet Union: The Pre-Stalin period." *Historia Mathematica* **26** (1999):107-124.

Webb, Judson Chambers. *Mechanism, Mentalism and Meta-mathematics: An Essay on Finitism.* Synthese Library. Vol. 137. Dordrecht and London: D. Reidel Publishing Co., 1980,

Welby, Lady Victoria. "Significs." *Encyclopaedia Britannica.* 11th Ed.

______. *What is Meaning? Studies in the Development of Significance.* London: Macmillan, 1903.

CHAPTER 12

The Emergence of the Japanese Mathematical Community in the Modern Western Style, 1855–1945

Chikara Sasaki*
University of Tokyo (Japan)

Japanese Mathematics from Traditional to Modern

Around the Meiji Restoration of 1868,[1] Japanese society underwent a dramatic period of modernization, resulting in a cultural transformation almost unparalleled in worldwide social history.[2] During this time, its long-established traditional culture became almost entirely Westernized, largely through the introduction of Western science, technology, and their institutional bases. The style of mathematics that the Japanese people pursued also changed from *wasan*—the indigenous, traditional, Japanese mathematics—to the more modern methods that had originated in sixteenth- and seventeenth-century Europe.

During this process of rapid Westernization or modernization, the traditional type of scholars virtually disappeared.[3] The community of modern scholars trained in Western science and technology began not only to emerge but also to dominate in almost all intellectual scenes. Even *kampoi*, or traditional Japanese doctors who used Chinese medical practices, failed to retain their intellectual and social status after 1869 when the new Meiji govenment decided that the new Japan's official style of medicine should be Western, particularly German, rather than Chinese. Such a radical change did not occur in China or Korea, where traditional medical practices continued to be respected. But by 1912, the last year of the Meiji period, most of the traditional type of scholars had all but died out in Japan, with most scientists and engineers having adopted the Western style. This chapter examines the similar transformation that occurred in the case of early modern Japanese mathematics.

---

*In this chapter, the names of Japanese are presented in the normal East Asian order, family name first and given name second, except, as here, where the author is writing in a Western language.

[1]The Meiji period, Japan's first modern period under the rule of the Meiji Emperor, lasted from 1868 through 1912.

[2]For general background on the history of nineteenth-century Japan, see, for example, Marius Berthus Jansen, ed., *The Cambridge History of Japan*, vol. 5, *The Nineteenth Century* (Cambridge: University Press, 1989).

[3]Here, "traditional scholars" are those possessing the matured knowledge originating in East Asia, mainly China, Korea, and Japan, for example, traditional *wasan* mathematicians, calendar makers, herbalists, and practitioners of Chinese medicine. On Japanese science before the Meiji Restoration, see Masayoshi Sugimoto and David L. Swain, *Science and Culture in Traditional Japan* (Rutland & Tokyo: Charles E. Tuttle, 1989).

Chinese Mathematics and Its Reform in Seventeenth-Century Japan

Before the introduction of Western methods in the second half of the nineteenth century, Japanese mathematics was the preserve of a group of mathematical practitioners engaged in a traditional subject called *wasan*.[4] *Wasan* was a branch of Chinese mathematics[5] that had been drastically reformed by Seki Takakazu and his followers in seventeenth-century Japan. Seki transformed the Chinese instrumental art named *tianyuan shu* (technique of the celestial element) of solving equations by manipulating calculating rods into a kind of symbolic algebra in the written form called *bosho-ho* (art of side-writing). This specific form of algebra represented both known and unknown quantities by symbols of Chinese ideographs and was later named *tenzan* (adding and deleting) algebra by Matsunaga Yoshisuke, one of Seki's talented followers. With this reformed symbolic algebra, *wasan* mathematicians produced a number of remarkable results. For example, in 1722, Takebe Katahiro, another disciple of Seki, succeeded in computing the value of π to forty-one digits in his masterpiece, *Tetsujutsu Sankei* (*Mathematical Canon on the Art of Linking*), dedicated to the Tokugawa family's eighth shogun.[6]

Following the seclusion policy adopted by the Tokugawa government in 1639, the study of any kind of Western knowledge was prohibited. But in 1720, the shogun had relaxed this policy slightly to allow the study of Dutch scientific works as a part of his program of calendrical reform. As a result, some *wasan* practitioners began to read foreign mathematical books, mainly in Chinese. Toward the end of the Tokugawa, or Bakumatsu, period, there also emerged mathematicians proficient in Dutch. For example, Uchida Itsumi, from the fifth generation of the Seki school, is known to have studied Dutch and to have called his private school of *wasan* "*matematika-juku*," or "school of mathematics," using the Latin word *mathematica*. But, even during the Bakumatsu period, most Japanese mathematicians satisfied themselves with their knowledge of traditional methods, which to them were far superior to their newer competitors.

---

[4] *Wa* was an old name for Japan, and *san* meant arithmetic or mathematics.

[5] On Chinese traditional mathematics, see Yoshio Mikami, *The Development of Mathematics in China and Japan*, 1st ed. (Leipzig: B. G. Teubner Verlag, 1913); 2nd ed. (New York: Chelsea Publishing Co., 1974); Li Yan and Du Shiran, *Chinese Mathematics: A Concise History*, trans. John N. Crossley and Anthony W.-C. Lun (Oxford: Clarendon Press, 1987); and Jean-Claude Martzloff, *A History of Chinese Mathematics*, trans. Stephen S. Wilson (Berlin-Heidelberg-New York: Springer-Verlag, 1997).

[6] On the history of *wasan*, see Mikami, *Development of Mathematics in China and Japan*; David Eugene Smith and Yoshio Mikami, *A History of Japanese Mathematics* (Chicago: Open Court Press, 1914); and Annick Horiuchi, *Les mathématiques japonaises à l'époque d'Edo* (Paris: J. Vrin, 1994). Detailed monographs in Japanese include Endo Toshisada, *Nippon Sugakushi* (*A History of Mathematics in Japan*), revised and enlarged by Mikami Yoshio and Hirayama Akira (Tokyo: Koseisha Koseikaku, 1980), and the Japan Academy, ed., *Meiji-zen Nippon Sugakushi* (*The History of Mathematics in Pre-Meiji Japan*), 5 vols. (Tokyo: Iwanami Shoten, 1954-1960), actually written by Fujiwara Matsusaburo. Mikami Yoshio's small 1922 book, *Bunkashijo yori mitaru Nippon no Sugaku* (*Japanese Mathematics from the Viewpoint of Cultural History*), ed. Sasaki Chikara (Tokyo: Iwanami Shoten, 1999), is quite illuminating. For similarities and dissimilarities between *wasan* algebra and the symbolic algebra of early modern Europe, see Chikara Sasaki, "The French and Japanese School of Algebra in the Seventeenth Century: A Comparative Study," *Historia Scientiarum* **9** (1) (1999):17-26.

Learning Western Mathematics as Military Science, 1855–1868

The type of knowledge gleaned from Dutch scholars and books during the Tokugawa period was called *rangaku* (meaning "Dutch studies" or "Dutch learning"), and, before 1840, the principal Japanese area of interest was medicine. This began to change after the Opium Wars between Qing China and Great Britain. Japan's isolation ended with the signing of the treaties of amity with the United States in 1854 and with the other main Western countries soon thereafter. As a result, feudal Japan opened some of its ports to the Western world, a move soon followed by the establishment of a national navy. To this end, in 1855 the *Nagasaki Kaigun Denshu-sho* (Nagasaki Naval Training Institute) was established, and an official request was made to the Dutch government to send their naval officers to train samurais as navy men.[7] The first group of twenty-two Dutch teachers and seventy samurais arrived in Nagasaki soon after. Altogether, about 200 samurais were trained between the years 1855 and 1859.

It was at the Nagasaki Institute that Western mathematics was systematically taught for the first time in Japan, although not directly by Japanese instructors. Japanese interpreters with a working knowledge of Dutch translated what the Dutch teachers taught, which at this time included arithmetic, the theory of algebraic equations of the second degree, plane and spherical trigonometry, logarithms, and an elementary part of differential and integral calculus.

At the same time, the Tokugawa government also opened an official institution for the translation, study, and teaching of Western languages and scientific subjects. This institution, initially called *Yogaku-sho* (Institute for Western Learning), was opened in Edo in 1856 as *Bansho Shirabe-sho* (Institute for the Investigation of Barbarian [Foreign] Books) amidst strong opposition from Confucianist scholars, who were then promoters of the Tokugawa shogunate's official ideology. The *Bansho Shirabe-sho* was the first central institution to study Western knowledge systematically and, consequently, was a forerunner of the University of Tokyo, the first Japanese university to be modeled in the modern Western style. In 1862, a *rangaku* expert named Kanda Takahira was appointed as instructor of Western mathematics at this institute. Although he certainly possessed an elementary knowledge of Western mathematics, it would be inaccurate to think of him as a real mathematician. The Institute hired him not for his mathematical expertise but for his general knowledge of Dutch studies.

Japanese textbooks containing Western arithmetic were first published in 1857. One was *Seisan Sokuchi* (*A Short Course on Western Arithmetic*) by Fukuda Riken, a *wasan* mathematician with a limited knowledge of Dutch. Another was *Yosan Yoho* (*Method of Western Arithmetic*) (see Figure 1) by Yanagawa Shunsan, a very talented scholar of Dutch studies. In the latter monograph, the Western way of performing the four elementary arithmetical operations and the rule of three in the elementary theory of proportion were explained in detail using the language of *wasan*. At the very beginning of his book's main text, Yanagawa wrote: "Western arithmetic is more or less similar to our art of *tenzan*." In the second column (i.e., vertical line) of the page shown in Figure 2, the author presented the Japanese

---

[7] See Fujii Tetsuhiro, *Nagasaki Kaigun Denshu-sho* (*Nagasaki Navy Training Institute*) (Tokyo: Chuo Koron, 1991). The first chapter of Komatsu Atsuo, *Bakumatsu Meiji-shoki Sugakusha Retsuden* (*Biographies of Mathematicians during the Bakumatsu and Early Meiji Periods*), vol. 1 (Kyoto: Yoshioka Shoten, 1990) is dedicated to the Nagasaki Institute.

(also Chinese) character for addition, its Western symbol +, a corresponding word in Dutch (*optelling*) in its Japanese form, and, finally, a notation in *tenzan* algebra equivalent to $a + b$ in modern European symbolic algebra.[8] Then, successively, subtraction − or *aftrekking*, multiplication × or *vermenigvuldiging*, and division ÷ or *verdeling* are explained by referring to the notation of *tenzan* algebra. In the sixth and final column, he wrote: " ':' this symbol is used for the so-called *evenredigheid*, or proportion." His statements indicate that Western arithmetic could be rather easily translated into the language of *tenzan* algebra for a scholar of Dutch studies during the Bakumatsu period. The only difference may have been the way of writing in addition to the letters; Dutch is written horizontally, while Japanese is vertical! Yanagawa later became a prominent journalist and, after the Meiji Restoration, a key fugure in the establishment of the institution which later became the University of Tokyo.

The Tokugawa shogunate had friendly diplomatic relations not only with the Netherlands but also with France. In the late 1860s, the Yokosuka Shipyard and Arsenal was built with French engineering assistance. After the fall of the Tokugawa regime in 1868, the new Meiji government took over the project, opening the *Yokosuka Kaigun Zosenjo Kosha* (School for the Yokosuka Navy Shipyard) in 1870. There, all instruction was given in French. For example, Paul Sarda, who had graduated from the *École centrale* in Paris, taught various courses in mathematics. Tatsumi Hajime, then a young boy from the domain of Kanazawa, studied there under French teachers including Sarda from 1870 through 1877. His surviving notebooks contain approximately 5,000 pages and include lectures on the theory of algebraic equations as well as on advanced calculus. After graduating from the Yokosuka School, Tasumi was sent to Cherbourg to study shipbuilding for a further four years. He later played an important role as a Japanese naval officer and engineer.[9]

It may seem remarkable that during the Bakumatsu and early Meiji periods, many navy men played leading roles in the introduction of Western mathematics to Japan. This was, however, a natural consequence of the militarization of Japanese learning as a whole at this critical juncture. The upshot was that, at this time, those who studied Western mathematics most seriously were naval officers, and although some civilian scholars such as Yanagawa were certainly familiar with it, their interest was, in general, quite amateur.[10]

Educational Reform during the Early Meiji Period, 1868–1877

During the Bakumatsu period, the Japanese people increasingly came to realize the need for knowledge of Western natural sciences and technology, especially military science, in order to defend their national independence. They had recognized

[8]A photographic edition with explanations of both *Seisan Sokuchi* and *Yosan Yoho* was published under the editorship of Ohya Shin'ichi (Tokyo: Kochi Shuppan, 1979). Figure 2 is on p. 153 (or f. 7r in the original edition).

[9]On the Yokosuka Shipyard, see Tomita Hitoshi and Nishibori Akira (supervised by Takahashi Kunitaro), *Yokosuka Zosenjo no Hitobito* (*People of the Yokosuka Shipyard*) (Yokohama: Yurindo, 1983). On Tatsumi's notebooks, which are now located at the University of Tokyo Library, see Sasaki Chikara, "Tasumi Hajime Monjo no Sugakushiteki-Igi (The Significance of the Documents of Tatsumi Hajime in the History of Mathematics)," *Sugaku Semina (Mathematics Seminar)* **37** (2) (1998):2-5.

[10]Compare the similar situation for China described by Joseph Dauben below.

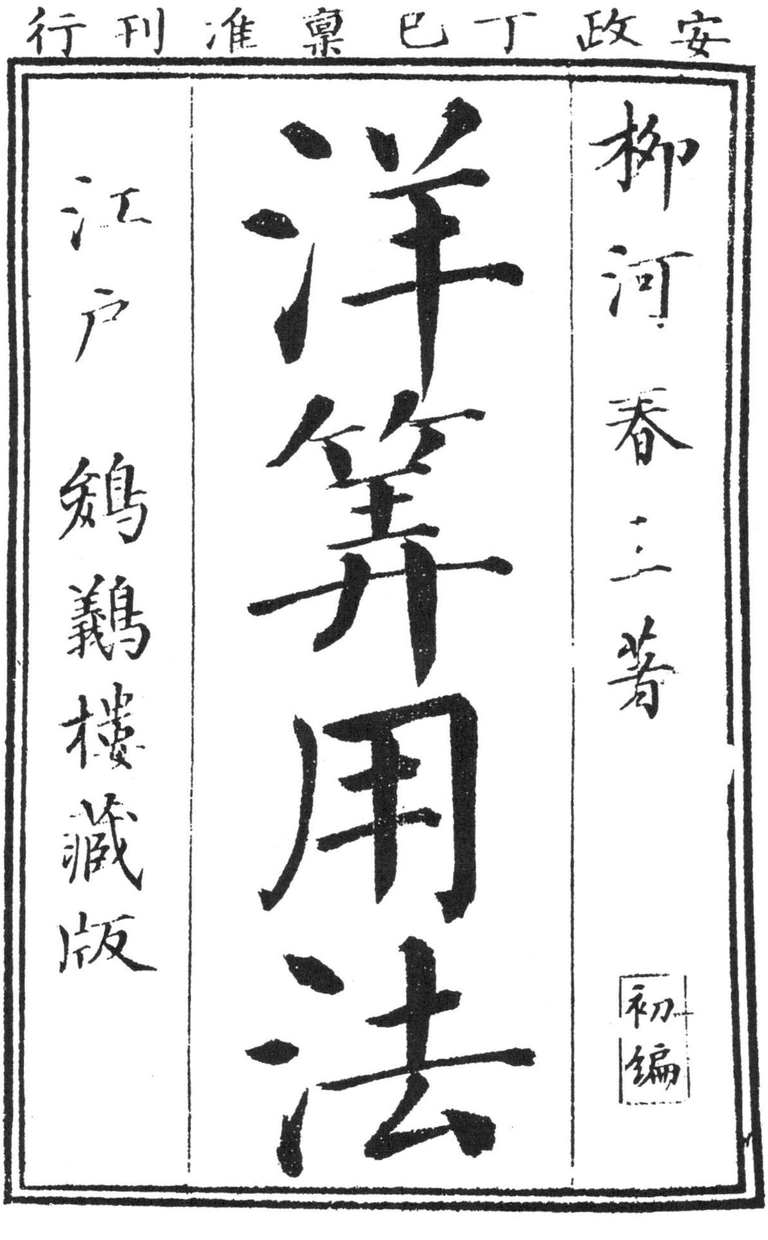

FIGURE 1. The Title Page of Yonagawa Shunsan's *Yôsan Yôhô* (1857)

the superiority of Western military technology both through Qing China's defeats in the first and second Opium Wars and through their own by the British Navy in 1863. The belief in "*toyo dotoku, seiyo geijutsu*" or "*wakon yosai*" (Eastern morality and Western techniques) as formulated by Sakuma Shozan was retained firmly among the feudal Japanese.

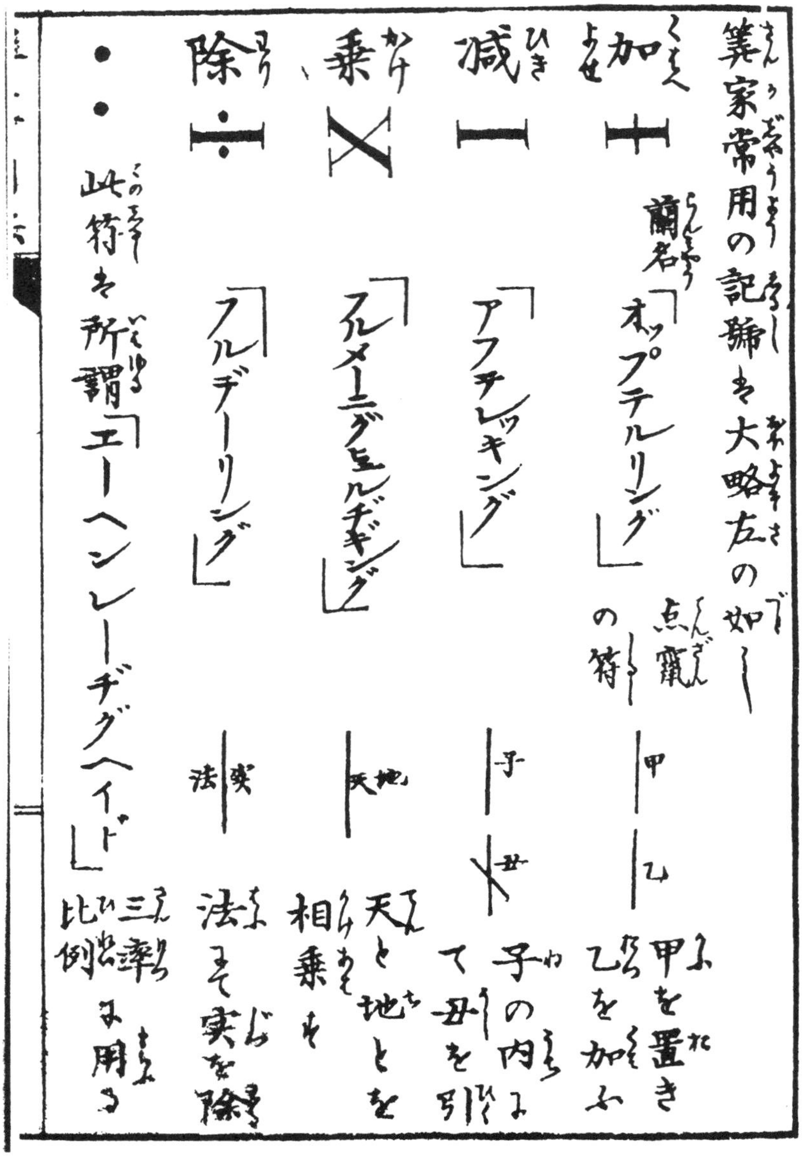
筭家常用の記號を大略左の如し
加 ＋ 蘭名「オップテルリング」 点竄の符 甲 乙 甲を置き乙を加ふ
減 − 「アフテレッキング」 子 丑 子の内より丑を引く
乗 × 「フルメーニグヒュルヂギング」 天 地 天と地とを相乗る
除 ÷ 「ヂーリング」 法 実 法を以て実を除る
∴ 此符を所謂「エーヘンレーヂグヘイド」 三率比例に用る

FIGURE 2. Yonagawa's Explanation of Four Fundamental Arithmetical Operations and of Proportion in *Yôsan Yôhô*

From the very beginning of the Meiji Restoration, the study of Western learning was generally encouraged, although the zeal for *kokugaku* (national learning) and for Neo-Confucian teaching was also still strong. The enthusiasm for Western learning can be sensed from the *Gokajo no Goseimon* (Imperial Oath on the Five Principles), read by the new Meiji Emperor on 14 March, 1868.[11] The fifth principle read:

[11]This is the date in the traditional Japanese calendar; it is April 6 in the European calendar.

"Intellect and learning shall be sought throughout the world, in order to establish the foundations of the Empire."[12] This idealisic tone inaugurated the modern era of *bunmei kaika* (civilization and enlightenment) in Japan.

A key figure in this new period was Fukuzawa Yukichi. In the first section of his *Gakumon no Susume* (*An Encouragement of Learning*) of 1871, one of the bestsellers during the first decade of the Meiji period, Fukuzawa stated: "It is only the person who has studied diligently, so that he has a mastery over things and events, who becomes noble and rich, while his opposite becomes base and poor."[13] In the same spirit of enlightenment, in 1872, the fifth year of Meiji, the central government decided that Western mathematics rather than traditional Japanese mathematics should be a part of the *Gakusei* (school system) for the new Japan. This decision was a crucial event for the Westernization of Japan as a whole and, in particular, for the eventual emergence of the Japanese mathematical community in the modern Western style.

At this point, there were basically two ways for Western culture and ideas to reach Japan: through Western visitors or through Japanese students visiting the Western world. In the first category, foreign employees (*oyatoi*) hired by the Japanese government were key. Examples include the American mathematician and educator, David Murray, who worked for the Meiji government's Ministry of Education between 1873 and 1878 and who proved crucial in implanting the Western educational system on Japanese soil; the German doctor, Erwin Bälz, who worked at the medical school of the University of Tokyo from its foundation in 1877 to 1906, becoming the father of modern Japanese medical science; and Julius Scriba, who, between 1881 and 1901, played a similar role in the case of surgery. Other examples include the Dutch naval officers at the Nagasaki Naval Training Institute and the French who worked at the Yokosuka Shipyard. In all, the number of hired Westerners in the early Meiji period is estimated to have been between 1,500 and 2,000.[14]

Both the Tokugawa and the new Meiji governments sent many students to advanced Western countries, mostly at the state's expense. Between 1853 and 1868, at least 152 people can be identified who went to the Western world for the purpose of study.[15] In 1866, for example, twelve students who had learned English at the Kaiseijo school (formerly the *Bansho Shirabe-sho*) were selected to go to England, since by that time the Tokugawa government had recognized that English-speaking nations were advanced both intellectually and politically. Among the students were Toyama Masakazu, then eighteen years old, who later became a sociologist and President of the University of Tokyo, and Kikuchi Dairoku, then the youngest at eleven years of age, who became the first professor of mathematics at, and later President of, the University of Tokyo.

---

[12] *Meiji Japan Through Contemporary Sources, 1844–1882* (Tokyo: The Centre for East Asian Cultural Studies, 1970), 2:73.

[13] Fukuzawa Yukichi, *An Encouragement of Learning*, trans. David A. Dilworth and Umeyo Hirano (Tokyo: Soshia University, 1969), p. 1.

[14] See Ardath W. Burks, ed., *The Modernizers: Overseas Students, Foreign Employees, and Meiji Japan* (Boulder & London: Westview, 1985); and Iwatsuki Minoru *et al.*, ed., *The Yatoi: A Comparative Study of Hired Foreigners* (Kyoto: Shibunkaku Shuppan, 1987).

[15] On overseas students, see Iwatsuki Minoru, *Kindai Nippon no Kaigai Ryugaku-shi* (*A History of Overseas Students in Modern Japan*) (Kyoto: Mierva Shobo, 1972; reprint ed., Tokyo: Chuokoron, 1992).

Cultural missionaries were also sent to the Western world. Fukuzawa was a representative samurai who went abroad for this purpose. Before the downfall of the Tokugawa regime, he visited the United States twice and Europe once. It was through his own experiences that he became one of the first full-fledged leaders of the enlightenment movement during the early Meiji period.

The University of Tokyo and the Tokyo Mathematical Society, 1877–1881

In the same spirit of enlightenment, in 1877, the tenth year after the Meiji Restoration, the University of Tokyo and the Tokyo Mathematical Society were established. The University, opened on 12 April, was the first national university in modern Japan. Although some foreign mathematics teachers were initially employed there, they were soon replaced by young Japanese professors who had studied overseas. The first such mathematician worthy of attention is Kikuchi Dairoku. Kikuchi was a grandson of Mitsukuri Bempo, a professor at the *Bansho Shirabe-sho*. As noted, he was sent by the Tokugawa shogunate to study in London in 1866. After returning in 1868, the Restoration government sent him back to England, where he enrolled at University College School in London in 1870, and subsequently at University College itself in January 1873. Later that year, he moved to St. John's College, Cambridge, where he studied under Isaac Todhunter, graduating as nineteenth Wrangler in the Mathematical Tripos of 1877. It was Kikuchi who, at the age of twenty-two, became the first professor of mathematics at the newly founded University of Tokyo in 1877.[16] Although not a research mathematician, he had a remarkable career as an educator and administrator of science, serving as Dean of the College of Science from 1881 until 1898 when he was named President of the University, and as Minister of Education of the national government in 1901. A baronage was bestowed on him in 1904 for his efforts as a science administrator and as a promotor of the Anglo-Japanese Alliance negotiated in 1902.

Also in Tokyo, the Tokyo Mathematical Society was one of the first learned societies in Japan to be molded in the modern Western style. Founded in September 1877 with 117 members, it quickly inaugurated a regular publication, the *Tokyo Sugaku Zasshi* (*Journal of the Tokyo Mathematical Society*). During its early years, the great majority of its members were *wasan* mathematicians, with only a few favoring the Western approach. For example, its first President, Kanda Takahira, had been an instructor of mathematics at the *Bansho Shirabe-sho*. As a consequence of the Society's constituency, the mathematical problems in its journal during the early years were in the *wasan* style. However, by about 1884, when it became the Tokyo Mathematico-Physical Society,[17] Western-style mathematicians had achieved dominance and had taken control of its operations. Not surprisingly perhaps, it was Kikuchi who figured crucially in this transformation.

The Germanization of the Political System and of Learning, 1881–1945

Around 1881, the growing Germanization or, more properly, Prussianization of the Japanese educational system became apparent. In fact, the Japanese had

---

[16] For Kikuchi's career, see Editorial Committee, ed., *Nippon no Sugaku Hyakunen-shi* (*A Hundred Years of Japanese Mathematics*), 2 vols. (Tokyo: Iwanami Shoten, 1983–1984), 1:117-122.

[17] See *ibid.*, 1:81-100.

begun to recognize the German superiority in the natural sciences as early as the beginning of the Meiji period. In 1869, two medical doctors, Sagara Tomoyasu and Iwasa Jun, had been ordered by the government to judge which Western system of medicine was superior. They discovered that Dutch medical books had mainly been translated from German and, on this basis, deduced the superiority of German medicine. A document drafted by Sagara stated clearly that "[t]eachers should be hired from Germany, those teachers who have reached manhood and who have mature knowledge of medicine. And they should be two in number for the first year, and afterwards, another three should be hired. Germany is the best country in the world and all other countries have based their medical knowledge on it."[18] From this point until 1945, the model of modern Japanse medicine shifted from Chinese and Dutch to German.

In 1871, the Iwakura diplomatic mission—consisting of Kido Takayoshi, Okubo Toshimichi, Ito Hirobumi, and the Ambassador extraordinary and plenipotentiary, Iwakura Tomomi—was dispatched to the United States and leading European countries. In Prussia, they found the perfect example of a late-developing modernizer like the new Japan. In fact, Otto von Bismarck is said to have candidly explained to the Japanese representatives the need for *Realpolitik* and hard work in their task of nation-building.[19] From this point, Bismarck's Prussia became an ideal model for the Meiji leaders, who wanted to reform Japan into an absolutist constitutional monarchy centered on an Emperor, like the new Germany.

But the Germanization really became apparent with a kind of political *coup d'état* attempted by Ito in 1881. Ito's mentor was the conservative thinker, Inoue Kowashi. On the eve of the *coup*, Inoue composed a memorandum entitled "Jinshin Kyodo Iken-an" (A Draft of My Opinion to Direct People's Mind). This document demonstrated clearly that, by 1881, Meiji Japan no longer regarded Western nations equally; Germany was now its role model for modernization. Inoue emphasized two aspects of education that he believed should be encouraged, namely, traditional Confucianist moral education, on the one hand, and the study of German, on the other. With regard to Chinese studies, he said that "[s]ince the Restoration, English and French studies have had high priority, and this has caused the sprouts of revolutionary thought to appear in our country for the first time. However, for teaching the Way of loyalty to ruler, love of country, and allegiance—values in danger of disappearing at present—nothing equals Chinese studies. We must revive these values and thereby maintain a balance."[20] In Inoue's opinion, German studies were to be encouraged instead of English and French:

> Under our present educational system, the only students who study the German language are found in medicine. Students studying law and related subjects all learn English and French. It is only natural that those who study English admire English ways, and that those who study French envy French

---

[18]Okubo Toshiaki, *Tokyo Teikoku Daigaku Gojunen-shi* (*Fifty Years of the Imperial University of Tokyo*) (Tokyo: The Imperial University of Tokyo, 1932), 1:375.

[19]See Morikawa Jun, *Doitsu Bunka no Ishoku Kiban* (*Bases to Implant German Culture*) (Tokyo: Yushodo Shuppan, 1997).

[20]The Committee for Editing the Biography of Inoue Kowashi, ed., *Inoue Kowashi Den* (*A Biography of Inoue Kowashi*), *Shiryo-hen* (*Historical Archives*) (Tokyo: Kokugakuin Daigaku Toshokan, 1966), 1:250. The quotation in English is from Hirakawa Sukehiro, "Japan's Turn to the West," trans. Bob Tadashi Wakabayashi, in Jansen, p. 494.

> government. But of all nations in present-day Europe, only Prussia is similar to us with regard to the circumstances of its unification .... If we want to make men throughout the land more conservative-minded, we should encourage the study of German and thereby allow it, several years hence, to overcome the dominance now enjoyed by English and French.[21]

For Inoue, and then Ito, liberalism, which was considered to be influenced by British constitutionalism, and radicalism, which was regarded as related to the French Revolution of 1789, were dangerous for the future of Japan. But these ideas of Inoue and Ito were opposed by, among others, Fukuzawa Yukichi, champion of the enlightenment movement, and leaders of the *Jiyu minken undo* (Freedom and People's Rights Movement). With the *coup* of 1881, the study of German began to be encouraged, while British-style liberalism became unfashionable.[22] Students also began to travel to Germany to study. It was in the spirit of Prussian absolutism that the University of Tokyo was reorganized into the Imperial University in 1886.[23]

After 1881, the intellectual and political mood in Japan became authoritarian and militaristic. From the point of view of international political circumstances, it was in the 1880s that the imperialist tempo of the Western powers quickened, with France, Britain, Germany, and other states extending their powers to regions hitherto either loosely tied to European powers or lying beyond their control. In 1881, for example, France established a protectorate over Tunis, Britain occupied Egypt in 1882, Germany began its colonial activities in Southeast Africa in 1883, and France and Britain extended their respective sways to Indochina and Burma between 1884 and 1885. Not surprisingly, Japan was eager to join the ranks of these empire-building nations.[24]

"For the Nation!" Fujisawa and Mathematical Research at Tokyo

The first mathematician to engage seriously in Western-style mathematical research was Fujisawa Rikitaro.[25] Fujisawa entered the University of Tokyo in 1877, where he was taught by Kikuchi. In 1882, acting on his professor's advice, he went to London to study mathematics, but the quality of the English mathematical education available did not prove satisfactory. So, on the urging of his professors there, he moved to Germany, where, at the University of Berlin, he attended lectures by Karl Weierstraß and Leopold Kronecker. Traveling on to Strasbourg in October 1884, he studied analysis with Theodor Reye and Elwin Bruno Christoffel. The University of Strasbourg, founded in 1872 as a "mission for German culture,"[26] had a politically ideal academician in Christoffel because of his loyalty toward the

---

[21] *Ibid.*, pp. 250-251. Compare Hirakawa.

[22] *Tokyo Teikoku Daigaku Gojunen-shi*, pp. 645-646.

[23] Imperial University was renamed the Imperial University of Tokyo with the establishment of the second national university, Kyoto Imperial University, in 1897.

[24] See Akira Irie, "Japan's Drive to Great-Power Status," in Jansen, p. 747.

[25] On Fujisawa's career, see *Nippon no Sugaku Hyakunen-shi*, 1:224-227.

[26] See John Eldon Craig, "A Mission for German Learning: The University of Strasbourg and Alsatian Society, 1870–1918," (Stanford University, unpublished Ph.D. dissertation, 1973).

German nation.[27] This profound sense of purpose proved to be a great influence on Fujisawa, both mathematically and politically.[28]

Fujisawa's period in Strasbourg ultimately played an extremely important role both in the history of mathematics in modern Japan and in his own career. Awarded a doctorate in July 1886 for a thesis on Fourier series, Fujisawa returned to Tokyo in 1887 and became the second professor of mathematics at his *alma mater*, ushering in a drastic change in the mathematical curriculum. Fujisawa seemed to have acquired both the mathematical acumen and the political characteristics of his teacher at Strasbourg. It is reported that when a student of his at the University of Tokyo asked why he studied mathematics, Fujisawa's answer was "For the nation!"[29] No words seem more appropriate to characterize not only why Fujisawa but also why Japanese mathematicians as a whole studied mathematics before 1945.[30]

Fujisawa essentially established a Western mathematical research tradition in Japan. The University of Tokyo had been reconstituted as the Imperial University in 1886 as part of the overall Germanization of Japan immediately before Fujisawa became a professor there. In this sense, Fujisawa's return from Germany was well-timed. Through his rigorous courses of lectures and research seminars, he began to implant a research imperative in the next generation of Japanese mathematicians.[31] Kikuchi had occasionally given lectures on higher mathematical areas such as non-Euclidean geometry and Hamilton's quaternions, but, essentially, his teaching did not go beyond analytic geometry and differential and integral calculus. Most of his lectures are known to have been delivered in English, using no particular textbooks, but they were clearly heavily influenced by the mathematical education he had received in England in the 1860s and 1870s. Fujisawa went far beyond these, delivering lectures on the theory of real and complex variables, the theory of differential equations, and an introduction to the theory of elliptic functions.

Even more important was Fujisawa's introduction of the German-style research seminar. In this forum, he deliberately emphasized algebra, which had witnessed a

---

[27]For a sense of Christoffel's personality and works, see Paul Leo Butzer and Franziska Feher, ed., *E. B. Christoffel: The Influence of His Work on Mathematics and the Physical Sciences* (Basel-Boston-Stuttgart: Birkhäuser Verlag, 1981), and Charles C. Gillispie, ed., *Dictionary of Scientific Biography*, 16 vols. and 2 supps. (New York: Charles Scribner's Sons, 1970–1990), s.v. "Christoffel, Elwin Bruno" by Dirk J. Struik.

[28]Wilhelm Lorey's comprehensive study of mathematical education in nineteenth-century Germany referred to Fujisawa as a product of Christoffel's seminar in this way: "Einer davon ist der japanische Mathematiker Fujisawa [One of them is the Japanese mathematician Fujisawa]." Wilhelm Lorey, *Das Studium der Mathematik an den deutschen Universitäten seit Anfang des 19. Jahrhunderts* (Leipzig: B. G. Teubner Verlag, 1916), p. 158.

[29]Suetuna Zyoiti, "Fujisawa Sensei" (Professor Fujisawa), in *Fujisawa Hakase Tsuiso-roku* (*Reminiscences of Dr. Fujisawa*), ed. The Memorial Committee of Dr. Fujisawa (Tokyo: The Memorial Committee of Dr. Fujisawa, 1938), p. 326. See also Suetuna, *Chosaku-shu* (*Collected Works*) (Tokyo: Nansosha, 1989), 2:133.

[30]Interestingly, Fujisawa is best remembered today as the father of the actuarial profession in Japan. In 1889, he published a book on life insurance designed to prevent workers from committing themselves to radical social movements such as socialism or anarchism. Fujisawa believed that life insurance guaranteed people of lower social status a stable life. His work on life insurance thus seems to have been a natural outcome of his strong nationalist sentiment.

[31]For the terminology "research imperative," see Roy Steven Turner, "The Growth of Professional Research in Prussia, 1818 to 1848: Causes and Context," *Historical Studies in the Physical Sciences* **3** (1971):137-182.

rapid development toward the end of the nineteenth century and in which the university had no specialists. (Kikuchi was essentially a geometer, Fujisawa himself an analyst.) Since records of Fujisawa's research seminars were sometimes published, they exerted an enormous influence on the more ambitious students. For instance, notes written for Fujisawa's seminars by his students were published as *Seminari Enshu-roku* (*Notes of Seminar Exercises*) in five successive volumes between the years 1896 and 1900.[32] Contributors comprised those mathematicians who would later play an important role in the institutionalization of mathematical research in Japanese universities.[33]

Takagi Teiji, the most brilliant student to come out of Fujisawa's research seminars, became the first Japanese mathematician to enjoy an international reputation.[34] It is no exaggeration to say that no history of modern Japanese mathematics would be complete without Takagi. He grew up in a rather poor family in a rural area of central Japan. Thanks to his brilliance, however, he succeeded in entering the Department of Mathematics at the Imperial University in 1894. What he studied can be ascertained from a talk he gave in 1940, entitled "Kaiko to Tenbo" (Reminiscences and Perspectives).[35] According to this, his teachers were Kikuchi and Fujisawa. In his first year, he studied analytic geometry and differential and integral calculus, reading textbooks on elliptic functions and algebraic curves the second year. In his third year, however, as Kikuchi became busy with his administrative work at the Ministry of Education, Takagi came almost completely under Fujisawa's influence.[36] To prepare himself for Fujisawa's research seminar, he read Joseph Serret's two-volume *Cours d'algèbre supérieure* and Heinrich Weber's *Lehrbuch der Algebra*, eventually contributing a paper on the theory of algebraic equations.

In 1898, after having spent one year as a graduate student in Tokyo, Takagi was sent to Germany by the Ministry of Education. The library cards kept at the University of Tokyo show that, before leaving Japan, Takagi borrowed David Hilbert's influential *Zahlbericht* of 1897. At the University of Berlin, he attended courses by Lazarus Fuchs, Hermann Amandus Schwarz, and Georg Frobenius. He

---

[32]The Editorial Committee of the Tokyo Mathematico-Physical Society, ed., *Fujisawa Kyoju Seminari Enshu-roku* (*Notes of Prof. Fujisawa's Seminar Exercises*), 5 vols. (Tokyo: Maruzen, 1896–1900).

[33]To mention only notable names and the titles of their essays (all in Japanese), Hayashi Tsuruichi wrote a paper "On the Transcendence of e and π," Yoshiye Takuzi on "Conformal Mappings," and Takagi Teiji "On Abelian Equations." While these essays, which appeared in volume 2 of 1897, cannot be said to be totally original, they were of a very high standard, especially considering the fact that they were all composed by undergraduate students. Hayashi later became the founder of the Department of Mathematics of Tohoku Imperial University, and both Yoshiye and Takagi became professors of mathematics at the Imperial University of Tokyo. See below.

[34]On Takagi's career, see Shokichi Iyananaga, "On the Life and Works of Teiji Takagi," in Teiji Takagi, *Collected Works*, ed. S. Iyagnaga, K. Iwasawa, K. Kodaira, and K. Yosida, 2nd ed. (Tokyo-Berlin-Heidelberg-New York: Springer-Verlag, 1990), pp. 354-376 (hereinafter cited as *Takagi: Collected Works*), and Kinya Honda, "Teiji Takagi, A Biography," *Commentarii mathematici Universitatis Sancti Pauli* **24** (1975):141-167.

[35]Takagi Teiji, "Kaiko to Tenbo" (Reminiscences and Perspectives), in Takagi Teiji, *Kinsei Sugaku-shi Dan* (*Talks on the History of Modern Mathematics*) (Tokyo: Iwanami Shoten, 1995), pp. 198-214.

[36]Some indication of his superb mathematical ability can be ascertained from a story concerning Takagi's undergraduate days. He scored 140 on an examination of Fujisawa's, its full mark being 100!

was most impressed by Frobenius's vivid lectures on the Galois theory of algebraic equations and on the theory of numbers. After three semesters in Berlin, he moved to Göttingen in 1900, studying under Felix Klein and Hilbert for three semesters. According to Takagi's reminiscences, he listened to Klein's popular lectures on geometry for less than the first six weeks, since they were free of charge and he had acquired sufficient general knowledge. He was attracted by Klein's talks because the Göttingen professor had occasionally criticized mathematical theories of "empty generality [*leere Allgemeinheit*]." But it was Hilbert with whom Takagi shared his passion for algebraic number theory, "especially the special case of the so-called Kronecker Jugendtraum where the ground field is the Gaussian field, that is, a complex multiplication of the lemniscate function."[37] While at Göttingen, Takagi had come to think of mathematical research in Japan as just half a century behind the frontline in advanced European countries such as Germany. But after his eighteen-month sojourn, he felt he had caught up to some extent. In fact, before leaving Germany for Tokyo, he succeeded in solving the problem he had proposed to Hilbert, a result that led to the receipt of his doctorate from the Imperial University of Tokyo in 1903.[38]

Takagi returned to Tokyo toward the end of 1901 and took a position in the Department of Mathematics there. His 1903 thesis provided the first indication that Japan had produced a world-class mathematician. He was promoted to titular professor in 1904, but by 1914, when World War I broke out, Takagi seems to have become rather inactive. For several years, books from Germany rarely reached Japan because of the war. This gave Takagi a chance to devote himself to his own research independently of others. A product was his general class field theory, published in 1920.[39] In September of the same year, Takagi visited Strasbourg in order to participate in the sixth International Congress of Mathematicians (ICM). He read a paper at the Congress on his newly discovered class field theory, which very few people could understand, since the German mathematicians were excluded from the meeting.[40] In fact, there was essentially no reaction to Takagi's lecture whatsoever.[41]

In early 1921, en route from the Congress, Takagi went to the University of Hamburg, where he saw a female assistant reading his monumental paper of 1920. Although it is said that Carl Ludwig Siegel was the first mathematician to recognize its importance, it was not until much later that Hilbert became interested in it and asked Takagi if it could be republished in the prestigious *Mathematische Annalen*. In 1927, Emil Artin, who had been told of the paper by Siegel, generalized Takagi's class field theory, and, consequently, their algebraic number theory became known

[37]Takagi, *Kinsei Sugaku-shi Dan* (*Talks on the History of Modern Mathematics*), p. 204.

[38]Teiji Takagi, "Über die im Berichte der rationalen complexen Zahlen Abel'schen Zahlkörper," *Journal of the College of Science, Imperial University of Tokyo* **19** (1903):1-42, or *Takagi: Collected Works*, pp. 13-39.

[39]Teiji Takagi, "Über eine Theorie des relativ Abel'schen Zahlkörpers," *Journal of the College of Science, Imperial University of Tokyo* **41** (1920):1-133, or *Tagaki: Collected Works*, pp. 73-167.

[40]Compare the chapters by Sanford Segal and Olli Lehto below on the circumstances surrounding the Strasbourg ICM.

[41]The paper Tagaki presented was a summary of the 1920 work. See Teiji Takagi, "Sur quelques théorèmes généraux de la théorie des nombres algébriques," *Comptes rendus du Congrès international des mathématiciens*, ed. Henri Villat (Toulouse: n.p., 1921), pp. 185-188, or *Tagaki: Collected Works*, pp. 168-171.

as Takagi-Artin class field theory.[42] Artin, in turn, advised Helmut Hasse to read Takagi's article, and it was Hasse's *Zahlbericht*, published from 1926 to 1930, that went a long way in introducing the Takagi-Artin theory to a wider audience. The theory was later extended and revised by such mathematicians as Claude Chevalley, Jacques Herbrand, and Takagi's own disciple, Iyanaga Shokichi.

In 1932, Takagi again traveled to Europe, this time to participate as a Vice-President at the ninth ICM held in Zürich. At its opening ceremony, President Karl Rudolf Fueter made a point of calling the audience's attention to the presence of the "Japanese mathematician Takagi."[43] By then Takagi had become rather well-known in the international community of mathematicians, and during this Congress, he was named a member of the Fields Prize committee together with Francesco Severi, Constantin Carathéodory, and Élie Cartan. Besides his mathematical eminence, a possible reason for his inclusion in the committee was his quiet, scholarly, but, above all, non-political nature. It is also likely that he attracted students for these same reasons.

During his last year of service at the Imperial University of Tokyo, in 1935–1936, Takagi gave a course on infinitesimal analysis. (Among the undergraduate students in attendance was Kodaira Kunihiko, who became Japan's first Fields medalist in 1954.) Takagi based the influential 1938 monograph, *Kaiseki Gairon* (*Introduction to Analysis*), on these lectures. This work can be compared to the various treatises entitled *Cours d'analyse* and written by professors of the *École polytechnique* from Cauchy's time to the early twentieth century. Takagi retired from the Imperial University of Tokyo in 1936 and received the Order of Culture from the Japanese government in 1940.

In addition to Takagi, the Department of Mathematics at the Imperial University of Tokyo had several other talented mathematicians,[44] of whom Yoshiye Takuzi and Nakagawa Senkichi are particularly worthy of attention. Yoshiye entered the University in the same year as Takagi and was another favorite student of Fujisawa. For three years from 1899, he also studied with Klein and Hilbert at Göttingen, specializing in the theory of ordinary differential equations and the application of the theory of partial differential equations to the calculus of variations; he obtained some remarkable results, especially in the latter field. Yoshiye became professor at his *alma mater* in 1909, where he taught analysis with special emphasis on the theory of differential equations. One notable feature of his teaching was that he permitted students to submit a report based on their own research instead of taking an examination. Under Yoshiye's guidance, Nagumo Mitio, Kunugui Kinjiro, Hukuhara Masuo, and Yosida Kosaku all became expert analysts and went on to make important contributions to the field. Yoshiye's colleague, Nakagawa, graduated from the Imperial University of Tokyo in 1898 and studied geometry with

---

[42]Katsuya Miyake, "The Establishment of the Takagi-Artin Class Field Theory," in *The Intersection of History and Mathematics*, ed. Chikara Sasaki *et al.* (Basel-Boston-Berlin: Birkhäuser Verlag, 1994), pp. 109-128.

[43]Shokichi Iyananaga, "On the Life and Works of Teiji Takagi," in *Takagi: Collected Works*, p. 364.

[44]On mathematical studies at the Imperial University of Tokyo, see History of Science Society of Japan, ed., *Suri-Kagaku* (*Mathematical Sciences*), vol. 12 of *Nippon Kagakugijutsu-shi Taikei* (*An Outline of the History of Science and Technology in Japan*) (Tokyo: Daiichi Hoki, 1969), pp. 91-96, and *Nippon no Sugaku Hyakunen-shi*, 1:161-175 and 252-258.

Schwarz at the University of Berlin for three years beginning in 1901. A full professor at Tokyo from 1914, he trained some notable geometers such as Yano Kentaro, who later became known in the field of differential geometry.

The Kyoto University School

Following the Sino-Japanese War of 1894–1895, a second imperial university was founded in 1897 in the ancient Japanese capital of Kyoto. The chief mathematicians at this university in its early years were Kawai Jittaro and Sono Masazo.[45] Kawai had been educated by Kikuchi and Fujisawa at Tokyo between 1886 and 1889. Following his graduation, he became professor at the Third High School in Kyoto, where he delivered well-organized lectures that attracted many young students including Hayashi, Takagi, and Yoshiye. Moving to Kyoto Imperial University, he was promoted to titular professor in 1903. He enjoyed a good reputation, although more as a mathematics instructor than as a researcher.

One of Kawai's Kyoto students was Sono Masazo, who, on his advice, became an algebraist. Sono graduated in 1910 and traveled to Europe in 1919, where he attended lectures at the *Collège de France*. Returning to Japan in 1920, he became a Full Professor at Kyoto Imperial University in 1921. His research earned him the reputation of a pioneer of abstract algebra in Japan, especially for his study of the theory of ideals. One of Sono's students, Akizuki Yasuo, also took a position as a professor at the Third High School and attracted students who, in turn, became leading Japanese mathematicians. Moving to Kyoto Imperial University, Akizuki became a professor there in 1948. He worked to advance mathematics through not only his research, but also his organizational skills. He was a close friend of Oka Kiyoshi, who had studied in France from 1929 to 1932 and whose research focused on the theory of functions of several complex variables. Oka's devotion to mathematical research had an almost religious quality to it in the sense that his Buddhist beliefs led him to contemplate difficult mathematical problems. He obtained remarkable results on three problems in the field, namely, Cousin's problem, the approximation problem, and Levi's problem.[46] Thanks to Sono and Akizuki, the Kyoto school produced several high-caliber researchers, particularly in algebraic geometry. For example, the two Fields medalists, Hironaka Heisuke and Mori Shigefumi, came from this school.

The Tohoku University School

In 1907, soon after the Japanese victory over Russia in the war of 1904–1905, a third national university, Tohoku Imperial University, was established in Sendai, the largest city in northeastern Japan. The most important mathematical figure to contribute to the foundation of Tohoku was Fujisawa, who nominated his students, Hayashi and Fujiwara Matsusaburo, as the new university's founding mathematics professors.[47] Hayashi and his younger colleague, Fujiwara, consciously decided to

[45] For the early years of mathematics in Kyoto, see *Suri-Kagaku*, pp. 121-133, and *Nippon no Sugaku Hyakunen-shi*, 1:175-178 and 253-265.

[46] Oka Kiyoshi's principal works were collected and published under the title, *Sur les fonctions analytiques de plusieurs variables* (Tokyo: Iwanami Shoten, 1961).

[47] On mathematics at Tohoku Imperial University, see *Suri-Kagaku*, pp. 143-152, and *Nippon no Sugaku Hyakunen-shi*, 1:239-252. See also Sasaki Shigeo, *Tohoku Daigaku Sugaku-kyoshitsu no Rekishi* (*A History of the Mathematical Institute of Tohoku University*) (Sendai: The Alumni Association of the Mathematical Institute of Tohoku Univeristy, 1984).

make their department a rival to that at Tokyo. Indeed, there was a general wish among the scientific faculty in Sendai that their home institution should become a Japanese replica of Göttingen. By this analogy, the Imperial University of Tokyo was equivalent to the University of Berlin, whose professors were at times distracted by administrative work. In fact, even today the slogan of Tohoku University maintains that "priority is placed on research."

Key institutional factors contributed to the activity of the Tohoku Department of Mathematics during the period before World War II. First, Hayashi began to publish the *Tohoku Mathematical Journal* privately in 1911 as a means for encouraging research activities. Up to this point, there had been no international journals in Japan devoted to mathematics, but the *Tohoku Mathematical Journal* functioned as a unique scholarly resource for Japanese as well as for foreign mathematicians, publishing papers mostly in European languages but occasionally in Japanese. Responsibility for the journal was transferred from Hayashi to the University when the journal became an official organ of the institution in 1916.

Second, the methods of mathematical instruction and research at the University were also unique. Professors and students regularly communicated the results of their mathematical studies with one another, not only through ordinary lectures and seminars, but also through colloquia and "meetings for readings." Colloquia were organized so that faculty members could transmit their own original results and the recent developments of others. "Meetings for readings" were for students to enlarge their stock of mathematical knowledge through reading foreign journals and books. As to mathematical content, the department encouraged researchers and students alike to study various fields instead of one narrow area. In fact, the Departments of Mathematics and of Physics regularly collaborated so that mathematics students ordinarily took basic physics courses and *vice versa*. Sometimes, several common fields or problems were posed and attacked. Consequently, the University became a world leader in differential geometry, particularly in studies of ovals, and in real analysis, especially in Fourier series.

Besides Hayashi and Fujiwara, eminent mathematicians who became faculty members in Sendai before World War II were Kubota Tadahiko, Kakeya Soichi, and Okada Yoshitomo. Kubota, considered the founder of modern geometry in Japan,[48] was succeeded by Sasaki Shigeo, a brilliant differential geometer. Kakeya obtained several highly significant results in the theories of algebraic and integral equations. He became the first mathematician to be awarded the Imperial Academy Prize in 1928 for his results on the theory of integral equations and afterwards moved to the Imperial University of Tokyo. Finally, Okada obtained original results in analysis, particularly concerning series. One of his students, Kakutani Shizuo, became an outstanding functional analyst and obtained a professorship at Yale University after World War II.

Both Hayashi and Fujiwara were known as erudite and prolific authors of high-quality mathematics books. For instance, Fujiwara published a two-volume treatise on advanced algebra called *Daisugaku* (*Algebra*) in 1928–1929, which may be compared to the nineteenth-century classics by Serret or Weber. This is said to have inspired the young Kodaira. Fujiwara also brought out another comprehensive and

[48]Here modern geometry refers to geometry other than Euclidean or elementary analytic geometry.

influential, two-volume book, *Bibun Sekibun-gaku* (*Differential and Integral Calculus*), between 1934 and 1939. One of Fujiwara's most talented disciples in the field of algebra, Tannaka Tadao, was another initiator of the Japanese tradition in abstract algebra. He obtained a remarkable result, known today as "Tannaka's duality theorem,"[49] which extended Pontrijagin's duality theorem on commutative topological groups to non-commutative compact groups.

From its inauguration, Tohoku Imperial University had a very liberal policy toward student admissions in that it was the first national university to admit students who had not graduated from normal high schools. In particular, it accepted female and foreign students, such as the Chinese mathematicians, Chen Kien-Kwong (Chen Jiangong) and Su Bu-Chin (Su Buchin).[50] Chen studied analysis, and especially Fourier series, under Fujiwara, while Su Bu-Chin studied differential geometry under Kubota. In 1929, Chen was the first international student to be awarded a doctorate by a Japanese university. He was offered positions at Beijing University, Wuhan University, and other leading institutions in China, but chose to return to Zhejiang University in Hangzhou to be near his parents. When Su took his doctorate in 1931, Chen invited him to come to Zhejiang as department chair so that Chen could devote himself to mathematical research. They thus established the Chen-Su school of mathematics in Hangzhou. Later, in 1952, both moved to Fudan University in Shanghai and continued their school of mathematics there. In a sense, then, Tohoku Imperial University was one of the birthplaces of modern Chinese mathematics.

From the very beginning, the Tohoku Department of Mathematics also emphasized the history of mathematics because its founders believed research, education, enlightenment, and history to be equally important.[51] Hayashi's interest in studying the history of mathematics, especially *wasan*, began with a suggestion from his teacher, Kikuchi, when he was a student at the Imperial University of Tokyo. His collection of papers on *wasan*, entitled *Wasan Kenkyu Shuroku* (*Collected Papers on the Old Japanese Mathematics*), was published posthumously in two volumes in 1937. Fujiwara began his serious study of *wasan* after Hayashi's death. This resulted in five volumes published (posthumously) from 1954 to 1960, *Meiji-zen Nippon Sugaku-shi* (*The History of Japanese Mathematics before the Meiji Period*). Ogura Kinnosuke, an assistant in the department during its early period, also wrote several informative books on the history of both Western and Japanese mathematics, in which, influenced by Marxist historiography, he took the social dimension of mathematics into consideration. It was from Tohoku University that the eminent historian of mathematics, Mikami Yoshio, took his doctoral degree in 1949. Kondo Yoitsu also studied mathematics and the history of mathematics at the same institution. After the war, he established himself as an historian of Western mathematics and published many influential works, particularly on the formation of non-Euclidean geometry.

---

[49]Tadao Tannaka, "Über den Dualitätssatz der nichtkommutativen Gruppen," *Tohoku Mathematical Journal* **45** (1938):1-12.

[50]The alternative spellings are the renderings into modern *pinyin*. For more on the influence of Japan on China, see Joseph Dauben's chapter that follows.

[51]On the study of the history of mathematics at Tohoku University with related bibliographical data, see Chikara Sasaki, "Historians of Mathematics in Modern Japan," *Historia Scientiarum* **6** (2) (1996):67-78.

The Other Colleges and Universities

Other than the imperial universities in Tokyo, Kyoto, and Sendai, there were few institutions in Japan teaching higher mathematics in the pre-war period: the Mathematics Departments of two teacher-training colleges in Tokyo and Hiroshima and four imperial universities in Sapporo (on Hokkaido Island), Osaka, Fukuoka (on Kyushu Island), and Nagoya. One prominent mathematician who flourished in this environment was Shoda Kenjiro of Osaka Imperial University. Shoda graduated from the Imperial University of Tokyo in 1925 and studied algebra for three years in Weimar Germany, first with Issai Schur at Berlin and then with Emmy Noether at Göttingen. While there, he belonged to a group of young mathematicians known as the "Noether boys" and, in 1932, published a monumental monograph, *Chusho Daisugaku* (*Abstract Algebra*). This work was a Japanese counterpart of the famous textbook, *Moderne Algebra*, published in 1930–1931 by Baertel van der Waerden, one of Noether's best known "boys." On returning to Japan in 1933, Shoda became the founding professor of mathematics at Osaka Imperial University.[52] Once again, the knowledge of abstract algebra, topology, and functional analysis, freshly acquired in Europe, was quickly implanted into the newly established Japanese universities.

One private institution that provided a unique mathematical education also merits mention. The Tokyo School of Physics,[53] established in 1881, was relatively liberal in politics and, in general, emphasized pedagogy over original research. Its founders, who had studied physics in France, actively resisted the increasing Germanization of Japanese culture and produced students of a generally high quality. Interestingly, historians of mathematics such as Mikami and Ogura also taught there.

At all of these institutions where higher mathematics was taught prior to the outbreak of World War II—the imperial universities as well as the teacher-training schools—mathematical research was an intellectual activity aimed at turning Japan into a first-class world power. This is one of the reasons why the German cultural ethos was so influential in modern Japan and why so many young Japanese mathematicians studied in Germany. Moreover, Fujisawa Rikitaro's urge to study mathematics "for the nation" was perfectly in tune with this general mood. It is therefore not surprising that, with a few exceptions, mathematicians were directly or indirectly mobilized for the war without any serious resistence. After all, some were already engaged in research that had direct military application. For them, the "research imperative" and the "war imperative" were almost identical in the sense that both were for the good of the Japanese Empire.

Towards Democratization and Internationalization after World War II

With Japan's surrender to the Allies in 1945 almost everything changed, and mathematical culture was no exception. German authoritarianism was abandoned, and American-style democracy began to prevail. This meant two things in particular: the nationalistic sentiment in the period after 1945 became much weaker

[52]On Shoda's career and works, see *Nippon no Sugaku Hyakunen-shi*, 2:31-32, and *Shoda Kenjiro Sensei: Essei to Omoide* (*Prof. Shoda Kenjiro: Essays and Reminiscences*) (Osaka: Keirinkan, 1978).

[53]On this school, see Kitsuoka Shigeyoshi, *Butsuri Gakko no Densetsu* (*A Legend of the School of Physics*) (Tokyo: Subaru Shobo, 1982).

than it had been during the pre-war period, and the "monopoly" of the imperial universities broke down.

The first important historical event for Japanese mathematicians after the war was the reorganization of their scholarly community. Before 1945, both mathematicians and physicists belonged to the Japanese Mathematico-Physical Society, which had succeeded the Tokyo Mathematico-Physical Society, founded in 1884. Since this arrangement had sometimes been viewed as hindering the autonomous development of the two disciplines, the drastic upheaval following the war provided a good opportunity for reorganization. Consequently, in 1946 the mathematicians set up an independent body called the Mathematical Society of Japan, with Shoda as its first President. The new society had several regional branches and seven subcommittees according to subfields of mathematics, namely, the theory of functions, topology, functional equations, algebra, geometry, probability and statistics, and applied mathematics. It also published two journals, one in Japanese and the other in European languages. The new mathematical society aimed to revitalize mathematical activities that had become stagnant because of the war, although this was not an easy task. It is generally considered that it took almost ten years for Japanese mathematicians to recover from the war's devastating effects.

With the establishment of the new mathematical society, the monopoly of the imperial universities also began to weaken, signifying a "democratization" of the Japanese intellectual community. Around 1949, the whole Japanese educational system underwent a major reform, with the formerly prestigious imperial universities reorganized into ordinary national universities with the same status as teachers' colleges or institutes of technology. This new system, which emphasized general liberal education, was influenced strongly by the American approach. It marked the beginning of a new period for modern, Western-style, Japanese mathematics. The mathematician who worked to promote this process most energetically was Iyanaga Shokichi, who had studied number theory under Takagi in Tokyo, under Emil Artin in Hamburg, and finally under Claude Chevalley in Paris.

Conclusion: General Characteristics of Mathematical Studies in Modern Japan

The history of mathematical culture in Japan since the Meiji Restoration of 1868 can be seen as the story of the transplantation of Western mathematics onto Japanese soil. This is why external or institutional considerations are important for describing its history. In Japan, Western mathematics began at an extremely elementary level, but not from nothing. The Japanese only needed to transform their traditional style of mathematics, *wasan*, into the Western style, since it had already become very sophisticated before the process of Westernization began. At first, the attempt was made to popularize the new mathematics at both the elementary and advanced levels, with Kikuchi Dairoku being the chief contributor to this process. The second step was to instill a research ethos among Japanese mathematicians. The most vibrant center for this kind of activity, the research seminar of Fujisawa Rikitaro at the Imperial University of Tokyo, produced Takagi Teiji, the first Japanese mathematician to become famous internationally. With Takagi's remarkable achievement in class field theory of 1920, Japan can be regarded as having established a research tradition in Western mathematics soon after World War I.

Before 1945, first the Imperial University of Tokyo, and later Kyoto, Tohoku, and Osaka Imperial Universities, became centers of Japanese mathematical research. The generations of mathematicians after Fujisawa diffused what they had learned both from him and from their European (particularly German) teachers. However, although international exchanges among mathematicians certainly existed, they were almost all unilateral. In other words, Japanese students visited European countries to study advanced mathematics and imported mathematical knowledge from there to Japan, but while some overseas mathematicians did come to Japan, such visits were sporadic.

After the chaotic period during and immediately after World War II, Japanese mathematics recovered through the efforts of its own people and with the help of several American mathematicians. The newly organized Mathematical Society of Japan, for example, started its activities as early as 1946. Moreover, the rapid industrialization on a national scale brought with it the expansion of the population of mathematical researchers and students of the mathematical sciences as well as of the associated institutions for mathematical studies. Ambitious Japanese mathematicians soon went to America or France to catch up with the frontiers of mathematical research, and beginning in 1955, international conferences were often held in Japan for the purpose of stimulating talented young researchers. The staging in 1990 of the International Congress of Mathematicians in Kyoto provided symbolic confirmation, if any were needed, of how international Japanese mathematics became during the twentieth century.

Since the days of Takagi, Japan has produced a host of internationally recognized, first-class mathematicians: Oka Kiyoshi, Kodaira Kunihiko, Iwasawa Kenkichi, Sato Mikio, Hironaka Heisuke, Mori Shigefumi, and many others. Interestingly, many have specialized in pure mathematics, a characteristic inherited from pre-modern *wasan*. But the Japanese tendency toward pure mathematics seems also to have been influenced by the source of their modern, Western mathematics. Before World War II, Japanese students studied mostly in the leading German universities, where idealism, in general, and pure mathematics, in particular, prevailed.[54] After the war, they went primarily to the United States, where the research tradition of pure mathematics had been implanted since the arrival of J. J. Sylvester at the Johns Hopkins University in 1876, exactly one year prior to the opening of the University of Tokyo.[55]

From the time Fujisawa brought the German mathematical research seminar to Tokyo, modern Japan has produced talented, creative mathematicians. As a general rule, they have tended to appreciate pure mathematics and seem to have been less skilled in fields connected with the other sciences. This does not mean that they have necessarily been uninterested in applied mathematics, but the distinction between pure and applied mathematics may have been sharper in Japan than in other countries. There have, for example, been few Japanese mathematicians in the mold of Norbert Wiener. Still, Japanese mathematicians have proved themselves adept at solving difficult problems presented by mathematicians overseas. They

[54]See Lewis Pyenson, *Neohumanism and the Persistence of Pure Mathematics in Wilhelmian Germany* (Philadelphia: American Philosophical Society, 1983).

[55]See Karen Hunger Parshall and David E. Rowe, *The Emergence of the American Mathematical Research Community, 1876-1900: J. J. Sylvester, Felix Klein, and E. H. Moore*, HMATH, vol. 8 (Providence: American Mathematical Society and London: London Mathematical Society, 1994), chapters 2 and 3.

have not only caught up with theories created outside Japan, but they have also proven themselves more than capable in the international mathematical arena, as the Taniyama-Shimura-Weil conjecture exemplifies.

Be that as it may, it was through the introduction of Western mathematics, beginning with the deliberate institutionalization of a naval training school and a research institute for Western studies in 1855 and following the gradual abandonment of the old, indigenous mathematics, that Japan began to stand at the frontline of modern, international, mathematical culture.

References

Akira Irie. "Japan's Drive to Great-Power Status." In *The Cambridge History of Japan.* Vol. 5. *The Nineteenth Century.* Ed. Marius B. Jansen. Cambridge: University Press, 1989, pp. 721-782.

Burks, Ardath W., Ed. *The Modernizers: Overseas Students, Foreign Employees, and Meiji Japan.* Boulder & London: Westview, 1985.

Butzer, Paul Leo and Feher, Franziska, Ed. *E. B. Christoffel: The Influence of His Work on Mathematics and the Physical Sciences.* Basel-Boston-Stuttgart: Birkhäuser Verlag, 1981.

The Committee for Editing the Biography of Inoue Kowashi, Ed. *Inoue Kowashi Den* (*A Biography of Inoue Kowashi*). *Shiryo-hen* (*Historical Archives*). Tokyo: Kokugakuin Daigaku Toshokan, 1966.

Craig, John Eldon. "A Mission for German Learning: The University of Strasbourg and Alsatian Society, 1870–1918." Stanford University. Unpublished Ph.D. dissertation, 1973.

The Editorial Committee, Ed. *Nippon no Sugaku Hyakunen-shi* (*A Hundred Years of Japanese Mathematics*). 2 Vols. Tokyo: Iwanami Shoten, 1983–1984.

The Editorial Committee of the Tokyo Mathematico-Physical Society, Ed. *Fujisawa Kyoju Seminari Enshu-roku* (*Notes of Prof. Fujisawa's Seminar Exercises*). 5 Vols. Tokyo: Maruzen, 1896–1900.

Endo Toshisada. *Nippon Sugakushi* (*A History of Mathematics in Japan*). Rev. and Enlarged by Mikami Yoshio and Hirayama Akira. Tokyo: Koseisha Koseikaku, 1980.

Fukuzawa Yukichi. *An Encouragement of Learning.* Trans. David A. Dilworth and Umeyo Hirano. Tokyo: Soshia University, 1969.

Fujii Tetsuhiro. *Nagasaki Kaigun Denshu-sho* (*Nagasaki Naval Training Institute*). Tokyo: Chuo Koron, 1991.

Gillispie, Charles C., Ed. *Dictionary of Scientific Biography.* 16 Vols. and 2 Supps. New York: Charles Scribner's Sons, 1970–1990.

The History of Science Society of Japan, Ed. *Suri-Kagaku* (*Mathematical Sciences*). Vol. 12 of *Nippon Kagakugijutsu-shi Taikei* (*An Outline of the History of Science and Technology in Japan*). Tokyo: Daiichi Hoki, 1969.

Hirakawa Sukehiro. "Japan's Turn to the West." Trans. Bob Tadashi Wakabayashi. In *The Cambridge History of Japan.* Vol. 5. *The Nineteenth Century.* Ed. Marius B. Jansen. Cambridge: University Press, 1989, pp. 432-498.

Honda Kinya. "Teiji Takagi, A Biography." *Commentarii mathematici Universitatis Sancti Pauli* **24** (1975):141-167.

Horiuchi, Annick. *Les mathématiques japonaises à l'epoque d'Edo.* Paris: J. Vrin, 1994.

Iwatsuki Minoru *et al.*, Ed. *The Yatoi: A Comparative Study of Hired Foreigners.* Kyoto: Shibunkaku Shuppan, 1987.

———. *Kindai Nippon no Kaigai Ryugaku-shi* (*A History of Overseas Students in Modern Japan*). Kyoto: Mierva Shobo, 1972. Reprint Ed. Tokyo: Chuokoron, 1992.

Iyanaga Shokichi. "Évolution des études mathématiques au Japon depuis l'ère Meiji." *Historia Scientiarum* **4** (3) (1995):181-206.

———. *Mémoires sur l'histoire des mathématiques contemporaines au Japon.* Tokyo: Maison franco-japonaise, 1996.

———. "On the Life and Works of Teiji Takagi." In Teiji Takagi. *Collected Works*, Ed. S. Iyagnaga, K. Iwasawa, K. Kodaira, and K. Yosida, 2nd Ed. Tokyo-Berlin-Heidelberg-New York: Springer-Verlag, 1990, pp. 354-376.

Jansen, Marius B., Ed. *The Cambridge History of Japan.* Vol. 5. *The Nineteenth Century.* Cambridge: University Press, 1989.

Japan Academy, Ed. *Meiji-zen Nippon Sugakushi* (*The History of Mathematics in Pre-Meiji Japan*). 5 Vols. Tokyo: Iwanami Shoten, 1954-1960. (Actually written by Fujiwara Matsusaburo).

Kitsuoka Shigeyoshi. *Butsuri Gakko no Densetsu* (*A Legend of the School of Physics*). Tokyo: Subaru Shobo, 1982.

Komatsu Atsuo. *Bakumatsu Meiji-shoki Sugakusha Retsuden* (*Biographies of Mathematicians during the Bakumatsu and Early Meiji Periods*). 2 Vols. Kyoto: Yoshioka Shoten, 1990–1991.

Li Yan and Du Shiran. *Chinese Mathematics: A Concise History.* Trans. John N. Crossley and Anthony W.-C. Lun. Oxford: Clarendon Press, 1987.

Lorey, Wilhelm. *Das Studium der Mathematik an den deutschen Universitäten seit Anfang des 19. Jahrhunderts.* Leipzig: B. G. Teubner Verlag, 1916.

Martzloff, Jean-Claude. *Histoire des mathématiques chinoises.* Paris: Masson, 1988.

———. *A History of Chinese Mathethematics.* Trans. Stephen S. Wilson. Berlin-Heidelberg-New York: Springer-Verlag, 1997.

Meiji Japan Through Contemporary Sources, 1844–1882. Tokyo: The Centre for East Asian Cultural Studies, 1970.

The Memorial Committee of Dr. Fujisawa, Ed. *Fujisawa Hakase Tsuiso-roku* (*Reminiscences of Dr. Fujisawa*). Tokyo: The Memorial Committee of Dr. Fujisawa, 1938.

Mikami Yoshio. *Bunkashijo yori mitaru Nippon no Sugaku* (*Japanese Mathematics from the Point of View of Cultural History*). Ed. Sasaki Chikara. Tokyo: Iwanami Shoten, 1999. (Originally published in 1922).

———. *The Development of Mathematics in China and Japan.* 1st Ed. Leipzig: B. G. Teubner Verlag, 1913. 2nd Ed. New York: Chelsea Publishing Co., 1974.

Mikami Yoshio and Smith, David Eugene. *A History of Japanese Mathematics.* Chicago: Open Court Press, 1914.

Miyake Katsuya. "The Establishment of the Takagi-Artin Class Field Theory." In *The Intersection of History and Mathematics.* Ed. Chikara Sasaki *et al.* Basel-Boston-Berlin: Birkhäuser Verlag, 1994, pp. 109-128.

Morikawa Jun. *Doitsu Bunka no Ishoku Kiban* (*Bases to Implant German Culture*). Tokyo: Yushodo Shuppan, 1997.

Oka Kiyoshi. *Sur les fonctions analytiques de plusieurs variables.* Tokyo: Iwanami Shoten, 1961.

Okubo Toshiaki. *Tokyo Teikoku Daigaku Gojunen-shi* (*Fifty Years of the Imperial University of Tokyo*). Tokyo: The Imperial University of Tokyo, 1932.

Ogura Kinnosuke. *Kindai Nippon no Sugaku* (*Mathematics in Modern Japan*). Vol. 2 of *Chosaku-shu* (*Collected Works*). Tokyo: Keiso Shobo, 1973.

Parshall, Karen Hunger, and Rowe, David E. *The Emergence of the American Mathematical Research Community, 1876-1900: J. J. Sylvester, Felix Klein, and E. H. Moore.* HMATH. Vol. 8. Providence: American Mathematical Society and London: London Mathematical Society, 1994.

Pyenson, Lewis. *Neohumanism and the Persistence of Pure Mathematics in Wilhelmian Germany.* Philadelphia: American Philosophical Society, 1983.

Sasaki Chikara. "The Adoption of Western Mathematics in Meiji Japan, 1853-1903." In *The Intersection of History and Mathematics.* Ed. Chikara Sasaki *et al.* Basel-Boston-Berlin: Birkhäuser Verlag, 1994, pp. 165-186.

———. "Asian Mathematics from Traditional to Modern." *Historia Scientiarum* **4** (2) (1994):69-77.

———. "The French and Japanese Schools of Algebra in the Seventeenth Century: A Comparative Study." *Historia Scientiarum* **9** (1) (1999):17-26.

———. "Historians of Mathematics in Modern Japan." *Historia Scientiarum* **6** (2) (1996):67-78.

———. "Tasumi Hajime Monjo no Sugakushiteki-Igi" (The Significance of the Documents of Tatsumi Hajime in the History of Mathematics). *Sugaku Semina* (*Mathematics Seminar*) **37** (2) (1998):2-5.

Sasaki Chikara *et al.*, Ed. *The Intersection of History and Mathematics.* Basel-Boston-Berlin: Birkhäuser Verlag, 1994.

Sasaki Shigeo. *Tohoku Daigaku Sugaku-kyoshitsu no Rekishi* (*A History of the Mathematical Institute of Tohoku University*). Sendai: The Alumni Association of the Mathematical Institute of Tohoku University, 1984.

Shoda Kenjiro Sensei: Essei to Omoide (*Prof. Shoda Kenjiro: Essays and Reminiscences*). Osaka: Keirinkan, 1978.

Suetuna Zyoiti. *Chosaku-shu* (*Collected Works*). 3 Vols. Tokyo: Nansosha, 1989.

Sugimoto Masayoshi and Swain, David L. *Science and Culture in Traditional Japan.* Rutland and Tokyo: Charles E. Tuttle, 1989.

Takagi Teiji. *Collected Works.* Ed. S. Iyagnaga, K. Iwasawa, K. Kodaira, and K. Yosida. 2nd Ed. Tokyo-Berlin-Heidelberg-New York: Springer-Verlag, 1990.

———. "Kaiko to Tenbo" (Reminiscences and Perspectives). In Takagi Teiji. *Kinsei Sugaku-shi Dan* (*Talks on the History of Modern Mathematics*). Tokyo: Iwanami Shoten, 1995, pp. 198-214.

———. *Kinsei Sugaku-shi Dan* (*Talks on the History of Modern Mathematics*). Tokyo: Iwanami Shoten, 1995.

———. "Sur quelques théorèmes généraux de la théorie des nombres algébriques." *Comptes rendus du Congrès international des mathématiciens.* Ed. Henri Villat. Toulouse: N.p., 1921, pp. 185-188.

———. "Über die im Berichte der rationalen complexen Zahlen Abel'schen Zahlkörper." *Journal of the College of Science, Imperial University of Tokyo* **19** (1903):1-42.

______. "Über eine Theorie des relativ Abel'schen Zahlkörpers." *Journal of the College of Science, Imperial University of Tokyo* **41** (1920):1-133.

Tannaka Tadao. "Über den Dualitätssatz der nichtkommutativen Gruppen," *Tohoku Mathematical Journal* **45** (1938):1-12.

Tomita Hitoshi and Nishibori Akira. *Yokosuka Zosenjo no Hitobito* (*People of the Yokosuka Shipyard*). Yokohama: Yurindo, 1983.

Turner, Roy Steven. "The Growth of Professional Research in Prussia, 1818 to 1848: Causes and Context." *Historical Studies in the Physical Sciences* **3** (1971):137-182.

CHAPTER 13

Internationalizing Mathematics East and West: Individuals and Institutions in the Emergence of a Modern Mathematical Community in China

Joseph W. Dauben
City University of New York (United States)

Introduction

Mathematics in the West has long been associated with teachers, schools, and a variety of institutions—including universities, academies, and journals—by means of which mathematical communities have been created and defined. In the course of the nineteenth century, mathematics in Europe became less parochial, less nationalistic. By the end of the century, the first truly international meetings for mathematicians were being held in Europe and North America. Only recently, however, has China come into the international picture, adopting in the process a basically Western model of teaching, publishing, and institutionalizing mathematics. Traditionally, China has supported a large class of practitioners who have used mathematics in a variety of ways for pragmatic ends. Interests that in the West would be considered didactic, theoretical, or even philosophical were basically foreign to the Chinese temperament for a variety of reasons. Thus, important questions about society and ideology arise in trying to account for what happened when Chinese scholars first encountered Euclid and the axiomatic method through the translation efforts of Jesuits and other missionaries in the late Ming and early Qing dynasties (basically, the seventeenth century).[1]

At the time, the Chinese cannot be described as a receptive audience, but this was to change. Two centuries later, following the Opium Wars that began to reopen China to the West after 1839, and somewhat later the "Self-Strengthening" or "Westernization Movement" that stimulated thoughts of reform, a series of events occurred that ultimately transformed the perceptions of Chinese intellectuals. China's defeat by France in 1885, its unsuccessful war with Japan in 1895, the Boxer Uprising of 1900, and finally, the Revolution of 1911, opened the eyes of reformers to the importance of adopting Western practice as well as theory, if China were to compete and survive in the modern world.

---

[1]For a discussion of the earliest reception of Euclid in China, refer to Peter M. Engelfriet, "The Chinese Euclid and its European Context," in *L'Europe en Chine: Interactions scientifiques, religieuses et culturelles aux XVIIe et XVIIIe siècles: Actes du colloque de la Fondation Hugot*, ed. Catherine Jami and Hubert Delahaye, Mémoires de l'Institut des hautes études chinoises, vol. 34 (Paris: Collège de France and Institut des hautes études chinoises, 1993), pp. 111-135, and *Euclid in China: The Genesis of the First Chinese Translation of Euclid's Elements Books I-VI (Jihe yuanben, Beijing, 1607) and Its Reception up to 1723*, Sinica Leidensia, vol. 40 (Leiden: Brill, 1998).

This point of view was eventually reflected in educational changes throughout China. The constitutional Reform Movement of 1898 under the Emperor Guang Xu resulted in new institutions and modernization of education, including the founding of the Capital University (later Beijing University). Although the Reform Movement of 1898 only lasted 100 days, by 1905 the central government had done away with the civil service examinations, and a new Education Department had been founded. After the Revolution in September of 1911, further changes in the educational system were made, and mathematics became a compulsory subject for all students.

In the course of these dramatic changes, Western mathematicians began to visit China, and for the first time, Chinese intellectuals encountered scholars who were not missionaries interested first in conversion and only incidentally in science. Similarly, for the first time, Chinese scholars in modest numbers began to travel and study abroad. In learning Western mathematics, they were also exposed to and quickly adopted more than just its content. As the mathematical community in China became increasingly Westernized, and correspondingly internationalized, schools, colleges, universities, mathematical societies, academies, and journals all began to fashion a growing Chinese research community. How this happened in concert with the international community to which it was indebted and on which it largely depended constitutes one of the main concerns of this chapter, which is devoted primarily to the emergence of a local, indigenous, mathematical community in China in the twentieth century.

Modern Science Emerges in China: The Self-Strengthening Movement

In the 250 years after the first translation of Euclid into Chinese was completed in 1610, China underwent profound and dramatic changes, the most important of which for the fate of modern mathematics at the end of the nineteenth century was what the Chinese called the *yangwu yundong* (Westernization movement) or the "self-strengthening movement." Moreover, the most significant reason why China was more receptive to Western science in the nineteenth century than it had been in the seventeenth century is directly related to the Opium Wars and the repeated military defeats China suffered throughout the rest of the century. The First Opium War (1839–1842), resulting in the treaties of Nanjing (1842) and the Bogue (1843), not only opened the ports of Canton (Guangzhou), Amoy (Xiamen), Fuzhou, Ningbo, and Shanghai to the British, but also ceded Hong Kong to Britain. Even more significant in opening China to foreign influence was the treaty of Tianjin signed in 1860 at the conclusion of the Second Opium War (1856–1860), in which the French also took part. This opened additional ports to foreigners and not only allowed them to travel in China but also provided for the admission of missionaries.[2]

One especially important provision of the Treaty of Tianjin was the requirement that thenceforth all official communications from Chinese authorities to representatives of the British and French governments be conveyed in English or French. In order to deal effectively with foreign envoys, the first Chinese Ministry of Foreign Affairs, the *Zongli Yamen*, was established in 1861, and this, in turn, created the

---

[2]For details, see Jack Beeching, *The Chinese Opium Wars* (London: Hutchinson, 1975), and Edgar Holt, *The Opium Wars in China* (London: Putnam, 1964). The creation of the Chinese Ministry of Foreign Affairs was also a result of the Treaty of Tianjin.

Tongwen Guan, or College of Languages, which was also authorized by the imperial government in Beijing in 1861.[3] Similar schools were established in Canton and Shanghai in 1863.

The language schools were not only necessary for training the diplomats and administrative go-betweens who would serve China's political and economic interests in dealing with foreigners, they were also a necessary prerequisite if China were serious about importing Western learning. Another and more important result of the Opium Wars, in addition to opening China to foreign influences, was the sudden awareness of the Chinese that to compete and survive in the modern world, China would have to adopt modern ways. The Chinese equated the success of the British armed forces with armaments and the building of ships. The latter were associated with engineering and a basic knowledge of science, which, in turn, were believed to be dependent upon mathematics as the ultimate foundation for all of Western science.[4] These sentiments were significantly reinforced in the aftermath of the Sino-Japanese War over Korea (1894–1895), two results of which were the independence of Korea and the loss of Taiwan to Japan. This time, it was Japan's success in Western modernization that made a strong impression on Chinese leaders.[5]

Thus, it was apparent as a result of repeated lessons learned in the latter half of the nineteenth century that in order to establish a solid, indigenous cadre of Chinese scientists and engineers, it would be necessary to learn both Western languages and Western sciences. Consequently, attempts were made not only to teach foreign languages but also to add mathematics and the sciences to the curricula at a variety of institutions, among the most important of which were schools associated with the *Tongwen Guan* (f. 1861–1862) in Beijing, the Shanghai *Tongwen Guan* (1863–1864), the Jiangnan Arsenal at Shanghai (1865), and the Fuzhou Shipyards School (1866).[6] The aim of all these institutions, to varying degrees, was "to impart specific

---

[3]In a memorial to the Emperor of 13 January, 1861, the *Zongli Yamen*, China's first foreign office, was proposed, along with creation of the *Tongwen Guan*, which was in operation a year later. For a detailed description of the school, see Knight Biggerstaff, *The Earliest Modern Government Schools in China* (New York: Cornell University Press, 1961), pp. 94-153. For the memorial to the Emperor on "The New Foreign Policy of January 1861," see Teng Ssu-Yü and John K. Fairbank, *China's Response to the West: A Documentary Survey, 1839–1923* (New York: Atheneum, 1970), pp. 47-49. Note that the names of Chinese institutions, like the *Tongwen Guan*, are given here in their *pinyin* romanization, although many variants will be found in the literature; Biggerstaff, for example, refers to the same institution as the *Tung-wen kuan*.

[4]This was similar to French feelings about Germany in the wake of the Franco-Prussian War of 1870–1871. For the fortunes of mathematics in France in the late nineteenth century, see the contributions to this volume by Thomas Archibald and Hélène Gispert.

[5]One immediate response, as the twentieth century began, was that large numbers of Chinese students were sent to Japan for their education, although most of these studied literature or military science. Few studied the sciences, and even fewer, mathematics. See Jean Chesneaux, *et al.*, *China from the Opium Wars to the 1911 Revolution*, trans. Anne Destenay (New York: Pantheon Books, 1976); John K. Fairbank, *The United States and China*, 4th ed. (Cambridge: Harvard University Press, 1979); Edwin O. Reischauer, John K. Fairbank, and Albert M. Craig, *East Asia: Tradition and Transformation* (Boston: Houghton Mifflin, 1973). Also relevant to the transmission of modern mathematics to China is the example of and China's interaction with Japan; see the contribution to this volume by Chikara Sasaki.

[6]Delia Davin, "Imperialism and the Diffusion of Liberal Thought: British Influences on Chinese Education," in *China's Education and the Industrialized World: Studies in Cultural Transfer*, ed. Ruth Hayhoe and Marianne Bastid (Armonk, NY: M. E. Sharpe, 1987), pp. 33-56 on p. 36. To this list add the Tianjin Telegraph School (1880, notable for its all-Danish staff) and the Viceroy's Hospital Medical School at Tianjin (1881). See as well Hiroshi Abe, "Borrowing from Japan: China's First Modern Education System," in *China's Education and the Industrialized*

skills in language and technology to a very small number of selected students."[7] By 1895, in fact, there were no fewer than twenty-two establishments of various kinds teaching diverse aspects of Western knowledge in China. Virtually all of these included mathematics in one form or another. There were also educational programs abroad, training carefully selected Chinese students in the United States, France, England, and Germany. But the numbers were small, and by 1894, on the eve of the Sino-Japanese War, barely 300 students had studied abroad.[8] Before analyzing the significance of these efforts for mathematics in China, it is necessary to examine four of the most important, local, teaching institutions and their impact on Chinese mathematics more closely.

The Beijing *Tongwen Guan* (1861–1862)

As already noted, a significant result of the British and French Treaties of Tianjin was the creation of a school to train young Chinese students in foreign languages, so that all official communications with English and French envoys might be conducted in their native languages. Regarded as the first "modern school" to be established by the Chinese government, the Beijing *Tongwen Guan* hoped to find competent Chinese instructors of French and English in either Canton or Shanghai. Failing this, they turned to foreigners. The earliest to be hired was an Englishman, John S. Burdon, who met with his first class of ten students in July of 1862.[9] In April of the folowing year, two classes of twenty students each had begun to study French and Russian. In 1863, Burdon was replaced by another Englishman, John Fryer, who left a year later to head the translation department of the Jiangnan Arsenal in Shanghai (see below) and was, in turn, replaced by an American, William A. P. Martin.[10]

In 1866, the *Zongli Yamen* added a scientific department to the language school, described by Knight Biggerstaff as "an even more radical step than the setting up of a foreign language school had been in the first place, for it appeared to imply recognition of shortcomings in the time-honored Chinese educational system."[11] In December, a proposal was submitted to the Throne calling for the creation of a Department of Astronomy and Mathematics, to which students would be admitted by special examination. It was not until 1869, however, that the first mathematician, Li Shanlan, was appointed to teach the new subject. Li, an outstanding Chinese mathematician recruited from Shanghai, was the only Chinese to teach in the "scientific department," as Martin referred to it. The students, while studying English and French, were otherwise taught by foreign science teachers.

Martin assumed the directorship of the *Tongwen Guan* full-time. As he wrote to Frederick F. Low, the American minister in Beijing, on 18 January, 1872:

World, ed. Ruth Hayhoe and Marianne Bastid (Armonk, NY: M.E. Sharpe, 1987) pp. 57-70 on p. 58. Abe also notes the Shanghai School of Engineering, the Tianjin Naval Academy, the Guandong Naval and Military Academy, and the Hubei Self-Strengthening School, among others.

[7] Davin, p. 36. See also Biggerstaff, *The Earliest Modern Government Schools in China*, especially pp. 65-68.

[8] Marianne Bastid, "Servitude or Liberation? The Introduction of Foreign Educational Practices and Systems to China from 1840 to the Present," in *China's Education and the Industrialized World: Studies in Cultural Transfer*, ed. Ruth Hayhoe and Marianne Bastid (Armonk, NY: M. E. Sharpe, 1987), pp. 3-20 on p. 8.

[9] Biggerstaff, *The Earliest Modern Government Schools in China*, p. 99.

[10] For more on the English and American influences in China, see below.

[11] Biggerstaff, p. 108.

> From the date of my appointment, a little more than two years ago, the College (as we prefer to call it) has been undergoing a gradual but thorough reorganization. ... Mathematics are taught by a native professor, whose influence does much to awaken a taste for such studies in the minds of his countrymen. Physics have been taught by the President (Mateer); and since the last Spring we have had a professor of Chemistry lecturing and giving experiments to intelligent classes. Within the month past, a chair of Anatomy and Physiology has been added; and the Professor (Dr. Dudgeon) is to enter on his duties with the opening of the coming Spring. ... As to students, we have but a handful, only ninety-two, all told; but this is twice as many as Mr. Seward found in attendance at the time of his visit a year ago. The best of these have been brought up from the schools at the open ports, which are henceforth to be regarded as feeders for the central institution and affiliated to it. ... [12]

Mathematics was taught in both parts of the Beijing *Tongwen Guan*, in the language school and in the scientific school, but more intensively in the five-year scientific program, which included arithmetic, algebra, geometry, plane and spherical trigonometry, differential and integral calculus, as well as navigation, astronomy, and practical mechanics. In 1879, out of a total of 103 students in the Beijing *Tongwen Guan*, thirty-three were studying mathematics; in 1898, out of 133 students, thirty were studying mathematics.

What Li Shanlan taught was traditional Chinese mathematics alongside European mathematics at the *Tongwen Guan*. He based the latter primarily on translations he made himself of Western works. When he died in 1882, having taught at the College for thirteen years, he was eventually replaced not by a foreign mathematician, but by a graduate of the *Tongwen Guan*, Hsi Kan (Xi Gan),[13] who taught mathematics from 1886 until 1898. Of graduates who did not go into government service as translators, many apparently applied their knowledge as teachers in modern schools that were beginning to appear throughout China.[14]

Because the primary aim of the *Tongwen Guan*, however, was to train young men for government service, the education given to those in the science school did not lead to great payoffs in terms of creative or sustained scientific activity. Perhaps the best that can be said for the *Tongwen Guan* is that it encouraged "a taste" for mathematics in the minds of its students, some of whom—like Xi Gan—went on to become teachers of mathematics themselves. The only long-term impact of the *Tongwen Guan* was linked to the translations that it published, which by the end

---

[12] *Ibid.*, p. 126, quoting from U.S. Department of State Legation Archives, China, CCXLIII, pp. 92-93.

[13] Chinese names may be transliterated in any number of ways, and names transliterated from dialectical pronunciations like Cantonese may diverge substantially from the standard Mandarin pronunciation of the same characters. When multiple versions of the same name appear in the text, the first is the name as it appears in whatever work is cited; the second in parentheses is the now-standard *pinyin* transliteration of the same name. Thus, Hsi Kan is the transliteration used by Biggerstaff; Xi Gan is the *pinyin* equivalent.

[14] Biggerstaff, p. 147.

of the century amounted to some two dozen books, ten or so of which were devoted to the sciences.[15]

Apart from its publications, the *Tongwen Guan's* most important role was as a model for the new universities that were just beginning to appear in China. As Biggerstaff notes: "Finally it was in the end the model for the new Peking University, which was to become within two decades the center of an epochal and vigorous cultural renaissance and within four decades one of the great educational institutions of the world."[16] Even earlier, however, the Beijing *Tongwen Guan* had served as a model for similar institutions in Shanghai and Canton, where Western mathematics and sciences were also introduced. For the future of mathematics in China, the Shanghai *Tongwen Guan* was to prove especially important.

The Shanghai *Tongwen Guan* (1863–1864) and the Jiangnan Arsenal (1865)

On 11 March, 1863, Li Hung-chang (Li Hongzhang), Superintendent of Trade for southern ports and Governor of Jiangsu Province, submitted a memorial to the Emperor urging the establishment of foreign language schools in Shanghai and Canton (Guangzhou), to be modeled on the *Tongwen Guan* in Beijing.[17] He also explained how "instruction in foreign languages would ultimately provide a key to Chinese mastery of Western mathematics, physics, and technology, preparing the way for the construction of steamships and the manufacture of firearms in China."[18]

The Shanghai *Tongwen Guan* was established immediately, with a curriculum including English, mathematics, Chinese classics, history, and writing. English and mathematics were to be studied every day, the other subjects in what time remained. As Biggerstaff emphasizes: "All subjects except foreign languages were to be taught by the Chinese professors. Particular emphasis was laid on the study of mathematics, which was recognized as the basis of Western technological achievement; specialization in mathematics was to be not only permitted but encouraged. ... Prizes of from four to eight taels would be awarded to the best students in order to encourage diligence."[19]

As the Shanghai *Tongwen Guan* was getting under way, plans for an arsenal that would also serve as a shipyard for the manufacture of arms and the construction of ships were being laid as early as 1865.[20] By 1869, it had been decided to

---

[15] *Ibid.*, p. 151. The question of the influence of the translations made by the Beijing *Tongwen Guan* remains in dispute, especially since the translations were distributed by the Ministry of Foreign Affairs; some, it seems, were reprinted by missionary newspapers or sold in bookstores. See Hu Mingjie, "Merging Chinese and Western Mathematics: The Introduction of Algebra and the Calculus in China, 1859–1903" (Princeton University, unpublished Ph.D. dissertation, 1998; Ann Arbor: University Microfilms #9833137, 1998), pp. 93-96, and Su Jing, *Qing ji Tongwen Guan* (*The Tongwen Guan of the Late Qing*) (Taipei: Shanghai Yinshuachang, 1977), p. 207.

[16] Biggerstaff, *The Earliest Modern Government Schools in China*, p. 153.

[17] Note that while in *pinyin* it is the Jiangnan Arsenal, other works often refer to this same institution as the Kiangnan Arsenal, but the two are identical. For a detailed study of the Jiangnan Arsenal, see Thomas L. Kennedy, *The Arms of Kiangnan: Modernization in the Chinese Ordnance Industry, 1860–1895* (Boulder: Westview Press, 1978).

[18] Biggerstaff, p. 156.

[19] *Ibid.*, p. 158.

[20] The Arsenal's official name, *Jiangnan jiqi zhizao ju* (Biggerstaff = *Kiangnan chi-ch'i chih-tsao chu*), may be directly translated as Jiangnan Machine Manufacturing Bureau. Here, it will be referred to in English by its popular designation, the Jiangnan Arsenal. On the founding of the Arsenal, see Teng and Fairbank, pp. 64-65. On the arsenals, in general, and science in China,

move the Shanghai *Tongwen Guan* to the Arsenal, where all teaching and translation activities could thenceforth be coordinated. Mathematics received special emphasis due to its status as one of the "six arts" of traditional Chinese study and as the foundation of Western technology. The Arsenal also sought to build up a comprehensive library of Western and Chinese works and even maintained a small collection of scientific instruments.

The translation department of the Jiangnan Arsenal was under the control of Hsü Shou (Xu Shou) and Hua Hengfang. Hua became noted for the translations he produced, together with the Englishman, John Fryer, of many basic scientific works that served as textbooks in "Westernized Schools" like those in Shanghai.[21] Fryer was a prolific translator and either translated or wrote nearly ninety works published by the Arsenal. Among other Westerners who joined the Arsenal's translating team, the Englishman, Alexander Wylie, was also interested in mathematics.[22]

In 1905, the foreign language school and translation department of the Shanghai Arsenal were merged into a new School of Technology, which soon thereafter was reorganized by the Ministry of War into a technical military school on three different levels. Although the Arsenal's translation department continued to operate until sometime after the fall of the Qing dynasty, it was unable to maintain the pace or quality of publications set by Fryer and Wylie.

For the advancement of mathematics in China, the Jiangnan Arsenal was clearly much more important for what it translated than for what it taught. As for its graduates, except for those from the language school who went on to more advanced work in Beijing's *Tongwen Guan*, most who graduated from the Shanghai school apparently had considerable trouble finding stable or productive employment. Demand for professional interpreters turned out to be much less than anticipated, although of twenty-five graduates who went on to the Beijing *Tongwen Guan* between 1867 and 1890, many rose to positions of considerable importance.

As for the Arsenal's publications, John Fryer, Hua Hengfang, Xu Shou, and Young Allen were primarily responsible for the translation department. In the single decade between 1870 and 1880, over 30,000 copies of its books were sold, not to mention numerous cheaper, pirated editions. After the Sino-Japanese War, when even greater emphasis was placed on the importance of Western scientific knowledge and mathematics, demand for publications from the Jiangnan Arsenal grew substantially. In addition to publishing translations of articles from the *Encyclopaedia Britannica* on "algebra," "fluxions," and "probability," as well as introductory texts on practical geometry and plane and spherical trigonometry, John Fryer translated

---

consult William J. Haas, *China Voyager* (Armonk, NY: M. E. Sharpe, 1996), pp. 60-63. See also Biggerstaff, *The Earliest Modern Government Schools in China*, p. 201.

[21]The translations of Hua and Fryer were generally regarded as easier to use by local Chinese than Li Shanlan's translations. Although John Fryer (1829–1928) is described by Meng Yue as a "sojourner," in addition to the position he held in the Jiangnan Arsenal, he also helped to popularize science and engineering through a popular science journal he founded, the *Gezhi huibian* (*Scientific and Industrial Magazine*, 1876–1892). Fryer not only edited the *Magazine*, he financed the journal, which was printed by the London Missionary Society Press (*Mohai shuguan*). See Meng Yue, "Hybrid Science versus Modernity: The Practice of the Jiangnan Arsenal, 1864–1897," *East Asian Science, Technology, and Medicine* **16** (1999):13-52, especially p. 28.

[22]For more about Fryer and Wylie, see below. Others on the staff, like Daniel J. Macgowan, had other interests; Macgowan, for example, helped Hua Hengfang translate Lyell's *Geology* into Chinese and wrote a book in Chinese on typhoons and marine meterology. See Biggerstaff, *The Earliest Modern Government Schools in China*, pp. 174-175.

Augustus De Morgan's *The Elements of Arithmetic* into Chinese, and produced works with such indicative titles as *Weiji xuzhi* (*What One Should Know About the Infinitesimal Calculus*) and *Quxian xuzhi* (*What One Should Know About Curved Lines*).[23]

Although the mathematical level of such works fell far short of what contemporary, European mathematicians were then producing, these works were more suitable to the fledgling abilities of Chinese students (and perhaps to the levels of their translators as well). And while it may seem strange that any effort should have been given to translating so dated a work as one on "fluxions," or even De Morgan's *Arithmetic*, which by the time Fryer got to it was nearly a half-century old, these choices may simply reflect what their English translators had themselves studied in school a generation earlier, or had available to them in Shanghai.

Despite the disappointingly low level of the works they produced, the translators of the Jiangnan Arsenal, and among them especially John Fryer, were quite successful in making basic mathematical texts available in Chinese. As Knight Biggerstaff surmises in his study of the Jiangnan Arsenal, "[t]here can be little doubt of the great contribution made by the translation department of the Kiangnan [Jiangnan] Arsenal to the introduction of Western knowledge into China—particularly of scientific and technical knowledge."[24] In fact, although the Jiangnan Arsenal launched only a "negligible" number of ships (nine in all between 1868 and 1885), and produced only limited amounts of gunpowder and small arms, its translation activities far surpassed what the other arsenals and language institutes combined managed to translate during roughly this same period.[25]

The Fuzhou Shipyard (1866)

The Fuzhou Shipyard was established in 1866, fully in the spirit of China's "Self-strengthening Movement."[26] Although Chinese craftsmen were experienced in building wooden ships, steamships were an entirely different matter, and the

---

[23]This inventory of mathematical texts is based upon forty-six titles of works translated into Chinese listed in Jean-Claude Martzloff, *A History of Chinese Mathematics*, trans. Stephen S. Wilson (Berlin: Springer-Verlag, 1997), pp. 372-389. Where the sources of the translations are known, they are given along with the edition upon which the translation was based. Note that it was in 1879 that Fryer translated De Morgan's *Arithmetic*, which had first appeared in 1830; Alexander Wylie had already translated De Morgan's *Algebra* of 1835 into Chinese in 1859. See below.

[24]Biggerstaff, *The Earliest Modern Government Schools in China*, p. 197.

[25]In addition to the data provided by Biggerstaff, see the list of publications in Adrian A. Bennett, *John Fryer: The Introduction of Western Science and Technology into Nineteenth-Century China* (Cambridge: East Asian Research Center, Harvard University, 1967), as well as the study by Tsien Tsuen-hsuin, "Western Impact on China Through Translation," *Far Eastern Quarterly* **13** (1954):305-327.

[26]This is also referred to as the Foochow Shipyard by Mary Clabaugh Wright, *The Last Stand of Chinese Conservatism: The T'ung-Chih Restoration, 1862–1874* (Stanford: University Press, 1957), pp. 212-213; as the Fuzhou Dockyard by Steven A. Leibo, *Transferring Technology to China: Prosper Giquel and the Self-strengthening Movement* (Berkeley: Institute of East Asian Studies, 1985), pp. 88-106; and as the Foochow Arsenal by Prosper Giquel himself, in *L'Arsenal de Fou-Tcheou, ses résultats* (Shanghai: A. M. de Carvalho, 1874). Zhang Lihua and Luo Rongqu refer to it as the Fujian Ship Management Office (see below), and Knight Biggerstaff calls it the "Foochow Navy Yard." Since the facility at Fuzhou, in Fujian Province, was not involved to any significant extent in the production of ordinance or ammunition (see Kennedy, p. 78), but was devoted almost exclusively to steamship production, it will be referred to here as the Fuzhou Shipyard.

earliest attempts to construct their own failed. According to Tso Tsung-t'ang (Zuo Zongtang), "[a]s a result, it was necessary to train large numbers of Chinese technicians and workers who would be able to master Western technology. Hence, when the Fuzhou Ship Management Office was set up, it was decided to employ foreign instructors to give the necessary modern technical education. The hiring of foreign specialists was regarded as an extraordinary measure."[27] The first advisor and deputy-advisor to the Fuzhou Shipyard were both French: Prosper-Marie Giquel (1835–1886) and Paul d'Aiguebelle (1831–1875). The two were commissioned to oversee the construction within five years of eleven large and five small ships, in the course of which they were also expected to teach their Chinese workers how to construct and maintain necessary equipment, as well as how to follow blueprints and how to navigate ships. However, when the first five-year contract for the French expired in 1874, their positions were not renewed, and most of the foreign technicians returned to their respective countries. Nevertheless, in the course of their five years in China, the foreigners trained some 2,000 Chinese apprentices, all of whom received a basic technical education.[28]

There were two schools in the Fuzhou Shipyard, both of which taught mathematics. In the "front" school, which concentrated on French techniques of naval design and construction, all teaching was in French and included, over a five-year period, arithmetic, elementary algebra, plane geometry, trigonometry, analytical geometry, calculus, perspective drawing, elementary physics, and mechanics.[29] In the "back" school, where practical navigation skills were emphasized, including offshore navigation and the actual handling of ships and the cannon they carried, all teaching was in English. The "back" school also included a considerable amount of mathematics; over a three-year period students covered arithmetic, elementary algebra, plane and spherical trigonometry, nautical astronomy, meteorology, theoretical navigation, and geography. In all, between 1867 and 1874, the Fuzhou Shipyard employed twenty-four foreign teachers, seventeen of whom only taught part-time, spending the rest of their time on ship construction.

When the contracts of the Shipyard's foreign staff were not renewed, it was decided that in order to keep abreast of the latest Western practices, select graduates should continue their education in Europe. As early as 1875, five students were sent to France and Britain, and beginning in 1877, marine engineering students were sent abroad on a regular basis. In fact, by 1897, eighty-six students from China had gone abroad, seventy-six of whom were drawn from the Fuzhou Shipyard School. Concentrated primarily in Britain, France, and Germany, these Chinese students studied or apprenticed in various schools, mines, and naval institutions. Meanwhile, back in China, over the course of thirty years some forty steamships, including warships, had been built, and Chinese students had studied mathematics and the sciences, learned the art of navigation, and expanded their

---

[27]Tso Tsung-tang, quoted by Zhang Lihua and Luo Rongqu, "Technical Education of the Fujian Ship Management Office and the Transfer of Modern Western Technology to China," in *The Transfer of Science and Technology Between Europe and Asia, 1780–1880*, ed. Yamade Keiji (Kyoto: International Research Center for Japanese Studies, 1994), pp. 229-239 on p. 229.

[28]Prosper Giquel, *The Foochow Arsenal and Its Results*, trans. H. Lang (Shanghai: Shanghai Evening Courier, 1874), quoted in Zhang and Luo, p. 231. For full details about the Fuzhou Shipyard School, see Biggerstaff, *The Earliest Modern Government Schools in China*, especially chapter four.

[29]Zhang and Luo, p. 234. Mathematics was taught by Léon Médard, according to Biggerstaff, *The Earliest Modern Government Schools in China*, p. 210.

engineering expertise to mining, smelting, communications, and the construction of railways. Unfortunately, due to labor disruptions and political corruption, China did not make as much progress as it might have.[30]

Nevertheless, the Fuzhou Shipyard was the "leading naval shipbuilding establishment in China, at least until the end of the Qing dynasty. Between the withdrawal in 1874 of the French engineers and master workmen who had built it and the arrival of five new French technicians in 1897, the yard was operated entirely by graduates of the French division of the school, many of whom had received additional training in France."[31] As a result of the unexpected defeat of China by Japan during the Sino-Japanese War in 1895, it was decided to reinforce the Fuzhou Shipyard, and thus a large contingent of French experts was sent to re-staff the school and oversee ship construction. Meanwhile, the English division of the Fuzhou Shipyard continued to supply the navigation expertise, engine-room personnel, and both deck and engine-room officers for the entire Chinese navy until the late 1880s, when newer naval academies in Tianjin, Whampoa, and Nanjing began to supply officers as well. But by the end of the century, responsibility for teaching mathematics to large numbers of Chinese students had moved from the specific goals of the shipyards and translation schools to China's earliest colleges and universities.

Educational Reform and the "Reign of One Hundred Days"

For about a decade following China's war with Great Britain and France in 1860 (the Second Opium War), diplomacy replaced warfare until the Tianjin Massacre of 1870. Between 1876 and 1885, the Chinese had to face confrontations on four fronts: with Russia over possession of a large portion of Xinjiang Province; with Japan over islands off the southeast coast of China; with its vassal state, Korea, when Japan began trade in 1876; and with France in the southwest, when the French occupation of Annam led to war in 1884–1885. When the failure of the early Self-Strengthening Movement led to calls for educational reform, among the "new self-strengtheners" was Chang Chih-tung (Zhang Zhidong), who after the Sino-French War "embraced similar views and turned to Westernization with a vengeance."[32]

As a result, yet another series of school reforms began in 1894 that were meant to bring China into the modern world, at least by European standards. It was decreed that every county had to have a school and that mathematics was to be taught in all schools, even in the countryside. This had a considerable effect upon mathematics because now it was necessary to train instructors specifically to teach mathematics in the newly created schools across all of China. Consequently, many normal universities were founded in this period. Most teachers went to country schools to teach, and mathematics suddenly began to experience considerable growth. Because everyone believed that it was the foundation for all of the sciences,

---

[30]As Zhang and Luo conclude: "However, because of the poor management and the corruption of the Ch'ing administration, China's modern shipbuilding industry was unable to develop any further." Quoted from Zhang and Luo, pp. 238-239.

[31]Biggerstaff, *The Earliest Modern Government Schools in China*, p. 244.

[32]William Ayers, *Chang Chih-tung and Educational Reform in China* (Cambridge: Harvard University Press, 1971), p. 98. For further details, see pp. 100-136.

mathematics was accorded a special place and given considerable prominence in the efforts to reform education in China.[33]

In 1898, the Emperor Guang Xu, in the course of his famous reign of 100 days, not only abolished the civil service examination system but also established new schools, colleges, and universities—with the stipulation that mathematics be taught throughout China.[34] It was also decided that students should be sent abroad to study, mainly to Japan.[35] Unfortunately, Guang Xu's reign was brief. The Dowager Empress, in a brief *coup*, regained control of the throne in September of 1898, and by the end of the year, China was again at war with the Europeans.

The so-called "Boxer Rebellion" or "Boxer Uprising" of 1900 was directed largely against foreigners and Chinese Christians. As a consequence of the threat to foreign lives and property, it served to bring the combined forces of Russia, Britain, Germany, France, Italy, Japan, Austria, and the United States to Beijing, where they occupied the city and sought to protect their locally vested interests. Once the uprising was over, a "Boxer Protocol" was eventually signed that levied a severe judgment of $322 million in punitive damages against the Chinese.[36]

In the wake of the Boxer Uprising, there were once again renewed calls for serious educational reforms throughout China. The most significant were proposed and adopted in 1904, when the *Zhouding xuetang zhangcheng* (Memorial Determining School Regulations) finally succeeded in establishing a modern school system throughout China. This was largely the work of Zhang Zhidong, who drew explicitly on the Japanese model of modern education, which embraced a combination of Confucian thought and modern science.[37]

Japan as an Early Role Model

At the turn of the century, China had to face both its defeat at the hands of the Japanese in the Sino-Japanese War of 1895 and the consequences of the Boxer Uprising a few years later. It was clear that earlier attempts to promote Western knowledge (and to import wholesale industrialized methods and techniques

---

[33] For a comprehensive study of the reform movement at the end of the nineteenth century, see Chien Po-tsan (Jian Bozan), *et al.*, *Wu-hsü pien-fa* (*Wuxu bianfa*) (*The Reform Movement of 1898*), 4 vols. (Shanghai: n.p., 1953). Volume two includes 190 memorials and 316 edicts from this period. Of the memorials to the Emperor, about a quarter deal with education; of the edicts, more than sixty in the period 1894–1898 are concerned with educational reform. See also Cyrus H. Peake, *Nationalism and Education in Modern China* (New York: Columbia University Press, 1932; reprint ed., New York: Howard Fertig, 1970); Wolfgang Franke, *The Reform and Abolition of the Traditional Chinese Examination System* (Cambridge: Harvard University Press, 1960), pp. 32-40; and William Ayers, "Education in the Hundred Days' Reform," in Ayers, pp. 173-178.

[34] All of this was outlined explicitly in Chang Chih-tung's widely-read book of 1898, *Ch'üan-hsueh-p'ien* (*Quanxue pian*) (*Exhortation to Learning*), which called for educational reforms throughout China. It was Chang who advocated the combination of "Chinese learning for fundamental principles, Western learning for practical applications [*Zhongxue wei ti, xixue wei yong*]." For translations of excerpts from Chang's book, see Teng and Fairbank, pp. 166-174. Ayers also discusses school reform at length; see especially pp. 160-173.

[35] See Abe, p. 59, and compare Sasaki's chapter above.

[36] For discussion of the initially exorbitant indemnity and efforts to ameliorate its effect to the positive benefit of the Chinese through education, see Michael H. Hunt, "The American Remission of the Boxer Indemnity: A Reappraisal," *Journal of Asian Studies* **31** (3) (1972):539-559. A general overview of the event is provided by Chester C. Tan, *The Boxer Catastrophe* (New York: Columbia University Press, 1955; reprint ed., New York: Octagon Books, 1967). Compare also Xu Yibao's chapter in the present volume.

[37] See Abe, pp. 60-62.

of shipbuilding and modern warfare) had not been as successful as originally hoped. In light of the patent success of the Japanese at modernization, Chinese reformers looked more carefully at the Japanese reorganization of education and at how Japan had succeeded in rapidly industrializing during the Meiji period.[38] The extent of the Japanese achievement impressed the Chinese not only because of Japan's success in the Sino-Japanese War but also because Japan, as a culture heavily influenced by China, was for many a role model preferred over the West. Japanese experience in translating texts, especially scientific and technological works into Japanese, was expected to help expedite the production of acceptable Chinese translations.[39]

Beginning in 1898, Chinese students went to Japan in remarkable numbers for formal study. By 1903, 1,000 students were studying in Japan, and by 1906, the number had jumped to 8,000. Most, however, studied literature, and few seem to have studied science or mathematics. Among those who did was Ma Zuju. Born in Xin Jiang in 1880, he studied at Tokyo Imperial University in 1914. In 1918, he joined the Department of Mathematics at Beijing University and was simultaneously made Chair of the Department of Mathematics at Northeast China Normal University. Chen Jiangong also studied in Japan from 1921 until 1929 and published one research paper there. Later, he had considerable influence in southern China, where he established the first mathematical seminar at Hangzhou.[40]

One important component of the development of modern mathematics in both China and Japan that bears further comparative study is the role of arsenals and shipyards in Japan as centers that also promoted the teaching of science, including mathematics, as well as scientific research. In the case of Japan's largest private shipyard, the Mitsubishi Nagasaki Shipyard, just as in China, the influence of both English and French engineers was of crucial importance, especially at the beginning, in establishing the shipyards as well as the schools and research facilities they incorporated. But the British and French were influential in the early development of mathematics in China for reasons that go well beyond their importance for the shipyards and arsenals in which they were involved.

England: Supporting the Rise of Mathematics in Modern China

The earliest foreign influence to have any significant impact upon Chinese mathematics in the nineteenth century was English, due at least indirectly to the activities of the London Missionary Society based in Shanghai. As noted above, among those helping the Society to establish a presence in China was Alexander Wylie, who first went to Shanghai in 1847 and spent thirteen years there working for the Society's Printing Establishment. He returned briefly to England for three years

[38]Bastid, p. 11. There is also a very interesting study by Yukiko Fukasaku, *Technology and Industrial Development in Pre-War Japan* (London: Routledge, 1992), where the history of the Mitsubishi Nagasaki Shipyard involves scientific education and research, including mathematics. It shows striking similarities with the experience of the Chinese in their various arsenal and shipyard projects. For more details on the emergence of modern mathematics in Japan, see Sasaki's chapter above.

[39]Hu, *Merging Chinese and Western Mathematics*, pp. 93-96.

[40]For further details about Chinese students and their study of mathematics in Japan, see Abe, pp. 57-80. For remarks on the experiences of mathematicians who studied in Japan, see Cheng Minde, "Chen Jiangong," in *Zhongguo jiandai shuxuejia zhuan* (*Biographies of Modern Chinese Mathematicians*), ed. Cheng Minde, 3 vols. (Huai Yin: Jiangsu Education Publishing House, 1995), 2:18-19.

in 1860, but then returned to China in 1863 for another fourteen years, which he spent working for the British Foreign Bible Society in Shanghai.

Wylie was especially important for the transmission of Western mathematics to China primarily as a result of his fruitful collaboration with the mathematician, Li Shanlan. Together they served to bring such basic works as De Morgan's *Algebra* as well as calculus texts by the American, Elias Loomis, into the hands of Chinese readers via carefully translated editions in Chinese.

As already discussed above, the British were also influential in supporting the leading institution for translation in China, the Beijing *Tongwen Guan*, where the first two Professors of English were British: John S. Burdon and John Fryer.[41] What made Fryer especially influential was his interest in science and technology and his insistence that these could be conceptualized and promoted in Chinese. Although there were those who continued to argue that the Chinese language was antithetical to scientific ideas, and that science education in China could only proceed successfully if it were taught in Western languages, Fryer thought otherwise. It was largely through his translations, as well as his efforts to create a suitable scientific vocabulary in Chinese, that he helped to refute such prejudice.[42] In the course of his career at the Jiangnan Arsenal, he published some seventy-five translations of works devoted to mathematics, pure and applied science, medicine, and the social sciences.

Fryer also helped to establish a public library (later known as the Shanghai Polytechnic Institution) and edited a popular scientific magazine that included extracts of articles from British and American educational and scientific journals.[43] The most remarkable effort of the Polytechnic Institute, however, was the regular series of lecture courses on mathematics it began in 1895. Two years later, the *Zongli Yamen* reported that the Shanghai Polytechnic Institution was training half of the students studying mathematics in the entire country. (Recall that at the time, mathematics was also being taught primarily at the Beijing *Tongwen Guan*, the Fuzhou Shipyard, and the Jiangnan Arsenal.) As a result of the growing number of missionary institutions, private colleges, and government reforms of public education throughout China over the next few decades, however, the need for the kind of instruction offered by the Shanghai Polytechnic Institution soon declined precipitously. By 1904, it discontinued classroom instruction altogether.[44]

As for British missionaries, they were not as influential in the realm of education as were Americans, due largely to the fact that most British missionaries were not university graduates themselves and were intent more on evangelical conversion than on using education as a means of converting Chinese, the latter a

---

[41] Both Burdon and Fryer left the *Tongwen Guan* after a year and were replaced by an American, W. A. P. Martin, who served the school for twenty-five years, and from 1869, as its director. See Davin, p. 37.

[42] *Ibid.*, p. 40.

[43] For a detailed study of the Shanghai Polytechnic Institution, see Knight Biggerstaff, "Shanghai Polytechnic Institution and Reading Room: An Attempt to Introduce Western Science and Technology to the Chinese," *The Pacific Historical Review* **25** (1956), and reprinted in Knight Biggerstaff, *Some Early Chinese Steps Toward Modernization* (San Francisco: Chinese Materials Center, 1975), pp. 69-91. One of the most successful projects of the Institution was launched in 1895, a preparatory class in arithmetic that started with twenty students, and soon grew to about fifty in 1896. This effort persisted for about a decade. See Biggerstaff, "Shanghai Polytechnic Institution," p. 82.

[44] Biggerstaff, "Shanghai Polytechnic Institution," p. 91.

strategy followed by American missionaries in China like Gist Gee and Calvin Mateer.[45] American missionaries were so convinced of the importance of science in the curriculum that they sought to introduce systematic instruction of mathematics as well as laboratory science (see below). Americans also brought greater financial support to China, especially with the Boxer Indemnity Scholarship Program to prepare students in China and then to pay for their further education in the United States.

By the time Great Britain launched its own Boxer Indemnity program in the 1930s, the often radical character of student movements persuaded many in Britain that "it was not the moment for a great expansion of education."[46] Initially, the Boxer Indemnity Bill as presented to Parliament was vaguely worded, saying only that the monies were to be spent "on purposes to the mutual advantage of Great Britain and China."[47] It did not initially insist that these purposes be educational. In a memorandum on this subject, Bertrand Russell urged that "it is of the utmost importance that an Amendment should be adopted specifying Chinese education as the sole purpose to which the money should be devoted."[48] Among his reasons: "That any other course would contrast altogether too unfavourably with the action of America, which long ago devoted all that remained of the American share of the Boxer indemnity to Chinese education."[49]

Russell addressed his memorandum to the Labour Party, with the following warning:

> The Bill in its present form opens the door to corruption, is not calculated to please Chinese public opinion, displays Great Britain as less enlightened than America and Japan, and therefore fails altogether to achieve its nominal objects. The Labour Party ought to make at least an attempt to prevent the possibility of the misapplication of public money to purposes of private enrichment. This will be secured by the insertion of the words "connected with education" in Clause 1, after the word "purposes."[50]

In the end, the Advisory Committee for the disbursement of Boxer Indemnity funds recommended that British monies be used to support British-operated secondary schools, Christian-run universities, and the University of Hong Kong. It set aside special funds for the endowment of chairs in English literature, philosophy, history, and political science in Chinese universities, which were to be occupied by British scholars whenever possible. Between 1933 and 1938, 148 students received "Boxer scholarships" to study in Britain, and in 1939, of 350 students from China studying in Britain, sixty-seven (about one fifth) were supported by the Indemnity fund.[51]

---

[45]Mateer's importance for the promotion of mathematics in China is discussed below. On Gist Gee, see the biography by Haas.

[46]Davin, pp. 54-55.

[47]*Ibid.*

[48]Bertrand Russell, *The Autobiography of Bertrand Russell*, 3 vols. (Boston: Little, Brown, 1951), 2:214.

[49]*Ibid.*

[50]*Ibid.*, p. 215.

[51]Davin, p. 55. Russell was invited by Prime Minister J. Ramsey MacDonald to represent education on the *ad hoc* committee to advise the government how best to disperse the British share of the China Boxer Indemnity. See his letter of 31 May, 1924, to Russell as reproduced in Russell's *Autobiography*, 2:221-222. To this Russell replied with a "Memorandum on the Boxer Indemnity"

Of these, Wu Tajen (Wu Daren) went in 1933 to study mathematics with Thomas L. Wren at University College London, under the auspices of the Foundation. Of others supported, the best-known was the analytical number theorist, Hua Loo-Keng (Hua Luogeng).[52]

In addition to training young, Chinese mathematicians, Britain also helped to stimulate interest in modern mathematics through distinguished figures who visited China. Among the earliest of these was Russell, who sailed for China from Marseilles in early autumn of 1920 and spent several months as a visiting professor at the National University of Beijing, where he was greatly impressed by the seriousness of his students and their determination to bring China into the twentieth century.[53]

In Beijing, Russell presented (among other topics) his basic ideas about set theory and logic. Although he covered the paradoxes and the axiom of choice, he did not attempt to offer a comprehensive introduction to set theory in general. Nevertheless, it was enough to create preliminary interest in Russell's mathematics. In 1924, Fu Zhongsun translated (with Zhang Bangming) Russell's *Introduction to Mathematical Philosophy* into Chinese as *Luosu suanli zhexue* (*Russell's Philosophy of Mathematics*). This was later retranslated, with the advantage of more than a half-century's understanding of mathematical logic and Western philosophy of mathematics, by Yan Chengshu as *Shuli zhexue daolun* (*Introduction to Philosophy of Mathematics*, 1982).[54]

Not only did Fu make the first translation of a Western book on the philosophy of mathematics into Chinese, he also made Oswald Veblen's geometric axioms widely known through an article "On the Foundations of Geometry."[55] But the point and methods of these early publications were not so clearly expressed, and the translations were often rough and not as easily understandable as they might have been had there been a longer history of Western mathematical logic in China. These early attempts by Fu Zhongsun were also made before China's first specialist on modern logic had appeared. Beginning in 1931, Xiao Wencan at Wuhan University began a systematic series of articles in *Like jikan* (*Science Quarterly*), which proved to be the first systematic introduction to set theory in Chinese.[56] Of special importance, Xiao carefully explained both the meaning and significance of

---

in which he strongly urged that the Boxer Indemnity Bill then in Committee before Parliament should be amended, "specifying Chinese education as the sole purpose to which the money should be devoted." Russell's memorandum is also reproduced in his *Autobiography*, 2:222-224. Russell was informed shortly after the fall of the Labour Government that his services on the Committee would no longer be needed by the succeeding Conservative Government. Russell, *Autobiography*, 2:199.

[52]On Wu Daren and Hua Luogeng, see Xu Yibao's chapter that follows.

[53]Russell, *Autobiography*, 2:191-192.

[54]For an analysis of Russell's *Introduction* and the impact of mathematical logic and its development in China, see the next chapter.

[55]This was published in two parts, "Jihe xue zhi jichu" (On the Foundations of Geometry), *Shuli zazhi* (*Mathematics and Physics Magazine*) **3** (4) (1922):1-40, and **4** (1) (1922):1-23. Fu Zhongsun also translated, with Han Guicong, Hilbert's *Foundations of Geometry*, which appeared as *Jihe Yuanli* (*Principles of Geometry*) (Beijing: Commercial Press, 1924).

[56]Xiao Wencan, "Wulishu zhi lilun" (The Theory of Irrational Numbers), *Like jikan* (*Science Quarterly*) **2** (4) (1932); "Wuqiongda zhi jie" (The Orders of Infinity), *op. cit.*, **3** (1) (1933); and "Jihe lun" (Set Theory), *op. cit.*, **4** (2) (1933); *op. cit.*, **4** (4) (1934); *op. cit.*, **5** (1) (1934); and *op. cit.*, **5** (2) (1934). The articles on set theory were published together as a booklet entitled *Jihe lun chubu* (*Basics of Set Theory*) (Shanghai: Commercial Press, 1939).

the new approaches to mathematics, particularly the role that set theory could play in numerous applications throughout modern mathematics.

Finally, in addition to its educational impact on China through students, teachers, and lecturers, Britain was also influential as the colonial administrator of Hong Kong, a city that served as a role model for what the Chinese could accomplish on Western terms, given the right means and circumstances. Britain's most emancipating legacy, perhaps, was the English that the citizens of Hong Kong learned in school. Yung Wing (Rong Hong), the first Chinese to attend an American college, had been at school in Hong Kong where he acquired a basic command of English, as had many who went abroad for further study. Basically, Hong Kong provided young men with enough fluency in English either to travel abroad for further study or to find profitable employment in China. Reform leaders like Kang Youwei and Liang Qichao certainly regarded the prosperity of Hong Kong as evidence that the poverty and inferior conditions in the rest of China must be a result of its political system. Sun Yat-Sen, who also received a British education, "contrasted the modernity of Hong Kong with the backwardness of the hinterland of China and found in it an inspiration to action."[57]

France: Contributing to the Transmission of Modern Mathematics to China

Meanwhile, the French had nearly as long an influence and virtually as much impact upon Chinese mathematics as did the English, beginning with the French presence at the Fuzhou Shipyard. But it was not until 1912 that Chinese students began going to France on a regular basis to obtain diplomas in mathematics, and only then did France begin to exert an influence on the development of advanced, research mathematics in China to any appreciable degree. Most often, Chinese students enrolled at the faculties of science in Paris, but also elsewhere in France.[58] This was due directly to the establishment of the Chinese Society for Affordable French Education, created in 1912 by Li Shizeng (whose given, family name was Li Yüying) and Cai Yuanpei.[59] The Association aimed to find positions for students in France as salaried workers, preferably in French industries. This enabled them to build up sufficient savings to finance their studies at the same time that it gave them the opportunity to learn the basics of the French language. Between 1913 and 1919, more than 2,000 Chinese students studied in France, many under the Society's auspices. Without proper facilities in China to prepare them for study abroad, students with backgrounds neither in mathematics nor French spent as many as seven years before earning their *licence* in mathematics.

At first, among the most successful and best-known of those who studied mathematics in France was Hiong King-Lai (Xiong Qinglai), who, upon his return to China, headed the Department of Mathematics at Nanjing University. In 1926, Xiong was invited to organize the department at Qinghua University in Beijing. Four years later, he returned to France to study for the doctoral degree he received

[57]Davin, p. 49.

[58]Li Wenlin and Jean-Claude Martzloff, "Aperçu sur les échanges mathématiques entre la Chine et la France (1880–1949)," *Archive for History of Exact Sciences* **53** (1998):181-200 on p. 183.

[59]*Ibid.*, p. 183. Cai was named Chinese Minister of Education in January of 1912, Rector of Beijing University in 1916, and President of the Academia Sinica in 1928.

in 1933 from the *Institut Henri Poincaré*.[60] This provided the opportunity for Xiong to attend the International Congress of Mathematicians (ICM) in Zürich, which he did as an official delegate representing the Chinese Mathematico-Physico Society. Hsu Koh-Pao also attended the Zürich Congress, representing Jiao Tong University in Shanghai. Together, the two constituted the first official delegation to represent China at an ICM, although neither Hsu nor Xiong presented a paper.

From France, Xiong went to Germany before returning to Qinghua University. In 1937, he became Rector of the University in exile during China's war with Japan when Beijing University, Qinghua University, and Nankai University all moved their faculties and students to safer territory in Yunan Province. After World War II, Xiong returned to France. A decade later, he was invited to return to China, which he did in order to take up a post at the Academia Sinica in Beijing in 1957.

Meanwhile, France had several occasions to influence Chinese mathematics directly through distinguished visitors to China. In 1920, Paul Painlevé and Émile Borel went to China at the invitation of the Chinese government, to study the reorganization of the Chinese railroad.[61] They also, implicitly, had a further mission: to "remedy the decline of the French language and French influence on intellectual and technical relations between China and France, and to 'attract' a notable part of the élite of young Chinese to French culture."[62] Unfortunately, it is clear neither how they did the latter nor whether they made any arrangements to meet specifically with mathematicians or to give any lectures on their mathematical research.

A year later, in 1921, when the Franco-Chinese University Association was established to help facilitate exchanges between France and China, Painlevé was a member of its Board of Directors. By 1933, a new tri-partite university was established with branches in Beijing, Canton, and Lyon. Lyon's long association with China and the manufacture of silk once again came into play, making it an obvious choice for the French headquarters of the new university. Up to 1937 (when yet another Sino-Japanese War broke out), seven Chinese students had earned their Ph.D.s at the university in Lyon. Meanwhile, the celebrated Jacques Hadamard visited China, where he had a decisive influence.[63] When Hadamard and his wife arrived in Beijing in April of 1936, their photograph was prominently featured on the cover of the monthly journal, *Beijing zheng wenbao* (*Peking Politics*).[64] Hadamard,

[60]Zhuang Jingtai and Zhu Dexiang, "Xiong Qinglai," in *Zhongguo jiandai shuxuejia zhuan* (*Biographies of Modern Chinese Mathematicians*), ed. Cheng Minde, 3 vols. (Huai Yin: Jiangsu Education Publishing House, 1995), 1:35-47.

[61]Painlevé, a graduate of the *École normale supérieure*, had a Ph.D. from Göttingen. In addition to teaching at the *École polytechnique*, the *Collège de France*, and the *École normale supérieure*, in 1904 he taught a course at the *École supérieure d'aeronautique*, and had even flown with one of the pioneers of modern aviation, Wilbur Wright. In 1910, Painlevé was elected a deputy from the 5th arrondissement, headed a naval and aeronautical commission, and in 1917 served his country as Minister of War. It was in 1920 that the Chinese government commissioned him to help reorganize the country's railroads.

[62]Archives nationales, dossier No. 313/AP/203 (1920), cited by Li and Martzloff, p. 188. In all, Borel and Painlevé were in China for five months.

[63]An account of Hadamard's visit to China was reported in *Kexue* (*Science*) as a news item: **20** (5) (1936):416-417. Norbert Wiener also paints a lively portrait of this visit in *I Am a Mathematician: The Later Life of a Prodigy* (Garden City: Doubleday & Company, Inc., 1956), pp. 199-200.

[64]*Peking Politics* appeared monthly in French; the issue that carried the photograph of Professor Hadamard and his wife was vol. 23, no. 16, dated 18 April, 1936.

who lectured at Qinghua University in the spring of 1936, had a particularly strong influence on Hua Luogeng and Wu Wenjun.

Germany: A Model for Developing Modern Mathematics in China

As calls for the reform of education in China mounted during the last quarter of the nineteenth century, government officials seriously studied foreign approaches in the spirit of the Self-Strengthening Movement with its conviction that Western knowledge was imperative to the modernization of China. One of the earliest and most influential studies appeared in 1873, *Deguo xuexiao lunlüe* (*An Overview of German Schools*), written by the German missionary, Ernst Faber. This influential work was prefaced with observations by Li Shanlan, who noted that "Prussia owed her recent military victories to the education received by her soldiers, which had inspired them to fight for an ideal and for certain principles."[65]

The German model of education was congenial to Chinese reformers for several reasons. Not only did they equate Germany's military success with its industrialization—which, in turn, was linked to education, especially to the *Technische Hochschulen* (Technical Colleges)—but the Chinese in general approved of the authoritarian elements that were a signature of German education. Germany's greatest influence upon mathematics in China, however, was not through the examples its pedagogical theorists set but more directly through its training of Chinese students. By far the majority of these went to Göttingen, where the combined efforts of Felix Klein and David Hilbert had created an international center for the study and promotion of mathematics early in the twentieth century. (Compare Figure 1.)

The first Chinese student in Germany was Wei Ciluan, who after a year in Frankfurt went on to Göttingen to study analysis with Richard Courant in 1920.[66] In 1924, Wei received his doctoral degree in applied mathematics for work related to the equi-distribution of pressure, using the calculus to determine various maximum/minimum problems. After returning to China, he taught first in Shanghai then at Sichuan University. Another student who studied with Courant in Göttingen, Zhu Gongjin, translated Courant's *Differential and Integral Calculus* into Chinese in 1934.[67] Among students who went to Göttingen in the 1930s, Ceng Jiongzhi received his Ph.D. under the guidance of Emmy Noether for a thesis (1936) on the Hasse-Brauer theorem and the "*Stufentheorie*" of quasi-algebraic closedness of commutative groups.[68]

German mathematicians also went to China. In 1910-1911, Konrad Knopp spent a year at the German-Chinese Academy in Qing Dao. Born in Berlin in 1882, Knopp had received his Ph.D. in 1907 before moving to the Commercial Academy in Nagasake, Japan. Although a specialist on the subject of generalized limits, Knopp taught only basic mathematics at Qing Dao, and while he could have offered more,

---

[65]Li Shanlan, as recounted by Bastid, p. 9. For the analogous French reaction, see Hélène Gispert's chapter above.

[66]Li Wenlin, "Getinggen shuxue de shijie yingxiang" (The International Influence of Göttingen Mathematics), *Shuxueshi yanjiu wenji* (*Collected Research Papers on History of Mathematics*) **6** (1995):117-123 especially on p. 121.

[67]*Ibid.*, p. 121.

[68]*Ibid.*, p. 121, and James W. Brewer and Martha K. Smith, ed., *Emmy Noether: A Tribute to Her Life and Work* (New York: Marcel Dekker, 1981), p. 42.

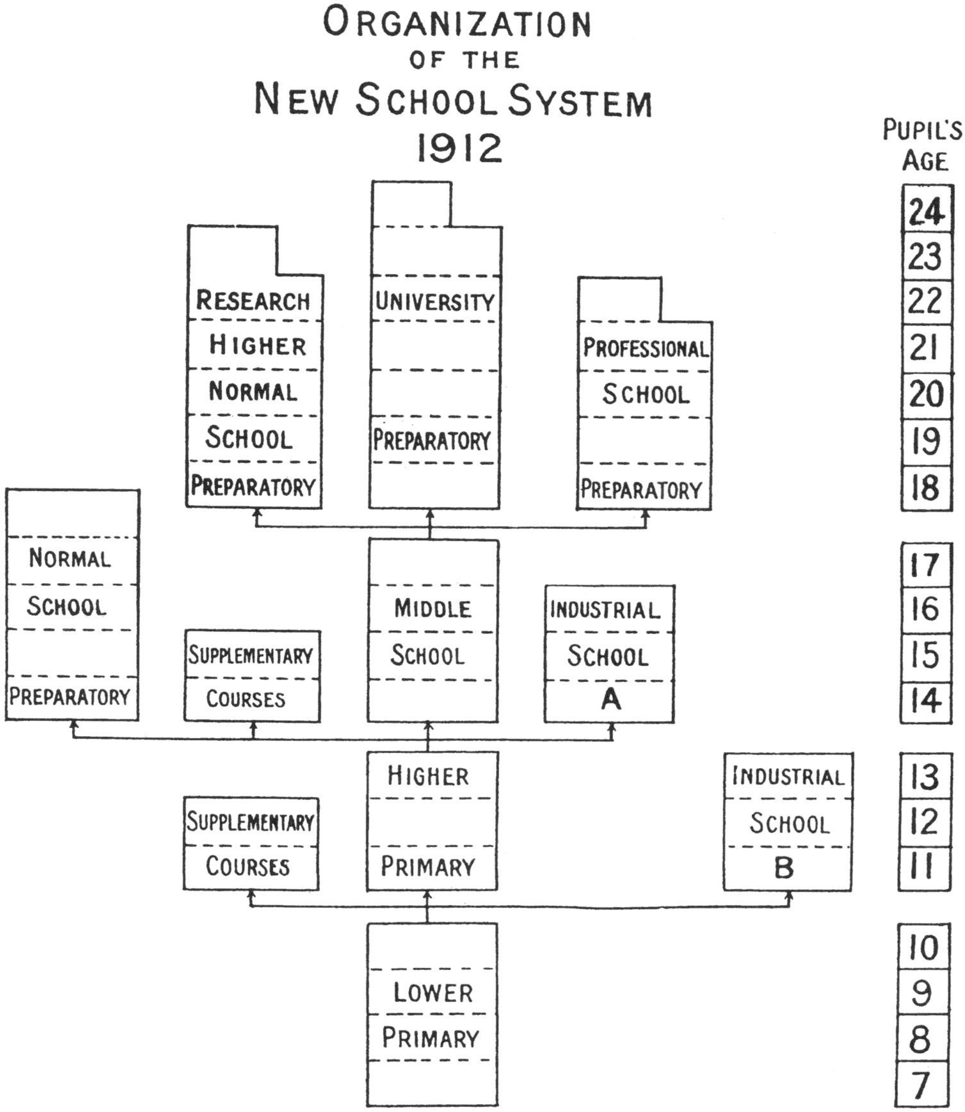

FIGURE 1. Organization of the New School System, 1912

"at the time the level of students was not advanced enough to take advantage of his guidance."[69]

Several decades later the German mathematician, Wilhelm Blaschke, also visited China, where he made a strong impression in Beijing lecturing on topology

[69]Li Wenlin, p. 121. In 1912, Knopp was back in Berlin, where he had completed his *Habilitation* in 1911 and then taught at the Berlin Military Academy. Later, he went on to teach at Königsberg and Tübingen.

and differential geometry in 1932.[70] Two years later, thanks to Blaschke's efforts, Emanuel Sperner, a student of George D. Birkhoff, spent the years 1932–1934 teaching primarily topology and foundations of geometry in China. It was Blaschke's influence that prompted Chen Shengshen (better known as S. S. Chern) to begin his study abroad in Hamburg, with support from the U.S. China Foundation (Blaschke's trip to China was also sponsored by the Foundation). Chern later went to France, where he studied further with Élie Cartan. Other Chinese students who wanted to study mathematics were also attracted to Germany. Among these, Wu Daren and Jiang Lifu both traveled together with Chern to visit Hamburg in 1936.[71]

The United States

The United States was a major early influence on the development of mathematics in China for several reasons, at first largely because American missionaries and philanthropies took a special interest in education. The most influential of the former for the introduction of mathematics to China was a young Presbyterian from Pennsylvania, Calvin W. Mateer.[72]

Upon graduating from Jefferson College (today Washington and Jefferson College in Canonsburg, Pennsylvania), Mateer went on to Western Theological Seminary in Allegheny, Pennsylvania. After completing his studies in 1861, he married Julia Ann Brown, and a year later, the two sailed for China, where Mateer began a decade of evangelizing in the countryside of Shandong province. At the same time, his wife opened a small school in their hometown of Dengzhou for peasant boys in a broken-down Buddhist temple "from which the Tengchow (Dengzhou) College grew later on."[73]

From the beginning, Mateer had sporadically taught arithmetic and over the years compiled a set of written lessons that he published in a three-volume, lithographed edition in 1877. Mateer's method was to introduce Arabic numerals and concentrate on problems from everyday life, including such useful mathematical matters as taxation, partnerships, customs duties, and banking charges, all cast in terms of Chinese currency, weights, and measures. The Mateer method soon became popular in China and was adopted by other mission schools.[74]

Mateer's method, as well as his books, also reflected his belief that educational works should be written in colloquial Chinese rather than in the more erudite, scholarly language that Li Shanlan had used in his translations, and which thereby made them unsuitable, Mateer believed, for poorly educated students in the provinces.

---

[70] *Ibid.* As Li Wenlin suggests, Blaschke provided "new channels" for Chinese mathematicians to consider, just as Hadamard had done a decade earlier. For more on Blaschke's involvement in China, see Xu Yibao's chapter below.

[71] Cheng Minde, ed., 2:iii.

[72] Wang Yüan-te (Wang Yuande), "Di kaowen xiansheng chuan" (Biography of Mr. Calvin Mateer), in Wang Yüan-te (Wang Yuande) and Liu Yü-feng, ed., *Wen huiguan zhi* (*Alumni History of Tengchow College*) (Wei-hsien: n.p., 1913), p. 1. See also Daniel Webster Fisher, *Calvin Wilson Mateer, Forty-Five Years a Missionary in Shantung Province, China: A Biography* (Philadelphia: The Westminster Press, 1911), pp. 20 and 30-31.

[73] Irwin T. Hyatt, Jr., *Our Ordered Lives Confess: Three Nineteenth-Century American Missionaries in East Shantung*, Harvard Studies in American-East Asian Relations, **8** (Cambridge: Harvard University Press, 1976), p. 147.

[74] See the review of Mateer's *Arithmetic* by W. A. P. Martin, "A Tribute to Dr. Mateer," *The Chinese Recorder* **10** (September-October 1879):397-398.

Mateer liked to stress that his preference for colloquial Chinese meant that he was writing for people in "their own spoken language," thus honoring a "cardinal principle of Protestantism."[75] His insistence that the Chinese vernacular was no handicap for teaching or learning science also led him into vigorous arguments with his contemporaries.[76] While some still believed Chinese was inherently unsuited for mathematics or science, others like John Fryer not only wrote in Mandarin Chinese but preferred traditional Chinese numeral characters over Arabic numbers in his translations, believing the latter would seem too foreign to Chinese students. No one seemed to agree about the best translations to use for basic, mathematical terminology, and Mateer, in fact, from 1893 to 1908 headed a special committee of the Educational Association of China devoted to developing scientific nomenclature. In the course of its work, the committee managed to identify and translate some 12,000 terms, "briefly the whole range of scientific and general terminology."[77]

Determined to give his students a solid foundation in basic mathematics, Mateer was convinced of the need for more "brain-energizing subjects" and thus introduced algebra to the Dengzhou School in 1873. The results were so encouraging that he was soon offering additional courses in geometry, trigonometry, astronomy, and chemistry. As the quality of Mateer's efforts became more widely known, applications to the school steadily increased. In February of 1877, on the occasion of the first graduation exercises held for three students in a two-day commencement, the school changed its name to the *Wen hui-kuan* (*Wen huiguan*) (*Literary Guild Hall*), and from a boys' boarding school it became Tengchow High School, regarded among the missionary community as "really a college ... and decidedly the best school in China."[78]

Four years later, the Boys' High School officially became Dengzhou College. Within a year, enrollments increased to seventy, and it was thanks to the courses it offered in mathematics and science that the College's reputation was second to none in China. All graduates after the mid-eighties had taken at least two years of geometry, a year of algebra, trigonometry, and calculus, along with courses in surveying and navigation. In the sciences, students took chemistry, physiology, astronomy, geology, and three years of physics.[79] In 1904, the College moved to the city of Wei-hsien (Weixian), where it merged with an institution operated by English Baptists to form Shandong Union College. This, in turn, moved in 1917 to the provincial capital at Jinan, where it formed the nucleus of Shandong Christian University.[80]

[75]Hyatt, p. 202. The *wenli* favored by the Mandarinate was a relic dependent upon the traditional Chinese examination system. When that was abolished in 1904, the reasonableness of Mateer's views on translation became even more apparent.

[76]Objections were also raised against Mateer's preference for writing in colloquial Chinese. W. A. P. Martin, for example, reflected the opinion of many other missionaries who found Mateer's "low kind of Mandarin" offensive to "the taste of the educated classes." See Martin as quoted from the *Chinese Recorder and Missionary Journal* **10** (September-October 1879):397-398. According to Hyatt, "Fryer's position started Mateer on a twenty-year search for a usable scientific vocabulary in Chinese. Hostility to the use of Mandarin stung him first into issuing a *wen-li* edition of his *Arithmetic*, and then to a general campaign on behalf of colloquial Chinese." See Hyatt, p. 195.

[77]See Hyatt, pp. 195-196. A detailed study of mathematical terminology as it figured in the Association's examination of technical nomenclature would be of considerable interest, but is beyond the scope of this chapter.

[78]Hyatt, p. 175.

[79]*Ibid.*, p. 185.

[80]*Ibid.*, p. 217.

Although Chinese interest in Western knowledge was essentially utilitarian, Mateer emphasized mathematics and the sciences because he was convinced of their use as tools of religious conversion. Science, he believed, was "in fact but an exposition of the unwritten laws of God."[81] It was also useful, as the Jesuits before him had believed, in showing the superiority of Western science and, by implication, of Western theology.[82]

Mateer also believed that science could be used to liberate the Chinese mind, which had been regimented—and thereby made rigid and uncreative—by traditional Chinese thought. As a result of Confucianism, he wrote, the typical Chinese remained "like a donkey, with eyes hooded and head tied fast to the centre of the mill he is compelled to turn ... as he stupidly and patiently trudges round and round the same old track." Mathematics and science, "in large doses," were devices to "lift the donkey's hood," to "develop the mental faculties symmetrically and stimulate men to think and act for themselves."[83]

Mateer always kept the interests of his Chinese students in mind and understood the importance of reaching out to them with an awareness of their own experience and the specifically Chinese context within which he was working. This philosophy was as much responsible for the success of Mateer's books as it was for the school itself. Mateer wrote his own texts with Chinese students specifically in mind; he did not simply translate American or European texts, as did, for example, Alexander Wylie and John Fryer. Mateer found their books "too elaborate and got up in too expensive a style," and he objected to the emphasis they placed on methods and principles "but having no practical problems."[84] As a result, Mateer did not believe they were well-suited for use in Chinese schools. Instead, he consciously tried to use examples from daily Chinese life in writing his book on arithmetic, *Bisuan shuxue* (*Mathematics of Computation with the Brush*) of 1892, which was a great success and sold "by tens of thousands" for over forty years.[85]

This was soon followed by *Xin xingxue* (*New Geometry*), published in Chinese in 1895 and described by one reviewer as not so "chattily Chinese" as Mateer's *Arithmetic*.[86] Unlike Alexander Wylie, who had translated Euclid without great embellishment, Mateer offered algebraic shortcuts, provided simple examples, and included modifications to Euclid based upon several modern texts. Mateer's last two books, an elementary and an advanced algebra, were written with a similar plan in mind. The emphasis Mateer placed on everyday problems from Chinese life was designed to draw the student into the world of "symbolic mathematics."[87] The advanced text was the product of twenty years and many revisions when it finally appeared in print in 1908.

American missionaries like Mateer were dedicated to laying the indigenous foundations upon which the Chinese could begin to produce their own generations of

---

[81] *Ibid.*, p. 187.

[82] *Ibid.*, p. 285, note 83.

[83] Calvin W. Mateer, *The Chinese Recorder* **14** (November-December 1883):463-469.

[84] Calvin W. Mateer, "Mathematics in Chinese," *Chinese Recorder and Missionary Journal* **9** (September-October 1878):372-378.

[85] *Ibid.*, p. 378.

[86] The term Mateer introduced for geometry, *hsing-hsüeh* (*xingxue*) meaning "form study," was greeted enthusiastically by Martin and his colleagues at the government's *Tongwen Guan* in Beijing. See the *Chinese Recorder and Missionary Journal* **17** (August 1886):314-316. See also Hyatt, p. 195.

[87] Hyatt, p. 195.

young students well versed in the Western sciences. Still, it was primarily through Chinese scholars trained in the United States under the auspices of the Boxer Indemnity fellowship program that mathematics was established in China's major universities.[88]

New Institutional Models

In 1914, Chinese science in general made a fundamental advance through the creation of the Science Society of China. Significantly, this was not done in China, but in Ithaca, New York. Patterned on the American Association for the Advancement of Science by a group of Chinese students at Cornell University, the Society was the product of its members, virtually all of whom had attended Western-style schools in China before going abroad. They firmly believed that "the unity and universality of modern science flows from its methods"[89] and were determined to bring the benefits of Western science back to China.

After the students began to return home from completing their studies in the United States, the Science Society was formally registered with the Chinese government in 1917 and immediately became the largest and most important scientific organization in China. It was also shaped by the environment of post-1911 revolutionary China, where the new regime was concerned with theories of nationalism, constitutional government, freedom, and democracy. The Science Society of China had its ideological dimensions as well, for its founders were "convinced that the institutions of American science were voluntary associations that would liberate a tyrannized people from centuries of Confucian oppression."[90] Not surprisingly, virtually all of the founding members of the Society at Cornell went on to become active in China in various prominent and influential positions in the 1920s and 1930s.[91]

What proved most significant for mathematicians was the fact that the Science Society of China provided the country with a good example of how a nationwide organization could serve not only to professionalize scientists but also to offer indispensable institutional support in the way of libraries, journals, and local meetings. This model was not lost on mathematicians a few decades later. As H. C. Zen (Ren Hongjun) put it, the Science Society of China aimed to "[bring about] innovations in thinking by means of literary agitation,"[92] namely, through the Society's journal established at Cornell as *Kexue* (*Science*). The students' founding intentions were

---

[88]For further details on the impact both of Chinese students studying in the United States and of American mathematicians—like William Fogg Osgood and Norbert Wiener—who visited China, see Xu Yibao's chapter below.

[89]Peter Buck, *American Science and Modern China* (New York: Cambridge University Press, 1980), p. 5. In fact, approximately one-third of the Chinese students at Cornell had received a modern education before the abolition of the Chinese Civil Service Examinations in 1905.

[90]*Ibid.*, p. 94.

[91]*Ibid.*, pp. 94-95. For example, H. C. Zen (Ren Hongjun), Executive Director of the China Foundation for the Promotion of Education and Culture, was a tireless supporter of the sciences in Republican China and an influential figure in administering the second phase of the American Boxer Indemnity Scholarship Program. Yang Ch'uan (Hsing-fo) became Secretary General of the Chinese Academia Sinica. Chou Jen (Tzu-ching) became first director of the Academy's Institute of Engineering. And Y. R. Chao (Chao Yuan-jen/Zhao Yuanren), who was trained in mathematics and physics at Cornell before taking his Ph.D. at Harvard, later became an internationally recognized linguist, holding many important positions and contributing substantially to the improvement of both education and the level of research in China.

[92]*Ibid.*, p. 95.

explicit: "To promote science, encourage industry, authorize terminologies, and spread knowledge."[93] As the Science Society took root back in China, basing itself in Nanjing, special sections were created (beginning in 1915) for agriculture and forestry, biology, engineering, chemistry, mining and metallurgy, mathematics and physics.[94] A local committee was appointed to edit *Science*. Some book translations were commissioned, and plans were laid to establish a science research library. Meanwhile, articles were beginning to appear in *Kexue* on terminology. This was an important preliminary to introducing Western science on any broad scale in China, where progress would continue to be slow if students had first to master Western languages before they could begin their study of any of the sciences. Among the first of the articles to appear on terminology was one devoted to mathematics.[95]

In the 1920s, a public lecture series was also launched, and committees were established to study and promote science education and industrial research. In 1922, annual *Transactions* of the Society began to appear, in addition to *Kexue*. These were published in Western languages and sent to Western libraries not only to provide some idea in the West of scientific activity in China but also to exchange publications for the Society's library.

In 1929, the Society founded the Chinese Scientific Books and Instruments Corporation in Shanghai, which was designed to publish scientific materials for use in schools and to provide an additional source of revenue for the Society. A year later, the Society created a Bureau for Scientific Information at the request of the Nanjing government, and this marked officially the Society's importance to the nation as an advisory body to the government. In 1931, the Society built a major new scientific research library in Shanghai, which by 1937, when the Sino-Japanese War broke out, had amassed over 10,000 volumes and had more than 100 foreign journals.

Universities

Meanwhile, the number of colleges and universities with departments of mathematics was steadily increasing, and more and more mathematicians were being trained in China. The beginnings of a national university in Beijing can be traced back to the founding in 1898 of the Imperial University. There were also diverse private and missionary colleges and universities in China, but consideration here will be limited primarily to those that became important national universities in the course of the first decades of the twentieth century. These included, prior to or just after declaration of the Republic in 1912: Beiyang Public School (Tianjin, 1895), Nanyang University (Shanghai, 1896), Imperial University (Beijing, 1898), Shanxi University (1901), Qinghua College (Beijing, 1911), and Beijing University

---

[93] *Ibid.* At first, *Kexue* was the primary, indeed the sole, tangible activity of the Society. Initially, "shares" were sold for $10, and the students at Cornell tried to operate on a business-like basis, hoping the journal would pay for itself. This proved unrealistic, and as the Society took hold back in China, it began to operate as a more traditional society relying on membership fees and private contributions.

[94] By 1930, a branch of the Society had been established in Canton, providing additional opportunities for membership and further development of the Society's activities.

[95] For details about efforts to translate Western mathematical works into Chinese, and the significance of discussions about terminology, see the contribution to this volume by Xu Yibao that follows.

(1912).[96] Of these, the most outstanding groups of mathematicians came to be associated with Beijing University and with what would later become known as Qinghua University. To give but one example of the quality and experience of the mathematicians teaching at China's best institutions in the 1920s, by 1930, when Qinghua University launched its graduate program in mathematics (it had been promoted from the status of college in 1925 and a Department of Mathematics had been founded in 1928), its solid faculty included Xiong Qinglai, who had been trained in France and who served as departmental chair, Zheng Zhifan (Cornell/Harvard), Sun Guangyuan (Chicago), and Yang Wuzhi (Chicago).

Universities were the most important institutions for the support of mathematics in China prior to the founding of the Chinese Academy of Sciences. In addition to the mathematics taught as a matter of course in the universities, there were additional ways in which the universities helped to institutionalize and professionalize mathematics in China. Among the highlights, in 1934, a Mathematical Association was established at Qinghua University. A year later, the most important step for the institutional success of mathematics in China was also taken.

The Chinese Mathematical Society

The formation of a Chinese society for mathematicians was surprisingly late, but in 1935, one came into existence under the leadership of Qin Fen (see Figure 2). Qin, a Harvard-trained mathematician, went on to become President of Southeast University in Nanjing after teaching at Shanghai's Jiao Tong University and at Beijing University. Among members of the Society's first Governing Board were Hu Dunfu, Zeng Zifan, Wang Renfu (also a Harvard graduate), and Feng Zuxun (who had studied in Japan).[97] The Society's most important early efforts were its journals; the *Journal of the Chinese Mathematical Society* was a serious research journal, while *Mathematics Magazine* was intended for more expository articles with an emphasis on teaching. The *Journal* published works only in Western languages, predominantly in English, thus intending to reach an international audience; the *Magazine* published only in Chinese and served more local purposes.

Unfortunately, the Sino-Japanese War that broke out in 1937 was followed all too soon by the outbreak of World War II, which put an end to this encouraging beginning for modern mathematics in China. Nevertheless, many of China's leading mathematicians managed to continue working at Southwest Associated University in Yunan, to which many scholars were evacuated from occupied parts of China.[98] The *Journal of the Chinese Mathematical Society* even managed to publish its second volume in 1940. A brief examination of the first two volumes of the journal will suffice to indicate the nature and level of work that Chinese mathematicians were capable of producing, even during this period of great duress.

---

[96]This list does not claim to be exhaustive and is the result of information compiled from a number of sources, including Ruth Hayhoe, *China's Universities, 1895–1995: A Century of Cultural Conflict* (New York: Garland, 1996), p. 266. It does not include institutions sponsored by foreign organizations, notably missionary societies and the colleges and universities they supported.

[97]See Ren Nanheng and Zhang Youyu, ed., *Zhongguo shuxue hui shiliao* (*Materials for a History of the Chinese Mathematical Society*) (Nanjing: Jiangsu jiaoyu chubanshe, 1995), p. 30.

[98]The Southwest Associated University was a combination of Qinghua, Peking, and Nankai Universities, located in Kunming. There was also a less well-known Northwest Associated University. See Hayhoe, p. 57.

FIGURE 2. Annual meeting of the Chinese Mathematical Society on the occasion of its 50th anniversary. Fudan University, Shanghai, December 1985

Volume one appeared in the spring of 1936. The lead article by Kuen-sen Hu (Hu Kunsheng) was on "A Problem with Variable End Points in the Calculus of Variations."[99] At the time, Hu was teaching at the National Central University in Nanking (Nanjing), and his results drew especially on Marston Morse's *The Calculus of Variations in the Large* (1934).[100]

The second article was by the American ex-prodigy, Norbert Wiener, who signed himself as a Research Professor at Tsing Hua (Quinghua) University in Beijing for 1935–1936. Wiener's paper, "A Tauberian Gap Theorem of Hardy and Littlewood," offered alternative hypotheses under which a converse of Abel's theorem on power series might be obtained in terms of a theorem first proved by Hardy and Littlewood in 1924. Wiener gave an alternative proof of the Hardy-Littlewood theorem linking it with the theory of interpolation.[101]

The third paper was by the soon-to-be-famous Loo-Keng Hua (Hua Luogeng), who listed himself as a "Research Fellow of the China Foundation for the Promotion

[99]Hu Kunsheng, "A Problem with Variable End Points in the Calculus of Variations," *Journal of the Chinese Mathematical Society* **1** (1936):1-14.

[100]Marston Morse, *The Calculus of Variations in the Large*, American Mathematical Society Colloquium Publications, vol. 18 (New York: American Mathematical Society, 1934).

[101]Norbert Wiener, "A Tauberian Gap Theorem of Hardy and Littlewood," *Journal of the Chinese Mathematical Society* **1** (1936):15-22; see also Godfrey H. Hardy and John E. Littlewood, "Abel's Theorem and its Converse II," *Proceedings of the London Mathematical Society* **22** (1924):254-269. For more on Wiener's stay in China, see Xu Yibao's chapter below.

of Education and Culture."[102] Like Wiener, he was at Qinghua University and offered a paper "On Waring's Problem with Polynomial Summands," which also drew from work of Hardy and Littlewood.[103]

Another paper, "Zur Stufentheorie der Quasi-algebraisch-abgeschlossenheit kommutativer Körper," was written by Chiungtze C. Tsen (Zeng Jiongzhi) and dedicated to the memory of Emmy Noether.[104] Zeng had studied in Hamburg as a Research Fellow of the China Foundation for Promotion of Education and Culture and was pleased to thank Emil Artin, in particular, for inspiration and "*Gastfreundlichkeit.*"[105] Zeng cited his own dissertation, *Algebren über Funktionenkörpern* (Göttingen, 1934), as well as work that he had already published in the *Göttinger Nachrichten* (1933) on division algebras.[106] As for his affiliation, however, he listed it as "*Mathematisches Sominar* [sic] *der Universität von Tschekiang*" in Zhejiang Province (May 1936).[107]

Yu-cheng Shen (Shen Youcheng) contributed a paper "On Interpolation and Approximation by Rational Functions with Preassigned Poles," which contained results from his Harvard dissertation (June 1934), and noted that some of his results had already been "incorporated" in Professor James Walsh's recent book, *Interpolation and Approximation by Rational Functions in the Complex Domain.*[108]

The first issue of volume two of the new journal appeared in February of 1937, and was even more international in its list of authors than volume one. It began with a paper by the Belgian mathematician, Lucien Godeaux, of the University of Liège, "Sur les quadriques associées aux points d'une surface."[109] This was followed by a contribution from the French mathematician, Jacques Hadamard, who was then at Qinghua University under the auspices of the *Commission des oeuvres franco-chinoises*. Hadamard had been lecturing in Beijing in the spring of 1936 and signed his paper "*Université Tsing-Hua, Pékin.*" In this work, Hadamard returned to a problem that he had considered earlier, namely, whether it was possible to determine a solution to a given hyperbolic partial differential equation under certain given conditions. This time, he considered the question from a geometric point of view, and it was this approach that he used to point out "some remarkable circumstances that had eluded [him] earlier."[110]

---

[102]Hua Luogeng, "On Waring's Problem with Polynomial Summands," *Journal of the Chinese Mathematical Society* **1** (1936):23-61 on p. 23.

[103]*Ibid.* Hua cited a series of papers by Hardy and Littlewood, beginning with "Some Problems of 'Partitio Numerorum': I. A New Solution of Waring's Problem," *Göttinger Nachrichten* (1920):33-54, and culminating with "The Number $\Gamma(k)$ in Waring's Problem," *Proceedings of the London Mathematical Society* **28** (1929):518-542.

[104]Zeng Jiongzhi, "Zur Stufentheorie der Quasi-algebraisch-abgeschlossenheit kommutativer Körper," *Journal of the Chinese Mathematical Society* **1** (1936):81-92.

[105]*Ibid.*, p. 81.

[106]Zeng Jiongzhi, *Algebren über Funktionenkörpern* (Göttingen: n.p., 1934), and "Divisionsalgebren über Funktionenkörpern," *Göttinger Nachrichten* (1933):335-339.

[107]Zeng, "Zur Stufentheorie," p. 92.

[108]Shen Youcheng, "On Interpolation and Approximation by Rational Functions with Preassigned Poles," *Journal of the Chinese Mathematical Society* **1** (1936):154-173, and James L. Walsh, *Interpolation and Approximation by Rational Functions in the Complex Domain*, Colloquium Publications of the American Mathematical Society, vol. 20 (New York: American Mathematical Society, 1935).

[109]Lucien Godeaux, "Sur les quadriques associées aux points d'une surface," *Journal of the Chinese Mathematical Society* **2** (1937):1-5.

[110]". . . quelques circonstances remarquables qui m'avaient échappé précédemment." Jacques Hadamard, "Le problème de Dirichlet pour les équations hyperboliques," *Journal of the Chinese Mathematical Society* **2** (1937):6-20 on p. 6.

Publication of the *Journal of the Chinese Mathematical Society* was interrupted in 1938 by the outbreak of the Sino-Japanese War, and for the next two years, mathematicians were among the intellectuals and scholars who were evacuated to Kunming in Yunan Province. The second issue of volume two, however, eventually appeared in February of 1940, beginning with page 139, but with a noticeable decline in the quality of paper. Nevertheless, the quality of research was impressive. Hua Luogeng contributed "On a Generalized Waring Problem II,"[111] while Su Buchin offered several papers on projective differential geometry and the projective theory of n-dimensional curves that drew on work he had just published in the *Annali di matematica pura ed applicata* (1939).[112]

Similarly, Ta-jen Wu (Wu Daren), in his paper on "Der Dual der Grundformel in Integralgeometrie," cited a work of his own, "Über elliptische Geometrie," that had recently appeared in the *Mathematische Zeitchrift*.[113] He had written the latter while studying with Blaschke in Hamburg in 1937. Chao Ko (Ke Zhao) contributed a "Note on the Diophantine Equation $x^x y^y = z^z$," devoted to an Erdös conjecture, namely, that $x^x y^y = z^z$ has no integer solutions if $|x| > 1$, $|y| > 1$, $|z| > 1$, and Shao-lien Chow (Zhou Shaolian), writing from Chongqing in 1939, offered a study of closed point sets that drew on work he had already published in French: *Questions de géométrie des ensembles*.[114] Zhou's theorem showed that given any closed point set E and two points not belonging to E, it was always possible to join the two points in question by a simple arc of limited length that did not intersect E.

Chern Shiing-shen (Chen Shengshen) dedicated his first publication in the *Journal of the Chinese Mathematical Society* to Élie Cartan (on the occasion of Cartan's seventieth birthday). Chern's paper, "The Geometry of Higher Path-Spaces," investigated systems of differential equations in terms of equivalence properties, using a method, as Chern acknowledged, pioneered by Cartan.[115] And Hua Loo-keng provided a paper "On an Exponential Sum" that managed to achieve a better upper bound than one he had published a year earlier in the *Journal of the London Mathematical Society*.[116] In all, the first two volumes of the *Journal of the Chinese Mathematical Society*, which appeared between 1937 and 1940, included forty-one works, among them contributions from three distinguished foreign mathematicians.

---

[111] Hua Luogeng, "On a Generalized Waring Problem II," *Journal of the Chinese Mathematical Society* **2** (1940):175-191.

[112] Su Buchin, "Plane Sections of the Tangent Surface of a Space Curve," *Annali di matematica pura ed applicata* **18** (1939):77-96.

[113] Wu Daren, "Der Dual der Grundformel in Integralgeometrie," *Journal of the Chinese Mathematical Society* **2** (1940):199-204, and "Über elliptische Geometrie," *Mathematische Zeitschrift* **43** (1938):495-521.

[114] Ke Zhao, "Note on the Diophantine Equation $x^x y^y = z^z$," *Journal of the Chinese Mathematical Society* **2** (1940):205-207; Zhou Shaolian, "Sur les ensembles fermées punctiformes," *Journal of the Chinese Mathematical Society* **2** (1940):235-237; and Zhou Shaolian, *Questions de géométrie des ensembles* (Paris: Vuibert, 1936).

[115] Shiing-shen Chern, "The Geometry of Higher Path-Spaces," *Journal of the Chinese Mathematical Society* **2** (1940):247-276.

[116] Hua Loo-keng, "On an Exponential Sum," *Journal of the London Mathematical Society* **13** (1938):54-61, and "On an Exponential Sum," *Journal of the Chinese Mathematical Society* **2** (1940):301-312.

Conclusion

Despite the constant political turmoil and economic uncertainties that have plagued China throughout the twentieth century, by the end of World War II and the creation of the People's Republic of China in 1949, Chinese mathematicians could look back on many accomplishments of which they could justifiably be proud. Not only had Chinese mathematics come of age during the first half of the twentieth century, taking its place among work done at the most advanced levels anywhere in the West, but it also achieved at least a preliminary independence and self-identity as well. By the end of the Second World War, Chinese mathematicians could be found at home as well as abroad—teaching, lecturing, publishing at the highest levels in the best journals anywhere in the world. Chinese mathematicians had formed their own national organization, established regular serial titles, and mobilized a retinue of writers, readers, and students eager to help in any way necessary to make indigenous Chinese mathematics a success. As Bertrand Russell said in reflecting on the spirit he felt alive and at work in Beijing when he was there in the early 1920s, "[t]he atmosphere was electric with the hope of a great awakening."[117]

And awaken it did. China, given the serious interests of its early reformers and the philosophy—if not the immediate success—of the "Self-Strengthening Movement," made remarkable progress against formidable odds by the end of the nineteenth century. From a handful of students who learned the most basic mathematics at missionary schools or in the classrooms of a few select arsenals and shipyards, the country slowly began to develop a widespread appreciation for modern mathematics that quickly became established in colleges and universities as the country underwent massive educational reforms at the beginning of the twentieth century.

At first, the success of modern mathematics in China was achieved by virtue of students who studied abroad, or thanks to visitors to Beijing and elsewhere like Russell, Blaschke, Osgood, Wiener, and Hadamard, through whom scholars in China were exposed to many of the newest mathematical ideas. The real test of Chinese mathematics, however, came during the 1930s when it established itself, professionally and institutionally, in ways that no longer depended upon foreign inspiration or support for its continuation and success. Remarkably, even at the beginning of the war with Japan—when many of the country's best faculty and students were evacuated to Southwest Associated University in Kunming—mathematics was still taught, research was still published. After World War II, mathematicians had much to contribute to the new People's Republic of China, and thanks to the foundations laid during the first half of the century, they were prepared to do so. Although many individuals and institutions played their parts, it was the colleges and universities, the Chinese Mathematical Society, and the journals it supported that gave modern mathematics in China a presence and an institutional stability that would ensure its persistence and enable it to withstand its greatest test to come later in the century—the Cultural Revolution.

[117]Russell, *Autobiography*, 2:191-192.

References

Abe, Hiroshi. "Borrowing from Japan: China's First Modern Education System." In *China's Education and the Industrialized World: Studies in Cultural Transfer*. Ed. Ruth Hayhoe and Marianne Bastid. Armonk, NY: M. E. Sharpe, 1987, pp. 57-80.

Ayers, William. *Chang Chih-tung and Educational Reform in China*. Cambridge: Harvard University Press, 1971.

Bastid, Marianne. "Servitude or Liberation? The Introduction of Foreign Educational Practices and Systems to China from 1840 to the Present." In *China's Education and the Industrialized World: Studies in Cultural Transfer*. Ed. Ruth Hayhoe and Marianne Bastid. Armonk, NY: M. E. Sharpe, 1987, pp. 3-20.

Beeching, Jack. *The Chinese Opium Wars*. London: Hutchinson, 1975.

Bennett, Adrian A. *John Fryer: The Introduction of Western Science and Technology into Nineteenth-Century China*. Cambridge: East Asian Research Center, Harvard University, 1967.

Biggerstaff, Knight. "Shanghai Polytechnic Institution and Reading Room: An Attempt to Introduce Western Science and Technology to the Chinese." *The Pacific Historical Reiew* **25** (1956); Reprinted in Biggerstaff, Knight. *Some Early Chinese Steps Toward Modernization*. San Francisco: Chinese Materials Center, 1975, pp. 69-91.

______. *The Earliest Modern Government Schools in China*. New York: Cornell University Press, 1961.

Brewer, James W. and Smith, Martha K., Ed. *Emmy Noether: A Tribute to Her Life and Work*. New York: Marcel Dekker, 1981.

Buck, Peter. *American Science and Modern China*. New York: Cambridge University Press, 1980.

Cheng Minde, Ed. *Zhongguo jiandai shuxuejia zhuan* (*Biographies of Modern Chinese Mathematicians*). 3 Vols. Huai Yin: Jiangsu Education Publishing House, 1995.

Chern Shiing-shen. "The Geometry of Higher Path-Spaces." *Journal of the Chinese Mathematical Society* **2** (1940):247-276.

Chesneaux, Jean; Bastid, Marianne; and Bergère, Marie-Claire, Ed. *China from the Opium Wars to the 1911 Revolution*. Trans. Ann Destenay. New York: Pantheon Books, 1976.

Davin, Delia. "Imperialism and the Diffusion of Liberal Thought: British Influences on Chinese Education." In *China's Education and the Industrialized World: Studies in Cultural Transfer*. Ed. Ruth Hayhoe and Marianne Bastid. Armonk, NY: M. E. Sharpe, 1987, pp. 33-56.

Engelfriet, Peter. "The Chinese Euclid and its European Context." In *L'Europe en Chine: Interactions scientifiques, religieuses et culturelles aux XVIIe et XVIIIe siècles. Actes du colloque de la Fondation Hugot*. Ed. Catherine Jami and Hubert Delahaye. Mémoires de l'Institut des hautes études chinoises. Vol. 34. Paris: Collège de France, Institut des hautes études chinoises, 1993, pp. 111-135.

______. *Euclid in China: The Genesis of the First Chinese Translation of Euclid's Elements Books I-VI (Jike yuanben; Beijing, 1607) and Its Reception up to 1723*. Sinica Leidensia. Vol. 40. Leiden: Brill, 1998.

Fairbank, John K. *The United States and China.* 4th Ed. Cambridge: Harvard University Press, 1979.

Fisher, Daniel Webster. *Calvin Wilson Mateer, Forty-Five Years a Missionary in Shantung Province, China: A Biography.* Philadelphia: The Westminster Press, 1911.

Franke, Wolfgang. *The Reform and Abolition of the Traditional Chinese Examination System.* Cambridge: Harvard University Press, 1960.

Fukasaku, Yukiko. *Technology and Industrial Development in Pre-War Japan.* London: Routledge, 1992.

Giquel, Prosper. *L'Arsenal de Fou-Tcheou, ses résultats.* Shanghai: A. M. de Carvalho, 1874.

———. *The Foochow Arsenal and Its Results.* Trans. H. Lang. Shanghai: Shanghai Evening Courier, 1874.

Godeaux, Lucien. "Sur les quadriques associées aux points d'une surface." *Journal of the Chinese Mathematical Society* **2** (1937):1-5.

Haas, William J. *China Voyager: Gist Gee's Life in Science.* Armonk, NY: M. E. Sharpe, 1996.

Hadamard, Jacques. "Le problème de Dirichlet pour les équations hyperboliques." *Journal of the Chinese Mathematical Society* **2** (1937):6-20.

Hardy, Godfrey H. and Littlewood, John E. "Some Problems of 'Partitio Numerorum': I. A New Solution of Waring's Problem." *Göttinger Nachrichten* (1920):33-54.

———. "Abel's Theorem and its Converse II." *Proceedings of the London Mathematical Society* **22** (1924):254-269.

———. "The Number $\Gamma(k)$ in Waring's Problem." *Proceedings of the London Mathematical Society* **28** (1929):518-542.

Hayhoe, Ruth *China's Universities, 1895–1995: A Century of Cultural Conflict.* New York: Garland, 1996.

Hayhoe, Ruth and Bastid, Marianne, Ed. *China's Education and the Industrialized World: Studies in Cultural Transfer.* Armonk, NY: M. E. Sharpe, 1987.

Holt, Edgar. *The Opium Wars in China.* London: Putnam, 1964.

Hu Kunsheng. "A Problem with Variable End Points in the Calculus of Variations." *Journal of the Chinese Mathematical Society* **1** (1936):1-14.

Hu Mingjie. "Merging Chinese and Western Mathematics: The Introduction of Algebra and the Calculus in China, 1859–1903." Princeton University. Unpublished Ph.D. dissertation, 1998; Ann Arbor: University Microfilms #9833137 (1998).

Hua Luogeng. "On Waring's Problem with Polynomial Summands." *Journal of the Chinese Mathematical Society* **1** (1936):23-61.

———. "On Waring's Problem." *The Quarterly Journal of Mathematics* **9** (1938): 199-202.

———. "On an Exponential Sum." *Journal of the London Mathematical Society* **13** (1938):54-61.

———. "On an Exponential Sum." *Journal of the Chinese Mathematical Society* **2** (1940):301-312.

———. "On a Generalized Waring Problem II." *Journal of the Chinese Mathematical Society* **2** (1940):175-191.

Hunt, Michael H. "The American Remission of the Boxer Indemnity: A Reappraisal." *Journal of Asian Studies* **31** (3) (1972):539-559.

Hyatt, Irwin T., Jr. *Our Ordered Lives Confess: Three Nineteenth-Century American Missionaries in East Shantung.* Harvard Studies in American-East Asian Relations. Vol. 8. Cambridge: Harvard University Press, 1976.

Jian Bozan (Chien Po-tsan) *et al.* *Wu-hsü pien-fa* (*Wuxu bianfa*) (*The Reform Movement of 1898*). 4 Vols. Shanghai: N.p., 1953.

Ke Zhao. "Note on the Diophantine Equation $x^x y^y = z^z$." *Journal of the Chinese Mathematical Society* **2** (1940):205-207.

Kennedy, Thomas L. *The Arms of Kiangnan: Modernization in the Chinese Ordnance Industry, 1860–1895.* Boulder: Westview Press, 1978.

Leibo, Steven A. *Transferring Technology to China: Prosper Giquel and the Self-strengthening Movement.* Berkeley: Institute of East Asian Studies, 1985.

Li Wenlin. "Getinggen shuxue de shijie yingxiang" (The International Influence of Göttingen Mathematics). *Shuxueshi yanjiu wenji* (*Collected Research Papers on History of Mathematics*) **6** (1995):117-123.

Li Wenlin and Martzloff, Jean-Claude. "Aperçu sur les échanges mathématiques entre la Chine et la France (1880–1949)." *Archive for History of Exact Sciences* **53** (1998):181-200.

Martin, William A. P. Review of Mateer's *Arithmetic*. *The Chinese Recorder and Missionary Journal* **10** (September-October 1879):397-398.

——. "A Tribute to Dr. Mateer." *The Chinese Recorder* **39** (December 1908):694-695.

Martzloff, Jean-Claude. *A History of Chinese Mathematics.* Trans. Stephen S. Wilson. Berlin: Springer-Verlag, 1997.

Mateer, Calvin W. "Mathematics in Chinese." *The Chinese Recorder and Missionary Journal* **9** (September-October 1878):372-378.

Morse, Marston. *The Calculus of Variations in the Large.* American Mathematical Society Colloquium Publications. Vol. 18. New York: American Mathematical Society, 1934.

Peake, Cyrus H. *Nationalism and Education in Modern China.* New York: Columbia University Press, 1932. Reprint Ed. New York: Howard Fertig, 1970.

Reischauer, Edwin O.; Fairbank, John K.; and Craig, Albert M. *East Asia: Tradition and Transformation.* Boston: Houghton Mifflin, 1973.

Ren Nanheng and Zhang Youyu, Ed. *Zhongguo shuxue hui shiliao* (*Materials for a History of the Chinese Mathematical Society*). Nanjing: Jiangsu jiaoyu chubanshe, 1995.

Russell, Bertrand. *Luosu suanli zhexue* (*Russell's Philosophy of Mathematics*). Trans. Fu Zhongsun and Zhang Bangming. Beijing: Shang Wu Press, 1922.

——. *Shuli zhexue daolun* (*Principles of Mathematical Philosophy*). Trans. Yan Chengshu. Beijing: Shang Wu Press, 1982.

——. *The Autobiography of Bertrand Russell.* 3 Vols. Boston: Little, Brown, 1951.

Shen Youcheng. "On Interpolation and Approximation by Rational Functions with Preassigned Poles." *Journal of the Chinese Mathematical Society* **1** (1936):154-173.

Su Buchin. "Plane Sections of the Tangent Surface of a Space Curve." *Annali di matematica pura ed applicata* **18** (1939):77-96.

Su Jing. *Qing ji Tongwen Guan* (*The Tongwen Guan of the Late Qing*). Taipei: Shanghai yinshuachang, 1977.

Tan, Chester C. *The Boxer Catastrophe.* New York: Columbia University Press, 1955.

Teng Ssu-Yü and Fairbank, John K., Ed. *China's Response to the West, A Documentary Survey, 1839-1923.* New York: Atheneum, 1970.

Tsien Tsuen-hsuin. "Western Impact on China Through Translation." *Far Eastern Quarterly* **13** (1954):305-327.

Walsh, James L. *Interpolation and Approximation by Rational Functions in the Complex Domain.* Colloquium Publications of the American Mathematical Society. Vol. 20. New York: American Mathematical Society, 1935.

Wang Yüan-te (Wang Yuande). "Di kaowen xiansheng chuan" (Biography of Mr. Calvin Mateer). In *Wen hui guan zhi* (*Alumni History of Tengchow College*). Ed. Wang Yuande and Liu Yü-feng. Wei-hsien: N.p., 1913, pp. 1-10.

Wang Yüan-te (Wang Yuande) and Liu Yü-feng, Ed. *Wen hui guan zhi* (*Alumni History of Tengchow College*). Wei-hsien: N.p., 1913.

Wiener, Norbert. "A Tauberian Gap Theorem of Hardy and Littlewood." *Journal of the Chinese Mathematical Society* **1** (1936):15-22.

———. *I Am a Mathematician: The Later Life of a Prodigy.* Garden City: Doubleday & Company, Inc., 1956.

Wright, Mary Clabaugh. *The Last Stand of Chinese Conservatism. The T'ung-Chih Restoration, 1862-1874.* Stanford: University Press, 1957.

Wu Daren. "Über elliptische Geometrie." *Mathematische Zeitschrift* **43** (1938):495-521.

———. "Der Dual der Grundformel in Integralgeometrie." *Journal of the Chinese Mathematical Society* **2** (1940):199-204.

Yamade Keiji, Ed. *The Transfer of Science and Technology Between Europe and Asia, 1780-1880.* Kyoto: International Research Center for Japanese Studies, 1994.

Yue Meng. "Hybrid Science versus Modernity: The Practice of the Jiangnan Arsenal, 1864–1897." *East Asian Science, Technology, and Medicine* **16** (1999):13-52.

Zeng Jiongzhi. "Zur Stufentheorie der Quasi-algebraisch-abgeschlossenheit kommutativer Körper." *Journal of the Chinese Mathematical Society* **1** (1936):81-92.

———. "Divisionsalgebren über Funktionenkörpern." *Göttinger Nachrichten* (1933):335-339.

———. *Algebren über Funktionenkörpern.* Göttingen: N.p., 1934.

Zhang Lihua and Luo Rongqu. "Technical Education of the Fujian Ship Management Office and the Transfer of Modern Western Technology to China." In *The Transfer of Science and Technology Between Europe and Asia, 1780-1880.* Ed. Yamade Keiji. Kyoto: International Research Center for Japanese Studies, 1994, pp. 229-239.

Zhou Shaolian. *Questions de géométrie des ensembles.* Paris: Vuibert, 1936.

———. "Sur les ensembles fermées punctiformes." *Journal of the Chinese Mathematical Society* **2** (1940):235-237.

Zhuang Jingtai and Zhu Dexiang. "Xiong Qinglai." In *Zhongguo jiandai shuxuejia zhuan* (*Biographies of Modern Chinese Mathematicians*). Ed. Cheng Minde. 3 Vols. Huai Yin: Jiangsu Education Publishing House, 1995, 1:35-47.

CHAPTER 14

Chinese–U.S. Mathematical Relations, 1859–1949

Yibao Xu*
City University of New York (United States)

Introduction

China's relationship with the United States dates back to the earliest years of the American Republic. In 1784, the American ship *Empress of China* sailed from New York harbor to Canton and, in so doing, inaugurated trading relations between the two countries. Over the next five decades, businessmen provided the sole connections between the United States and China. This began to change when the missionary, Elijah Coleman Bridgman, sent by the American Board of Commissioners for Foreign Missions, arrived in China and began attempts to convert the Chinese to Christianity in 1830. Following his example, more missionaries from different denominations entered into China's five treaty ports after the first Sino-American treaty, the Wangxia Treaty, was signed in 1844. After China was defeated in the Opium War (1856–1860) by the British and French, restrictions on missionary activities in the treaty ports were totally removed. As a result, missionaries from Europe as well as from the United States began to flood into China in the mid-nineteenth century. It was through the missionaries, primarily English and American, that modern Western mathematics was again introduced into China after the termination of the first wave in the 1720s.

American missionaries exerted considerable effort to transform mathematical education in China from a traditional style to a modern, Western one. The Presbyterian, Calvin Wilson Mateer, is a good example.[1] He, together with his associates, compiled a series of elementary mathematics textbooks for Chinese students. In these textbooks, Mateer paid considerable attention to mathematical terminology

*The author wishes to thank Karen Hunger Parshall and Adrian C. Rice for inviting him to participate in the conference, "Mathematics Unbound," and for suggesting the topic of the paper from which this chapter has developed. Over the course of its preparation, I have been greatly indebted to Joseph W. Dauben for his guidance and for polishing my English. I also want to express my sincere thanks to Lisa R. Coats for providing relevant documents preserved at the Historical Studies–Social Science Library of the Institute for Advanced Study at Princeton as well as to the Institute for permission to publish from its archives. Similar thanks go to William M. Roberts at the Bancroft Library of the University of California, Berkeley. Last, but not least, I am grateful to the two referees for the criticisms, comments, and suggestions that have certainly made this chapter stronger. (My name is presented here in Western order.)

[1]For detailed information about Mateer and his mission in China, see Daniel W. Fisher, *Calvin Wilson Mateer: Forty-Five Years a Missionary in Shantung, China: A Biography* (Philadelphia: The Westminster Press, 1911), and Irwin T. Hyatt, Jr., *Our Ordered Lives Confess: Three Nineteenth-Century American Missionaries in East Shantung* (Cambridge: Harvard University Press, 1976).

as well as to the form of presentation. Mateer was also the first person to study systematically the correspondence of Chinese-English mathematical terminology.

Through the hands of missionaries like Mateer in China, mathematical books, especially American textbooks, prevailed in China. The first such book was Elias Loomis's *The Elements of Analytical Geometry and That of Differential and Integral Calculus*,[2] which also happened to be the first book on calculus to be translated into Chinese (1859). Because of the great success of this translation, it had an immediate and lasting impact on the Chinese mathematical community. Loomis, who was a professor of natural philosophy and astronomy at Yale from 1860 to 1889, immediately became a well-known mathematician in China.[3] As a consequence, all but four of his fourteen mathematics textbooks were subsequently translated into Chinese.[4] Moreover, the second edition of his calculus text was actually rendered twice into Chinese by two different groups. It was in large measure through Loomis's textbooks that America exerted its earliest influence on the transformation from traditional to modern mathematics in China.

It is against this backdrop of the modernization of mathematics in nineteenth-century China that this chapter will analyze the influence of the United States in shaping the course of Chinese mathematics in the first half of the twentieth century. By focusing on the role of the Boxer Indemnity in the opening decades of the twentieth century and of specific American institutions, including Harvard University and the Institute for Advanced Study at Princeton, among others, it will present a case study of the effects of increased contact with scholars from abroad on the development of a particular mathematical community.

The Boxer Indemnity and the Modernization of Chinese Mathematics

About one hundred years ago, a movement against foreigners and Christians, known as the Boxer Uprising, began in Shandong Province and rapidly spread to other parts of northern China. Conservatives at the Qing court, represented by the Empress Dowager, wanted to rid China of foreign forces and therefore sought to exploit the uprising. As it gained strength, eight Western countries, namely, Austria-Hungary, Britain, Germany, France, Italy, Japan, Russia, and the United States, all sent forces to China in order to protect their various interests there. The uprising was soon put down, and as a result, representatives from fourteen countries—the abovementioned eight together with Belgium, Denmark, the Netherlands, Portugal, Spain, and Sweden—forced the Qing government to sign the Boxer Protocol in 1901. The provisions of Article VI of the Protocol demanded a substantial indemnity totaling 450,000,000 taels, equivalent to approximately $334,000,000, to be paid over thirty-nine years with an annual interest of 4%. The United States was apportioned about 5.4% of the total or $24,440,000. Fortunately, through the lobbying of Chinese diplomats and some American political activists, the U.S. Congress and

---

[2]Elias Loomis, *The Elements of Analytical Geometry and That of Differential and Integral Calculus* (New York: Hapers & Brothers, 1851). This book was very popular in the United States; it had sold 25,000 copies by 1874, when the second edition appeared. The Chinese translation was done by the English missionary, Alexander Wylie, and Li Shanlan.

[3]For Loomis's biography, see Hubert A. Newton, *Elias Loomis, L.L.D., 1811–1889: Memorial Address* (New Haven: Tuttle, Morehouse & Taylor, 1890).

[4]For a list of Loomis's textbooks, see Zhang Dianzhou and Joseph W. Dauben, "Mathematical Exchanges Between the United States and China: A Concise Overview (1850–1950)," in *The History of Modern Mathematics*, vol. 3, ed. Eberhard Knobloch and David E. Rowe (Boston: Academic Press, Inc., 1994), pp. 263-297 on pp. 288-289.

President Theodore Roosevelt agreed in 1908 that after paying for property and other damages, the rest of the indemnity—in the amount of $10,784,507.71—was to be remitted to China, mainly for the purpose of educating Chinese students in both the United States and China.[5]

Prior to this time, there were few Chinese students studying in the U.S. The first Chinese to graduate from an American college was Yung Wing, who was brought to the U.S. by the Protestant missionary, Samuel Robbins Brown. After receiving his Bachelor's degree from Yale University in 1854, he returned to China. Due to Yung's persistent efforts, the Qing government later sent about 120 boys in four successive years, beginning in 1872, to study in the Connecticut Valley as part of the Self-Strengthening Movement. This movement, as noted in the previous chapter, was one of the major consequences of the Opium Wars. Its primary goal was to strengthen China by acquiring and teaching advanced Western military technologies, particularly in the areas of machinery- and steamship-building. Unfortunately, this educational mission to the U.S. was abandoned by the Qing government in 1881, basically because of differences of opinion between Yung Wing and various of his Chinese colleagues over how the students should be educated, and due to objections that some of the students had been converted to Christianity.[6] When the boys were ordered back to China in 1881, only a handful had as yet graduated from college. Although most of them were still in high school, the students who returned nevertheless made considerable contributions to the modernization of China. None, however, had influence on the development of modern Chinese mathematics.[7] The Chinese imperial government did not resume sending specially selected students to study in the U.S. until 1890, and even then there were few.[8] Thereafter, the numbers steadily grew, however; in 1907, 155 Chinese were reported at American institutions, studying on funds provided either by the

[5]The data presented here were drawn from George Marvin, "An Act of International Friendship," *The Outlook* (November 14, 1908), pp. 582-586. For a detailed account of the Boxer Indemnity, besides Marvin's articles, see Terence Eldo Brockhausen, "The Boxer Indemnity: Five Decades of Sino-American Dissension" (Texas Christian University, unpublished Ph.D. dissertation, 1981), and Wang Shuhuai, *Gengzi peikuan* (*The Boxer Indemnity*) (Taipei: Institute of Modern History, Academia Sinica, 1974). For the Boxer Indemnity and modern Chinese mathematics, see Zhang Dianzhou, "Gengzi peikuan he zhongguo xiandai shuxue de fazhang" (The Boxer Indemnity and the Rise of Modern Chinese Mathematics), in *Keshi xinchuan* (*Passing the Torch of the History of Science*) (Shengyang: Liaonin jiaoyu chubanshe, 1997), pp. 179-184.

[6]Lai Hui-lan, *Qingmo liumei youtong zhi yanjiu* (*Study on the Chinese Education Mission in the United States in the Late Qing Dynasty*) (unpublished Master's thesis, The Chinese Cultural University, Taipei, 1984), pp. 134-46, and Xu Yibao, "The Chinese Educational Mission: Hopes, Failures and Reasons for its Recall by the Qing Government in 1881," in press. See also Yung Wing, *My Life in China and America* (New York: Henry Holt and Company, 1909).

[7]Arthur G. Robinson, "The Senior Returned Students: A Brief Account of the Chinese Educational Commission 1872–1881 under Dr. Yung Wing," *P. & T. Times* (June 24, 1932); Thomas F. LaFargue, *China's First Hundred* (Seattle: Washington State University Press, 1942; reprinted in 1987); and Gao Zonglu, *Zhongguo liumei youtong shuxinji* (*Collected Correspondence of Former Students of the Chinese Educational Commission*) (Taipei: Zhuangji wenxue congshu, 1986).

[8]Chen Xuexun and Tian Zhengping, *Zhongguo jindai jiaoyushi ziliao huibian: liuxue jiaoyu* (*Collected Sources for the History of Modern Chinese Education: Studying Abroad*) (Shanghai: Shanghai jiaoyu chubanshe, 1991), p. 8.

imperial or provincial governments. Of these, about one half were in either colleges or universities, but none of them opted to study mathematics.[9]

The Boxer Indemnity Scholarships opened the floodgates for Chinese students studying in the U.S. With the help of the Boxer Indemnity Fund, the Office for Studying in the United States was established in Beijing in 1909. That year, through nationwide examinations, the Office selected forty-seven students to go to the U.S., followed by seventy in 1910, and sixty-three in 1911. In the meantime, additional students—sponsored by local Chinese provincial governments, institutions, or by their own families—also found their way into American colleges and universities. In 1911, the Chinese Students' Alliance of America was founded and claimed 800 members. The figure soared to 1,300 in 1914 and to 1,500 in 1917.[10]

Of the 180 students selected between 1909 and 1911 by the Office for Studying in the U.S., seven were assigned to study mathematics: Wang Renfu (1909), Yan Jiazou (1909), Chao Yuanren (1910), Zhu Lu (1910), Hu Mingfu (1910), Ma Xianjiao (1910), and Jiang Lifu (1911).[11] Wang went to Harvard University directly. Yan, Chao, Hu, and Jiang eventually found their separate ways to Harvard (see below). Zhu and Ma were admitted to the University of Wisconsin where they received their B.A.s. They both then moved on to Columbia University, earning their Master's degrees in mathematics in 1915 and 1916, respectively. Zhu returned to China and subsequently became a high school mathematics teacher in Shanghai.[12]

In 1911, the Office for Studying in the U.S. was replaced by a preparatory school for studying abroad. The school, named the Qinghua School, recruited boys and girls at the age of about fifteen and then trained them for several years before sending them to American colleges and universities. Between 1913 and 1929, the year the Preparatory Department for Studying in America was abandoned, nineteen students had been chosen to study either mathematics or physics.[13] Among the mathematics students were He Yunhuang, Wen Yuqing, Deng Encong, and Ceng Yuanrong. He Yunhuang and Wen were sent to Cornell and Harvard, respectively, in 1914, and both of them received their B.A.s in mathematics in 1917. He Yunhuang continued his graduate study in Ithaca, New York, but unfortunately died of fever in February 1920 before finishing his Ph.D. Deng went to Princeton University in 1915, but withdrew in 1918 because of illness.[14] Ceng was recruited by the Preparatory Department in 1919 and was subsequently sent to the University of Chicago, where he received a Master's degree and then a Ph.D. in mathematics in 1933 under the guidance of Eliakim Hastings Moore. Parts of his dissertation were then published in the *Bulletin of the American Mathematical Society* under

---

[9]George Marvin, "The American Spirit in Chinese Education," *The Outlook* (November 28, 1908), pp. 670-72.

[10]Chen and Tian, pp. 214-221.

[11]Wang Huanchen, ed., *Liuxiu jiaoyu: zhongguo liuxiu jiaoyu shiliao* (*Studying Abroad: Historical Documents of Chinese Students Studying Abroad*), vol. 1 (Taipei: Guoli bianyiguan, 1980), pp. 218-225. See also Chen and Tian, pp. 188-202.

[12]John B. Powell, "Directory of American Returned Students," *Supplement to the Third Edition of Who's Who in China* (Shanghai: The Weekly Review, 1928).

[13]The Writing Group of the History of Qinghua University, ed., *Qinghua daxue xiaoshigao* (*A Draft History of Qinghua University*) (Beijing: Zhonghua shuju, 1981), pp. 68-69.

[14]Wang Huanchen, pp. 1115 and 1119.

the title "Expansions According to a Given System of Functions."[15] Ceng went on to pioneer functional analysis in China.

In May of 1924, the U.S. Senate passed a bill to remit a second Boxer Indemnity of $12,545,437 to China.[16] Thanks to these funds, the China Foundation for the Promotion of Education and Culture was created in September 1924. With support from the Foundation, Qinghua began to send the students it had recruited directly to America to pursue graduate study. Among those who won Boxer Indemnity Scholarships for mathematics were Liu Jin-nian (1925), Jiang Zehan (1927), and Hu Kunsheng (1929).[17] Both Liu and Jiang went to Harvard. Their education and the roles they played in the development of modern Chinese mathematics will be discussed below. Hu went to the University of Chicago to study mathematics, where he became especially interested in the calculus of variations. In 1930, he wrote a thesis on "Improper Double Integrals Depending Upon a Parameter," for which he was awarded a Master's degree. Over the next two years, he studied under the guidance of Gilbert A. Bliss and Lawrence M. Graves and received his Ph.D. in 1932. After teaching briefly at Qinghua, Hu moved to the Central University at Nanjing, then to Chongqing University in 1946, and finally to Sichuan University in 1953. He died in the latter post in 1959 at the early age of fifty-eight. Hu was a longtime Dean of the Mathematics Departments of both the Central University and Chongqing University. He was also a pioneer in the field of the calculus of variations in China, and some of his early research results were published in the first issue of the *Journal of the Chinese Mathematical Society.*[18]

The China Foundation also helped upgrade the Qinghua School to the status of a university in 1925. After the Mathematics Department was established at Qinghua three years later, it immediately became a center for teaching and research in mathematics. Its early faculty included Xiong Qinglai (Dean), Zheng Zhifan, Sun Guangyuan, and Yang Wuzhi. With the exception of Xiong, all were educated in the U.S. Zheng studied mathematics at Cornell, where he received his B.A. in 1910 before moving on to Harvard for one more year of study. Beginning in 1920, he taught mathematics at the Qinghua School. Sun and Yang were both Chicago graduates, receiving their Ph.D.s in 1928. Sun's interest lay in projective differential geometry, while Yang's was in number theory.

The China Foundation likewise understood the importance of sending Chinese professors abroad to do their own research.[19] One mathematician who received such support was Hua Luogeng. Hua never went to college but showed his mathematical talent very early. After publishing two highly original papers in 1929 and 1930 in

---

[15]Ceng Yuanrong, "Expansions According to a Given System of Functions," *Bulletin of the American Mathematical Society* **39** (1933):26-27.

[16]Yang Tsuihua, *Zhongjihui dui kexue de zanzhu* (*Patronage of Sciences: The China Foundation for the Promotion of Education and Culture*) (Taipei: Institute of Modern History, Academia Sinica, 1991), p. 6.

[17]Wang Huanchen, pp. 1107-1109.

[18]Hu Kunsheng, "A Problem with Variable End Points in the Calculus of Variations," *Journal of the Chinese Mathematical Society* **1** (1937):1-14, and recall the discussion in the previous chapter. For Hu's life and his mathematical contributions, see The Department of Mathematics of Sichuan University, ed., *Hu Kunsheng yizhu* (*Posthumous Works of Hu Kunsheng*) (Beijing: Renmin jiaoyu chubanshe, 1960).

[19]From 1928 to 1945, it provided thirty-two fellowships. See Sun E-tu Zen, "The Growth of the Academic Community, 1912-49," in *The Cambridge History of China*, vol. 13, pt. 2 (New York: Cambridge University Press, 1986), p. 406.

Kexue (*Science*), Hua came to the attention of Xiong Qinglai. Shortly thereafter, Xiong brought Hua to Qinghua. In Beijing, Hua studied Waring problems under the guidance of Yang Wuzhi. In order to encourage his potential, the China Foundation awarded Hua Luogeng $1,200 to visit Cambridge University in Great Britain for the 1936–1937 academic year. It also gave him a research scholarship for 1937–1938 to study with Godfrey H. Hardy. During his stay in Cambridge, Hua focused on analytical number theory, particularly on the circular method and the estimation of trigonometric sums. In this short period, he made a number of substantial discoveries, and the various articles he published then have subsequently become classics in these fields.[20]

Thanks to the financial support made possible by the China Foundation, Qinghua University created China's first graduate school in mathematics in 1930. Its first two graduate students were Chern Shiing-shen and Wu Daren. Before his graduation, Wu received the first British Boxer Indemnity Scholarship in 1933 and went to University College London, where he studied with Thomas L. Wren. Chern graduated from Qinghua in 1934 and then received an American Boxer Indemnity Scholarship. As a consequence of lectures Chern had heard by the German mathematician, Wilhelm Blaschke, who had visited Beijing in April 1932, the China Foundation made a special arrangement for Chern to go to work with him at the University of Hamburg. Two years later, Chern's Boxer Indemnity Scholarship also made it possible for him to do postdoctoral work in Paris with Élie Cartan.

Between 1931 and 1944, six students completed Master's degrees at the Qinghua graduate school, namely, Chern (1934), Shi Xianglin (1935), Zhuang Qitai (1936), Chung Kai-lai (1942), and Wang Hsien-chung (1944). Each later became a world-renowned mathematician.[21] Shi and Chung came to the U.S. for further study and received Ph.D.s from Harvard and Princeton in 1942 and 1947, respectively. Zhuang went to the University of Paris, earning a doctoral degree there in 1938 for his work in functional analysis. On the faculty at Beijing University by 1946, he trained a new generation of Chinese functional analysts. Wang earned his doctorate from Victoria University in Manchester, England and, in 1949, moved to the United States to establish what became his distinguished career.[22]

The China Foundation also provided financial support for foreign mathematicians to lecture in China. The first to be invited by the Foundation to visit China was Wilhelm Blaschke. Blaschke gave a series of lectures on topological questions in differential geometry in Beijing in the spring of 1932. His lectures had a profound impact on Chern's career, in particular, and upon the development of modern Chinese mathematics, in general. It was Blaschke who convinced Chern to study in Germany rather than in the U.S. His lectures also helped inspire an enduring interest in topology and differential geometry among Chinese mathematicians. Moreover, on Blaschke's recommendation, Beijing University invited the young German, Emanuel Sperner, with funding from the China Foundation, to spend two years

[20]For Hua's biography, see Wang Yuan, *Hua Loo-Keng: A Biography*, trans. Peter Man-Kit Shiu (Singapore: Springer-Verlag, 1999).

[21]The Writing Group of the History of Qinghua University, ed., *Qinghua daxue shiliao xunpian* (*Selected Historical Documents Related to Qinghua University*), vol. 3 (Beijing: Qinghua daxue chubanshe, 1994), pp. 102-07.

[22]About Wang's life and mathematical contributions, see Hu Shizhen T., "Hsien Chung Wang 1918-1978," *Bulletin of the Institute of Mathematics, Academia Sinica (Taipei)* **8** (1980):x-xii, and William M. Boothby, Shiing-shen Chern, and S. P. Wang, "The Mathematical Work of H. C. Wang," *op. cit.*, pp. xiii-xxiv.

in China. Sperner, a specialist in topology and geometrical foundations, came to China during the academic years 1932–1934. His lectures attracted students from both Qinghua and Beijing Universities. Sperner's "careful and detailed account of Erhard Schmidt's proof of the Jordan curve theorem" made a lasting impression on the young Chinese mathematics students.[23]

After Sperner left China, the newly retired Harvard professor, William Fogg Osgood, was invited to lecture at Beijing University for the academic years 1934–1936. Over those years, Osgood offered two courses, one entitled "Functions of Real Variables," the other "Functions of a Complex Variable." Subsequently, the University Press published Osgood's manuscripts of these two courses in 1936. Both became the most popular textbooks on these subjects in China, and both were popular in the United States.

While Osgood was still in Beijing, Qinghua invited his fellow American, Norbert Wiener, to come to China for the 1935–1936 academic year. Wiener's main work at Beijing, as he stated, "was lecturing on generalized harmonic analysis and on the material [Raymond] Paley and I had included in our book. I also was engaged in new pure-mathematics research on the so-called problem of quasi-analytic functions."[24] Wiener's classes, like those of other invited foreign lecturers, were attended by all mathematics students and faculty members both at Qinghua and Beijing Universities. It was from Wiener that Hua Luogeng, Shu Shien-siu, and Zhuang Qitai learned the Fourier analysis that proved such a useful tool in their future researches. Under Wiener's guidance, Hua and Shu co-authored a paper, entitled "On Fourier Transforms in L_p in the Complex Domain," which was published in 1936 in the *Journal of Mathematics and Physics*; Zhuang's paper, "On the Normalization of the Set of Functions X^n Sech x," appeared in the same issue. In his acknowledgment, Zhuang wrote: "The writer wishes to thank Prof. Wiener for suggesting the problem and his kind guidance during the course of the work."[25] Wiener's lectures at Qinghua University clearly had a strong impact on the introduction and development of harmonic analysis in China.[26]

The China Foundation was instrumental in sponsoring lectures given by eminent foreign mathematicians in China, and these, in turn, not only gave Chinese students a wider scope of mathematical knowledge but, above all, inspiration. Foreign mathematicians encouraged promising Chinese students to go abroad to pursue the mathematical sciences. This was, and still is, one of the most effective means of modernizing and internationalizing Chinese mathematics. Unfortunately, invitations to a number of prominent American mathematicians, like John von Neumann,

---

[23]Chern Shiing-shen, "My Mathematical Education," in *S. S. Chern: A Great Geometer of the Twentieth Century*, ed. Yau Singtung (Hong Kong: International Press, 1992), pp. 1-14 on p. 3.

[24]Norbert Wiener, *I Am a Mathematician* (Garden City: Doubleday & Company, Inc., 1956), p. 190. For details about this invitation and Wiener's activity in China, see Li Xuhui, "Sanshi niandai N. Wiener fangwen qinghua daxue handian shimo" (The Correspondence Concerning N. Wiener's Visiting China in the 1930s), *Zhongguo keji shiliao* (*China Historical Materials of Science and Technology*) **19** (1) (1998):42-51, and Wiener, chapter 10 (China and Around the World), pp. 183-207.

[25]Hua Loo-keng and Shien Shien-siu, "On Fourier Transforms in L_p in the Complex Domain," *Journal of Mathematics and Physics* **15** (1936):249-263, and Chuang Chi-tai, "On the Normalization of the Set of Functions X^n Sech x," *op. cit.*, pp. 264-267.

[26]Wang Yuan, "Hua Luogeng," in *Zhongguo kexue jishu zuangjia zhuanglue: shuxue* (*Brief Biographies of Chinese Scientific and Technological Specialists: Mathematicians*), vol. 1 (Shijiazhuang: Hebei jiaoyu chubanshe, 1996), p. 233.

were precluded by the Sino-Japanese War, which broke out in 1937. The immediate result of the conflict with Japan was that the Chinese mathematical community was totally isolated from the rest of the world.

Harvard: An Educational Center for Chinese Mathematicians

As noted, when Boxer Indemnity Scholarships first became available, some Chinese students began to come to the United States to study advanced mathematics. Thereafter, other Chinese students found alternative ways to come to the U.S. As a result, in the first half of the twentieth century, Chinese mathematics students were scattered across the entire country, but with significant concentrations at Harvard University, the University of Chicago, the University of Michigan, Princeton University, and the University of California at Berkeley.[27] On their return to China, most of the mathematics graduates from these universities (particularly those who received Ph.D.s and Master's degrees) became the driving force for the development of modern, Chinese mathematics. Most became professors of mathematics at either Beijing University, Nankai University, Qinghua University, the National Central University, or other prestigious institutions. To get a sense of their importance to the mathematical endeavor in China, consider those Chinese students who graduated from Harvard University as a case in point.

The earliest Harvard graduate of note here was probably Qin Fen. In 1906, he went to Harvard and studied astronomy and mathematics supported financially by the Tianjin Government. He earned his Bachelor's degree three years later, and then his Master's degree in astronomy in 1913. After returning to China, Qin became one of the three founders of the Department of Mathematics at Beijing University. He was also one of the nine founding members of the Chinese Mathematical Society in 1935 and was elected a member of the Society's governing Board.[28]

Wang Renfu was the first Chinese student picked to study mathematics with the support of the Boxer Indemnity Scholarship. He went to Harvard to study mathematics in 1909, where he received his B.A. in 1913. After a short stay at Beijing Teachers College, he became a professor of mathematics at Beijing University in the Department of Mathematics, which was the first mathematics department in China. Wang, together with Qin Fen and Feng Zuxun, the latter a Japan-educated mathematician, exerted considerable early influence upon mathematical teaching at Beijing University, in particular, and across the whole of China, in general.[29] Like Qin, Wang was also a member of the Board of the Chinese Mathematical Society.[30]

Hu Mingfu arrived at Harvard in 1914, the year after Qin and Wang had left; Chao Yuanren and Jiang Lifu followed in 1915. Hu and Chao were two of the three selected to study mathematics in the second dispatch of the Boxer Indemnity students. They first went to Cornell, where they proved to be excellent students. After receiving his B.A. in 1914, Hu transferred to Harvard to pursue graduate

---

[27] For a list of Chinese students who received their Ph.D.s in mathematics or related fields from American institutions, as well as their dissertation titles, see Yuan Tung-li, *A Guide to Doctoral Dissertations by Chinese Students in America 1905–1960* (Washington D.C.: Sino-American Cultural Society, Inc., 1961), and the Dissertation CD-ROM by A. Bell & Howell Company.

[28] Yang, pp. 124-125. See also Ren Nanheng and Zhang Youyu, *Zhongguo shuxuehui shiliao* (*Historical Materials of the Chinese Mathematical Society*) (Nanjing: Jiangshu jiaoyu chubanshe, 1995), pp. 117-118.

[29] On the mathematical training of Chinese students in Japan in the early twentieth century, compare Chikara Sasaki's chapter in the present volume.

[30] Ren and Zhang, pp. 120-121.

study, where he developed an interest in integral equations. Under the guidance of Maxime Bôcher, Hu wrote his thesis, "Linear Integro-differential Equations with a Boundary Condition," and became the first Chinese ever to receive a Ph.D. in mathematics. His thesis was an extension of Vito Volterra's work on the first and second classes of integral equations. The major results of his dissertation were published in the *Transactions of the American Mathematical Society*, becoming the first research paper on modern mathematics to be published in English by a Chinese mathematician.[31] After returning to China, Hu devoted himself to teaching mathematics at Da Tong (Utopia) University in Shanghai. Tragically, before he could exert any considerable influence on the development of Chinese mathematics, he drowned on 12 June, 1927 while swimming in a river in his hometown. It was a major loss for the entire Chinese mathematical community.

Unlike Hu, Chao Yuanren enrolled in the Department of Philosophy at Harvard. There, his interests centered on mathematical logic, and he became the first Chinese to study this subject. In 1918, he received his Ph.D. under the guidance of Henry M. Sheffer.[32] When the British logician and philosopher, Bertrand Russell, lectured in China from October 1920 to June 1921, Chao served as his principal personal translator. It was thus through Chao that Russell's Chinese audiences got their first introduction to Russell's philosophy and his technical work on mathematical logic. For better or for worse, Chao's interests soon turned. He pioneered the use of Western linguistic theories to study the Chinese language, and it is for this that he is now best remembered.

Jiang Lifu's career took a course different from that of Chao; he devoted the rest of his life to teaching and promoting Western mathematics in China. After receiving his B.A. in mathematics from the University of California at Berkeley in 1915, Jiang went to Harvard for graduate study. There, he developed a strong interest in geometry. His thesis, "The Geometry of a Non-Euclidean Line-Sphere Transformation," was written under the supervision of Julian L. Coolidge. In 1920, Jiang was appointed professor of mathematics at the newly founded Nankai University in Tianjin, where he soon built up a highly successful Department of Mathematics and educated soon-to-be eminent mathematicians such as Chern, Jiang Zehan, Liu Jin-nian, Shen Youcheng, and Wu Daren, among others. Jiang, Liu, and Shen followed in their mentor's footsteps, went to Harvard, and received their doctorates in 1930, 1930, and 1935, respectively. Jiang Lifu trained his pupils in a very strict way, which built a solid foundation for their future careers. Jiang Zehan later wrote: "The tests [at Harvard] were not really as hard as those in Nankai .... According to Mr. Liu Jingnan [Liu Jin-nian], who was my former classmate at Nankai, and who also studies mathematics here, there is no danger of my failing the examination even if I were to play around here for the next three years."[33]

---

[31]Hu Mingfu, "Linear Integro-differential Equations with a Boundary Condition," *Transactions of the American Mathematical Society* **19** (1918):363-407.

[32]See Yuan Ren Chao, *Life With Chaos: Yuen Ren Chao's Autobiography, First 30 Years, 1892-1921* (Ithaca: Spoken Language Services, 1975), Part III, Ten Years in America.

[33]Jiang Zehan to Hu Shih, 2 February, 1928, written from Harvard. See Geng Yunzhi, ed., *Hu Shih Yigao ji michang shuxin* (*Hu Shih's Manuscripts and his Secretly Treasured Correspondence*), vol. 25 (Hefei: Huangshang shushe, 1994), pp. 146-47. This volume includes Jiang Zehan's letters to Hu Shih on pp. 84-196. On the nature of these letters, see Xu Yibao, "Zhongguo xiandai shuxue de zhongyao shiliao: Jiang Zehan zhi Hu Shih de xinhan" (An Important Document for the History of Modern Chinese Mathematics: The Letters of Jiang Zehan to Hu Shih), *Shuxue chuanbo* (*Dissemination of Mathematics*) (*Taipei*) **23** (3) (1999):34-48.

Besides educating his students at Nankai, Jiang Lifu was also a central figure in the formation of the Chinese Mathematical Society, as well as the national mathematical institute, the Institute of Mathematics, Academia Sinica. For the latter, he was designated Director of the Preparatory Committee in 1941. Through both the Society and the Institute, Jiang further exerted his influence upon the course of mathematics in China.

Following the example set by his mentor Jiang Lifu, Jiang Zehan also devoted his life to mathematics. After finishing his thesis, "Existence of Critical Points of Harmonic Functions of Three Variables," under the supervision of Marston Morse at Harvard, Jiang Zehan went to Princeton in the summer of 1930, where he spent a year studying with Solomon Lefschetz. Jiang Zehan's research at both Harvard and Princeton was devoted to topology. Accordingly, he became the first Chinese to introduce the subject into his country, and through his efforts, the American topological tradition was transmitted directly to China. Topology was to become one of the strongest areas of Chinese mathematics, with valuable contributions to the field being made by Chinese topologists.

Jiang Zehan, like his mentor Jiang Lifu, had considerable administrative ability. Three years after he returned to China, he was elected Dean of the Department of Mathematics at Beijing University, a post he held until 1952. As Dean, he successfully maintained close connections with his former teachers at Harvard, as well as with mathematicians in Europe, by inviting them to lecture at Beijing University. He also encouraged and helped his students and junior colleagues to go abroad for further study.

Between 1909 and 1949, dozens of Chinese mathematics students, including eleven Ph.D.s, graduated from Harvard University. After they returned to China, the majority of them secured jobs as professors of mathematics at various universities. Thanks to their efforts, particularly those of Jiang Lifu, Jiang Zehan, and Shen Youcheng, modern American mathematics was transmitted directly to China, where a new indigenous generation of mathematicians took root. The universities, however, were not the only American institutions to have a strong influence on Chinese mathematicians.

The Institute for Advanced Study: A Bridge Between the U.S. and China

The Institute for Advanced Study (IAS) was founded in 1930. Mathematics was soon created as the first of the Institute's Schools.[34] The earliest faculty at the School of Mathematics included such eminent figures as Oswald Veblen, Hermann Weyl, John von Neumann, James Alexander, and Marston Morse. Given its excellent faculty, ideal location first within then adjacent to Princeton University, and stable funding, among other things, the Institute for Advanced Study immediately became a world center for mathematical research. The fortunes of the Institute were also affected by Nazi anti-Semitism in Axis countries and the exodus of many Jewish mathematicians to the United States prior to World War II. Among the latter, some of the best found their way to Princeton.[35]

[34] Armand Borel, "The School of Mathematics at the Institute for Advanced Study," in *A Century of Mathematics in America—Part III*, ed. Peter L. Duren *et al.* (Providence: American Mathematical Society, 1989), pp. 119-147.

[35] On the effects of the Nazi regime on mathematics, see the chapters below by Reinhard Siegmund-Schultze and Sanford Segal.

The first Chinese mathematician to become a member of the Institute was Jiang Zehan, who was awarded membership for the academic year 1936–1937. During his stay, besides working part-time as an assistant to his former Harvard thesis adviser, Marston Morse, Jiang carried out his own investigations into Lefschetz's fixed point theorem.[36] In a letter dated 12 November, 1936 to Hu Shih, his cousin's husband and John Dewey's champion in China, Jiang confided that "[r]ecently, I have made some progress in my studies, and am now becoming more interested in [mathematical research]."[37]

The academic atmosphere at the Institute not only inspired Jiang's intellectual work but also made him realize the important role research institutes could play in the advancement of mathematics. While still in Princeton, he envisioned his own mathematics institute and began to recruit young Chinese mathematicians who had recently received their doctoral degrees or who had been well-trained in mathematics abroad, particularly in the United States. Tragically, the Japanese declared war against China on 7 July, 1937, just one day before Jiang went back to Beijing University. As a result, Jiang's plan for an advanced mathematical institute in China died in embryo. He expressed his sadness in a letter on 7 October, 1937 to Hu Shih, who by then was Ambassador to the United States: "Last year I dreamed for the whole year of founding a Mathematical Institute within the Department of Mathematics. I invited [several] mathematicians and had also purchased the necessary books and references. Now not only has the dream blown up, but the things we already had have gone with it. It is really sad!"[38]

In 1947, when Jiang had a second opportunity to go abroad, sponsored by the Chinese Education Ministry, he again applied for a position at the Institute for Advanced Study. Once again, he was admitted, this time as a member for the 1947-1948 academic year. But Jiang soon changed his mind and decided instead to visit the topologist, Heinz Hopf, at the *Technische Hochschule* in Zürich, Switzerland. On his way to Zürich, Jiang wrote a letter dated 11 September, 1947, to Frank Aydelotte, then Director of the IAS, expressing both his regret for having changed his plan and his wish to visit Princeton the following year. Accordingly, the Institute held Jiang's membership until the end of 1948.[39] Ultimately, Jiang did not go to Princeton but remained in Zürich until May 1949, when he returned to the mainland by way of Hong Kong, only a few months before Mao announced the creation of the People's Republic of China on 1 October.[40]

The second Chinese mathematician to be admitted as a member of the Institute for Advanced Study (in the Fall of 1936) was Wong Yue-kei, who was also admitted again in the years 1948–1950. Wong was educated at the University of Chicago, where he received a B.S. with honors in 1927; he earned his M.S. in 1929 and his Ph.D. in 1931. Notably, Wong was elected a member of both the honor societies, Phi Beta Kappa and Sigma Xi.[41] According to Saunders Mac Lane, "[o]f the six students of [E. H.] Moore and [R. W.] Barnard during this period [1927–1941], only

[36]Geng, p. 162.

[37]*Ibid.*, p. 163.

[38]*Ibid.*, p. 184.

[39]Jiang Zehan, Administrative File, Historical Studies–Social Sciences Library, IAS, Princeton.

[40]Jiang Zehan, "Huiyi Hu Shih de jijianshi" (Reminiscences of Hu Shih), *World Journal* (*East Edition*) (December 19-20, 1990).

[41]Wong Yue-kei, Administrative File, Historical Studies–Social Sciences Library, IAS, Princeton.

Y. K. Wong (Ph.D. 1931) continued substantial activity. With Moore, he had studied matrices and their reciprocals; in his later research (at the University of North Carolina) he was concerned with the use of Minkowski-Leontief matrices in economics."[42] During his stay at Princeton in 1936, Wong became friends with Jiang Zehan, who later tried unsuccessfully to recruit him to serve at Beijing University.[43] Wong went on to pursue a distinguished mathematical career in the U.S.

The third Chinese mathematician to undertake substantial research at the IAS was Chern. He visited the Institute on three occasions, first as a member from August 1943 to February 1946; second, in the Spring Semester of 1949; and then again, as a member for the academic year 1954–1955. How Chern first came to the IAS is one of the most celebrated stories in Sino-American mathematical relations.

In a memorandum dated 22 April, 1942 to the Institute's Director, Frank Aydelotte, Oswald Veblen explained how Chern had approached him, how highly he regarded Chern, and how it might be possible to handle the financial problem of bringing Chern to the IAS:

> He wrote me on May 8, 1941, asking whether he could get a stipend in the Institute in 1942. With his letter he enclosed an account of his researches, and a paper which he offered for publication in one of the American mathematical journals. The paper struck me as being extremely good, and the referee's report pronounces it first class. His work altogether seems to establish that Chern is the most promising Chinese mathematician who has thus far come to our attention. Our mathematical group would therefore, except for the difficult situation in which we find ourselves, have recommended him for a stipend from the Institute.
>
> In the present circumstances our recommendation is that an attempt should be made to bring him to the Institute for a couple of years. The problem of bringing him here and of returning him to China at the end of his period might be referred to the Chinese Embassy in Washington, and the funds might be sought from one of the Foundations. We feel that Chern seems to be a man of such unusual quality, and the need of China for the development of such men so immediate, that there should be a good chance of carrying out a program of this sort.[44]

Although a request to the Rockefeller Foundation for funding was denied, Aydelotte approached the Chinese Embassy in Washington, D.C. as Veblen had suggested. In a letter dated 23 May, 1942 to Ambassador Hu Shih, Aydelotte wrote:

> It seems to us quite clear that Professor Chern himself and Chinese mathematics in general would be advanced if it could be made possible for him to come to the Institute for a year or, still better, for two. At the end of the war Professor Chern

---

[42]Saunders Mac Lane, "Mathematics at the University of Chicago: A Brief History," in *A Century of Mathematics in America—Part II*, ed. Peter L. Duren *et al.* (Providence: American Mathematical Society, 1988), pp. 127-154 on p. 140.

[43]Geng, p. 191.

[44]Chern Shiing-shen, Administrative File, Historical Studies–Social Sciences Library, IAS, Princeton.

> could be immensely useful in his own professorship at National Tsing Hua University and indeed throughout all the universities of China.
>
> We should be most happy to have Professor Chern as a member of the Institute if it could be made possible for him to come. I regret to say that we have no funds out of which we could offer to defray the cost of his travel or of his expenses in Princeton. I am writing you in the hope that it may be possible for the Chinese government to make Professor Chern's visit to Princeton financially possible. His own high qualifications and the importance of the subject would seem to us to justify the Chinese government in undertaking this expense.[45]

Unfortunately, not even Hu Shih could find a way to help support Chern financially. The IAS then turned to other channels but, by January 1943, decided to offer a stipend of $1,500 to Chern from its own already tight budget. The cost of his transportation, however, both to Princeton and later for his return to China, was still an open question. On 1 February, 1943, Veblen wrote to the Division of Cultural Relations of the U.S. Department of State, requesting its support by pointing out the importance both to China of developing leaders in mathematics and for the U.S. to cooperate in this process. Once again, the reply was negative. In the meantime, Chern had applied for travel grants from the Chinese Ministry of Education and Qinghua University, where he was then posted. In a letter to Veblen dated 4 March, 1943, Chern explained that he would make the trip by boat thanks to a subsidy from his University, even if his other application for a travel grant should fail.

In order to relieve most of Chern's financial burdens for traveling and to expedite his trip to Princeton, Aydelotte and his colleagues did everything they could to help prepare for Chern's long journey from China. In July 1943, Aydelotte wrote letters to the Office of the Secretary of War in Washington, D.C., to the Commanding Officer of the U.S. Military Transport Service in Cairo, Egypt, and to the Chinese Embassy, again requesting an allowance and assistance for Chern to travel on U.S. Army transport planes. As a result of all these carefully made arrangements, Chern finally arrived at the IAS safely on 14 August.[46]

Chern very soon proved that he was worthy of the extraordinary effort that had been made on his behalf. Over the next two years, he made important discoveries in both differential geometry and topology. His most influential work, a proof of the generalized Gauss-Bonnet formula, involving what are now known as Chern characteristic classes, was done during this period.[47] Chern's excellent work at the IAS very much impressed Veblen, Weyl, Morse, and other mathematicians there. "It is a great pleasure to have you here at the Institute and I hear with satisfaction

---

[45] *Ibid.*

[46] For Chern's own account of the trip, see his letter of 25 August, 1943 to Mei I-ch'I (Y. C. Mei), then President of Qinghua University, in The Writing Group of the History of Qinghua University, ed., *Qinghua daxue shiliao xunpian* (*Selected Documents on Qinghua University*), pp. 309-310.

[47] Phillip A. Griffiths, "Some Reflections on the Mathematical Contributions of S. S. Chern," in *Shiing-shen Chern Selected Papers* (New York: Springer-Verlag, 1978), pp. xiii-xix. See also *Shiing-shen Chern: A Mathematician and His Mathematical Work* (Singapore: World Scientific Publishing Co., 1996), pp. 76-82.

fine reports of the work you are doing," said Aydelotte in a letter to Chern dated 17 March, 1945, when he extended Chern's membership for the subsequent 1945–1946 academic year.[48] Chern himself also regarded his stay at Princeton as a crucial time in his long and distinguished career. On his way back to China, he wrote to Aydelotte from San Francisco on 23 February, 1946, expressing his gratitude: "The years 1943–45 will undoubtedly be decisive in my career, and I have profited not only on the mathematical side. I am inclined to think that among the people who have stayed at the Institute I was the [one] who has profited the most, but other people may think the same way."[49]

Chern returned to China in March 1946. Due to his outstanding work at the IAS in the previous two-and-a-half years, he was immediately appointed as Acting Director of the Institute of Mathematics, Academia Sinica, founded in 1947 after six years of preparation. Soon after Chern assumed this position, however, the political situation in China began to change dramatically.

In July 1946, the Nationalists and the Communists abandoned their cooperative alliance—the so-called Second United Front formed during the Anti-Japanese War—and China was again plunged into civil war. Initially, the Nationalists under Chiang Kai-Shek took the advantage in military action, but by November 1948, they had been driven out of Manchuria as a result of the Communists' Liaoxi-Shenyang campaign. Despite the civil war, Chern wrote to J. Robert Oppenheimer, who had succeeded Aydelotte as Director of the Institute, that "the conditions in Nanking have been quieter recently and the work at Academia Sinica is going on as usual."[50] Nevertheless, Chern's American friends at the IAS had already begun to worry about his situation. In a memorandum of 11 November, 1948, Veblen suggested to Oppenheimer that they should offer some sort of help to Chern. On 19 November, Oppenheimer sent an historic cable to Chern, which read as follows:

> If there should be any steps that you would like to have us take in the next months to facilitate your coming to this country, please let us know.
> Robert Oppenheimer[51]

Chern made up his mind to return to the United States after very careful consideration. This time he would not come alone but planned to bring his wife, the daughter of the American-educated mathematician, Zheng Zhifan, and their two children. This, however, raised an additional difficulty. According to Nanking Foreign Ministry regulations, passports could not be issued for the entire family unless Chern was given a three-year appointment. In order to meet this requirement, Oppenheimer made an exceptional decision and offered Chern a three-year membership at the Institute effective 1 January, 1949, through 31 December, 1951. Like Aydelotte, Oppenheimer also requested the Chinese Embassy and the U.S. Department of State to help get the Cherns their passports and visas. In a letter dated 30 November, 1948, to the Chinese Embassy, Oppenheimer wrote: "The School of Mathematics here is most anxious to facilitate the coming of Professor Chern to this country, and in present circumstances, believes that prompt action is necessary.

---

[48] Chern Shiing-shen, Administrative File, Historical Studies–Social Sciences Library, IAS, Princeton.

[49] *Ibid.*

[50] Chern to J. Robert Oppenheimer, 22 November, 1948 in *ibid.*

[51] J. Robert Oppenheimer to Chern, 19 November, 1948, in *ibid.*

We would appreciate your sending a certified cable embodying this appointment to membership at the Institute for Advanced Study immediately."[52] Thanks to all of these efforts, the Cherns were able to leave Shanghai for San Francisco on New Year's Eve 1948. Chern spent the spring of 1949 at the IAS, but then moved to the University of Chicago. In 1954, he came back for the third time to Princeton as a member for 1954–1955 and, in a letter to Oppenheimer, expressed once again his gratitude that the IAS had "played such a great role in my scientific career."[53]

Partly because of Chern's great success at the Institute for Advanced Study, and partly because of the growing number of well-qualified Chinese mathematicians, the Institute began to offer memberships increasingly often to eminent Chinese mathematicians after World War II. Between the years 1945 and 1949, although China was still in turmoil and the general situation was unstable for research, the following eight Chinese found their way to Princeton, either supported by the Chinese Government, by the Institute, or by the mathematicians themselves:[54]

- Fan Ky, 1945-1947: Functional Analysis.
- Jiang Lifu, 1946-1947: Geometry, Organization of Mathematical Research.
- Hua Luogeng, 1946-1948: Number Theory, Automorphic Functions, Geometry of Matrices, and Extended Spaces of Complex Variables.
- Chow Wei-liang, 1947-1948: Algebraic Geometry. (Admitted again in 1954-1955).
- Min Sihe, 1947-1948: Riemann Zeta Function, Analytic Number Theory.
- Shu Shien-siu, 1947-1948: Applied Mathematics (Fluid Mechanics).
- Zhang Zongsui, Fall 1947: Field Theory, Theory of Elementary Particles in Quantum Mechanics.
- Chen Yu-why, 1949-1950: Analysis; Differential Equations, Calculus of Variation. (Admitted again in Fall 1955, and Spring 1972).
- Zhang Shixun, Spring 1949: Integral Equations.

Fan Ky, Chow Wei-liang, Shu Shien-siu, and Chen Yu-why remained in the United States, where they made important contributions to their own respective fields. Jiang Lifu, Hua Luogeng, Min Sihe, Zhang Zhongsui, and Zhang Shixun all eventually went back to mainland China and made further contributions there, with those of Hua Luogeng being particularly important. After a period at the University of Illinois, Hua returned to China in 1950, where he was soon appointed Director of the Institute of Mathematics of the Chinese Academy of Sciences and President of the Chinese Mathematical Society. Holding both positions until 1983, he exerted tremendous influence on the development of mathematics in mainland China.

Two Unsuccessful Invitations

Chinese mathematical educators realized that one of the most effective ways to help the development of mathematics in China, especially to benefit students

[52] J. Robert Oppenheimer to the Chinese Embassy, 30 November, 1948 in *ibid.*

[53] Chern to J. Robert Oppenheimer, 16 May, 1954 in *ibid.*

[54] See the Administrative Files of Fan Ky, Jiang Lifu, Hua Luogeng, Chow Wei-liang, Min Sihe, Shu Shien-siu, Chang Zongsui, Chen Yu-why, and of Chang Shixun at the Historical Studies–Social Sciences Library, IAS, Princeton. See also The Institute for Advanced Study, ed., *A Community of Scholars: The Institute for Advanced Study Faculty and Members, 1930-1980* (Princeton: The Institute for Advanced Study Press, 1980).

and instructors who could not travel abroad, was to invite eminent foreign mathematicians to lecture in China. In the 1920s and 1930s, among those who came to China as guest lecturers were the Englishman Bertrand Russell, the Germans Wilhelm Blaschke and Emanuel Sperner, the Frenchman Jacques Hadamard,[55] and the Americans William F. Osgood and Norbert Wiener, among others. As already mentioned, these mathematicians exerted a great influence on the younger generation of Chinese students and faculty alike. In the late 1930s and 1940s, however, Beijing University and the newly founded Institute of Mathematics at the Academia Sinica unsuccessfully attempted to bring two eminent American-based mathematicians to China. One was Witold Hurewicz, the other Hermann Weyl.[56]

In 1937, Beijing University, with the support of the China Foundation, sent an invitation to Hurewicz, which he accepted. Unfortunately, he was prevented from making the trip due to the outbreak of the Sino-Japanese War. Interest in Hurewicz's visit was not forgotten, however, and after the war, the mathematicians at Beijing University again approached him about coming to China. In a letter to Hu Shih dated 3 September, 1945, Jiang Zehan explained that

> I wrote two letters to my former teacher Yao Yu-Tai and asked him to invite Witold Hurewicz. We are eager to invite an excellent foreign mathematician to help us to run the Institute. If Hurewicz cannot come, would you please ask my former teacher Yao, together with [Chern] Shiing-shen, to consult Lefschetz, Morse, and Veblen, to find another one? [We understand] it is not easy to find so perfect a man as Hurewicz.[57]

Three months later, Jiang explained in detail why he thought Hurewicz was such a "perfect" choice:

> We invited him in 1937, but he could not make the trip to Peiping because of the war [with Japan]. His contributions to mathematics are substantial. And more important to us is that our professors such as Sheng [You-cheng], Cheng [Yu-why], Fan Ky, including me, are all interested in his work, and junior faculty members such as Wang Xiang-hao, Sun Shu-ben, Liao Shang-tao, Leng Sheng-ming could do research under his guidance. The most important thing in choosing foreign mathematicians is that we can understand their work, have interests in that, and do research immediately with them. Hurewicz is someone who satisfies these conditions. Shiing-shen, in a letter three months ago, told me he [Hurewicz] was doing war-time

---

[55] For Hadamard's activities in China, see Vladimir G. Mazia and Tatyana Q. Shaposhnikova, *Jacques Hadamard: A Universal Mathematician*, HMATH, vol. 14 (Providence: American Mathematical Society and London: London Mathematical Society, 1998), pp. 221-224, and Wiener, pp. 198-200.

[56] Hurewicz was Polish-Russian. Born in 1904 in Lodz, Poland, he was educated in Vienna, where he received his Ph.D. in 1926. A decade later, in 1936, he emigrated to America. His major contributions were related to dimension theory and homotopy groups. In 1940, together with Henry Wallman, Hurewicz published his famous book on dimension theory. He died accidentally on 6 September, 1956 in Mexico City. For more on Hurewicz's life and his mathematical contributions, see Solomon Lefschetz, "Witold Hurewicz, in Memoriam," *Bulletin of the American Mathematical Society* **63** (1957):77-82.

[57] Geng, p. 189.

> work at MIT, and it seems to him that Hurewicz has not yet found a very good position. Now the biggest problem for us is how to pay him. Our currency is still devalued. According to Meng-zheng [Fu Shi-nian], the salaries of foreign professors must be the same as those of our home professors. [If so], then there will be no hope of inviting foreign professors unless we have a special stipend for them. Meng-zheng will discuss these things with the Education Ministry. He also asked if it might be possible for you to raise some foreign currency to hire several foreign professors first. Therefore, please weigh whether or not we might hire Hurewicz, and then let my former teacher Shu-ren, and Shiing-shen, know what you think.[58]

Unfortunately, the invitation to Hurewicz never materialized, apparently only for lack of funds. Another invitation that was financially supported by a solid grant also failed due to the civil war in China.

In March 1946, Chern returned to Shanghai and was invited to organize the Institute of Mathematics of the Academia Sinica. In order to promote the mathematical relationship with the United States and to build an international reputation for the Institute as quickly as possible, Chern, with the support of the Chinese Education Ministry, invited Hermann Weyl to visit China in June 1946. In a memorandum dated 15 October, 1946 to Frank Aydelotte, Weyl reported:

> In July I received a letter from our good friend Professor Shiing-Shen Chern by which he invited me on behalf of the Academia Sinica (the Chinese National Academy) to spend the academic year 1947–1948 in China, if feasible one term in Peiping and one term in Nanking, and to help them start work at the newly created Institute of Mathematics of that Academy.
>
> To help in establishing close and friendly relations between this country and China, and more particularly between the new Chinese research Institute of Mathematics and our School of Mathematics, appears to me a task of some significance. Were this not so, I should not easily be persuaded to interrupt my quiet life and work in Princeton.[59]

Chern's letter was reinforced by a formal invitation from Dr. Zhu Jiahua, the Minister of Education and President of the Academia Sinica. With the invitation, it also provided a check for $10,000 to defray Weyl's travel expense and for his lectures in China. Nevertheless, Weyl postponed the trip because of the Chinese civil war and the illness of his wife, eventually canceling the trip altogether after the Communist takeover.

Ironically, the Communists, like the Nationalists, showed an interest in Weyl and invited him to China. Hua Luogeng, later head of the Chinese Mathematical Society, wrote to Weyl on 23 December, 1949, just before he decided to return to the newly created People's Republic of China from the University of Illinois:

---

[58]Jiang Zehan to Hu Shih, 5 November, 1945 in Geng, p. 192.

[59]Hermann Weyl, Administrative File, Historical Studies–Social Sciences Library, IAS, Princeton. See also Chern, "My Mathematical Education," p. 8.

> [Although] I have now decided to go back to China, I must express my deep appreciation of the kindness which you extended to me, while I have been in [the] U.S.A.
>
> The most regrettable thing is that Chinese mathematicians did not have the good fortune to receive your broad inspiration. But I daresay that the Chinese communists will welcome you as well as the nationalists, since according to our tradition, we always respect scholars .... Any help [from you] will be appreciated by Chinese mathematicians.[60]

Although Weyl did not go to China to lecture to Chinese audiences as he had done at the Institute for Advanced Study, both his works and Hurewicz's nurtured and inspired Chinese mathematicians. Hurewicz's *Dimension Theory* was one of the few new mathematical books available for mathematicians to study in China during World War II. Chern, Jiang Zehan, and his students such as Wang Xianghao all profited from it. Chern even copied the whole book, minus the last chapter, by hand.[61]

Conclusion

From 1859, when the calculus textbook by Elias Loomis was first translated into Chinese, to the present, over 140 years have elapsed. Over this period, modern Chinese mathematics has followed a tortuous path in its quest to achieve international recognition. One of two major indications of this recognition is that in 1986 the International Mathematical Association advanced China to membership in its fifth and highest group, on the same level as the United States, Russia, France, Great Britain, Germany, and Japan. The other is that China has been chosen to host the International Congress of Mathematicians, the first of the new millennium, in 2002. Such achievements, to a great extent, are attributable to China's mathematical contacts with foreign countries, particularly with the United States. The Boxer Indemnity Funds, Harvard University, and the Institute for Advanced Study are only a few of the most important means by which the United States has influenced the development of modern Chinese mathematics.

In October 1949, when the People's Republic of China was founded, diplomatic relations between the two countries came to an end, but not the American influence on Chinese mathematics. Some Chinese mathematicians who had come to the United States following World War II returned to the mainland in response to a call from the Communists. These mathematicians were instrumental in extending American influence upon mathematics throughout mainland China. Those who remained in the United States eventually became Chinese-Americans. Some of them subsequently made substantial contributions of their own to American mathematics. These Chinese-American mathematicians, in turn, educated new generations of American students and eventually became a bridge between the U.S. and China.

One excellent example is Chern Shiing-shen, who enjoyed a distinguished career in the U.S. Besides his own important mathematical contributions to differential

[60] Hermann Weyl's Administrative File at the Historical Studies–Social Sciences Library, IAS, Princeton.

[61] See Jiang Zehan, "Hui yi Hu Shih de ji jian shi" (Reminiscences on Hu Shih), *World Journal* (*New York Edition*) (December 20, 1990); and Chern, "My Mathematical Education," p. 8.

geometry, he educated a generation of young geometers. Forty-one students, both American and Chinese, received their Ph.D.s under his supervision.[62] Chern also lectured in Taiwan, Hong Kong, and in mainland China on numerous occasions, becoming a strong link between the American and Chinese mathematical communities. Furthermore, after his retirement from the University of California at Berkeley, he created the Nankai Institute of Mathematics, following the model of the Institute for Advanced Study at Princeton.

It was through Chinese-American mathematicians like Chern that the latest advancements in mathematics were introduced to China—or from China to the U.S. It was also through them that thousands of China's brightest students came to the U.S. to pursue advanced studies in mathematics. Between 1990 and 1998, American institutions have granted roughly 9,000 Ph.D.s in mathematics, of which some 2,000—or slightly more than one-fifth—have gone to Chinese or Chinese-Americans.[63] This figure is about fifty times greater than the total before 1949. Consequently, we have every reason to expect that, as Zhang Dianzhou and Joseph W. Dauben concluded in their study of mathematical exchange between the United States and China from 1850 to 1950, "Chinese-American relations will continue to dominate the future of mathematics in China, and mutually enrich scientific exchanges between both countries well into the next century."[64]

References

Archival Sources

Chang Shixun. Administrative File. Historical Studies–Social Science Library, IAS, Princeton.

Chen Yu-why. Administrative File. Historical Studies–Social Science Library, IAS, Princeton.

Chern Shiing-shen. Administrative File. Historical Studies–Social Science Library, IAS, Princeton.

Chow Wei-liang. Administrative File. Historical Studies–Social Science Library, IAS, Princeton.

Fan Ky. Administrative File. Historical Studies–Social Science Library, IAS, Princeton.

Hua Luogeng. Administrative File. Historical Studies–Social Science Library, IAS, Princeton.

Jiang Lifu. Administrative File. Historical Studies–Social Science Library, IAS, Princeton.

Jiang Zehan. Administrative File. Historical Studies–Social Science Library, IAS, Princeton.

---

[62] Cheng S.-Y., Li Peter, and Tian Gang ed., *Shiing-shen Chern: A Mathematician and His Mathematical Work*, (Singapore: World Scientific Publishing Co., 1996), pp. 697-698.

[63] John D. Fulton, "1995 Annual AMS-IMS-MAA Survey, Second Report," *Notices of the American Mathematical Society* **43** (1996):848-859 on p. 849; Paul W. Davis, "1996 AMS-IMS-MAA Annual Survey, Second Report," *Notices of the American Mathematical Society* **44** (1997):911-921 on p. 913; and Paul W. Davis, James W. Maxwell, and Kinda M. Remick, "1997 AMS-IMS-MAA Annual Survey, Second Report," *Notices of the American Mathematical Society* **45** (1998):1158-1171 on p. 1160. Also see the Dissertation Abstracts CD-ROM by A. Bell & Howell Company.

[64] Zhang and Dauben, p. 287.

Min Sihe. Administrative File. Historical Studies–Social Science Library, IAS, Princeton.

Shu Siu-shien. Administrative File. Historical Studies–Social Science Library, IAS, Princeton.

Weyl, Hermann. Administrative File. Historical Studies–Social Science Library, IAS, Princeton.

Wong Yue-kei. Administrative File. Historical Studies–Social Science Library, IAS, Princeton.

Zhang Shixun. Administrative File. Historical Studies–Social Science Library, IAS, Princeton.

Zhang Zongsui. Administrative File. Historical Studies–Social Science Library, IAS, Princeton.

Printed Sources

Boothby, William M., Chern, Shiing-shen, and Wang, S. P. "The Mathematical Work of H. C. Wang." *Bulletin of the Institute of Mathematics, Academia Sinica (Taipei)* **8** (1980):xiii-xxiv.

Borel, Armand. "The School of Mathematics at the Institute for Advanced Study." In *A Century of Mathematics in America—Part III.* Ed. Peter L. Duren *et al.* Providence: American Mathematical Society, 1989, pp. 119-147.

Brockhausen, Terence Eldo. "The Boxer Indemnity: Five Decades of Sino-American Dissension." Texas Christian University. Unpublished Ph.D. dissertation, 1981.

Ceng Yuanrong. "Expansions According to a Given System of Functions." *Bulletin of the American Mathematical Society* **39** (1933):26-27.

Chao Yuan Ren. *Life With Chaos: Yuen Ren Chao's Autobiography, First 30 Years, 1892-1921.* Ithaca: Spoken Language Services, 1975.

Chen Xuexun and Tian Zhengping. *Zhongguo jindai jiaoyushi ziliao huibian: liuxue jiaoyu (Collected Sources for the History of Modern Chinese Education: Studying Abroad).* Shanghai: Shanghai jiaoyu chubanshe, 1991.

Chern Shiing-shen. "My Mathematical Education." In *S. S. Chern: A Great Geometer of the Twentieth Century.* Ed. Sing-tung Yau. Hong Kong: International Press, 1992, pp. 1-14.

———. *Shiing-shen Chern: Selected Papers.* New York: Springer-Verlag, 1979.

Cheng S.-Y., Li Peter, and Tian Gang Ed. *Shiing-shen Chern: A Mathematician and His Mathematical Work.* Singapore: World Scientific Publishing Co., 1996.

Chuang Chi-tai. "On the Normalization of the Set of Functions X^nSech$x$." *Journal of Mathematics and Physics* **15** (1936):264-67.

Davis, Paul W. "1996 AMS-IMS-MAA Annual Survey, Second Report." *Notices of the American Mathematical Society* **44** (1997):911-921.

Davis, Paul W.; Maxwell, James W.; and Remick, Kinda M. "1997 AMS-IMS-MAA Annual Survey, Second Report." *Notices of the American Mathematical Society* **45** (1998):1158-1171.

Department of Mathematics of Sichuan University, Ed. *Hu Kunsheng yizhu (Posthumous Works of Hu Kunsheng).* Beijing: Renmin jiaoyu chubanshe, 1960.

Fisher, Daniel W. *Calvin Wilson Mateer: Forty-Five Years a Missionary in Shantung, China. A Biography.* Philadelphia: The Westminster Press, 1911.

Fulton, John D. "1995 Annual AMS-IMS-MAA Survey, Second Report." *Notices of the American Mathematical Society* **43** (1996):848-859.

Gao Zonglu. *Zhongguo liumei youtong shuxinji* (*Collected Correspondence of the Former Students of the Chinese Educational Commission*). Taipei: Zhuangji wenxue congshu, 1986.

Geng Yunzhi, Ed. "Jiang Ze-han's Letters to Hu Shih." In *Hu Shih Yigao ji michang shuxin* (*Hu Shih's Manuscripts and his Secretly Treasured Correspondence*). Hefei: Huangshang shushe, 1994, 25:84-196.

Griffiths, Phillip A. "Some Reflections on the Mathematical Contributions of S. S. Chern." In *Shiing-shen Chern Selected Papers.* New York: Springer-Verlag, 1978, pp. xii-xix.

Hu Kuensen. "A Problem with Variable and Points in the Calculus of Variations." *Journal of the Chinese Mathematical Society* **1** (1937):1-14.

Hu Mingfu. "Linear Integro-differential Equations with a Boundary Condition." *Transactions of the American Mathematical Society* **19** (1918):363-407.

Hu S. T. "Hsien Chung Wang 1918-1978." *Bulletin of the Institute of Mathematics, Academia Sinica* (*Taipei*) **8** (1980):x-xii.

Hua Loo-keng and Shien Shien-siu. "On Fourier Transforms in L_p in the Complex Domain." *Journal of Mathematics and Physics* **15** (1936):249-63.

Hyatt, Irwin T. Jr. *Our Ordered Lives Confess: Three Nineteenth-Century American Missionaries in East Shantung.* Cambridge, MA: Harvard University Press, 1976.

Institute for Advanced Study, Ed. *A Community of Scholars: The Institute for Advanced Study Faculty and members, 1930-1980.* Princeton: The Institute for Advanced Study Press, 1980.

Jiang Zehan. "Huiyi Hu Shih de jijianshi" (Reminiscences on Hu Shih). *World Journal* (*New York Edition*) (December 19-20, 1990).

LaFargue, Thomas F. *China's First Hundred.* Seattle: Washington State University Press, 1942. Reprint Ed. 1987.

Lai Hui-lan. *Qingmo liumei youtong zhi yanjiu* (*Study on the Chinese Education Mission in the United States in the Late Qing Dynasty*). Unpublished Master's thesis. The Chinese Culture University, Taipei, 1984.

Lefschetz, Solomon. "Witold Hurewicz, in Memoriam." *Bulletin of the American Mathematical Society* **63** (1957):77-82.

Li Xuhui. "Sanshi niandai N. Wiener fangwen qinghua daxue handian shimo" (The Correspondence Concerning N. Wiener's Visiting China in the 1930s). *Zhongguo keji shiliao* (*China Historical Materials of Science and Technology*) **19** (1) (1998):42-51.

Mac Lane, Saunders. "Mathematics at the University of Chicago: A Brief History." In *A Century of Mathematics in America—Part II.* Ed. Peter L. Duren *et al.* Providence: American Mathematical Society, 1988, pp. 127-154.

Marvin, George. "An Act of International Friendship." *The Outlook* (November 14, 1908):582-86.

———. "The American Spirit in Chinese Education." *The Outlook* (November 28, 1908):667-72.

Mazia, Vladimir G. and Shaposhnikova, Tatyana O. *Jacques Hadamard: A Universal Mathematician.* HMATH. Vol. 14. Providence: American Mathematical Society and London: London Mathematical Society, 1998.

Newton, Hubert A. *Elias Loomis, LL.D., 1811-1889: Memorial Address.* New Haven: Tuttle, Morehouse & Taylor, 1890.

Powell, John Benjamin. "Directory of American Returned Students." *Supplement to Third Edition of Who's Who in China.* Shanghai: The Weekly Review, 1928.

Ren Nanheng and Zhang Youyu. *Zhongguo shuxuehui shiliao* (*Historical Materials of the Chinese Mathematical Society*). Nanjing: Jiangshu jiaoyu chubanshe, 1995.

Wang Huanchen, Ed. *Liuxiu jiaoyu: zhongguo liuxiu jiaoyu shiliao* (*Studying Abroad: Historical Documents of Chinese Students Studying Abroad*). 6 Vols. Taipei: Guoli bianyiguan, 1980.

Wang Shuhuai, *Gengzi peikuan* (*The Boxer Indemnity*). Taipei: Institute of Modern History, Academia Sinica, 1974.

Wang Yuan. "Hua Luo-geng." In *Zhongguo kexue jishu zuangjia zhuanglue: shuxue* (*Brief Biographies of Chinese Scientific and Technological Specialists: Mathematicians*). Shijiazhuang: Hebei jiaoyu chubanshe, 1996, 1:230-248.

———. *Hua Loo-Keng: A Biography.* Trans. Peter Shiu Man-Kit. Singapore: Springer-Verlag, 1999.

Weil, André. "S. S. Chern as Geometer and Friend." In *Shiing-shen Chern: Selected Papers.* New York: Springer-Verlag, 1978, pp. ix-xii.

Wiener, Norbert. *I Am a Mathematician.* Garden City: Doubleday & Company, Inc., 1956.

Wilson, W. Stephen; Chern, Shiing-shen.; Abhyankar, Sheerem S.; Lang, Serge; and Igusa Jun-ichi. "Wei-Liang Chow." *Notices of the American Mathematical Society* **43** (1996):1117-24.

The Writing Group of the History of Qinghua University. *Qinghua daxue xiaoshigao* (*A Draft History of Qinghua University*). Beijing: Zhonghua shuju, 1981.

———. *Qinghua daxue shiliao xunpian* (*Selected Historical Documents on Qinghua University*). Vol. 3. Beijing: Qinghua daxue chubanshe, 1994.

Xu Yibao. "Zhongguo xiandai shuxue de zhongyao shiliao: Jiang Zehan zhi Hu Shih de xinhan" (An Important Document for the History of Modern Chinese Mathematics: The Letters of Jiang Zehan to Hu Shih). *Shuxue chuanbo* (*Dissemination of Mathematics*) (*Taipei*) **23** (1999):34-48.

———. "The Chinese Educational Mission: Hopes, Failures, and Reasons for its Recall by the Qing Government in 1881." In press.

Yang Tsuihua. *Zhongjihui dui kexue de zanzhu* (*Patronage of Sciences: The China Foundation for the Promotion of Education and Culture*). Taipei: Institute of Modern History, Academia Sinica, 1991.

Yau Sing-tung, Ed. *S. S. Chern: A Great Geometer of the Twentieth Century.* Hong Kong: International Press, 1992.

Yuan Ren Chao. *Life with Chaos: Yuen Ren Chao's Autobiography, First 30 Years, 1892–1921.* Ithaca: Spoken Language Services, 1975.

Yuan Tung-li. *A Guide to Doctoral Dissertations by Chinese Students in America 1905–1960.* Washington D.C.: Sino-American Cultural Society, Inc., 1961.

Yung Wing. *My Life in China and America.* New York: Henry Holt and Company, 1909.

Zen Sun E-tu. "The Growth of the Academic Community, 1912–49." In *The Cambridge History of China.* 15 Vols. New York: Cambridge University Press, 1986, 13:361-420.

Zhang Dianzhou and Dauben, Joseph W. "Mathematical Exchanges Between the United States and China: A Concise Overview (1850-1950)." In *The History of Modern Mathematics*. Vol. 3. Ed. Eberhard Knobloch and David E. Rowe. Boston: Academic Press, Inc., 1994, pp. 263-297.

Zhang Dianzhou. "Gengzi peikuan he zhongguo xiandai shuxue de fazhang" (The Boxer Indemnity and the Rise of Modern Chinese Mathematics). In *Keshi xinchuan* (*Passing the Torch of the History of Science*). Shengyang: Liaonin jiaoyu chubanshe, 1997, pp. 179-184.

A Glossary of Chinese Names

Ceng Yuanrong 曾遠榮

Chao Yuenren (or Chao Yuan Ren) 趙元任

Chen Yu-why 程毓淮

Chen Xuexun 陳學恂

Chern Shiing-shen 陳省身

Chow Wei-liang 周煒良

Chung Kai-lai 鍾開萊

Deng Encong 鄧恩聰

Fan Ky 樊畿

Gao Zonglu 高宗魯

Geng Yunzhi 耿云志

He Yunhuang 何運煌

Hu Kunseng (or Hu Kuen-sen) 胡坤陞

Hu Minfu 胡明復

Hu Shizhen (or Hu S. T.) 胡世楨

Hu Shih 胡適

Hua Luogeng (or Hua Loo-geng) 華羅庚

Jiang Lifu 姜立夫

Jiang Zehan 江澤涵 (or Kiang Tsai-han)

Lai Huilan 賴惠蘭

Li Shanlan 李善蘭

Li Xuhui 李旭輝

Liu Jinnian 劉晉年

Ma Xianjiao 馬仙嶠

Min Sihe 閔嗣鶴

Qin Fen 秦汾

Qin Yuanxun 秦元勳

Ren Nanheng 任南衡

Shen Youcheng 申又棖

Shi Xianglin 施祥林

Shu Shien-siu 徐賢修

Tian Zhengping 田正平

Sun Guangyuan (or Sun Dan) 孫光遠

Wang Huanchen 王煥琛

Wang Renfu 王仁輔

Wang Shuhuai 王樹槐

Wang Xianghao王湘浩

Wang Yuan 王元

Wen Yuqing 溫毓慶

Wong Yue-kei 黃汝琪

Wu Daren 吳大任

Xiong Qinglai 熊慶來

Xu Yibao 徐義保

Yan Sing-tung 邱成桐

Yan Jiazou 嚴家騶

Yang Tsuihua 楊翠華

Yang Wuzhi (or Yang Ko-chuen) 楊武之

Yu Tawei (or David Yule) 俞大維

Yuan Tung-li 袁同禮

Yung Wing 容閎

Zhang Dianzhou 張奠宙

Zhang Shixun 張世勛

Zhang Youyu 張友余

Zhang Zongsui 張宗燧

Zheng Zhifan 鄭之蕃

Zhu Jiahua (or Chu Chia-hua) 朱家驊

Zhu Lu 朱籙

Zhuang Qitai (Chuang Chi-tai) 庄圻泰

CHAPTER 15

American Initiatives Toward Internationalization: The Case of Leonard Dickson

Della Dumbaugh Fenster
University of Richmond (United States)

Introduction

In their book, *The Emergence of the American Mathematical Research Community, 1876-1900: J. J. Sylvester, Felix Klein, and E. H. Moore*, Karen Parshall and David Rowe suggest the notion of periodization (as opposed to continuity) as a means of historically investigating American mathematics. They characterize four developmental periods as follows:[1]

- 1776-1876: Mathematics in the general context of scientific development
- 1876-1900: Emergence of a research community
- 1900-1933: Consolidation and growth of research traditions and institutions
- 1933-1960: Influx of European mathematicians; Government funding.

At the turn of the twentieth century, this newly emerged mathematical research community, like its counterparts in other nations, came to view an active mathematical career as one that involved pursuing scholarly research in increasingly specialized fields, training future mathematicians in areas ripe with open questions, participating in mathematical societies with people of common interests, and contributing to specialized mathematical journals.[2] With mathematicians in various national contexts following this model—one dependent on open lines of communication—the internationalization of the field was well under way.[3]

These concurrent developments raise some intriguing questions relative to the American scene. How, for instance, did other countries influence the American mathematical research community in the late nineteenth and early twentieth centuries? What impact did American contributions make on mathematics internationally? Using the career of Leonard Eugene Dickson (1874-1954) as a point of departure, this study explores the international influences on American mathematics as well as the American role in an increasingly international community of mathematicians.

[1]Karen Hunger Parshall and David E. Rowe, *The Emergence of the American Mathematical Research Community, 1876-1900: J. J. Sylvester, Felix Klein, and E. H. Moore*, HMATH, vol. 8 (Providence: American Mathematical Society and London: London Mathematical Society, 1994), pp. 427-428.

[2]Karen Hunger Parshall, "How We Got Where We Are: An International Overview of Mathematics in National Contexts (1875-1900)," *Notices of the American Mathematical Society* **43** (1996):287-296 on p. 294. See also Karen Hunger Parshall, "Mathematics in National Contexts (1875-1900): An International Overview," in *Proceedings of the International Congress of Mathematicians, Zürich, Switzerland 1994* (Basel: Birkhäuser Verlag, 1995), pp. 1581-1591 on p. 1589.

[3]Parshall, "How We Got Where We Are," pp. 294-295.

FIGURE 1. Sketch made of L. E. Dickson during his lecture at the Strasbourg International Congress of Mathematicians

Why Dickson? From a purely chronological perspective, Dickson's mathematical pursuits span the second and third periods of American mathematics as characterized by Parshall and Rowe. He received his education while the American mathematical research community emerged and did almost all of his original research after 1900. In fact, Dickson's work in the theory of algebras serves as a case study supporting Parshall and Rowe's characterization of American mathematics in the first forty years of the twentieth century as a period of the consolidation and growth of research traditions within the American mathematical scene.[4] Dickson was also the most active member of the American mathematical research community from 1891 to 1906,[5] establishing a distinguished career early on and maintaining it for more than forty years. He authored eighteen books and more than 300 research papers; served as editor of the *Transactions of the American Mathematical Society* and the *American Mathematical Monthly*; guided sixty-seven doctoral students, eighteen of whom were women, to Ph.D.s in mathematics; led

[4]Parshall and Rowe, pp. 427-428. See Della Dumbaugh Fenster, "Leonard Eugene Dickson and His Work in the Arithmetics of Algebras," *Archive for History of Exact Sciences* **52** (1998):119-159.

[5]Della Dumbaugh Fenster and Karen Hunger Parshall, "A Profile of the American Mathematical Research Community: 1891-1906," in *The History of Modern Mathematics*, vol. 3, ed. Eberhard Knobloch and David E. Rowe (Boston: Academic Press Inc., 1994), pp. 179-227 on p. 185.

the American Mathematical Society (AMS) as its President from 1916 to 1918; and represented American mathematics twice at International Congresses of Mathematicians (ICM).[6] By viewing Dickson through the lenses of graduate education, scholarly research, and publication efforts, a detailed picture emerges both of how American mathematics matured in the late nineteenth and early twentieth centuries and of the complex interplay of international influences in that process.

Dickson in the Emergent Period of American Mathematics

Born in 1874, Dickson attended the University of Texas for his undergraduate and master's education and, while there, fell under the influence of George Bruce Halsted. In 1894, with his Master's degree in hand and two years of teaching experience to his credit, he chose the research-oriented University of Chicago over the up-and-coming (in terms of graduate training and research) Harvard as the place to pursue his doctorate in mathematics. If Dickson had arrived on the mathematical scene much earlier, he would probably have traveled to Europe, most likely to Germany, for his doctoral training in mathematics. Thus, whether he realized it at the time or not, Dickson's decision to go to Chicago placed him among the first generation of American-trained, research mathematicians.

Dickson and other aspiring, American mathematicians in the late nineteenth and early twentieth centuries largely owed their opportunity to earn American Ph.D.s in mathematics to a complex sequence of events that took place in the closing quarter of the nineteenth century and that relied on the efforts of three men, the Englishman James Joseph Sylvester, the German Felix Klein, and the American Eliakim Hastings Moore.[7] English and Continental influences thus crucially affected the development of research-level mathematics in the United States.

Sylvester had arrived in 1876 to head the Mathematics Department at the Johns Hopkins University, an institution focused not only on teaching but also on graduate studies and research. His department soon produced research-level work that caught the attention of mathematicians abroad.[8] Moreover, in 1878, Sylvester helped launch the *American Journal of Mathematics*. He solicited European contributions to the journal, especially from his British colleagues, and roughly one-third of the submissions came from non-Americans. The *American Journal* served as a silent advertisement to mathematical Europe of the awakening of mathematics in the United States. When Sylvester returned to England in 1883 to occupy the Savilian Chair of Geometry at New College, Oxford, however, the other American colleges could not adequately fill the void he left; they simply did not have faculties equipped to pursue mathematics at the research level.

Americans next turned primarily to Felix Klein in Germany for their training in mathematics. The extent of Klein's influence on American mathematics was made clear in 1893, when he participated—both at the Americans' invitation and as the official emissary of the Prussian government—in the Mathematical Congress

---

[6]The standard biographical sources on Dickson are A. Adrian Albert, "Leonard Eugene Dickson, 1874-1954," *Bulletin of the American Mathematical Society* **61** (1955):331-345, and Raymond C. Archibald, *A Semicentennial History of the American Mathematical Society, 1888-1938* (New York: American Mathematical Society, 1938), pp. 183-194.

[7]Parshall and Rowe.

[8]Karen Hunger Parshall, "America's First School of Mathematical Research: James Joseph Sylvester at The Johns Hopkins University 1876-1883," *Archive for History of Exact Sciences* **38** (1988):153-196.

held in Chicago in conjunction with the World's Columbian Exposition. He followed this with a two-week-long colloquium in nearby Evanston. The timing of the Congress and the colloquium lectures proved crucial to the American mathematical community, in general, and to the University of Chicago, in particular. The "notable and inspiring group"[9] of Chicago mathematicians helped pave the way for the fledgling New York Mathematical Society (founded in 1888) to grow into the American Mathematical Society in 1894. Moreover, Klein's visit "clearly signaled his desire to bow out gracefully after a decade of involvement in the American mathematical scene."[10] The University of Chicago and its Department of Mathematics, with E. H. Moore serving as its chair, soon emerged "not only as the leading center for mathematics in the United States but also as the first American institution of higher education to offer mathematical training comparable to that available at leading European universities."[11] But the German influence remained. With Moore's commitment to the highly successful German model of mathematical training and the first-hand experience of his German colleagues, Oskar Bolza and Heinrich Maschke, within that system as students of Klein at Göttingen University, the Chicago Mathematics Department set its sights on adapting the German model to its American setting.[12] In particular, that meant emphasizing pure research and implementing the unique German teaching tool, the seminar. Moore, Bolza, and Maschke established their successful department through effective classroom teaching, the organization of the Mathematical Club for both the review of current literature and the presentation of original research, and the production of top-quality published work.[13]

Owing to its successful implementation and continuation of a commitment to high research standards, the University of Chicago held a unique position among American institutions in the closing decade of the nineteenth century. Moore, Bolza, and Maschke sought to build and succeeded in forming a Mathematics Department that promoted original research, quality publications, and a broad view of the American mathematical community. The strong institutional and departmental philosophy inherent in these goals thus made the University of Chicago a viable option for Leonard Dickson as well as other aspiring American mathematicians, including Oswald Veblen, Gilbert Bliss, George D. Birkhoff, and Robert L. Moore.

When Dickson arrived at Chicago in the fall of 1894, he found a solid mathematics curriculum comparable to that offered at Göttingen.[14] He also discovered an E. H. Moore with newly acquired mathematical interests in pure group theory rather than group-theoretic approaches to geometry. In Chicago Mathematical Club meetings during the previous year, Moore had both revealed this shift in his

---

[9]George D. Birkhoff, "Fifty Years of American Mathematics," in *Semicentennial Addresses of the American Mathematical Society*, ed. Raymond C. Archibald (New York: American Mathematical Society, 1938), pp. 270-315 on p. 273.

[10]Parshall and Rowe, pp. 360-361.

[11]*Ibid.*, p. 361.

[12]For the interesting story of the genesis of the first Mathematics Department at the University of Chicago, see Karen Hunger Parshall, "The 100th Anniversary of Mathematics at the University of Chicago," *The Mathematical Intelligencer* **14** (2) (1992):39-44, and Parshall and Rowe, pp. 279-294.

[13]Parshall and Rowe, p. 371.

[14]*Ibid.*, p. 367.

research and exposed the Chicago students and faculty to finite simple groups, an active area of research on both sides of the Atlantic.[15]

Moreover, Moore drew from the work of Camille Jordan, Joseph Serret, Otto Hölder, and Eugen Netto to present "A Doubly-Infinite System of Simple Groups" at the Mathematical Congress held in Chicago in 1893; he published this as his first paper in abstract group theory. In this work, Moore both demonstrated his interest in the foundations of mathematics by stating his postulates for a field and exhibited his mathematical prowess by proving that every finite field is the abstract form of a Galois field of prime-power order.[16] More importantly, perhaps, "the great American protagonist" of the abstract point of view ushered abstract algebra into America with this work.[17] Through the Mathematical Club and the Chicago Congress, Moore thus demonstrated to the Chicago students and faculty not only the cutting-edge topics in mathematical research but also the importance of incorporating extant results—primarily European, in this case—into American work.

This environment prevailed when Dickson arrived at Chicago in 1894 on the heels of Moore's 1893 Congress presentation and just in time for his lectures on group theory. Following the lead of his adviser, Dickson acquainted himself with recent mathematical developments in Europe, particularly those of Jordan. In his 1896 dissertation, "The Analytic Representation of Substitutions on a Power of a Prime Number of Letters with a Discussion of the Linear Group," Dickson extended the theory of finite fields and established further its connections with group theory.[18]

Dickson spent the academic year from 1896 to 1897 studying with Sophus Lie in Leipzig and with Jordan, Émile Picard, and Charles Hermite in Paris.[19] On his return to America, and, most likely en route to his first job as an Instructor at the University of California, Dickson spoke on his year abroad at the Chicago Mathematics Club on 30 July, 1897. His lecture, "The Influence of Galois on Recent Mathematics," displayed his true expository potential and his overall flattering view of his contributions to mathematics (a rather bold approach given his youth and his audience!).[20] But this obviously confident twenty-three-year-old also offered a unique assessment of Chicago's school of mathematics. "It would thus appear," the young American concluded his talk,

---

[15] *Ibid.*, p. 378. Having been inspired at Yale by the geometer-turned-astronomer, Hubert Anson Newton, Moore's original research interests (from roughly 1885 to 1892) lay in the area of geometry. For more on Moore's research, see *ibid.*, pp. 372-379.

[16] Eliakim Hastings Moore, "A Doubly-Infinite System of Simple Groups," in *Mathematical Papers Read at the International Mathematics Congress Held in Connection with the World's Columbian Exposition: Chicago 1893*, ed. E. H. Moore, Oskar Bolza, Heinrich Maschke, and Henry S. White (New York: Macmillan & Co., 1896), pp. 208-242.

[17] George Birkhoff characterized Moore in this way in "Fifty Years of American Mathematics," p. 284.

[18] Leonard E. Dickson, "The Analytic Representation of Substitutions on a Power of a Prime Number of Letters with a Discussion of the Linear Group," *Annals of Mathematics* **11** (1897):65-143. See Parshall and Rowe, pp. 379-381 for an overview of Dickson's dissertation.

[19] *The University of Texas Record*, vol. 1, no. 3, August, 1899, University of Texas, The Center for American History, University of Texas Memorabilia Collection, James Benjamin Clark file.

[20] "The Influence of Galois on Recent Mathematics," 10 July, 1897, University of Chicago, Department of Special Collections, Mathematical Club Records 1893-1921. Dickson recorded his talk (by hand) in the Mathematical Club Records.

> that the brief products of the genius of Galois [were] still at work to unify and magnify the things that form the essence of many fields of mathematics. Along with these forces at Paris, with both Picard and Jordan in the group lines, also at Göttingen with Klein and Hilbert, and at Berlin with Frobenius (whose recent work on solvable groups has attracted so much attention), I am proud to feel that in the investigation as well as the teaching of the fields of mathematics traceable back to the influence of Galois, Chicago holds a high place, with Prof. Moore in both the group field and the field of Algebraic and number Corpora and Professors Bolza and Maschke in the group and substitution field. It is thus natural that my enthusiasm having been so strongly aroused for the lines opened by Galois as most fertile fields for investigation, I should still greatly admire those fundamental ideas; while my semester with Lie and that at Paris have shown me how remarkably those ideas have penetrated into and unified distant branches of mathematics.[21]

Dickson was one of the few people in the world (at that time) who could compare the training provided by this new American university with that of the older, established European schools. From his observation, the five-year-old Chicago Mathematics Department stood shoulder to shoulder with the great European institutions they aspired to emulate. The triumvirate at Chicago was capable of both producing the highest level of research mathematics and training others to do the same. On this occasion, Dickson recognized himself as one of the beneficiaries.

In a circuitous series of events, the University of Chicago Mathematics Department profited from its educational efforts when it invited Dickson to join its faculty as an Assistant Professor in 1900. In the 1900–1901 academic year, Dickson taught six undergraduate courses and lectured in the Graduate School of Mathematics. For the latter, he drew from his postdoctoral year and spoke on Lie's theory of differential equations and continuous groups.[22] Like his predecessors, Dickson incorporated the most recent, European mathematics into his graduate courses.

But Dickson imparted more than mathematical ideas to the next generation of Chicago mathematics students. He wrote prolifically, coming to Chicago in 1900 with a collection of roughly thirty-five mathematical papers on various lines of group-theoretic research inspired, in large part, by his dissertation. Moreover, these articles had appeared on both sides of the Atlantic in American, English, French, and German journals. In his first decade at Chicago, he turned out another 115 papers and advised five Ph.D. students. In terms of producing original mathematical results, he had firmly embraced—as had his teachers, Moore, Bolza, and Maschke—the production of original research, and he passed on this ethos to his students.

There were other, more subtle influences, however. Given Chicago's open policy relative to women at all levels, Moore and his department welcomed women to their

[21] *Ibid.*, p. 16.

[22] "Courses in Mathematics," *Register of the University of Chicago*, pp. 260-266, University of Chicago, Department of Special Collections.

graduate program from the first year of operation, the fall of 1892.[23] Mary Francis "May" Winston (later Newson), for example, came to Chicago as an "Honorary Fellow" for the 1892–1893 academic year and then traveled to Göttingen to complete her Ph.D. in mathematics under Felix Klein. Four years later, a group of AMS members in the Chicago area, led by Moore and Bolza among others, sent a circular announcing a conference to consider the possibility of a Chicago Section of the AMS. When the three-day conference convened on 31 December, 1896, at least one woman was in attendance.[24] When Dickson came to Chicago as a student, this favorable view towards women's presence both in the graduate program and at mathematical functions in the larger community was in place. When he returned as a faculty member, he helped advance the cause of women in mathematics by advising eighteen women doctoral students, as many as the entire faculty at Cornell in the same time period.[25]

Highlighting the role of women in American mathematics in the late nineteenth and early twentieth centuries reveals how the American mathematical research community made a mark on the international scene. Specifically, American women played a crucial role in advancing opportunities in higher education for women in Germany. This came about as somewhat of a by-product of efforts to open universities in the United States to American women for post-graduate study. Fueled by the energy of Christine Ladd-Franklin, the Association of Collegiate Alumnæ (ACA) introduced the ACA European Fellowship in 1888 to provide funds for American women to pursue graduate studies abroad. Their simple plan had far-reaching implications: by gaining ground at foreign universities, and particularly at the prominent and influential German ones, American women set a precedent for the universities at home to follow. The persistence of Ruth Gentry, the first ACA European Fellow, led directly to advances in higher education for women (both German and American) in Germany.[26] On an even broader scale, the American women had the international support of Sylvester, Klein, and influential AMS members, not the least of whom were E. H. Moore and, just after the turn of the century, Leonard Dickson.[27] Thus, an international effort—with a distinct American contribution—gained women entry into research-level mathematics.

Dickson's graduate education in the last decade of the nineteenth century thus brings into focus the unique training available to aspiring American mathematicians at the University of Chicago. Spearheaded by Moore, the Chicago Mathematics Department raised a generation of American-trained mathematicians with an international awareness of current mathematical ideas. Moore stretched and extended European results and encouraged Dickson as his advisee to do the same. In the process of producing original mathematical results in fields found valuable by the Europeans, Moore and Dickson gained ground for American mathematics

[23]Della Dumbaugh Fenster and Karen Hunger Parshall, "Women in the American Mathematical Research Community: 1891-1906," in *The History of Modern Mathematics*, vol. 3, ed. Eberhard Knobloch and David E. Rowe (Boston: Academic Press, Inc., 1994), pp. 229-261 on p. 244. See note 66 of "Women" for the behind-the-scenes tension facing women in mathematics.

[24]Archibald, *History*, p. 75.

[25]Judy Green and Jeanne LaDuke, "Women in the American Mathematical Community: The Pre-1940 Ph.D.'s," *The Mathematical Intelligencer* **9** (1) (1987):11-23.

[26]For the details of Gentry's experiences, see Fenster and Parshall, "Women," p. 236, and James C. Albisetti, *Schooling German Girls and Women: Secondary and Higher Education in the Nineteenth Century* (Princeton: University Press, 1988), p. 225.

[27]Fenster and Parshall, "Women," p. 244.

within the broader international arena. Moreover, as the issues surrounding the opening of graduate schools to American women illustrate, this advancement included not only mathematical results but also progress for women in mathematics at home and abroad.

Dickson's Research: The International Exchange of Mathematical Ideas

The University of Chicago furthered American initiatives toward internationalization in somewhat unexpected ways, too. Moore, Bolza, and Maschke simply could not have predicted the (indirect) contributions of the industrialist Andrew Carnegie to Chicago—and, consequently, American—mathematics. In the spring of 1901, with roughly $300,000,000 from the sale of his steel businesses, Carnegie announced that his next philanthropic effort would be in the field of higher education.[28] He first established a Scottish Trust to strengthen the universities of his boyhood country. The Trust included competitive, post-graduate scholarships and fellowships. During the 1904–1905 academic year, the young Scot Joseph Henry Maclagan Wedderburn traveled to the University of Chicago as a visiting Carnegie fellow.[29] On the heels of a winter semester at the University of Leipzig and a summer at the University of Berlin, Wedderburn arrived at Chicago in the fall of 1904, drawn to the University by its strong algebraic activity. At the time, Chicago "ranked undisputedly as the center of algebraic research in the United States."[30]

By January of 1905, both Wedderburn and Dickson were working to determine the nature of all finite division algebras.[31] Certainly, a finite field represented a finite division algebra with commutative multiplication, but were all finite division algebras fields? This question challenged both Dickson and Wedderburn in the early months of 1905.[32] On 20 January, they monopolized the Mathematics Club discussion at Chicago with their respective presentations of a "Proof of the Commutativity of Addition and Multiplication in Finite Fields" and a "Theorem on Non-Commutative Finite Systems."[33] The published papers which grew out of these talks seemed to reveal the two mathematicians' individual approaches to the query. Apparently, Wedderburn attacked the problem directly while Dickson looked for a counterexample.[34] Three months later, with the finished results in hand, Wedderburn presented what we now term the finite division algebra theorem, that is, every finite division algebra is a field, to the Chicago Section of the

---

[28]Joseph Wall, *Andrew Carnegie* (Pittsburgh: University of Pittsburgh Press, 1989), p. 836.

[29]Karen Hunger Parshall, "New Light on the Life and Work of Joseph Henry Maclagan Wedderburn (1882-1948)," in *Amphora: Festschrift für Hans Wussing zu seinem 65. Geburtstag*, ed. Menso Folkerts *et al.* (Basel/Boston/Berlin: Birkhäuser Verlag, 1992), pp. 523-537 on p. 527.

[30]Karen Hunger Parshall, "In Pursuit of the Finite Division Algebra Theorem and Beyond: Joseph H. M. Wedderburn, Leonard E. Dickson, and Oswald Veblen," *Archives internationales d'histoires des sciences* **33** (1983):274-299 on p. 278.

[31]This is in modern terms. In January of 1905, the concept of an algebra still included division. A finite algebra referred to a finite dimensional algebra over a finite field. See, for example, James Byrnie Shaw, *Synopsis of Linear Associative Algebra* (Washington, D.C.: Carnegie Institution, 1907), p. 58.

[32]Parshall, "In Pursuit of the Finite Division Algebra Theorem and Beyond," pp. 288-289. The subsequent discussion here follows that given in this source.

[33]Department of Mathematics, "Logbook of the Mathematical Club of the University of Chicago," Chicago, 1903–1954, p. 8. (Handwritten.)

[34]Parshall, "Pursuit of the Finite Division Algebra Theorem," p. 289.

American Mathematical Society, while Dickson discussed the class of finite algebras with inverses he had discovered.[35]

When viewed from the perspective of the internationalization of American mathematics, this episode from Dickson's professional career highlights a confluence of factors. The general setting is the University of Chicago, the first sustained American effort to incorporate the German ideals of post-graduate education. The specific setting is the Mathematics Department, built primarily on the strength of one American (with some German training) and two German mathematicians who each wholeheartedly embraced the German model of specialized graduate training, research, and the preparation of future researchers. Leonard Dickson, a first-generation, American-born and American-trained mathematician occupies a pivotal position, along with the Scot Wedderburn, who was supported by Carnegie's new Scottish Trust funds to spend a year at Chicago. An international research effort took place in the heart of Chicago and, perhaps more importantly, in the Mathematics Club, where the second-generation of American-trained mathematicians directly witnessed the benefits of international competition and collaboration and indirectly observed the efforts of a new Scottish organization committed, in part, to the advancement of research.

Dickson's scholarly pursuits showcase other philanthropic efforts to further American mathematics in the early decades of the twentieth century. Dickson was in his second year as an assistant professor at the University of Chicago when the Carnegie Institution of Washington came into existence in 1902. Although Carnegie initially considered establishing a national university in Washington, D.C., he ultimately settled on a foundation as a means of strengthening American universities through the support of basic research.[36] From virtually the time of its inception, the young Dickson pursued various funding opportunities from the Carnegie Institution.[37] E. H. Moore may have encouraged Dickson to seek out Carnegie funds. Moore served as president of the American Mathematical Society (AMS) in 1901 and 1902, chair of the committee advising the Carnegie Institution regarding original research in mathematics in 1902, and, closer to home for Dickson, chair of the Mathematics Department at the University of Chicago from 1896 to 1931.[38]

In 1911, just over a decade into a thriving career in pure mathematics, Dickson appealed to the Carnegie Institution of Washington as he began his "monumental" historical study of the theory of numbers.[39] In his initial letter to the Carnegie Institution seeking interest in the project, Dickson outlined that "[i]t would seem desirable to have undertaken in this country something of the kind done by the British Association, the Deutsche Mathematiker-Vereinigung, etc., in the preparation by specialists of note of extensive Reports each covering an important branch

[35]For the published versions of these works, see Joseph H. M. Wedderburn, "A Theorem on Finite Algebras," *Transactions of the American Mathematical Society* **6** (1905):349-352, and Leonard E. Dickson, "On Finite Algebras," *Nachrichten der Gesellschaft der Wissenschaften zu Göttingen, Mathematisch-Physicalische Abteilung* (1905):358-393.

[36]Wall, pp. 858-860.

[37]Dickson made his first appeal to the Carnegie Institution for support for his work in March, 1903.

[38]Archibald, *History*, p. 144.

[39]Albert, p. 333.

of science .... I have already given a solid year's work to such an expository Report on the theory of numbers (integral and algebraic) ... ."[40] By soliciting the Carnegie Institution, Dickson strove to gain outside support for mathematics. The British and German examples thus inspired Dickson to write his own compendium on the subject of number theory and in so doing to establish both himself solidly as a "specialist of note" and the United States as a competitive force in mathematics.

These were not all of the project's international implications. As Dickson explained in a letter a year after his initial proposal:

> The project of a history of the Theory of Numbers, to cover the ground in as complete a manner as possible has turned out to be a very formidable task. After a year and a half, I have secured probably absolutely complete data on the 19th century and fairly complete references on [the] earlier period ....
>
> But there remain several hundred references which it is necessary for me to examine before going further. This of course makes it necessary for me to spend some time in the libraries of Europe.
>
> I can secure leave of absence beginning shortly after April 1st—this part is easier than the securement of funds to cover *part* of the *extra* expenses involved and which I am not in a position to defray.[41]

Thus, in Dickson's view, a solid compendium of number theory required a trip overseas to visit European libraries and to collect various number-theoretic references. The University of Chicago, in granting Dickson a leave of absence, supported his international research agenda. The Carnegie Institution, in supplementing Dickson's funding, lent its support to the development of mathematics at a high level in the United States.

By the end of the second decade of the twentieth century, Dickson's career had advanced to the point where he received an invitation to deliver a plenary lecture at the 1920 International Congress in Strasbourg.[42] On this occasion, he selected a topic that grew, in part, from his historical acquaintance with the theory of numbers and from his work in the theory of algebras.[43] In particular, he demonstrated that

---

[40]Leonard Dickson to R. S. Woodward, 11 February, 1911, Carnegie Institution Archives, Washington, D.C., Dickson Papers. This and all subsequent quotes from the Carnegie Institution archives are published courtesy of the Carnegie Institution of Washington. I gratefully thank them both for their hospitality and for granting this permission.

[41]Leonard Dickson to R. S. Woodward, 3 March, 1912, Carnegie Institution Archives, Washington, D.C., Dickson Papers. Dickson's emphasis.

[42]At the fourth International Congress of Mathematicians in Rome in 1908, Simon Newcomb became the first American to deliver a plenary address. War completely extinguished the International Congress planned for 1916 in Stockholm. In 1920, the Congress excluded mathematicians from the Central Powers and admitted those from neutral countries only after a two-thirds vote from the executive committee. Consequently, the Strasbourg Congress was never designated as the sixth Congress, and subsequent Congresses have not been numbered. See Olli Lehto, *Mathematics Without Borders: A History of the International Mathematical Union* (New York: Springer-Verlag, 1998) as well as his contribution to the present volume.

[43]For the Congress paper, see Leonard E. Dickson, "Some Relations Between the Theory of Numbers and Other Branches of Mathematics," in *Comptes rendus du Congrès international des mathématiciens, Strasbourg*, ed. Henri Villat (Toulouse, 1921; reprint ed., Nendeln/Liechtenstein: Kraus Reprint Limited, 1967), pp. 41-56. Dickson's historical study resulted in his *History of the Theory of Numbers*, 3 vols. (New York: Chelsea Publishing Company, 1919, 1920, 1923). For the

the arithmetics of the complex numbers and the quaternions provide effective tools for studying the arithmetics of algebras. This result marked both his and his country's point of entry into the arithmetics of algebras.

The theory of the arithmetics of algebras hinged on the determination of a set of integral elements which led to an arithmetic analogous to that of the ordinary integers.[44] Although the Germans Carl Friedrich Gauss, Ernst Eduard Kummer, Richard Dedekind, Leopold Kronecker, and Rudolf Lipschitz had considered the arithmetic of complex integers, it was the German-born, Swiss professor of mathematics, Adolf Hurwitz (1859-1919), who first articulated the goal of constructing arithmetics resembling that of the ordinary integers for other mathematical systems. Lipschitz's proposed system of integral quaternions lacked several of the key, desirable properties (such as unique factorization), apparently prompting Hurwitz to recognize that the set of integral quaternions held the key to an arithmetic like that of the integers. Hurwitz thus shifted his focus from the coefficients of an integral quaternion to the structure of a set of integral quaternions. From this vantage point, he obtained an un-integer-like integral quaternion, but one that yielded the desired arithmetic.[45]

Drawing from the work of his adviser, Hurwitz, Louis Gustave Du Pasquier (1876-1957) considered the arithmetic of non-commutative number systems, beginning with matrices with rational entries and, roughly twenty years later, ending with all rational hypercomplex number systems. He formulated his definition of a set of integral elements based on the structure of the set of ordinary integers. Both the strengths and weaknesses of his theory resulted from this particular construction. On the one hand, he had so much confidence in his definition that he applied it to all rational hypercomplex number systems; prior to Du Pasquier, mathematicians had only considered the arithmetic of specific algebras. On the other hand, he clung so tightly to his definition that he seemingly lost sight of his objective. When he could not obtain unique factorization into primes, for example, he attempted to alter the concept of integer to include what he termed absolute, fractional, and conditional integers. With the ordinary integers supposedly serving as his guide, this idea of three types of integers in a single number system diverted him from his intended course.

Dickson more than likely heard Du Pasquier present this work at the Strasbourg Congress.[46] Soon after, Dickson established Du Pasquier's unsubstantiated claims

---

details of the genesis of the Strasbourg lecture, see Della Dumbaugh Fenster, "Leonard Dickson's History of the Theory of Numbers: A Historical Study with Mathematical Implications," *Revue d'histoire des mathématiques* **5** (2) (1999):159-179.

[44]The following general discussion of this development is intended to highlight the international exchange of mathematical ideas. For the technical details, see Fenster, "Dickson and His Work in the Arithmetics of Algebras."

[45]Hurwitz gave the general expression for an integral quaternion as $\frac{1}{2}(g_0 + g_1 i + g_2 j + g_3 k)$ where the g_i's are all even or all odd. Of course, it is when the g_i's are all odd that an integral quaternion takes on an un-integer-like appearance.

[46]Du Pasquier chose to discuss his ideas on the arithmetics of hypercomplex number systems in his "Sur les nombres complexes généraux" at the 1920 Congress where, along with Dickson, he presided over the sessions on "Arithmetic. Algebra. Analysis." For the written version of this talk, see L. Gustave Du Pasquier, "Sur les nombres complexes généraux," in *Comptes rendus du Congres international des mathématiciens, Strasbourg*, pp. 164-175. Since both men maintained somewhat of a high profile at the Congress and since each included the arithmetics of algebras in his talk, it seems highly probable that they heard one another speak.

regarding the use of ideals in certain algebras to obtain unique factorization.[47] In so doing, the American not only demonstrated his commitment to rigorous mathematics but also categorically demonstrated the superiority of his methods. He did not discreetly build the arguments Du Pasquier lacked; he proclaimed both the defects of Du Pasquier's ideas and the incompleteness of his proofs. Dickson went on to construct a set of integral elements, first in a rational algebra and later in an arbitrary algebra, which led to a rich theory (as opposed to case-by-case studies) of the arithmetics of algebras.[48]

As the evolution of this concept makes clear, international scientific exchange played a crucial role in the dynamic development of mathematics. Before Dickson ever brought the arithmetics of algebras to America, the subject had already advanced and retreated across Swiss and German borders for more than 100 years. Each mathematician who joined this pursuit drew from the work of his predecessors. Naturally, published works served to transmit ideas from one mathematician to another. The International Congress of 1920 may have also aided the progress of this arithmetic for it more than likely brought Dickson and Du Pasquier within earshot of one another. They met again four years later at the International Congress in Toronto.

In 1924, Dickson gave two addresses on the arithmetics of algebras. The first summarized the development of the arithmetics of algebras and showcased his definition of an integral element. The second extended the arithmetics of rational algebras to algebras over an arbitrary field.[49] His "remarkable theory" for rational algebras, as he liked to describe it, now held true for algebras over any field. He also proved that every rational algebra contained a set of integral elements. This result quite naturally marshaled even more evidence for his—as opposed to Du Pasquier's—postulational formulation of the definition of an integral element.

Du Pasquier did not have quite the same enthusiasm for Dickson's definition. He presented his views of the development of the concept of a hypercomplex number at the Toronto Congress.[50] Specifically, he described the "evolution" as a six-stage process which, not surprisingly, emphasized his first proposed definition of an integral hypercomplex number and presented Dickson's as one which (merely) introduced a "norm" postulate requiring every element in the system to possess an (ordinary) integral norm. (Dickson proposed this as a weaker assumption than the one he finally adopted. He did not make use of it in his version of the arithmetic of algebras that appeared in his celebrated *Algebras and Their Arithmetics.*)

The 1924 International Congress thus witnessed Dickson's overt criticisms of Du Pasquier's work and Du Pasquier's attempts to vindicate himself. It served as

---

[47] Leonard E. Dickson, "Impossibility of Restoring Unique Factorization in a Hypercomplex Arithmetic," *Bulletin of the American Mathematical Society* **28** (1922):438-442.

[48] For a complete discussion of this work, see Leonard E. Dickson, *Algebras and Their Arithmetics* (Chicago: University of Chicago Press, 1923).

[49] Leonard E. Dickson, "Outline of the Theory to Date of the Arithmetics of Algebras," in *Proceedings of the International Mathematical Congress held in Toronto, August 11-16, 1924*, ed. J. C. Fields, 2 vols. (Toronto: The University of Toronto Press, 1928), 1:95-102, and "Further Development of the Theory of the Arithmetics of Algebras," in *ibid.*, 1:173-184. Olive Hazlett, Dickson's former student, independently extended the arithmetics of rational algebras to algebras over an arbitrary field at the same Congress. See Olive C. Hazlett, "On the Arithmetic of a General Associative Algebra," in *ibid.*, 1:185-191.

[50] L. Gustave Du Pasquier, "L'Évolution du concept de nombre hypercomplexe entier," in *ibid.*, 1:193-205.

a forum for this exchange of mathematical ideas between Dickson, the American, and Du Pasquier, the Swiss. Both men brought their ideas and approaches before an international audience in an effort to gain support for their points of view and to cement further their reputations. In light of Dickson as a case study of American initiatives toward internationalization, moreover, the competition with Du Pasquier reflects the active engagement of American mathematics beyond the political boundaries of the United States. It represents a means toward the consolidation and growth of the American mathematical research community. A national agenda was being played out on the international stage.

Dickson and the Publication of Manuscripts and Book-length Treatises

A closer look at Dickson's professional career reveals various venues available to American mathematicians for the distribution of their mathematical work at home and abroad. The histories of these diverse publications provide additional insight into how American mathematicians both benefitted from and contributed to the internationalization of the field.

Dickson's textbooks, for example, promoted further consolidation and growth of the Chicago—and American—algebraic tradition. His eighteen books had, as A. Adrian Albert claimed, "a worldwide influence in stimulating research."[51] Perhaps Aubrey Kempner best captured Dickson's contributions as an author in his review of Dickson's *Studies in the Theory of Numbers* (1930). Kempner offered that

> [o]ne can but admire the courage of an author who will undertake to rebuild the whole structure rather than to patch up the unsound portions. One can only guess at the amount of labor covered by the modest words of the preface; 'It was no small task to write a satisfactory exposition.' On the other hand, we know, in this country as well as in Europe, how much the theory of numbers owes to the insistence of Dickson on precision in the statement of theorems and to his uncanny ability to detect, and to mend, unsound arguments; it seems therefore only fair that to him and his students should belong the credit of writing the first reliable treatment of the arithmetic of the theory.[52]

Thus, in addition to wielding influence in his own country, Dickson's texts began to repay a portion of the debt incurred by American mathematicians in the late nineteenth and early twentieth centuries when they borrowed extensively from the European mathematicians.[53]

Although few could deny Dickson's contribution to research-level teaching and training in the United States, some colleagues and students found the style of his texts less than satisfying. William Graustein described Dickson's *Linear Algebras* as a book with "a desert of statement and proof with not a refreshing oasis in sight,

[51] Albert, p. 332.

[52] Aubrey J. Kempner, "Review of *Studies in the Theory of Numbers*, by Leonard E. Dickson," *American Mathematical Monthly* **40** (1933):40-42 on p. 40.

[53] Moreover, Dickson's *Algebren und ihre Zahlentheorie* (Zürich: Orell Füssli, 1927) apparently provided German algebraists with an extensive development of hypercomplex number systems. Saunders Mac Lane, "History of Abstract Algebra: Origin, Rise, and Decline of a Movement," in *American Mathematical Heritage: Algebra and Applied Mathematics* (Lubbock, TX: Texas Tech University Press, 1981), pp. 3-35 on p. 14, and Birkhoff, "Fifty Years of American Mathematics," p. 287.

where the reader may pause to rest and take account of stock;" Saunders Mac Lane rather diplomatically referred to *Studies in the Theory of Numbers* as a book written with "sparse precision;" Paul Halmos "both feared and respected Dickson's *Modern Algebraic Theories*."[54] Relative to the latter work, Halmos also wrote that "[t]he exposition is brutal—correct but compressed, unambiguously decipherable but far from easy to read."[55] Although his style may have generated some criticism, his thoroughness and his ability to write so many texts on such a wide variety of subjects in algebra and number theory brought coherence to these areas and made advanced mathematics available to aspiring mathematicians across the country and around the globe.

The appearance of Dickson's first book, *Linear Groups with an Exposition of the Galois Field Theory*, perhaps reveals his most strategic maneuvering in the international publishing world.[56] In part one of this revised and expanded version of his dissertation, Dickson gave the first extensive presentation of the theory of finite fields entitled "Introduction to the Galois Field Theory." Having pulled together some of the loose ends of abstract field theory and Galois theory, Dickson devoted the second part of his book to a general theory of linear groups over $GF[p^n]$, a generalization of Camille Jordan's work over the prime field $GF[p]$. Dickson concluded his book with a summary of the known simple groups, a list Albert found "still valuable" in 1955.[57]

Although Dickson had originally intended to approach an American press regarding the publication of this work, E. H. Moore advised him to enlist the help of Felix Klein in the hopes of securing a German publisher.[58] Dickson wrote to Klein on 7 April, 1900 that "[i]t is probable that I shall publish it [*Theory of Linear Groups with an Exposition of the Galois Field Theory*] this summer in America; but, at the suggestion of Professor Moore, it has occurred to me that, if your sympathies could be enlisted, it might be accepted by Herr Teubner for his Sammlung."[59] Dickson must have persuaded Klein to take up his cause since his book appeared with Teubner the following year. "By thus securing the Teubner seal of approval," Parshall noted, "Dickson's work not only came before one of the most important segments of the international mathematical community at the turn of the century, but it also served to build its author's reputation at home and abroad."[60] Dickson's book with its "German seal of approval" also brought the international community an announcement they could not ignore: the Americans, particularly the American algebraists, had arrived on the mathematical scene.[61]

---

[54]William C. Graustein, "Dickson's Linear Algebras," *Bulletin of the American Mathematical Society* **10** (1915):511-522 on p. 513. Compare also Saunders Mac Lane, "Mathematics at the University of Chicago: A Brief History," in *A Century of Mathematics in America—Part II*, ed. Peter Duren *et al.* (Providence: American Mathematical Society, 1989), pp. 127-154 on p. 134, and Paul Halmos, "Some Books of Auld Lang Syne," in *A Century of Mathematics in America—Part I*, ed. Peter Duren *et al.* (Providence: American Mathematical Society, 1988), pp. 131-174 on p. 143.

[55]Halmos, p. 143.

[56]Leonard E. Dickson, *Linear Groups with an Exposition of the Galois Field Theory* (Leipzig: B. G. Teubner Verlag, 1901; reprint ed., New York: Dover Publications, Inc., 1958).

[57]Albert, p. 332.

[58]Karen Hunger Parshall, "A Study in Group Theory: Leonard Eugene Dickson's *Linear Groups*," *The Mathematical Intelligencer* **13** (1) (1991):7-11 on p. 9.

[59]*Ibid.*

[60]*Ibid.*, p. 7.

[61]*Ibid.*, p. 10.

The Teubner label may have also freed Dickson from what could have been a difficult search for an American publisher at the turn of the twentieth century. As Reinhard Siegmund-Schultze described the situation at that time, "[i]nsufficient public and private support regularly forced the American mathematical community to take on all of the financial responsibilities for its publishing."[62] Later, the American mathematicians would appeal to philanthropic organizations (see below) for, among other requests, funds for publication. These new national organizations would ultimately aid the Americans as they sought to gain independence from European publishers, an idea which would have full steam two decades after the appearance of Dickson's *Linear Groups*.[63] In 1901, however, Dickson—and Moore—desired a German publisher that would link American mathematics with its strong European counterparts.

And Dickson virtually waged a battle to keep that association solid. In particular, he remained emphatic about high-quality written work. Trusting his memory many years later, Kenneth Ghent, a 1935 Dickson doctorate, recalled that his adviser

> valued highly elegant proofs which avoided unnecessary verbiage. As a journal editor he was ruthless about rejecting shoddy work and about insisting on both clarity and brevity. He believed that it was a point of national honor to require that a paper by an American mathematician be one that would be respected by mathematicians abroad. He recounted in class his rejection of a journal paper by an individual who later succeeded in having the paper published abroad without making suggested revisions. Dickson believed that the foreign editor had accepted the paper as a courtesy to U.S. mathematicians. But to Dickson the author betrayed U.S. mathematicians. ... He believed that he was increasing the status of American mathematics by his emphasis on rigor and on clear, concise proofs.[64]

Dickson's standards for and pride in American mathematical publications, especially relative to European work, reflected a general attitude shared by many of his colleagues. American mathematicians had originally sought to emulate the European traditions in research mathematics and publications, and, in the early years of their development as a community, they recognized the superiority of their counterparts across the Atlantic. As this group moved into the twentieth century, however, they strove to produce American mathematical research on a par with, if not superior to, that of the European standard-bearers.

In order to turn out consistently high-quality work, even the most prominent American (and foreign) mathematicians allowed their ideas to undergo careful scrutiny. While Dickson served as editor of the *Transactions*, for example, he had this to say to R. L. Moore concerning a paper he had submitted for publication. "All that I expect you to do is to try to condense slightly. You are one of the staff—and it should not be said that you would not take the medicine that you

[62] Reinhard Siegmund-Schultze, "The Emancipation of Mathematical Research Publishing in the United States from German Dominance (1878-1945)," *Historia Mathematica* **24** (1997):135-166 on p. 138.

[63] *Ibid.*, p. 141.

[64] Kenneth Ghent to Della D. Fenster, 14 May, 1992.

and the others on staff are now obliged to offer outside authors."[65] If an American failed to meet the new standards, some, like Dickson, perceived it as a setback for the entire American endeavor.

Moreover, Dickson's correspondence with the Carnegie Institution in Washington reveals his concerted effort to secure publication for his work and so for American research-level mathematics. For example, on 25 November, 1904, Dickson wrote to Charles Wolcott, secretary of the Carnegie Institution, regarding work he had completed while a "Research Assistant of the Carnegie Institution" from 22 March to 1 October, 1904. "While the tenure of my appointment as research assistant to the Institution has terminated in one sense," Dickson began,

> the fiscal year's appointment has not. So I thought I might include the present paper ["Definitions of a Group and a Field by Independent Postulates"]—the outcome of considerable study the past two months—rather muchly "boiled down."
>
> There is being manifested considerable interest in America and Europe just now on the logical foundations of mathematics.
>
> While many formal definitions of a group have been offered, the one I hit up (and partially by Prof. Moore. His note is to appear simultaneously) is by all means the simplest and smoothest yet proposed.
>
> The brief paper will I think be a permanent contribution to group theory and so I should like to include it in the series of Carnegie Institution papers.[66]

Once again, this letter highlights Dickson's sense of competing in an international mathematical arena and of wanting American results effectively publicized.

A letter to Wolcott earlier in November also brings into focus Dickson's competitive spirit as well as his candor. He wrote that

> [t]he present little paper ["Determination of the Ternary Modular Groups"] treats without the limitations made by Burnside (the present Secretary of the London Math Soc.) a subject to which he devoted 50 pages—but with so many errors as to make his work really valueless—He acknowledges his errors but says he is not able to correct them. My attack is on wholly different lines.[67]

This brief summary makes it easy to understand why Dickson earned a reputation as "hard-bitten" and "gruff."[68] At the same time, it calls attention to Dickson's mathematical standards. When he published his own mathematical results, when

---

[65]Leonard E. Dickson to R. L. Moore, 18 December, no year given, University of Texas Archives, The Center for American History, R. L. Moore Papers. Dickson also served as editor of the *American Mathematical Monthly* from 1902 to 1908. For more on the early effort to promote the "best" publications, see Parshall and Rowe, pp. 411-415.

[66]Leonard E. Dickson to Charles D. Wolcott, 25 November, 1904, Carnegie Institution Archives, Washington, D.C., Dickson Papers.

[67]Leonard E. Dickson to Charles D. Wolcott, 3 November, 1904, Carnegie Institution Archives, Washington, D.C., Dickson Papers. As noted above, two decades later, Dickson expressed a nearly identical criticism of Du Pasquier's work in the arithmetic of algebras.

[68]Saunders Mac Lane, interview by author, 5–6 March, 1992, Charlottesville, Virginia, tape recording, and Donald J. Albers and Gerald L. Alexanderson, "A Conversation with Ivan Niven," *College Mathematics Journal* **22** (1991):370-402 on p. 377.

he edited the work of others for the *Transactions* or the *American Mathematical Monthly*, and when he trained future American mathematicians, he, like Moore before him, insisted on concise, high-quality publications with (preferably) general results.[69] This proved essential to the consolidation and growth of American mathematics in the early decades of the twentieth century.

Moreover, when promoting his work to the Carnegie Institution, Dickson situated it within the international mathematical context. In his view, this clearly lent credibility to his work and so should serve as a major selling point to a funding institution like the Carnegie. Although Dickson's motives here may appear self-serving at first glance, by pushing mathematics before the Carnegie Institution, he helped carve out a wedge of support for American mathematics.[70]

But Dickson was not the only mathematician soliciting the Carnegie Institution for funds for mathematical publications. Regarding a book Dickson submitted to the Carnegie Institution in March of 1903, E. H. Moore wrote to Wolcott that

> It was the feeling of the advisory committee, as expressed in the report, that the interests of mathematics would be effectively furthered by the Carnegie Institution, if, like the governments and the academies of Europe and the great presses of England, the Institution would undertake, directly or indirectly, the publication of treatises and collected works. We thought, and I still feel strongly, that in promoting the interests of mathematics the Institution might make considerable use of a society which, like the American Mathematical Society, is engaged in advancing those interests so far as its financial strength will permit.
>
> We understand, however, that for the present the Executive Committee has decided against any subventions to societies and journals. Accordingly, in writing six weeks ago with respect to the book of Mr. Dickson, the manuscript of which has been submitted to you, my recommendation was that the Carnegie Institution undertake the publication of a series of treatises on mathematics, to which Mr. Dickson's book should belong, and to which other books might be added as suitable manuscripts appear. Mr. Dickson's book can hardly be undertaken by one of the publishing houses, and it is too much of the nature of a treatise to appear in any of the journals. At any rate under present financial conditions, the journals cannot publish pages enough to include so extensive memoirs. The book is distinctly a strong one, and is entirely suitable for publication by the Institution.[71]

---

[69] Parshall and Rowe, p. 413.

[70] Loren Butler Feffer examined the significant effort Oswald Veblen exerted in the early 1920's to secure funds for American mathematical research through the AMS and Princeton University. Although Veblen's first-hand wartime experience and diplomatic personality contributed to his success, no single characteristic advanced his cause more than his strong assertion of the centrality of mathematics to all of the sciences. See Loren Butler Feffer, "Oswald Veblen and the Capitalization of American Mathematics: Raising Money for Research, 1923-1928," *Isis* **89** (1998):474-497.

[71] E. H. Moore to Charles D. Wolcott, 23 May, 1903, Carnegie Institution Archives, Washington, D.C., Dickson Papers.

Thus, from his position on the Carnegie advisory committee and as the immediate past-president of the AMS, Moore spoke authoritatively on American opportunities for publishing mathematical treatises. Once again, he was present at a crucial juncture in American mathematics. As the Carnegie Institution began to define its support of research, Moore was there using European models and standards to encourage the Institution to publish research-level mathematical treatises. Over the course of the next three decades, Dickson continued Moore's efforts to gain Carnegie support for mathematics.

Dickson's efforts to secure publishers for his manuscripts highlight the various avenues for publishing American mathematics and underscore, in particular, the importance of earning the German seal of approval early in one's career in the first decade of the twentieth century. A close examination of his publication record also reveals the absence of American publishers for research-level treatises and the initiatives of Moore and Dickson to fill this void through the newly formed Carnegie Institution by citing the European example. Moreover, Dickson's textbooks, especially his more advanced monographs, began to make a dent in "the German tradition in mathematical research which was still influential and reached American graduate students as late as in the 1930s without intermediate translation or commentary."[72] Thus, when examined through the lens of his publication contributions, Dickson serves as a case study to support Siegmund-Schultze's overall conclusion that "[o]n the one hand, the German example was the main point of reference for American mathematicians in their efforts to improve the national system of mathematical research publishing and to reach national independence in the field. On the other hand, the lack of governmental support and the undeveloped state of commercial scientific publishing up until the end of the 1930s forced American mathematicians to look for other sources of support, especially philanthropic foundations."[73]

American Mathematics: Unbound

The span of Dickson's career, commencing with his graduate training in 1894 and concluding with his retirement in 1939, demonstrates that America's research-minded mathematics community evolved in the midst of a constant exchange of American and international ideas. The traditional American metaphor applies to American mathematics in the late nineteenth and early twentieth centuries, that is, American mathematics was a melting pot of its own, with more than a sprinkling of German ingredients.

As the twentieth century progressed, Americans not only sought to emulate the Europeans, the international trendsetters as it were, but also to surpass them. Inspired in part by E. H. Moore, Dickson and other leaders in the American mathematical community prompted their colleagues to do and to publish mathematics on a par with—if not superior to—that of their European counterparts. Of particular importance, Dickson simultaneously held his Ph.D. students to these high American standards.[74] His sixty-seven students alone dispersed to no fewer than forty-five

---

[72]Siegmund-Schultze, p. 148.

[73]*Ibid.*, pp. 159-160.

[74]Della Dumbaugh Fenster, "Role Modeling in Mathematics: The Case of Leonard Eugene Dickson (1874-1954)," *Historia Mathematica* **24** (1997):7-24.

academic institutions in twenty-two states and three foreign countries with the professional ideas and ideals they had learned at the University of Chicago, thereby ensuring the successful continuation and expansion of a national community in an international arena.

But as Dickson's career makes clear, the advancement of American mathematics depended not only on the "production" of mathematical ideas, but also their "reception."[75] "Reception" hinged on publications. The histories of even a few of Dickson's manuscripts and book-length treatises make clear the often substantial effort involved in obtaining publishers for research-level mathematics. These histories also uncover steady attempts by E. H. Moore and Dickson to secure funds for mathematics from external (American) sources very early in the twentieth century. Using Dickson as a case study thus reveals that American mathematicians sought to carve out a wedge of the philanthropic pie more than a full decade before World War I. These advances, in turn, aided Americans in their quest for independence from European publishers. Dickson's professional pursuits show how American mathematics came full-circle in this regard. In 1901, in the infancy of his career, Dickson desired a German publisher for his first book to validate his career. From 1903 to 1935, but particularly for the quarter century beginning in 1910, he invested considerable energy in lobbying the Carnegie Institution for their American support of mathematical publications.

Examining these few aspects of the professional life of a significant American mathematician in the late nineteenth and early twentieth century shows that, even for the most distinguished mathematicians, mathematics involved a great deal more than proving theorems. As this case study of Dickson illustrates, a commitment to the German-inspired American training of aspiring mathematicians, a keen knowledge of mathematical results at home and especially abroad, a willingness to take advantage of new external support for mathematical research, and an awareness of how to advance national concerns in an increasingly international enterprise all became part-and-parcel of the American mathematician's professional life. These "historical origins" suggest a characterization of American mathematics as "internationalized science" as it emerged, consolidated, and strengthened in the opening decades of the twentieth century.[76]

References

Archival Sources

Department of Mathematics. "Logbook of the Mathematical Club of the University of Chicago." Chicago, 1903-1954. (Handwritten.)

Dickson, Leonard E. "The Influence of Galois on Recent Mathematics," July 10, 1897. University of Chicago. Department of Special Collections. Mathematical Club Records 1893-1921. (Handwritten.)

Leonard Dickson to Publication Committee of the Carnegie Institution. 3 March, 1903. Carnegie Institution Archives. Washington, D.C. Dickson Papers.

Leonard Dickson to President Daniel C. Gilman. 30 March, 1903. Carnegie Institution Archives. Washington, D.C. Dickson Papers.

[75] Siegmund-Schultze, p. 136.

[76] *Ibid.*, p. 160.

Leonard E. Dickson to R. L. Moore. 18 December, no year given. University of Texas Archives. The Center for American History. R. L. Moore Papers.

Leonard E. Dickson to Charles D. Wolcott. 3 November, 1904. Carnegie Institution Archives. Washington, D.C. Dickson Papers.

Leonard E. Dickson to Charles D. Wolcott. 25 November, 1904. Carnegie Institution Archives. Washington, D.C. Dickson Papers.

Leonard E. Dickson to Robert S. Woodward. 11 February, 1911. Carnegie Institution Archives. Washington D.C. Dickson Papers.

Leonard E. Dickson to Robert S. Woodward. 3 March, 1912. Carnegie Institution Archives. Washington D.C. Dickson Papers.

Leonard E. Dickson to Robert S. Woodward. 11 December, 1916. Carnegie Institution Archives. Washington D.C. Dickson Papers.

William Duren. Interview by author. 10 December, 1992. Charlottesville, Virginia. Tape Recording.

Kenneth Ghent to Della D. Fenster. 14 May, 1992.

M. Gweneth Humphreys to Della D. Fenster. 23 May, 1992.

Saunders Mac Lane. Interview by author. 5-6 March, 1992. Charlottesville, Virginia. Tape Recording.

E. H. Moore to Charles D. Wolcott. 23 May, 1903. Carnegie Institution Archives. Washington, D.C. Dickson Papers.

Frank Morley to Daniel C. Gilman. 18 April, 1903. Carnegie Institution Archives. Washington, D.C. Dickson Papers.

Ivan Niven to Della D. Fenster. 10 September, 1992.

Printed Sources

Albers, Donald J. and Alexanderson, G. L. "A Conversation with Ivan Niven." *College Mathematics Journal* **22** (1991):370-402.

Albert, A. Adrian. "Leonard Eugene Dickson 1874-1954." *Bulletin of the American Mathematical Society* **61** (1955):331-345.

Albisetti, James C. *Schooling German Girls and Women: Secondary and Higher Education in the Nineteenth Century.* Princeton: Princeton University Press, 1988.

Archibald, Raymond C. *A Semicentennial History of the American Mathematical Society, 1888-1938.* New York: American Mathematical Society, 1938.

______. Ed. *Semicentennial Addresses of the American Mathematical Society 1888-1938.* New York: American Mathematical Society, 1938.

Birkhoff, George D. "Fifty Years of American Mathematics." In *Semicentennial Addresses of the American Mathematical Society.* Ed. Raymond C. Archibald. New York: American Mathematical Society, 1938, pp. 270-315.

"Courses in Mathematics." *Register of the University of Chicago*, pp. 260-266. University of Chicago. Department of Special Collections.

Dickson, Leonard E. *Algebras and Their Arithmetics.* Chicago: University of Chicago Press, 1923.

______. *Algebren und ihre Zahlentheorie.* Zurich: Orell Füssli, 1927.

______. "The Analytic Representation of Substitutions on a Power of a Prime Number of Letters with a Discussion of the Linear Group." *Annals of Mathematics* **11** (1897):65-143.

______. *The Collected Mathematical Papers of Leonard Eugene Dickson.* Ed. A. Adrian Albert. 6 Vols. New York: Chelsea Publishing Co., 1975, 1983.

———. "Further Development of the Theory of the Arithmetics of Algebras." In *Proceedings of the International Mathematical Congress held in Toronto, August 11-16, 1924*. Ed. J. C. Fields. Toronto: The University of Toronto Press, 1928, 1:173-184.

———. *History of the Theory of Numbers.* 3 Vols. New York: Chelsea Publishing Company, 1919, 1920, 1923.

———. *Linear Groups with an Exposition of the Galois Field Theory.* Leipzig: B. G. Teubner Verlag, 1901. Reprint Ed., New York: Dover Publications, Inc., 1958.

———. "On Finite Algebras." *Nachrichten der Gesellschaft der Wissenschaften zu Göttingen, Mathematisch-Physicalische Abteilung* (1905):358-393.

———. "Outline of the Theory to Date of the Arithmetics of Algebras." In *Proceedings of the International Mathematical Congress held in Toronto, August 11-16, 1924*. Ed. J. C. Fields. Toronto: The University of Toronto Press, 1928, 1:95-102.

———. "Some Relations Between the Theory of Numbers and Other Branches of Mathematics." In *Conférence générale, Comptes rendus du Congrès international des mathématiciens, Strasbourg.* Ed. Henri Villat. Toulouse, 1921; Reprint Ed., Nendeln/Liechtenstein: Kraus Reprint Limited, 1967, pp. 41-56.

Du Pasquier, L. Gustave. "L'Évolution du concept de nombre hypercomplexe entier." In *Proceedings of the International Mathematical Congress held in Toronto, August 11-16, 1924*. Ed. J. C. Fields. Toronto: The University of Toronto Press, 1928, 1:193-205.

———. "Sur les nombres complexes généraux." In *Conférence générale, Comptes rendus du Congrès international des mathématiciens, Strasbourg.* Ed. Henri Villat. Toulouse, 1921; Reprint Ed., Nendeln/Liechtenstein: Kraus Reprint Limited, 1967, pp. 164-175.

Duren, William L. "Graduate Student at Chicago in the Twenties." In *A Century of Mathematics in America—Part II.* Ed. Peter Duren *et al.* Providence: American Mathematical Society, 1989, pp. 177-182.

Feffer, Loren Butler. "Oswald Veblen and the Capitalization of American Mathematics: Raising Money for Research, 1923-1928." *Isis* **89** (1998):474-497.

Fenster, Della D. "Leonard Dickson's History of the Theory of Numbers: A Historical Study with Mathematical Implications." *Revue d'histoire des mathématiques* **5** (2) (1999):159-179.

———. "Leonard Eugene Dickson and His Work in the Arithmetics of Algebras." *Archive for History of Exact Sciences* **52** (1998):119-159.

———. "Role Modeling in Mathematics: The Case of Leonard Eugene Dickson (1874-1954)." *Historia Mathematica* **24** (1997):7-24.

Fenster, Della D. and Parshall, Karen Hunger. "A Profile of the American Mathematical Research Community: 1891-1906." In *The History of Modern Mathematics.* Vol. 3. Ed. Eberhard Knobloch and David E. Rowe. San Diego: Academic Press, 1994, pp. 179-227.

———. "Women in the American Mathematical Research Community: 1891-1906." In *The History of Modern Mathematics.* Vol. 3. Ed. Eberhard Knobloch and David E. Rowe. San Diego: Academic Press, 1994, pp. 229-261.

Halmos, Paul. "Some Books of Auld Lang Syne." In *A Century of Mathematics in America—Part I.* Ed. Peter Duren *et al.* Providence: American Mathematical Society, 1988, pp. 131-174.

Hazlett, Olive C. "On the Arithmetic of a General Associative Algebra." In *Proceedings of the International Mathematical Congress held in Toronto, August 11-16, 1924.* 2 Vols. Ed. J. C. Fields. Toronto: The University of Toronto Press, 1928, 1:185-191.

Graustein, William C. "Dickson's Linear Algebras." *Bulletin of the American Mathematical Society* **10** (1915):511-522.

Green, Judy and LaDuke, Jeanne. "Women in the American Mathematical Community: The Pre-1940 Ph.D.'s." *The Mathematical Intelligencer* **9** (1) (1987): 11-23.

Kempner, Aubrey J. "Review of *Studies in the Theory of Numbers*, by Leonard E. Dickson." *American Mathematical Monthly* **40** (1933):40-42.

Koblitz, Ann Hibner. "Sofia Vasilevna Kovalevskaia (1850-1891)." In *Women of Mathematics: A Biobibliographic Sourcebook.* Ed. Louise S. Grinstein and Paul J. Campbell. Westport, CN: Greenwood Press, 1987, pp. 103-113.

Lehto, Olli. *Mathematics Without Borders: A History of the International Mathematical Union.* New York: Springer-Verlag, 1998.

Mac Lane, Saunders. "History of Abstract Algebra: Origin, Rise, and Decline of a Movement." In *American Mathematical Heritage: Algebra and Applied Mathematics.* Lubbock, TX: Texas Tech University Press, 1981, pp. 3-35.

______. "Mathematics at the University of Chicago: A Brief History." In *A Century of Mathematics in America—Part II.* Ed. Peter Duren *et al.* Providence: American Mathematical Society, 1989, pp. 127-154.

Moore, Eliakim Hastings. "A Doubly-Infinite System of Simple Groups." In *Mathematical Papers Read at the International Mathematics Congress Held in Connection with the World's Columbian Exposition: Chicago 1893.* Ed. E. H. Moore, Oskar Bolza, Heinrich Maschke, and Henry S. White. New York: Macmillan & Co., 1896, pp. 208-242.

Parshall, Karen Hunger. "America's First School of Mathematical Research: James Joseph Sylvester at The Johns Hopkins University 1876-1883." *Archive for History of Exact Sciences* **38** (1988):153-196.

______. "How We Got Where We Are: An International Overview of Mathematics in National Contexts (1875-1900)." *Notices of the American Mathematical Society* **43** (1996):287-296.

______. "In Pursuit of the Finite Division Algebra Theorem and Beyond: Joseph H. M. Wedderburn, Leonard E. Dickson, and Oswald Veblen." *Archives internationales d'histoires des sciences* **33** (1983):274-299.

______. "Mathematics in National Contexts (1875-1900): An International Overview." In *Proceedings of the International Congress of Mathematicians, Zürich, Switzerland 1994.* 2 Vols. Basel: Birkhuser Verlag, 1995, 2:1581-1591.

______. "New Light on the Life and Work of Joseph Henry Maclagan Wedderburn (1882-1948)." In *Amphora: Festschrift für Hans Wussing zu seinem 65. Geburtstag.* Ed. Menso Folkerts *et al.* Basel/Boston/Berlin: Birkhäuser Verlag, 1992, pp. 523-537.

______. "The 100th Anniversary of Mathematics at the University of Chicago." *The Mathematical Intelligencer* **14** (2) (1992):39-44.

———. "A Study in Group Theory: Leonard Eugene Dickson's *Linear Groups.*" *The Mathematical Intelligencer* **13** (1) (1991):7-11.

Parshall, Karen H. and Rowe, David E. *The Emergence of an American Mathematical Research Community: J. J. Sylvester, Felix Klein, and E. H. Moore.* HMATH. Vol. 8. Providence: American Mathematical Society and London: London Mathematical Society, 1994.

Rossiter, Margaret W. *Women Scientists in America: Struggles and Strategies to 1940.* Baltimore: The Johns Hopkins University Press, 1982.

Shaw, James Byrnie. *Synopsis of Linear Associative Algebra.* Washington, D.C.: Carnegie Institution, 1907.

Siegmund-Schultze, Reinhard. "The Emancipation of Mathematical Research Publishing in the United States from German Dominance (1878-1945)." *Historia Mathematica* **24** (1997):135-166.

The University of Texas Record. Vol. 1, No. 3, August, 1899. University of Texas. The Center for American History. University of Texas Memorabilia Collection. James Benjamin Clark file.

Wall, Joseph. *Andrew Carnegie.* Pittsburgh: University of Pittsburgh Press, 1989.

Wedderburn, Joseph H. M. "A Theorem on Finite Algebras." *Transactions of the American Mathematical Society* **6** (1905):349-352.

CHAPTER 16

The Effects of Nazi Rule on the International Participation of German Mathematicians: An Overview and Two Case Studies

Reinhard Siegmund-Schultze
Adger University College (Norway)

Introduction

The new political conditions in Germany after the Nazi seizure of power on 30 January, 1933 had obvious and deep implications for science and mathematics. The shockwave of expulsions of mostly Jewish scholars from Germany gravely disturbed Germany's international scientific relations. The latter had started to recover around 1930; the return of Germans to the International Congress of Mathematicians (ICM) in Bologna in 1928 and in Zürich in 1932 were clear signs of an easing of former political tensions.[1] After 1933, however, relations with emigrants—for example, the mere fact of having been a student of emigrants and/or of "Jews"—mattered greatly. Such relations tended to hamper careers, especially in the early years from 1933 to 1935 when there were still positions to fill. The course towards economic autarky, the shortage of foreign currency, and, finally, the war in 1939, all created an atmosphere of increasing self-isolation that militated against the international participation of German mathematicians.

However, the political conditions in Germany after 1933 are only part of the full story. That would have to include the larger pattern of international relations in science at least since World War I. In fact, it was in 1933 that the aftereffects of that catastrophe—and of the resulting poisoning of international relations—came fully and dramatically to the fore.[2]

This chapter examines some of the factors affecting and influencing the involvement of German mathematicians on the international stage from 1933 through World War II. It opens with an overview of the complicated notion of internationalization and of its manifestation, particularly in Germany, after World War I. It then explores briefly—via the examples of the mathematical abstracting journal, *Zentralblatt*, and the International Applied Mechanics Congress held in 1938—the conditions in Germany for international mathematical collaboration after 1933. Stress

[1]Compare Olli Lehto, *Mathematics Without Borders: A History of the International Mathematical Union* (New York: Springer-Verlag, 1998) as well as the chapters by both Sanford Segal and Olli Lehto in the present volume. On the gloomy atmosphere of suspicion and denunciation that had reigned in the period of the first expulsions, see Heinrich Behnke, *Semesterberichte: Ein Leben an deutschen Universitäten im Wandel der Zeit* (Göttingen: Vandenhoeck & Ruprecht, 1978).

[2]This also applies, as is well-known, to the overall political development in Nazi Germany, which must be explained against the backdrop of World War I.

is laid, in particular, on the effects of the insurmountable dogma of antisemitism in German politics on that interaction. The analysis then moves to two case studies. The first deals with one central event of international mathematical communication in the 1930s, namely, the International Congress of Mathematicians held in Oslo in 1936, and with the reactions of German mathematicians to it as revealed in a report by Walter Lietzmann. The second briefly describes Nazi strategies to "restructure" European mathematics during World War II and the role of mathematicians, Harald Geppert and Wilhelm Süss, in that endeavor.[3] The chapter closes with tentative remarks about the effects—both political and cognitive—of the restriction of the international participation of German mathematicians in the 1930s and 1940s. Finally, in order to provide an orientation for the analysis, an appendix containing four informational tables is included. The first table gives the dates of decisive events that determined the international participation of German mathematicians. The next two provide short characterizations of leading German mathematicians remaining in the country and their involvement in international communication. The fourth lists important international conferences and congresses in which German mathematicians took part during the Nazi era.

Internationalization: A Complex of Factors

In a letter to the Danish mathematician, Harald Bohr, written in June 1934, the Secretary of the American Mathematical Society (AMS), Roland G. D. Richardson (1878–1949), described the rise of national sentiment that had resulted in the wake of the First World War. In 1934, as refugees fled Europe, Richardson was still undecided whether to support or to contain the immigration of German mathematicians to the United States. As he put it, "[s]ince the war, we have been constantly compelled to think of colleagues as nationals and not as citizens in the international domain."[4] Paul Forman noted a similar attitude with respect to physicists in his study of the Weimar Republic:

> [T]he peculiarly interesting feature of the period following World War I is that while in contrast to the war years the political contribution of science was once again measured primarily in terms of prestige, with formal allegiance to the ideology of scientific internationalism, the German scientists—and in some measure the Allied as well—no longer conceived of their political role in the classical passive terms. Rather, they regarded themselves as agents, or even as bearers, of the foreign policy interests of their nation and as such were often obliged to sacrifice the interests

---

[3]While the first case study relies almost exclusively on unpublished material, the second draws from work already published in German. See, especially, Reinhard Siegmund-Schultze, "Faschistische Pläne zur 'Neuordnung' der europäischen Wissenschaft: Das Beispiel Mathematik," *NTM–Schriftenreihe für Geschichte der Naturwissenschaften, Technik und Medizin* **23** (2) (1986):1-17, and *Mathematische Berichterstattung in Hitlerdeutschland: Der Niedergang des Jahrbuchs über die Fortschritte der Mathematik (1869–1945)* (Göttingen: Vandenhoeck & Ruprecht, 1993).

[4]Roland G. D. Richardson to Harald Bohr, 26 June, 1934, R. G. D. Richardson Papers, Brown University Archives, Box Correspondence 1933 (German-Jewish Situation), f.:H. Bohr. Ultimately, Richardson became instrumental in helping immigrants who would figure prominently in his center for applied mathematics at Brown University. Compare Nathan Reingold, "Refugee Mathematicians in the United States of America, 1933–1941," *Annals of Science* **38** (1981):313-338.

> of German *science*, and their personal interests as scientists, for the sake of patriotic political posturing.[5]

Under the even more politicized conditions of the dictatorial regime of the Nazis, racist ideology occasionally intruded into what Forman called "the leading proposition in the ideology of scientific internationalism: the assertion of the universality of science."[6] To be sure, derivatives of Nazi racism such as *Deutsche Physik* and *Deutsche Mathematik* did not have much appeal to either scientists or politicians in Germany after the first waves of expulsions, since they were not in accordance with their long-term interests.[7] But even then, the prevailing "formal allegiance to the ideology of scientific internationalism"[8] left ample room for interpretation, depending on the actual power structure in the particular field of science or society. "Internationalism" could reveal a potential for resistance against the regime, as the tardy process of a political "coordination" (*Gleichschaltung*) of the *Deutsche Mathematiker-Vereinigung* (DMV) would show.[9] "Internationalism" could likewise be used for propaganda purposes, for example, taking a "German responsibility" for international science by urging a collaboration with French mathematicians during the German occupation of France.[10]

In addition to the political conditions stemming from World War I or originating in 1933, the material, technical, and cognitive infrastructure of international mathematical communication, the state of the 1920s, and the new developments in the 1930s must all be taken into account in order fully to appreciate the degree of international participation of Germans during Nazi rule. As of the late 1920s and early 1930s, mathematics in Germany had been the most "internationalized" of all national mathematical cultures in the world. This was true relative to a number of metrics: the nationality and origin of mathematicians teaching and studying at German universities; the number of German mathematicians sent abroad either as postgraduate students via, for example, Rockefeller's International Education Board or as guest professors like Wilhelm Süss, Richard Courant, and Wilhelm Blaschke;[11] national origins of authors of articles in German journals; the international importance of the German publication system in mathematics; and the variety of topics discussed.

Still, the degree of internationalization in mathematics was relative and restricted, even in Germany. There were clear distinctions between the internationalized mathematics at Göttingen University, which had received a new building from the American International Education Board in 1926, and the relatively more self-contained mathematics in Berlin and especially elsewhere in Germany. There were

---

[5]Paul Forman, "Scientific Internationalism and the Weimar Physicists: The Ideology and Its Manipulation in Germany after World War I," *Isis* **64** (1973):151-180 on p. 156 (his emphasis).

[6]*Ibid.*

[7]To substantiate this assertion, compare Herbert Mehrtens, "Ludwig Bieberbach and 'Deutsche Mathematik'," in *Studies in the History of Mathematics*, ed. Esther Phillips, MAA Studies in Mathematics, vol. 26 (Washington, D.C.: Mathematical Association of America, 1987), pp. 195-241, and Siegmund-Schultze, *Mathematische Berichterstattung*.

[8]Forman, p. 156.

[9]Herbert Mehrtens, "The Gleichschaltung of Mathematical Societies in Nazi Germany," *The Mathematical Intelligencer* **11** (3) (1989):48-60.

[10]See Case Study 2 on Geppert and Süss below. On the shades of meaning of the term "internationalism," recall the chapter by Parshall and Rice above.

[11]On Blaschke, in particular, see Xu Yibao's contribution to the present volume.

also thematic restrictions in German mathematics in fields such as topology, functional analysis,[12] and stochastics that resulted partly from traditions, but that were partly reinforced by the political alienation between countries after World War I. There is no doubt, for example, that anti-Polish resentments supported the partial abstention of the Germans from functional analysis and some aspects of foundational research. There was decidedly less reserve in this respect in Vienna, where a completely different political and philosophical atmosphere encouraged the collaboration of logical positivists and mathematicians and supported contacts with Polish scientists such as Alfred Tarski and Waclaw Sierpinski.[13] Relative to topology, the initiatives of individual mathematicians prevented a total international isolation of German mathematicians before 1933. For example, the Austrian-born geometer, Wilhelm Blaschke (1885–1962) of Hamburg, sent several of his students to Vienna, while mathematicians like Courant, Hilbert, and Emmy Noether in Göttingen invited Russians such as Pavel Aleksandrov. Still, one of the most brilliant young German topologists, Heinz Hopf, left for Switzerland in 1931.

The 1920s and 1930s also witnessed some rather new phenomena relative to the structures for the international communication of mathematics: specialized journals in, for example, statistics, number theory, and logic[14] and improvements in mathematical reviewing as evidenced by the *Zentralblatt für Mathematik*.[15] Nevertheless, the material conditions for individual mathematicians worldwide in the 1930s had not improved considerably compared to the preceding decade. If anything, overall conditions worsened. There were still few positions for assistants in the universities, various barriers for the appointment of foreigners were in place, and academic unemployment was growing. The latter was true even in the United States, despite the existence since 1923 of National Research Council fellowships.

If financial concerns militated against international contacts between mathematicians so, too, did the fact that there was, of course, no regular international civil air traffic yet. The participation of German mathematicians and other scientists in international conferences, which was supervised by the *Deutsche Kongreßzentrale* of Goebbels's propaganda ministry,[16] was heavily restricted by both political concerns and the shortage of foreign currency. What further complicated the situation for the Germans, however, was the fact that the Rockefeller fellowship program sponsored by the International Education Board, a program that had been so important in the late 1920s, came to a halt in the early 1930s. The German mathematician, Georg Aumann, was one of the last foreign Rockefeller fellows. After spending the 1934–1935 academic year in the United States, he compared the situation in

---

[12]Reinhard Siegmund-Schultze, "Die Anfänge der Funktionalanalysis und ihr Platz im Umwälzungsprozeß der Mathematik um 1900," *Archive for History of Exact Sciences* **26** (1982):13-71.

[13]Karl Menger, *Reminiscences of the Vienna Circle and the Mathematical Colloquium*, ed. Louise Golland, Brian McGuinness, and Abe Sklar (Dordrecht: Kluwer Academic Publishers, 1994), pp. 143ff.

[14]The Polish journals for the foundations of mathematics, *Fundamenta Mathematicæ*, and for functional analysis, *Studia Mathematica*, had been founded in 1920 and 1929, respectively.

[15]Reinhard Siegmund-Schultze, "The Emancipation of Mathematical Research Publishing in the United States from German Dominance (1878–1945)," *Historia Mathematica* **24** (1997):135-166.

[16]See the as yet almost unexplored materials of the *Deutsche Kongreßzentrale*, which was founded in December 1934. These materials are now housed at the Hoover Institution at Stanford University.

Germany in 1938 unfavorably to what he had experienced abroad. In his view, "[t]here are plenty of fellowships available in the United States with ideal conditions for research for younger men and, in addition, many guest lectures which are the best instrument to keep research going."[17] Beginning in May 1933, the Rockefeller Foundation had reoriented its support to emergency programs aimed at taking care of refugees fleeing Europe.[18] This exodus was, to be sure, a source of a tremendous upsurge in the internationalization of mathematics, especially in the sense of new and unexpected personal encounters and oral communication. Still, this type of internationalization was shaped in a peculiar way by emigration patterns. It was not necessarily healthy or natural when compared to the secular, long-term internationalization of mathematics that had been well under way in the decades before. Without entering into the foggy field of counterfactual history, it is important to focus on the *losses* for the various national cultures in mathematics in Europe that were brought about by the expulsions not just in Germany but also in other countries such as Poland, Hungary, and Austria. These losses were more than the sufferings of the refugees and the deaths of the victims.[19]

Germany Immediately after 1933: The Dogma of Antisemitism

When the explusions started in April 1933, several mathematicians such as Wilhelm Blaschke[20] and, most notoriously, Ludwig Bieberbach (1886–1982)[21] defended them. Others, like Helmut Hasse (1898–1979), who had recently been called to Göttingen as successor to emigré, Hermann Weyl, became overly anxious. When Hasse was asked by the Polish-Jewish mathematician, Arnold Walfisz, in a letter dated 4 June, 1934 whether he was willing to join the board of the new number-theoretic journal, *Acta Arithmetica*, Hasse deemed it wise to ask permission of the authorities.[22] He wrote to the *Kurator* of Göttingen University, the person responsible for contacts between the University and the Ministry of Education. After indicating his interest in joining the board, Hasse added, "[a]s to the factual matter I remark that the three managing editors are Polish mathematicians, two of them (Walfisz and Dickstein) presumably non-Aryans. Among the members of the editorial board there are two dismissed German professors, Landau and Rademacher, the first a Jew, the second not. Furthermore, certainly Mordell and presumably Ostrowski are Jews, all others are foreigners in any case."[23] Needless to say, the

[17] "durch Gewährung zahlreicher Stipendien mit idealen Forschungsbedingungen ein großzügige Förderung des wissenschaftlichen Nachwuchses und darüber hinaus die Veranstaltung vieler ... Gastvorträge, die ja das beste Mittel sind, die Forschung in Fluß zu halten." Siegmund-Schultze, *Mathematische Berichterstattung*, p. 141.

[18] Reinhard Siegmund-Schultze, *Rockefeller and the Internationalization of Mathematics Between the World Wars* (Basel: Birkhäuser Verlag, 2001).

[19] For more details, see Reinhard Siegmund-Schultze, *Mathematiker auf der Flucht vor Hitler: Quellen und Studien zur Emigration einer Wissenschaft* (Braunschweig: Vieweg Verlag, 1998).

[20] See Blaschke's correspondence with Oswald Veblen, in the Oswald Veblen Papers, Library of Congress, Washington, D.C.

[21] Compare Mehrtens, "Ludwig Bieberbach."

[22] Personal file H. Hasse, Göttingen University Archives, K XVI.V.Aa 53.

[23] "Zur Sache bemerke ich, daß die drei geschäftsführenden Redakteure polnische Mathematiker sind, von denen vermutlich zwei (Walfisz und Dickstein) nicht-arischer Herkunft sind. Unter den in Aussicht genommenen Mitgliedern des Redaktionsausschusses sind Landau und Rademacher entlassene deutsche Professoren, der erste Jude, der letztere nicht, ferner bestimmt Mordell und vermutlich auch Ostrowski Jude, die übrigen jedenfalls Ausländer." *Ibid.*, 5 June, 1934. (All translations from the German are the author's own.)

responsible functionary in the Nazi Ministry of Education, mathematician Theodor Vahlen, declined permission.[24] Even *Compositio Mathematica*, which was edited in Holland by Bieberbach's ally in the German-nationalist spirit, L. E. J. Brouwer, was not acceptable for "real" Germans after its appearance in 1935; its board contained expelled Germans, several of whom were "Jews" according to the arbitrary Nazi definition.[25]

Of course, many German mathematicians kept in contact with their expelled colleagues, even if much of their communication was overshadowed by misunderstandings, disappointment, and self-censorship. Letters were frequently sent during occasional trips abroad. Heinrich Behnke (1898–1979), for example, received and accepted many invitations, especially to France and to Switzerland, and used these opportunities to pass on correspondence.[26] There was also extensive correspondence between German mathematicians and emigré, Richard Courant (1888–1972), first from Cambridge, England and, after 1934, from New York City. In particular, Courant continued his collaboration with Kurt Friedrichs, who was in Braunschweig, on the second volume of *Methods of Mathematical Physics* (otherwise known as *Courant-Hilbert*). The volume eventually appeared in 1937.

There was, however, early Nazi interference into international work on mathematical publications, as in the case of the second edition of the *Enzyklopädie der mathematischen Wissenschaften* edited by B. G. Teubner Verlag in Leipzig.[27] Moreover, Bieberbach's partly racist journal, *Deutsche Mathematik*, appeared in 1936 in Gothic type, which certainly contributed to its international isolation.[28] Still, the Springer-dominated mathematical publication system—and especially the mathematical reviewing by *Zentralblatt*, which was edited from Copenhagen beginning in 1934 by the emigré, Otto Neugebauer (1899–1990)—secured a considerable amount of international contacts for German mathematicians at least until 1938.[29]

If the communication structure of mathematics within Germany stabilized after the shockwave of expulsions, international communication, especially with Western countries and the Soviet Union, deteriorated in the years to come. Two main

---

[24] *Ibid.*, 18 July, 1934. On Vahlen, see Reinhard Siegmund-Schultze, "Theodor Vahlen: Zum Schuldanteil eines deutschen Mathematikers am faschistischen Mißbrauch der Wissenschaft," *NTM-Schriftreihe für Geschichte der Naturwissenschaften, Technik und Medizin* **21** (1) (1984):17-32.

[25] Compare Mehrtens, "Ludwig Bieberbach," pp. 220-221.

[26] See Behnke, *Semesterberichte*, p. 135.

[27] Renate Tobies, "Mathematik als Bestandteil der Kultur: Zur Geschichte des Unternehmens 'Enzyklopädie der mathematischen Wissenschaften mit Einschluß ihrer Anwendungen'," *Mitteilungen österreichische Gesellschaft für Wissenschaftsgeschichte* **14** (1994):1-90 on pp. 52-56. Richard Brauer was forced to withdraw his contribution after the Nazi intervention of 1935.

[28] It was only in the war years that the publisher changed the print, when, incidentally, even schools in Nazi Germany discontinued the teaching of non-Latin letters. The international reception of the work of Oswald Teichmüller in function theory, which was to a considerable extent published in *Deutsche Mathematik*, was undoubtedly hampered by its place of publication.

[29] Michael Knoche, "Wissenschaftliche Zeitschriften im nationalsozialistischen Deutschland," in *Von Göschen bis Rowohlt: Beiträge zur Geschichte des deutschen Verlagswesens*, ed. Monika Estermann and Michael Knoche (Wiesbaden: Harrassowitz, 1990), pp. 260-281; Reinhard Siegmund-Schultze, "'Scientific Control' in Mathematical Reviewing and German-U.S.-American Relations Between the Two World Wars," *Historia Mathematica* **21** (1994):306-329; and Siegmund-Schultze, "The Emancipation of Mathematical Research Publishing."

reasons, which were, in a certain sense, intertwined contributed to this: the insurmountable dogma of antisemitism and the preparation for war. The affair surrounding the dismissal of the Italian-Jewish editor of the *Zentralblatt*, Tullio Levi-Civita, in 1938 and the ensuing foundation of the *Mathematical Reviews* in the United States have been much discussed.[30] Suffice it to say, here, that even mathematicians like Helmut Hasse, who were no old comrades of the Nazis, supported the Nazi measure to discontinue reviewing the work of "German authors" by Jewish mathematicians.[31] Like Bieberbach in 1934,[32] Hasse promoted an apartheid-notion of internationalism. Consider his letter to the American, Marshall Harvey Stone, on 15 March, 1939:

> Looking at the situation from a practical point of view, one must admit that there is a state of war between the Germans and the Jews. Given this, it seems to me absolutely reasonable and highly sensible that an attempt was made to separate within the domain of the Zentralblatt the members of the two opposite sides of this war. I do not understand why the American mathematicians found it necessary there on [sic] to withdraw their collaboration in bulk. I do not know whether it was the intention, but it certainly has the appearance of taking decidedly and emphatically one of the two sides, and thus deviating from a truly impartial and hence genuinely international course.[33]

Similarly, international congresses for pure and applied mathematics were increasingly hampered by the Nazi dogma of antisemitism. There had already been plans in 1934 to invite the International Congress for Applied Mechanics to Germany. In preparation for the 1938 congress, which ultimately took place in the United States, Ludwig Prandtl (1875–1953) of Göttingen had explored the question of whether a German invitation had a chance. He had the complete support of Göring's Ministry of Aviation in Germany, and mechanics scientists internationally did not appear averse to coming to Germany, provided no distinction was made between Jews and non-Jews. Despite intensive efforts, however, Prandtl could not guarantee the latter condition. The Reich's Ministry of Education, which was responsible for such matters, communicated to Prandtl on 20 August, 1938 that "Jews of foreign citizenship" who took part in the congress would not be regarded "as Jews here," but "German science has to be represented at international congresses by German scientists and not by Jewish ones."[34] The decision also destroyed the hope for a congress in Germany in 1942, which, due to the war, would not have taken place anyway.

---

[30]See Reingold and Siegmund-Schultze, *Mathematische Berichterstattung*.

[31]On Hasse's behavior in the Third Reich, see Sanford Segal, "Helmut Hasse in 1934," *Historia Mathematica* **7** (1980):46-56.

[32]In his article, "The Gleichschaltung of Mathematical Societies in Nazi Germany," Mehrtens described Bieberbach's attack on the Danish mathematician, Harald Bohr, as a "damager of all international cooperative work." See p. 52.

[33]Siegmund-Schultze, *Mathematische Berichterstattung*, p. 164.

[34]Mehrtens, "The Gleichschaltung of Mathematical Societies in Nazi Germany," p. 57.

Case Study 1: German Participation in the Oslo ICM, July 1936

German preparations for the International Congress of Mathematicians in Oslo were overshadowed by political caution on the part of the Ministry of Education and by shortages of foreign currency.[35] The first problem was to find an appropriate leader (*Führer*) for the German delegation. The official responsible for making this decision, Rudolf Mentzel, remarked in November 1935 that "[t]here are no old comrades of the party among the proposed mathematicians. Lietzmann appears the most adroit to me. He is, however, only a high school teacher. Nevertheless, I would propose him as the Führer of the delegation."[36] Lietzmann, the Göttingen professor of mathematical pedagogy and the German representative to the International Commission on Mathematical Education (ICME), seems to have divided his job in Oslo with Erhard Schmidt of Berlin. The latter was then the President of the DMV and represented the Germans on the mathematical side. His political naïveté, however, was apparent in his introductory address before the Congress. Schmidt recounted to the assemblage that "[w]hen I left Berlin, the air was filled with jubilations about the forthcoming Olympic Games. Everybody was prepared for a solemn reception of the gathering nations and for a celebration of the performance of their peaceful competition."[37] His remark was soon criticized by the emigré, Emil Gumbel, in a collection published in 1938 advocating freedom in science.[38] Still, Lietzmann and Schmidt tried to show due respect for their host nation. As Lietzmann reported to the Nazi Ministry, they laid a wreath at the tomb of Sophus Lie in the presence of Lie's family, and the Norwegian relatives, together with the German mathematicians, raised their hands in the German salute.[39] The shortage of Norwegian crowns led both to a reduction of the size of the delegation and of the means available for each participant (185 Reichmark instead of the 300 originally planned).

The following twenty-two German mathematicians formed the official delegation in Oslo: Behnke, Blaschke, Constantin Carathéodory, Harald Geppert, Helmut Grunsky, Gerhard Haenzel, Georg Hamel, Hasse, Erich Hecke, Paul Koebe, Gottfried Köthe, Lietzmann, Ernst Peschl, Paul Riebesell, Schmidt, Herbert Seifert, Carl Ludwig Siegel, Ernst Sperner, Egon Ullrich, Kurt Vogel, Aloys Timpe, and Karl

---

[35] The main sources for the material that follows are the official congress reports, *Comptes rendus du Congrès international des mathématiciens, Oslo 1936*, 2 vols. (Oslo: A. W. Brøggers Boktrykkeri A/S, 1937); unpublished material on the German participation in the Bundesarchiv Berlin, Reichserziehungsministerium (BAB-REM), 2905, particularly fol. 384-398, viz., Lietzmann's report to the Ministry, undated, August 1936; and the Lietzmann Papers in the Municipal Archives in Göttingen = Stadtarchiv Göttingen, Depositum 89 [MAG-WL]. The latter show that Lietzmann's report was informed, at least with respect to the mathematical part, by written contributions from several participants. On the Oslo Congress, compare also Sanford Segal's chapter in the present volume.

[36] "Alte Nationalsozialisten sind unter den Genannten nicht. Am gewandtesten erscheint mir Lietzmann. Dieser ist aber Gymnasiallehrer. Trotzdem würde ich ihn zum Führer der Delegation vorschlagen." BAB-REM 2905, fol. 54.

[37] "Als ich Berlin verließ, hallte und schallte dort alles von den freudigen Vorbereitungen für die Olympischen Festspiele. Alles schickte sich an, die zusammenströmenden Völker würdig zu empfangen und das Schauspiel ihres friedlichen Wettbewerbs zu feiern." *Comptes rendus ... Oslo*, 1:52.

[38] Emil J. Gumbel, ed., *Freie Wissenschaft: Ein Sammelbuch aus der deutschen Emigration* (Strasbourg: Sebastian Brant, 1938), p. 253.

[39] "erhoben mit uns drei Vertretern der Delegation die Hand zum deutschen Gruß." BAB-REM 2905, fol. 385.

Reinhardt. Lietzmann wrote in his report that he had observed several German mathematicians in Oslo who did not belong to the delegation, among them Max Dehn and Erich Reissner. The latter were considered Jewish in the Third Reich and had obviously financed their own trips. Dutifully, Lietzmann also mentioned several emigrants among the participants: Courant, Weyl, Neugebauer, Fritz Noether, Abraham Fraenkel, and, as Lietzmann called him, "the infamous Herr Gumbel."[40] The Italians were also absent due to Mussolini's negative reaction to political sanctions that followed in the wake of his Fascist occupation of Abbysinia (modern-day Ethiopia). France was represented by few, but influential, mathematicians: Émile Borel, Élie Cartan, and Gaston Julia.

In his report, Lietzmann described the American delegation as "very strong." The strong participation of the Americans may have resulted in part from political concerns. Over a year before the Congress, AMS Secretary Richardson had written to Stanford's Hans Blichfeldt: "I understand that some of the people in Scandinavia are concerned about the possible large Nazi contingent and eager to have Americans and British in considerable numbers to offset in order to keep politics and racial questions out of our meetings."[41] The very fact that Lietzmann had to report to the Ministry can be interpreted as a slight criticism of ideological interference into mathematics in Germany. As he put it, "[t]here is one thing that foreigners have a real problem understanding: our notion of the national pecularity of science, regardless of the universal validity of its results."[42]

Lietzmann commented likewise on the issue of the languages used at the Congress. "If the three official languages [English, French, and German] were enjoying the same rights," he reported, " ... German as a language came obviously more to the fore than originally planned. Of course, all of the members of the German delegation were obliged to speak German even in discussions."[43] He had to admit, however, that "today in Norway, English is becoming increasingly dominant over German."[44]

The language that most dominated Lietzmann's report was mathematics itself, and he was careful to point out his perception of Germany's place on the international research scene. Alluding to the three invited lectures by the Germans, Hasse, Hecke, and Siegel, Lietzmann remarked that "[t]he leading position of Germany in number-theoretic research has been maintained from Gauss to the present."[45] Referring specifically to the Congress's section on geometry, he further observed that

---

[40] "der berüchtigte Herr Gumbel." *Ibid.*

[41] Roland G. D. Richardson to Hans Blichfeldt, 12 February, 1935, Richardson Papers, Brown University Archives, Box Correspondence 1933 (German-Jewish Situation), f.:Miscellaneous Letters.

[42] "Eines ist allerdings für Ausländer ausserordentlich schwer zu begreifen: unsere Auffassung von der nationalen Eigenart einer Wissenschaft, unbeschadet der internationalen Bedeutung jeder Forschung." BAB-REM 2905, fol. 387.

[43] "Wenn auch bei den offiziellen Verlautbarungen all drei Sprachen gleichwertig auftraten ... trat doch die deutsche Sprache stärker, als ursprünglich beabsichtigt, in den Vordergrund. Selbstverständlich waren alle Mitglieder der deutschen Delegation gehalten, auch in der Diskussion deutsch zu sprechen." *Ibid.*, fol. 388.

[44] "weil sonst gerade in Norwegen heute das Englische gegenüber Deutsch stark in den Vordergrund tritt." *Ibid.*, fol. 394.

[45] "Die führende Stellung Deutschlands in der zahlentheoretischen Forschung ist von Gauss bis auf den heutigen Tag erhalten geblieben." *Ibid.*, fol. 389.

"topology was clearly dominant ... the absence of the Russians causing serious gaps."[46]

Lietzmann's remarks on the various talks in the mathematical sections also contained many interesting observations on the state of international mathematics, several of them obviously with overtones implicitly criticizing the actual state of official contacts. He reported that "at a social gathering it was stressed, relative to the Polish mathematician S[tefan] Banach (Lemberg) who was in attendance at the Congress, how admirable it was that Poland had developed such a broad and strong mathematical school of such a particular orientation in so short a time."[47] Lietzmann raised another politically sensitive point when he noted that "Professor Behnke (Münster) lectured impressively and lucidly on the theory of functions of several variables: this field has been cultivated for a decade in a friendly and exemplary manner in collaboration between German and French mathematicians."[48] Obviously to stress his political reliability in matters Polish and French, however, Lietzmann hastened to add that

> It would be very desirable indeed to cultivate close scientific relations with the southeastern European countries. It is remarkable that the French—obviously quite deliberately—are making 'scientific cultural propaganda' toward the Balkans. The French give regular lectures in that region and attend congresses in the Balkans. On the German side, we have neglected this so far (except for a trip by Herr Blaschke [Hamburg]). Perhaps they consider us too proud to go there.[49]

Lietzmann concluded these observations with one final judgment relative to the need to promote contacts with the Poles that was colored by the political situation in Germany. "In a similar way," he stated, "we should improve our contacts with Poland which is still largely oriented towards France relative to science. However, certain problems may arise, since many Polish mathematicians seem to be non-Aryans."[50]

---

[46] "Sachlich herrschte in den Arbeiten die Topologie weit vor ... das Fehlen der Russen bedingte wesentliche Lücken." *Ibid.*, fol. 393-394.

[47] "Bei einer gesellschaftlichen Zusammenkunft wurde anlässlich der Anwesenheit des polnischen Mathematikers S. Banach–Lemberg hervorgehoben, wie hoch es anzuerkennen sei, dass in Polen in so kurzer Zeit eine so grosse und leistungsreiche mathematische Schule einer ganz spezifischen Ausprägung entstanden sei." *Ibid.*, fol. 391.

[48] "Ferner sprach Prof. Behnke–Münster in einem einprägsamen und durchsichtigen Vortrag über Funktionen mehrerer komplexer Veränderlichen; dieses Gebiet wird seit einem Jahrzehnt in vorbildlicher, freundschaftlicher Zusammenarbeit zwischen deutschen und französischen Mathematikern bearbeitet." *Ibid.*, fol. 392.

[49] "Es mag bei dieser Gelegenheit hervorgehoben werden, wie wichtig es wäre, engere wissenschaftliche Beziehungen zu den südosteuropäischen Ländern zu pflegen. Es fällt auf, dass von französischer Seite—offenbar mit Bedacht—'wissenschaftliche Kulturpropaganda' nach den Balkanstaaten hin gemacht wird. Franzosen halten regelmässig Gastvorträge in jener Gegend und besuchen die Balkanischen Kongresse; von deutscher Seite wurde dem bisher keine Beachtung geschenkt (abgesehen von einer Reise des Herrn Blaschke–Hamburg), und man hält uns—vielleicht—für zu stolz, dorthin zu gehen." *Ibid.*, fol. 392-393.

[50] "Ähnlich müssten–wie mir scheint—unsere Verbindungen zu Polen verbessert werden, das wissenschaftlich zur Zeit noch ganz französisch eingestellt scheint. Freilich ergeben sich da besondere Schwierigkeiten, da offenbar viele polnische Mathematiker Nichtarier sind." *Ibid.*, fol. 393.

Case Study 2: WWII and German International Participation: Harald Geppert and Wilhelm Süss

Harald Geppert (1902–1945) was a second-rate differential geometer from Giessen, who had an Italian mother and wide-ranging international contacts partly due to the Rockfeller fellowship he took to Rome in 1928–1929. Politically active for the Nazis very soon after 1933, in 1938 he published, together with the statistician, Siegfried Koller, a book entitled *Hereditary Mathematics* (*Erbmathematik*) that was partly tainted by National Socialist racism. The Geppert/Koller book gave Berlin mathematician and fanatical Nazi, Ludwig Bieberbach, the chance to call Geppert to Berlin, purportedly to cultivate the "mathematical problems most important for our future."[51] Bieberbach's real intention, however, was to make Geppert the general editor of both German mathematical reviewing journals, the *Zentralblatt* and the *Jahrbuch über die Fortschritte der Mathematik*, in 1940. After the occupation of France by the Nazis in June 1940, Geppert offered to use reviewing as a nucleus for European mathematical collaboration under the German lead. On 12 November, 1940, Geppert represented the field of mathematics at a meeting of the Nazi Ministry of Education for the "new foundation of international scientific relations." Geppert proposed to "cooperate with the Academies and give them as an international task to review and control scientific literature, since they have had the most experience in this field for sixty to seventy years."[52]

In December of the same year, Geppert was sent on a trip to Paris to inquire about the *Institut Henri Poincaré*, which, it was thought, would continue the job of the International Mathematical Union (IMU) under French dominance. This was, of course, nonsense, since the IMU was long dead.[53] The former head of the IMU, Émile Picard, was very old (he died one year later in 1941) and had little influence at the *Institut Henri Poincaré* directed by Émile Borel. In his report on his Paris trip, Geppert stated that there was no political danger coming from the Institute, but he proposed to neutralize certain French mathematical journals by founding parallel German publications.[54] Geppert envisioned "a restoration of international mathematical communication in terms of smaller international meetings in special disciplines, which would be most appropriately organized by the DMV, preferably in connection with a larger institute. The financial means should be provided by the Ministry of Education."[55] As this suggestion makes clear, the Oberwolfach plan was already looming large, at least in the minds of mathematicians.

Geppert also used the trip to Paris to reach out to French mathematicians who might be willing to collaborate with the Germans. He met the French analyst, Gaston Julia (1893–1978), and tried to lure him into collaboration using Julia's

---

[51] "zukunftswichtigste Probleme." Siegmund-Schultze, "Faschistische Pläne," p. 6.

[52] "mit den Akademien zusammenzuarbeiten und ihnen als internationale Aufgabe die Literaturbesprechung und überwachung zu übertragen, da sie seit 60/70 Jahren über die größten Erfahrungen und Erfolge auf diesem Gebiete verfügen." Siegmund-Schultze, *Mathematische Berichterstattung*, p. 178.

[53] Lehto, *Mathematics Without Borders*. Compare, too, Lehto's chapter in the present volume on the early history of the IMU.

[54] "durch Gründung entsprechender deutscher Parallelunternehmen unschädlich gemacht werden." Siegmund-Schultze, *Mathematische Berichterstattung*, p. 183. Geppert particularly stressed the need for an equivalent to the short reports in the *Comptes rendus des séances de l'Académie des Sciences de Paris*, which did not exist in Germany at this time.

[55] *Ibid.*, p. 180.

longstanding relationship with the *Zentralblatt* and the fact that Julia's brother was a prisoner of war in Germany.[56] At about the same time, Helmut Hasse was in Paris as an officer in the German Navy, and he also tried to use his old connections with Julia as well as the latter's convictions that England had betrayed France and that the future of France was with that of Germany.[57] Relative to reviewing, Geppert and Hasse succeeded in persuading several French prisoners of war to cooperate. A few of them, like Frédéric Roger and Christian Pauc, were also prepared to leave the camps and become civil collaborators first in Berlin in Geppert's office for mathematical reviewing and later at several universities. Pauc, for example, went to the university in Erlangen. For some, such as the Alsacian Charles Pisot, this conduct caused political problems in France after the war. Others, like Jean Favard and the noted analyst, Jean Leray, did write mathematical reviews for Geppert, but insisted on staying in the camps. This unwillingness to collaborate formally with the Germans was not without danger for them.[58]

Geppert and Wilhelm Süss (1895–1958), the latter the President of the DMV during the war years, became spokesmen for German mathematics at the Ministry in matters of international relations. Süss prepared the foundation of the institute in Oberwolfach in 1944, to which Geppert had alluded in his report to the Ministry on his Paris trip,[59] and organized war research. Several mathematicians from southeastern European and from Scandinavian countries were involved in the German attempts to remodel European science in the years to come.[60] Others refused, and some, like Émile Borel in Paris, even went to jail.

In March 1942, both Geppert and Süss took part in a meeting at the Ministry of Education that aimed to prepare the overall strategies of German science and mathematics for the "restructuring" of European science, since "it is very likely that after the War, world science will split into one of the European continent and another of the American continent."[61] The mathematicians also mused whether America might take over the lead in the IMU after the war, and they were curiously not far from the truth.[62] The history of mathematical communication, and especially of the IMU and the boycott against German science in the 1920s, were likewise major points in what was a rather propagandistic discussion. Cooperation and competition with Italy, which was more advanced than Germany with respect to mathematical institutes, were discussed as well.

When the German mathematicians gathered in Berlin in March 1942, Nazi Germany was at the peak of its international power. As a consequence of the German military defeats in the following years, especially the decisive one of Stalingrad

[56] *Ibid.*, p. 185.

[57] *Ibid.*, p. 187.

[58] *Ibid*, p. 189. On Leray as a prisoner in a German camp, see Reinhard Siegmund-Schultze, "An Autobiographical Document (1953) by Jean Leray on His Time as Rector of the 'Université en captivité' and Prisoner of War in Austria, 1940–1945," *Gazette des mathématiciens* (*Supplément*) **84** (2000):11-15.

[59] On Oberwolfach, see Marjorie Senechal, "Oberwolfach, 1944–1964," *The Mathematical Intelligencer* **20** (4) (1998):17-24.

[60] For example, Rolf Nevanlinna gave a friendly radio address when he visited Jena in 1941, and Julia visited Berlin in 1942. Compare Table 4 in the Appendix.

[61] "damit zu rechnen ist, daß nach diesem Kriege die Wissenschaft der Welt in eine solche des eurpäischen Festlandes und des amerikanischen Kontinents auseinanderfällt." Siegmund-Schultze, *Mathematische Berichterstattung*, p. 181. See also Siegmund-Schultze, "Faschistische Pläne." The German term for "restructuring" was "*Neuordnung*."

[62] Siegmund-Schultze, "Faschistische Pläne," p. 10.

early in 1943, all plans for German dominance in science finally—and fortunately—collapsed. Süss gave a rather sober speech at the conference of German university rectors in Salzburg in August 1943, in which he stated that even in the number of *mathematical* publications, Germany had been outdistanced by the Americans. "From the totality of all the world's mathematical journals in 1937," he went on, "we have counted the citations and ordered them according to countries and years. Of all works written until 1870 and cited in 1937, 46% of the papers are by German authors, 20% by the English, and just 1% by the Americans. For the years 1931–1935, the figures are 28%, 13%, and 25% [respectively], and meanwhile the development has presumably become even worse for us."[63]

Physicist, Carl Ramsauer, made similar observations relative to physics. In a letter to Süss, he noted that Süss's results refuted the Nazi fiction of the alleged "qualitative superiority of German science" over allied science. After all, mathematics had generally been considered the quintessential, theoretical, "qualitative science," in the sense that it could not be easily promoted by simple material investment in personnel or technical equipment. Thus, Ramsauer felt compelled to acknowledge that "I find your results most important for my own activities at the moment. The situation of German mathematics may not be considered as important for the leading authorities as is the situation of physics. However, a surpassing of German mathematics by the Americans might be even more impressive as a general symptom."[64]

One year later, Süss had to acknowledge that he lacked any contacts with allied science whatsoever. In a letter to the German Research Council dated 4 July, 1944, he wrote that "[u]nfortunately, I do not know any titles of mathematical papers in American and English journals of recent years. So I would greatly welcome photocopies of the tables of contents of those journals for the years 1943 and 1944."[65] The ardent Nazi, Geppert, committed suicide in 1945 together with his family. Süss, who, although a member of the Nazi party, had not been a fanatical Nazi, served as director of the mathematical institute in Oberwolfach after the war. This became a center of restoration of international contacts for German mathematicians.

Conclusions

Nazi rule profoundly affected the ability of mathematicians working within Nazi Germany to engage in international communication after 1933. On the one hand, there were restrictions due to political isolation and militarization, for example,

---

[63] "Aus der Gesamtheit der mathematischen Fachzeitschriften der Welt des Jahrgangs 1937 ... sind alle Zitate ausgezählt und nach Ländern und Jahren geordnet worden. Von allen 1937 zitierten Arbeiten aus den Jahren bis 1870 entfallen z.B. 46% auf Arbeiten deutscher Verfasser, 20% auf englische und nur 1% auf amerikanische Arbeiten. Für 1931–1935 sind die Zahlen 28, 13 und 25 und inzwischen wird die Entwicklung für uns noch ungünstiger geworden sein." Siegmund-Schultze, *Mathematische Berichterstattung*, p. 5.

[64] "Ich lege bei meiner jetzigen Aktion größten Wert auf Ihre Feststellungen; denn wenn auch die Lage der deutschen Mathematik den maßgebenden Stellen sachlich nicht solchen Eindruck macht wie die Lage der deutschen Physik, so würde doch die Überflügelung der deutschen Mathematik durch die USA als ein Symptom von wesentlich stärker alarmierender Wirkung empfunden werden." Carl Ramsauer to Wilhelm Süss, 16 June, 1943, Freiburg University Archives, Süss Archives, C 89/12.

[65] "Leider kenne ich keine Titel von mathematischen Arbeiten in amerikanischen oder englischen Zeitschriften der letzten Jahre. Deshalb wären mir Fotocopien von Inhaltsverzeichnissen solcher Zeitschriften der Jahrgänge 1943 und 1944 doppelt erwünscht." Siegmund-Schultze, "Emancipation," p. 159, note 57.

secrecy regulations. There were also restrictions due to ideological resentments that were further sharpened by economic circumstances such as autarky and valuta shortage. These restrictions were reflected, above all, in the participation in international conferences—like the ICM in Oslo in 1936 and the Applied Mechanics Congress in Cambridge in 1938—and in publications—such as the *Zentralblatt*, the second edition of the *Enzyklopädie der mathematischen Wissenschaften*, and journals. On the other hand, however, there was also a tendency toward the "internationalization" of mathematics under German hegemony. Several German mathematicians and politicians not only reached out for influence in southeastern European mathematics but also worked to induce French mathematicians to collaborate with the Germans after the occupation in 1940.

Given these ambiguities, this chapter has aimed to show that, relative to mathematics in Nazi Germany, the notions of "international participation" and, more importantly, of "internationalism" in science and mathematics have to be understood in a very broad sense. Not only must the purely scientific interests of scientists and politicians be taken into account but also their political goals and imperialist aims. Interest in the continuation of the scientific enterprise, in particular, interest in international communication, sometimes led German mathematicians to criticize official, cultural policies. As the unpublished report by Walter Lietzmann on the Oslo ICM shows, this criticism did not penetrate the absolute barrier formed by the dogma of antisemitism, which had already made the expulsions an indisputable fact in the international arena. At the same time, Lietzmann's report reflects his unconditional support for the "German side" and for "scientific cultural propaganda" in Germany's cultural policies toward other nations. Moreover, Geppert, Süss, and others clearly "regarded themselves as agents," to use Paul Forman's words,[66] of the German cause and tried, in accordance with the Nazi Ministry, to instrumentalize the traumatic experiences of scientists during the First World War for their policies in World War II.

Nazi rule, however, affected mathematics not only in Germany but also in the postwar world. The flood of emigrations and the war finally made the United States the world center for mathematics. The postponed 1940 International Congress of Mathematicians ultimately took place in Cambridge, Massachusetts in 1950, still without German participation. The United States took the lead in the reorganization of the International Mathematical Union,[67] as Geppert, Süss, and others in Germany had suspected as they laid their own imperialist plans in the early 1940s.

The political aftereffects of the alienation of German mathematicians from foreign mathematicians were felt well into the 1960s,[68] despite efforts by mathematicians from allied countries immediately after the war to involve the Germans once again in international communication.[69] Developments in mathematics, like the French *Bourbaki* movement that had been very much inspired by German algebra of around 1930, had to be "re-imported" into Germany in the 1950s and 1960s, enriched with newer concepts such as topological ones. The overall scientific dominance of the United States after the war, which was also economically grounded,

[66]Forman, p. 156.

[67]Lehto, *Mathematics Without Borders*.

[68]Siegmund-Schultze, *Mathematiker auf der Flucht*, pp. 274ff.

[69]For example, John Todd, F. Joachim Weyl, and others were invited to visit Oberwolfach in the late 1940s. See Senechal.

was expressed by Garrett Birkhoff, son of the erstwhile leading American mathematician, George D. Birkhoff. "After the defeat of Hitler," the younger Birkhoff wrote,

> the U.S. (along with Canada) was standing triumphant, virtually unscathed in a world in which most advanced countries were prostrate and in ashes. Its scientific preeminence was taken for granted, and most Americans thought of computing machines as almost a national monopoly. They had largely forgotten the tradition of European scientific superiority that had been accepted only a decade earlier. What is now known about ENIGMA and the Zuse machines should help dispel this illusion.[70]

As this quote suggests, the isolation of German mathematicians during dictatorship and war had consequences not only for Germany. It also prevented foreigners from becoming aware of promising German scientific developments. This further underscores the complexity of any analysis of international communication and participation.

References

Archival Sources

AAMS: American Mathematical Society Archives. John Hay Library. Providence, R.I.

BAB-REM: Bundesarchiv Berlin. Reichserziehungsministerium.

BrUA: Roland G. D. Richardson Papers. Brown University Archives. Providence, R.I.

BUA: Berlin University Archives. Humboldt Universität.

DKZ: Deutsche Kongreßzentrale. Hoover Institution. Stanford University.

FUA-DMV: Deutsche Mathematiker-Vereinigung Archives. Freiburg University Archives.

FUA-WS: Wilhelm Süss Papers. Freiburg University Archives.

GSA: Geheimes Staatsarchiv Preußischer Kulturbesitz. Berlin.

GUA: Göttingen University Archives.

MAG-WL: Walter Lietzmann Papers. Municipal Archives Göttingen = Stadtarchiv Göttingen, Depositum 89.

HUA: George David Birkhoff. Papers. Harvard University Archives

LC: Oswald Veblen Papers. Library of Congress, Washington D.C.

MUA: Munich University Archives. Personal file: Constantin Carathéodory.

Printed Sources

Behnke, Heinrich. *Semesterberichte: Ein Leben an deutschen Universitäten im Wandel der Zeit.* Göttingen: Vandenhoeck & Ruprecht, 1978.

Birkhoff, Garrett. "Computer Developments 1935–1955, As Seen from Cambridge, U.S.A." In *A History of Computing in the Twentieth Century: A Collection*

[70]Garrett Birkhoff, "Computer Developments 1935–1955, As Seen from Cambridge, U.S.A.," in *A History of Computing in the Twentieth Century: A Collection of Essays*, ed. Nicholas Metropolis, Jack Howlett, and Gian-Carlo Rota (New York: Academic Press, Inc., 1980), pp. 21-30.

of Essays. Ed. Nicholas Metropolis, Jack Howlett, and Gian-Carlo Rota. New York: Academic Press, Inc., 1980, pp. 21-30.

Busemann, Adolf. "Compressible Fluids in the Thirties." *Annual Review of Fluid Mechanics* **3** (1971):1-12.

Comptes rendus du Congrès international des mathématiciens, Oslo 1936. 2 Vols. Oslo: A. W. Broggers Boktrykkeri A/S, 1937.

Deutsche Wissenschaft, Erziehung und Volksbildung: Amtsblatt des Reichsministeriums für Wissenschaft, Erziehung und Volksbildung **1** (1935).

Duren, Peter, Ed. *A Century of Mathematics in America—Parts I-III.* Providence: American Mathematical Society, 1988–1989.

Forman, Paul. "Scientific Internationalism and the Weimar Physicists: The Ideology and Its Manipulation in Germany after World War I." *Isis* **64** (1973):151-180.

Gumbel, Emil Julius, Ed. *Freie Wissenschaft: Ein Sammelbuch aus der deutschen Emigration.* Strasbourg: Sebastian Brant 1938.

Juhasz, Stephen, Ed. *IUTAM: A Short History.* Berlin: Springer-Verlag, 1988.

Kasper, G.; Huber, H.; Kaebsch K.; and Senger F., Ed. *Die deutsche Hochschulverwaltung.* 2 Vols. Berlin: N.p., 1942–1943.

Knoche, Michael. "Wissenschaftliche Zeitschriften im nationalsozialistischen Deutschland." In *Von Göschen bis Rowohlt: Beiträge zur Geschichte des deutschen Verlagswesens.* Ed. Monika Estermann and Michael Knoche. Wiesbaden: Harrassowitz, 1990, pp. 260-281.

Laitenberger, Volker. *Akademischer Austausch und auswärtige Kulturpolitik: Der Deutsche Akademische Austauschdienst (DAAD) 1923–1945.* Göttingen: Musterschmidt, 1976.

Lehto, Olli. *Mathematics Without Borders: A History of the International Mathematical Union.* New York: Springer-Verlag, 1998.

Mehrtens, Herbert. "The Gleichschaltung of Mathematical Societies in Nazi Germany." *The Mathematical Intelligencer* **11** (3) (1989):48-60.

______. "Ludwig Bieberbach and 'Deutsche Mathematik'." In *Studies in the History of Mathematics.* Ed. Esther R. Phillips. MAA Studies in Mathematics. Vol. 26. Washington: Mathematical Association of America, 1987, pp. 195-241.

Menger, Karl. *Reminiscences of the Vienna Circle and the Mathematical Colloquium.* Ed. Louise Golland, Brian McGuinness, and Abe Sklar. Dordrecht: Kluwer Academic Publishing, 1994.

Metropolis, Nicholas; Howlett, Jack; and Rota, Gian-Carlo, Ed. *A History of Computing in the Twentieth Century: A Collection of Essays.* New York: Academic Press, Inc., 1980.

Niven, Ivan. "The Threadbare Thirties." In *A Century of Mathematics in America–Part I.* Ed. Peter Duren *et al.* Providence: American Mathematical Society, 1988, pp. 209-229.

Reingold, Nathan. "Refugee Mathematicians in the United States of America 1933–1941." *Annals of Science* **38** (1981):313-338.

Scholz, Heinrich. "Die neue logistische Logik und Wissenschaftslehre." *Forschungen und Fortschritte* **11** (13) (1935):169-171.

Segal, Sanford. "Helmut Hasse in 1934." *Historia Mathematica* **7** (1980):46-56.

Senechal, Marjorie. "Oberwolfach, 1944–1964." *The Mathematical Intelligencer* **20** (4) (1998):17-24.

Siegmund-Schultze, Reinhard. "An Autobiographical Document (1953) by Jean Leray on His Time as Rector of the 'Université en captivité' and Prisoner of War in Austria 1940-1945." *Gazette des mathématiciens* (*Supplément*) **84** (2000):11-15.

———. "The Emancipation of Mathematical Research Publishing in the United States from German Dominance (1878–1945)." *Historia Mathematica* **24** (1997):135-166.

———. "Faschistische Pläne zur 'Neuordnung' der europäischen Wissenschaft: Das Beispiel Mathematik." *NTM–Schriftreihe für Geschichte der Naturwissenschaften, Technik und Medizin* **23** (2) (1986)1-17.

———. *Mathematiker auf der Flucht vor Hitler: Quellen und Studien zur Emigration einer Wissenschaft.* Braunschweig/Wiesbaden: Vieweg Verlag, 1998.

———. *Mathematische Berichterstattung in Hitlerdeutschland: Der Niedergang des Jahrbuchs über die Fortschritte der Mathematik (1869–1945).* Göttingen: Vandenhoeck & Ruprecht, 1993.

———. *Rockefeller and the Internationalization of Mathematics Between the World Wars.* Basel: Birkhäuser Verlag, 2001.

———. "'Scientific Control' in Mathematical Reviewing and German-U.S.-American Relations Between the Two World Wars." *Historia Mathematica* **21** (1994): 306-329.

———. "Theodor Vahlen: Zum Schuldanteil eines deutschen Mathematikers am faschistischen Mißbrauch der Wissenschaft." *NTM–Schriftenreihe für Geschichte der Naturwissenschaften, Technik und Medizin* **21** (1) (1984):17-32.

Timoshenko, Stephen P. *As I Remember.* Princeton: Van Nostrand, 1968.

Tobies, Renate. "Mathematik als Bestandteil der Kultur: Zur Geschichte des Unternehmens 'Encyklopädie der Mathematischen Wissenschaften mit Einschluß ihrer Anwendungen'." *Mitteilungen österreichische Gesellschaft für Wissenschaftsgeschichte* **14** (1994):1-90.

Trefftz, Erich. "IV. Internationaler Kongreß für angewandte Mechanik." *Zeitschrift für angewandte Mathematik und Mechanik* **14** (1934):255-256.

Appendix

Table 1: Important Events and Dates for International Communication of German Mathematics During the Third Reich, 1933–1945[71]

1933	**Apr. 7**	"Law for the restoration of Civil Service": Pseudolegal rationale for the majority of expulsions of Jewish scholars
1934	**Oct. 5**	Harvard President James B. Conant declines acceptance of scholarship for American students in Munich, offered by Hitler's press attaché, Ernst Hanfstaengl, a former Harvard student
	Dec. 22	Foundation of the German Congress Office (*Deutsche Kongreßzentrale*) with the Propaganda Ministry under Goebbels supervising international congresses, allocating foreign currency, nominating "leaders" of delegations, gathering reports, etc. [DKZ]
1935	**Sept. 9**	New regulation for export of German scientific books and journals: 25% subsidy on sale by the Propaganda Ministry; Bieberbach opposed, since his *Deutsche Mathematik* does not sell equally well abroad [Kn]
1936	**Mar. 9**	Collaboration with scientific organizations of the League of Nations is forbidden, after Germany left the League
	July	International Congress of Mathematicians in Oslo, Norway: Bureaucratic and political selection of German participants (Compare Case Study 1 above.)
1937	**Feb. 10**	Any contacts with Soviet science forbidden for German scientists by decree of the Ministry of Education; temporarily lifted during Hitler-Stalin Pact of 1939–1941
1938	**May 11**	Professorships of German Jews abroad generally declared undesirable, although Jewish emigration is still supported by the regime
	Summer	252 American students remain in German universities (800 in 1932, 540 in 1933); at the same time, there were 35 stipends for German students in America, mostly in the humanities (100 in 1930)
1939	**Aug. 22**	All trips to England, France, Poland, and the U.S. confidentially prohibited by the Ministry of Education

[71]For the symbols used, see the list of archival sources in the "References." Information in this table has also come from Knoche (abbreviated [Kn]) and Behnke (abbreviated [Beh]). See, in addition, *Deutsche Wissenschaft, Erziehung und Volksbildung: Amtsblatt des Reichsministeriums für Wissenschaft, Erziehung, und Volksbildung* **1** (1935); G. Kasper *et al.*, ed., *Die deutsche Hochschulverwaltung*, 2 vols (Berlin: 1942–1943); and Volker Laitenberger, *Akademischer Austausch und auswärtige Kulturpolitik: Der Deutsche Akademische Austauschdienst (DAAD) 1923–1945* (Göttingen: Musterschmidt, 1976).

Table 1 (cont.)

1940	**Jan.**	91 Chinese and 18 American guest professors in German universities in all fields; no French, British, or Polish due to the war
	Sept.	Beginning of German plans for a restructuring (*Neuordnung*) of European science, in particular, mathematics. Harald Geppert and Helmut Hasse consider reviewing as a core for international "collaboration" under German rule (See Case Study 2 above.)
	Dec.	Harald Geppert's trip to Paris (Compare Case Study 2 above.)
1941		Siemens engineer, Johannes Rasch, warns against falling behind the U.S. in applied science
	June 7	Ministry asks Carathéodory whether he is willing to contribute to the 75th volume of *Matematicheskii Sbornik*, two weeks before the attack on the Soviet Union [MUA]
1942	**Mar. 23**	Meeting at Ministry of Education on restructuring European science; most plans dropped in the years to come (Compare Case Study 2 above.)
1943	**Aug. 26**	DMV President, Wilhelm Süss, acknowledges surpassing of German mathematics by allies with respect to number of publications; contradicts the Nazi principle of "qualitative superiority" (See Case Study 2 above.)
	Nov.	Last of several trips to Switzerland by Heinrich Behnke due to German-Swiss professor exchange program [Beh]
1944	**Sept.**	Foundation of the research institute in Oberwolfach; nucleus of international collaboration after the war (See Case Study 2 above.)

Table 2: German Mathematicians Leading in International Communication, 1933–1945

Heinrich Behnke (1898–1979): cofounder of multidimensional complex function theory; German most frequently invited to foreign countries, however, because of his half-Jewish son, he was timid and without much official influence

Wilhelm Blaschke (1885–1962): differential geometry in the tradition of Felix Klein's *Erlanger Programm*; of Austrian origin, international experience, nationalist convictions, and particularly strong sense of opportunism

Constantin Carathéodory (1873–1950): fundamental contributions to variational calculus; although of Greek origin, well-liked abroad as "ambassador" of German mathematics in the traditional, apolitical sense

Harald Geppert (1902–1945): second-rate differential geometer and statistician; half-Italian background, proponent of "restructuring" of European mathematics under German hegemony (1940), tried to use mathematical reviewing for that purpose (See Case Study 2 above.)

Helmut Hasse (1898–1979): leading number theorist known for the "local-global principle;" half-American, typical German nationalist, overly loyal to authorities, engaged in Nazi activity during the German occupation of France

Eberhard Hopf (1902–1983): cofounder of ergodic theory; until 1936, professor at the Massachusetts Institute of Technology, returned to Leipzig as successor to expelled L. Lichtenstein, politically naïve and focused on research

Walter Lietzmann (1880–1950): mathematical pedagogy and ICME member; due to the Nazi bureaucracy, he was the "leader" (*Führer*) of the German delegation to the Oslo ICM in 1936 (See Case Study 1 above.)

Ludwig Prandtl (1875–1953): leading applied mathematician and aerodynamicist; failed in his effort to draw the International Mathematical Congress to Germany due to the dogma of antisemitism; supported Hitler's policies in a correspondence with the Englishman, Geoffrey I. Taylor

Heinrich Scholz (1884–1956): leading logician of the semantic approach; although politically conservative, was a decided internationalist with contacts to Poland, Bertrand Russell, and the Vienna Circle; limited influence due most likely to the limited influence of his subject within mathematics

Wilhelm Süss (1895–1958): second-rate geometer; due to Nazi bureaucracy, was the "leader" (*Führer*) of the DMV beginning in 1939, founded the institute in Oberwolfach (1944), which became the source of survival for German mathematics (See Case Study 2 above.)

Table 3:
Influential German Mathematicians with Hampered Potential in International Communication, 1933–1945

Emil Artin (1898–1962): leading algebraist; of Austrian origin, Jewish spouse, emigration to the U.S. in 1937

Ludwig Bieberbach (1886–1982): noted function theorist; principal proponent of "German (Aryan) mathematics," for this reason internationally isolated, no active foreign languages, did not participate in the Oslo ICM in 1936

Erich Hecke (1887–1947): leader in number theory and its relations to function theory; unable and unwilling to compromise with the regime, talked in the U.S. in 1936

Erich Kamke (1890–1961): expert, but not very original, in differential equations; persecuted because of Jewish spouse, barred from Oslo ICM in 1936, dismissed in 1937, after 1945 leading representative of Germany

Kurt Reidemeister (1893–1971): leader in topology and foundations; politically "unreliable" due to Russian contacts in the 1920s and to élitist behavior

Erhard Schmidt (1876–1959): cofounder of functional analysis; politically naïve and conservative, somewhat lazy, no Western languages other than German

Carl Ludwig Siegel (1896–1981): most important and universal mathematician; similar to Hecke, unable and unwilling to compromise, idiosyncratic behavior, returned from Princeton to Frankfurt (1935) allegedly to "protect" Jewish mathematician, Max Dehn

Theodor Vahlen (1869–1945): second-rate applied mathematician; leading Nazi functionary, did not participate in the Oslo ICM 1936

Baertel L. van der Waerden (1903–1996): leading algebraist; of Dutch origin was not an "ambassador" (in the sense of Carathédory) because of his youth and his relationships to Emmy Noether and to Springer, remained in Germany, the latter a cause of perplexity among emigrants

Table 4:
Participation or Planned Participation of German Mathematicians in International Conferences, 1933–1945[72]

1934	**Rome**
	Artin, Blaschke, Bieberbach, Hecke, Perron proposed as participants [GSA]
	Cambridge (UK): International Applied Mechanics Congress
	Participants from Germany; Erich Trefftz's report (1934) stresses Bush's differential analyzer [BAB-REM; Tref]
	Prague: Conference of Slavic Countries
	Bieberbach proposes participation by geometer, Blaschke [GSA]
1935	**Rome: Volta Congress on High Velocities in Aviation**
	Participation of Prandtl, von Kármán (formerly from Germany); supersonic speeds, international importance due to arms race [Buse; DKZ]
	Moscow: Topology
	Reidemeister stresses importance "from a purely scientific point of view" [BAB-REM]
	Paris: First Congress of Unity of Science (Vienna Circle)
	Participation by politically conservative logician, Scholz [BAB-REM 73; Sch]
1936	**Oslo: International Congress of Mathematicians**
	22 official participants from Germany; problems involving currency exchange and leaders of delegation—Lietzmann more "adroit" than Schmidt, but the latter propagandizes for the Olympic Games in Berlin; Kamke not allowed to participate due to Jewish wife and his politics on behalf of Springer-Verlag
	Cambridge (US): Harvard Tercentenary
	Official participation refused by German Ministry due to Hanfstaengl affair (unofficially, Carathéodory in attendance) [BUA]
	Heidelberg: 550th Anniversary of the University
	Official American delegation includes G. D. Birkhoff, who was criticized for his participation on his return to the U.S. [HUA]
1937	**Tbilisi: Soviet Mathematics Conference**
	Erhard Tornier was urged to refuse invitation "due to personal reasons" because of general ban on contacts with Soviet science [BAB]
	Göttingen: 200th Anniversary of the University
	No official American delegation in spite of reverence for Göttingen tradition, due partly to the experience in Heidelberg; Timoshenko takes part but is disillusioned [AAMS; Tim]

[72]Conferences with years in brackets did not take place. For the symbols used, see the list of archival sources in the "References." Information in this table has also been drawn from Adolf Busemann, "Compressible Fluids in the Thirties," *Annual Review of Fluid Mechanics* **3** (1971):1-12 (abbreviated [Buse]); Mehrtens, "Gleichschaltung" (abbreviated [Mehr]); Heinrich Scholz, "Die neue logistische Logik und Wissenschaftslehre," *Forschungen und Fortschritte* **11** (13) (1935):169-171 (abbreviated [Sch]); Stephen Timoshenko, *As I Remember* (Princeton: Van Nostrand, 1968) (abbreviated [Tim]); and Erich Trefftz, "IV. Internationaler Kongreß für angewandte Mathematik," *Zeitschrift für angewandte Mathematik und Mechanik* **14** (1934):255-256 (abbreviated [Tref]).

Table 4 (cont.)

	Geneva: Conference on Probability Only invited Germans (for cognitive reasons) were Heisenberg and Hopf [BAB-REM]
	Paris: International Congress of Actuaries Report by Paul Riebesell [BAB-REM]
1938	**Zürich: Conference on Foundations** Wilhelm Ackermann, Gerhard Gentzen have to refuse participation because conference connected to League of Nations [BAB-REM]
	Cambridge (US): 50th Anniversary of the AMS German mathematicians ask G. D. Birkhoff, who is considered "deutschfreundlich," to represent Germany [FUA-DMV]
	Cambridge (US): International Applied Mechanics Congress 22 participants from Germany; invitation for Germany in 1942 impossible since there were no guarantees for Jewish participants [Mehr]
	Helsinki: Scandinavian Congress Participation by Egon Ullrich, complex function theorist
[1939]	**Rome: Volta Congress** Hasse nominated leader; Congress postponed (canceled) due to war [FUA]
1940	**Bologna: Italian mathematics conference** Blaschke takes part; he also takes part in the inauguration of the Institute of Higher Mathematics in Rome with Severi; Duce Mussolini in attendance [DKZ: Blaschke's report]
[1940]	**Cambridge (US): International Congress of Mathematicians** German participation in political accordance with Italians; Süss as DMV President inquires about "race" of organizer, William Graustein, threatens boycott because of *Zentralblatt* affair; Congress canceled due to war (postponed to 1950) [AAMS; BAB-REM; FUA-DMV]
1941	**Jena: International DMV conference** Many foreigners, especially from Scandinavia; Nevanlinna gives radio speech; strategies under way for making DMV nucleus of European mathematics [DKZ; FUA-WS; BAB]
1942	**Rome: International mathematical congress (November)** Participation by Gröbner, Blaschke, Hasse, Carathéodory; Germans suspicious of Italian strategies for dominance given the cancellation of the DMV conference that had been scheduled for October with 60 foreign participants [DKZ; BAB]
1943	**Würzburg: "War Conference"** Strong security regulations; participation of Charles Pisot, French Alsacian, in "open session" [BAB-REM 73]
[1943]	**Leipzig or Köln: Southeastern European Actuarial Conference** Apparently canceled; during preparation there was fear of Italian dominance; stress on actuarial science, namely, economics [BAB-REM]

CHAPTER 17

War, Refugees, and the Creation of an International Mathematical Community

Sanford L. Segal
University of Rochester (United States)

Introduction

National mathematical communities date to before the turn of the twentieth century. The London Mathematical Society was founded in 1865, one year after the establishment of the Moscow Mathematical Society. Comparable institutions in France and Italy followed in 1872 and 1884, respectively, while the eventual American Mathematical Society made its appearance in 1888, with the German Mathematical Society following two years later. As national societies, these were basically interested in promoting a national mathematics. Karen Parshall has pointed out that they had common interests in mathematics education and generally adopted what was thought of as a German model of education.[1] However, mathematics was still largely perceived as a national ornament.

The initial effort toward an international or, more accurately, a Franco-German mathematical gathering was taken by Georg Cantor as early as 1888.[2] This came to naught at the time, but by 1897, Zürich, a neutral, German-speaking location, had hosted the first International Congress of Mathematicians (ICM). What started in Zürich and seemed to be gaining both structure and momentum was suddenly stopped by World War I and its aftermath. Germany, the leading mathematical country by 1900, continued to progress after World War I, but it was excluded for political reasons from international participation. Though formal efforts at mathematical internationalism began again in 1920, they were really pseudo-international and actually promoted nationalism under the guise of internationalism.[3]

In 1928, Germany was readmitted to the international mathematical family. Despite this, by 1933, all international mathematical efforts other than congresses had foundered. In 1933, Hitler became Chancellor of Germany, and this resulted in the expulsion of a large amount of mathematical talent from Germany and the

---

[1]Karen Hunger Parshall, "Mathematics in National Contexts (1875-1900): An International Overview," in *Proceedings of the International Congress of Mathematicians, Zürich, Switzerland, 1994*, 2 vols. (Basel: Birkhäuser Verlag, 1995), 2:1581-1591.

[2]Olli Lehto, *Mathematics Without Borders: A History of the International Mathematical Union* (Berlin: Springer-Verlag, 1991), p. 3.

[3]Brigitte Schröder-Gudehus and Paul Forman have both made this point, although their work does not deal principally with mathematicians. See Brigitte Schröder, "Caractéristiques des relations scientifiques internationales: 1870–1914," *Cahiers d'histoire mondiale* **10** (1966):161-177; and Paul Forman, "Scientific Internationalism and the Weimar Physicists: The Ideology and Its Manipulation in Germany after World War I," *Isis* **64** (1973):151-180.

German-dominated countries. When, after World War II, mathematical internationalism was again attempted, it was on a different basis than in 1920. The United States was the leader with the refugees from Hitler's Germany, mostly Jewish, key players. This chapter will assess these efforts as it explores both the extent to which true internationalism, as opposed to masked nationalism, underlay them and the role of the two World Wars in retarding or accelerating the process.

Before World War I

"National mathematics" manifested itself in a number of different ways in the decades before World War I. The competition in the early 1880s between Felix Klein in Germany and Henri Poincaré in France in the theory of automorphic functions came on the heels of the Franco-Prussian War; at its outbreak, Klein had been studying in Paris and had rushed home to join in the Prussian cause.[4] "National mathematics" was also apparent in the different styles of invariant theory that had developed in Britain and in Germany from the middle of the nineteenth century up to the 1890s and David Hilbert's transformative work.[5] At a more sociological level, "national mathematics" was also evident in the lengthy European—and particularly German—study tours of American mathematicians in the decades both before and after 1900. These mathematicians learned advanced material as students and brought it back home for further development.[6] They even published in foreign languages from time to time—like William Fogg Osgood and his *Lehrbuch der Funktionentheorie*—in their efforts to acquire recognition.[7] Such efforts, rather than indicative of internationalism, were examples of an attempt to associate professionally with a foreign nation and thereby to enhance one's own professional reputation at home.[8] This strategy, in fact, had been used throughout the nineteenth century.

During that century's first four decades, the French mathematicians were unquestionably the dominant professionals. Peter Lejeune-Dirichlet, who studied in Bonn and Cologne, went to Paris in 1822 because he felt there was insufficient mathematics in Germany; the solitary Gauss was certainly *sui generis*.[9] Even later in the century, Klein continued to recommend a trip to Paris for gifted young German mathematicians such as Hilbert.[10] In 1885, Hilbert's friend, Adolf Hurwitz, reinforced Klein's position, writing to Hilbert that young French mathematical talent

---

[4]See, for example, Renate Tobies, *Felix Klein* (Leipzig: B. G. Teubner Verlag, 1981), pp. 50-55.

[5]Karen Hunger Parshall, "Towards a History of Nineteenth-Century Invariant Theory," in *The History of Modern Mathematics*, ed. David E. Rowe and John McCleary, 2 vols. (Boston: Academic Press, Inc., 1989), 1:157-206.

[6]On America's mathematical pilgrims, see Karen Hunger Parshall and David E. Rowe, *The Emergence of the American Mathematical Research Community, 1876–1900: J. J. Sylvester, Felix Klein, and E. H. Moore*, HMATH, vol. 8 (Providence: American Mathematical Society and London: London Mathematical Society, 1994), pp. 189–259.

[7]William Fogg Osgood, *Lehrbuch der Funktionentheorie* (Leipzig: B. G. Teubner Verlag, 1906). Successive editions appeared in 1912 and 1928.

[8]On the distinction between "internationalism" and "internationalization," recall the discussion in Parshall and Rice's chapter above.

[9]Charles C. Gillispie, *Dictionary of Scientific Biography*, 16 vols. and 2 supps. (New York: Charles Scribner's Sons, 1970–1990), s.v. "Dirichlet, Peter Lejeune" by Oystein Øre; and Kurt R. Biermann, *Die Mathematik und ihre Dozenten an der Berliner Universität 1810–1933* (Berlin: Akademie-Verlag, 1988), p. 41.

[10]Constance Reid, *Hilbert* (New York: Springer-Verlag, 1970), pp. 20-21.

was "more intensive" than in Germany and counseling him that it "was necessary to master their results in order to go beyond them."[11] By 1900, however, the tables had turned. Germany was the new center of the mathematical world, even though Paris was the location of the second International Congress of Mathematicians and the site of Hilbert's famous address. The unification of the German states in 1871 on the heels of the Franco-Prussian War and the subsequent German political dominance in central Europe marked the beginning of a growing German intellectual dominance in many areas including mathematics. Germany, especially from the 1890s, spread the mathematical and scientific products of its educational system by a sort of cultural imperialism.[12]

Even earlier, German educational successes had led to the encouragement by other countries of the higher-level, German educational system with national modifications. For example, in the 1870s, Matthew Arnold, a school inspector as well as Professor of Poetry at Oxford, spent seven months studying European systems of higher education at the behest of the British government. He clearly admired the German universities both for their systematic pursuit of knowledge, or *Wissenschaft*, and for *Lehr-und Lernfreiheit*, the intellectual freedom of both professors and students. Arnold returned to Britain to urge his countrymen to consider a Germanic style of organization for their educational system: "It is in science that we have most need to borrow from the German universities. The French University has no liberty, and the English universities have no science; the German universities have both."[13]

A principal German educational innovation was the idea that teaching and research should inform one another by working in tandem. Relative to mathematics, this was supplemented in a critical way by the development of the research seminar. Invented by Carl G. J. Jacobi and Franz Neumann at Königsberg in 1834, the mathematics research seminar spread to all of the German universities over the next forty years.[14] A particularly successful practitioner of this training tool was Felix Klein, and among the students who learned from him in that context were Americans as well as Germans. Klein's German students, Oskar Bolza and Heinrich Maschke, were part of the first University of Chicago mathematics faculty in 1892. At Harvard, Osgood and Maxime Bôcher had also studied with Klein, although Osgood actually wrote his doctorate under the supervision of Klein's good friend at Erlangen, Max Noether. These and many other Americans had learned their mathematics in Germany, and the impact of that German training was not lost on leaders in the developing American mathematical research community. On 23 March, 1904, E. H. Moore, the chair of the Department of Mathematics at Chicago and the recipient himself of German training, wrote to Felix Klein acknowledging the debt American mathematics owed to Germany and to Klein:

---

[11] *Ibid.*

[12] David E. Rowe, "Die Wirkung deutscher Mathematiker auf die amerikanische Mathematik, 1815–1900," *Mitteilungen der Mathematischen Gesellschaft DDR* **3-4** (1988):72-96. See also Lewis Pyenson, *Cultural Imperialism and Exact Sciences: German Expansion Overseas, 1900–1930* (New York: P. Lang, 1985).

[13] Matthew Arnold, *Higher Schools in Germany* (London: Macmillan, 1874), p. 166. The epigraph to this report is from Wilhelm von Humboldt and speaks to using schools and universities as a means for raising national culture.

[14] Winfried Scharlau, ed., *Mathematische Institute in Deutschland 1800–1945* (Braunschweig: Vieweg Verlag, 1990).

> Certainly in the domain of mathematics German scholars in general and yourself in particular have played, by way of example and counsel and direct and indirect inspiration, quite the leading role in the development of creative mathematicians in this country, and on behalf of my colleagues here I wish to express our most grateful recognition and appreciation of our profound debt.[15]

This praise aside, Klein believed in varying national excellences in mathematics, but he viewed them as complementary. He also believed in the importance of moving beyond national mathematical communities to the formation of an international union.[16] Indeed, as the plenary speaker at the Mathematical Congress held in Chicago in connection with the World's Columbian Exposition in 1893 and organized by Moore, Bolza, Maschke, and their colleague at Northwestern, Henry Seely White, Klein declared that "our mathematicians must ... form international unions, and I trust that this present World's Congress at Chicago will be a step in this direction."[17]

Klein's attitudes exemplify that, at the turn of the twentieth century, there was in mathematics an incipient international movement with a sort of German focus. This was further reflected in the facts that the German *Enzyclopädie der mathematischen Wissenschaften* had contributors of other nationalities and that international congresses continued every four years after Paris successively in Heidelberg, Rome, and Cambridge, England, becoming increasingly international. Klein's idea of an international union also resurfaced. Not only Klein but other mathematicians of various nationalities pushed for this, despite the fact that some felt that national unions were quite adequate; as early as 1890, Klein's student, Walther von Dyck, had written to him wondering if international congresses were really necessary.[18] This tension between the promotion of excellence nationally and the desires of a growing community of mathematicians for an international audience for its work characterize the first inchoate tendencies towards an international community.

World War I and After

Whatever the general movement among mathematicians toward internationalism had been, it stopped and even reversed in the highly charged political climate that followed World War I. Within the victorious nations, widely believed but highly embroidered false stories of German atrocities in their invasion of neutral Belgium at the beginning of the war abounded and shaped popular sentiment.[19] These persisted throughout the war abetted by such facts as the German use of poison gas at Ypres. Various German intellectuals made a "Declaration to the Cultural World" justifying the invasion of Belgium and offering reasons for the war.

---

[15] Rowe, "Die Wirkung deutschen Mathematiker," p. 91.

[16] Lehto, p. 3.

[17] *Ibid.*, p. 6.

[18] *Ibid.*, p. 3.

[19] On the German terrorization of the Belgian population, see Barbara Tuchman, *The Guns of August* (New York: Macmillan & Co., 1962), see index entries for "Belgium—atrocities." For samples of the embroidery, see J. Selden Willmore, *The Great Crime and Its Moral* (New York: George H. Doran, 1917), chapters 6-7. For a brief analysis of the most notorious of these embroideries, the "Bryce Report," see Phillip Knightley, *The First Casualty* (New York: Harcourt, Brace, Jovanovich, 1975), pp. 82-85.

The original ninety-three signatories formed a "who's who" of the German academic world.[20] Among them, Felix Klein was the only prominent mathematician; Einstein and Hilbert had refused to sign.[21] In all, some 4,000 intellectuals affixed their names. At mid-war, five mathematicians—none of them prominent—were among the many German intellectuals who had signed a declaration for an ultra-annexationist, non-negotiated *Siegfrieden.*[22] To these and many similar German declarations, the British and French responded in kind. Many declared repeatedly that after the war's conclusion it would be impossible to resume the old relations that had been tending in an international direction in many scientific areas.

Among the mathematicians, two voices for sanity sounded. One was G. H. Hardy's, the other Gösta Mittag-Leffler's. Hardy had co-authored his *General Theory of Dirichlet Series* with Hungarian-born Marcel Riesz in 1915. Their book bore a Latin epigraph—Latin having been the universal language—that decried how the society of mathematicians was being ripped apart and that called for its reintegration. It ended with the signature "Auctores Hostes Idemque Amici," that is, "The authors, enemies, and all the same, friends."[23] Hardy's attempts immediately after the war to reestablish scientific relations between the war's enemies, at least among mathematicians, and his defense of Bertrand Russell's right to be a pacifist during the war, have led some to suppose that Hardy himself was a pacifist. That this was not so is evident in a letter he wrote to Russell in July of 1916. Apparently, Hardy, who was thirty-seven when the war broke out, applied for service not once, but three times and was refused.[24] Nevertheless, Hardy did think in September 1914 that the war was "monstrous" and that it was "impossible" to want to "crush" and "humiliate" Germany; in his view, what was wanted was peace on reasonable terms.[25] Both in 1914 and in 1934, Hardy thought that national passions often excited otherwise reasonable people to "imbecilities."[26] Right after the war, as a gesture of comity, he and Littlewood considered publishing in English in the *Göttingen Nachrichten.*[27]

As for Mittag-Leffler, he viewed *Acta Mathematica*, the journal he had founded in Sweden in 1882, as a vehicle for bringing mathematicians back together after the

---

[20] An English text and list of signatories can be found in John Jay Chapman, *Deutschland über Alles or Germany Speaks* (New York and London: C. P. Putnam, 1914), pp. 37-41. See also Klaus Böhme, *Aufrufe and Reden deutscher Professoren im Ersten Weltkrieg* (Stuttgart: Phillip Reclam, 1975).

[21] On Einstein, see Philipp Frank, *Einstein, His Life and Times* (New York: Alfred Knopf, 1947), p. 120; and Ronald Clark, *Einstein: The Life and Times* (New York: World Publishing, 1971), pp. 172 and 180. On Hilbert, see Reid, *Hilbert,* pp. 137-138.

[22] A complete list of the signatories can be found in the Nachlaß Schemann available at the Handschriftenabteilung, Universitätsbibliothek Freiburg im Breisgau.

[23] Godfrey H. Hardy and Marcel Riesz, *General Theory of Dirichlet Series*, Cambridge Mathematical Tracts, vol. 26 (Cambridge: University Press, 1915), preface. It is interesting to note that Hardy also praised the German, Edmund Landau, in this 1915 preface.

[24] Godfrey H. Hardy to Bertrand Russell, July 1916, Russell Correspondence, Russell Archive, McMaster University.

[25] Godfrey H. Hardy to Bertrand Russell, 16 September, 1914 in *ibid.*

[26] Godfrey H. Hardy, "The *J*-Type and the *S*-Type among Mathematicians," *Nature* **134** (1934):250. Compare, also, Lehto, pp. 30-31.

[27] Joseph W. Dauben, "Mathematicians and World War I: The International Diplomacy of G. H. Hardy and Gösta Mittag-Leffler as Reflected in their Personal Correspondence," *Historia Mathematica* **7** (1980):261-288 on p. 264.

war.[28] He also worked personally to reestablish relations between mathematicians from previously warring nations. While Mittag-Leffler may have been an "organizing" type of personality, he seems not to have been well-liked by many of his contemporaries, although he acquired and maintained substantial influence in the Swedish Academy of Sciences. This contemporary dislike may have stemmed partly from his support of female scientists, not only the Russian mathematician, Sonya Kovalevskaya, whom he helped to become a professor at Stockholm University but also Marie Curie, whom he twice helped acquire the credit due her from the Nobel Prize committee.[29] Whether or not Mittag-Leffler's support of female scientists had anything to do with his colleagues' dislike, no doubt many of his contemporaries saw his internationalism, although stemming from a citizen of a neutral country who had studied with both Charles Hermite in France and Karl Weierstraß in Germany, as an additional discomforting aspect of his uncomfortable personality.

In the immediate postwar climate, however, Hardy and Mittag-Leffler were literally voices crying in the wilderness. Those who thought the reestablishment of international relations an impossibility, or those who felt that it might take place only in the dim future, held sway in all of the victorious allied countries. In France, the principal exponents of this view were Émile Picard (who had lost a son in the war) and Gabriel Koenigs. Their nationalism found expression in Strasbourg in 1920 both in the creation of the International Mathematical Union (IMU) and in the International Congress of Mathematicians held there, a congress Mittag-Leffler termed "une affaire française."[30] The former Central Powers were not invited to the congress and they were barred from IMU membership; in fact, the French insistence on holding the congress in Strasbourg reflected their revanchist temper. Strasbourg was, after all, in Alsace, which had just returned to France as a consequence of the war. Apparently, though, all the other victorious allies agreed with France on the exclusion of Germany, Austria, Hungary, and Bulgaria. Internationalism on this occasion was thus restricted to a select group of countries, a circumstance that did not pass without vigorous protest from Hardy, then secretary of the London Mathematical Society.[31] The purportedly *International* Congress of Mathematicians in Strasbourg was largely a sham.

Opposition to the exclusion policy quickly grew, however, particularly in the United States. In 1920, the University of Chicago's Leonard E. Dickson, always internationally minded,[32] and Princeton's Luther P. Eisenhart had (without the prior approval of the American Mathematical Society) invited the next ICM to meet in New York City. At this time, there was a substantial American mathematical community,[33] and a New York gathering would have given it considerable cachet in the (largely European) world of mathematics.[34] It seems to have been the exclusion policy that prompted the withdrawal of the offer. Finally, the 1924 ICM

[28]On Mittag-Leffler and the founding of *Acta*, see June Barrow-Green's chapter in the present volume.

[29]On Kovalevskaya, see Ann Hibner Koblitz, *A Convergence of Lives: Sofia Kovalevskaia, Scientist, Writer, Revolutionary* (Boston: Birkhäuser Verlag, 1983). In fact, Kovalevskaya was the first woman to hold a professorship at a European university. On Marie Curie, see Susan Quinn, *Marie Curie: A Life* (New York: Simon & Schuster, 1995), pp. 187-189 and 327-329.

[30]Lehto, p. 24; "a French affair." Compare also Lehto's chapter in the present volume.

[31]*Ibid.*, p. 30.

[32]Compare Della Fenster's chapter on Dickson above.

[33]Parshall and Rowe.

[34]*Ibid.*, pp. 33-34.

did showcase North American mathematics, but in Toronto thanks to the efforts of John C. Fields. The circumstances behind the American and Canadian offers serve not only to illustrate internationalism in the service of nationalism but also to underscore the force of national principles. By 1924, mathematicians in Italy, Denmark, Holland, Sweden, Norway, and Great Britain had joined those in the United States in their opposition to the exclusionary clause. In June of 1926 that clause was finally lifted.

This spirit of international inclusivity mirrored broader political changes. In 1925, the German Foreign Minister, Gustav Stresemann, had negotiated the Locarno Pact, a set of treaties in which Germany voluntarily guaranteed its borders with France and Belgium and agreed to settle border disputes with Czechoslovakia and Poland by arbitration only. France, for its part, signed military aid treaties with Poland and Czechoslovakia, while Great Britain promised military aid in the event of German violation of its French and Belgian borders. (Significantly, in hindsight, Britain did not do the same for Poland and Czechoslovakia.) In the mid-1920s a new internationalist "spirit of Locarno" was abroad politically. In 1926, the same year the exclusionary clause was lifted, Germany joined the League of Nations and Stresemann shared the Nobel Peace Prize with the Frenchman, Aristide Briand. Germany was reentering the community of nations.

All was not well, however. In France, Picard and Koenigs were adamant that Germany not be permitted to participate in the ICM to be held in Bologna in 1928. Other French mathematicians took a more conciliatory view. Paul Painlevé, who, as the French Minister of War in 1917, had been responsible both for the appointment of Ferdinand Foch as head of the allied armies and for the negotiations with Woodrow Wilson that eventually brought the U.S. into the war, believed by 1926 that German mathematicians should be included in international efforts.[35] For their part, conservative German mathematicians including Erhard Schmidt, Hellmuth Kneser, and Ludwig Bieberbach argued against accepting the invitation to Bologna. In a view shared by the Dutch mathematician and ardent Germanophile, L. E. J. Brouwer, a new mathematical organization was needed, but it should be one with prominent German leadership.

The debate in Germany about rejoining the international mathematical community sparked a fierce exchange of opposing circulars between Bieberbach and his former teacher, Hilbert. In response to an inquiry from the *Rektor* at Halle, Bieberbach wrote a letter, which was widely circulated in universities and secondary schools and which attacked the notion of Germans attending the Bologna congress. Hilbert learned of this from his *Rektor* and wrote an equally widespread letter expressing his belief in the Italian dedication to true internationalism. Hilbert felt that it was in "the interest of German science and German respect" that everyone invited should attend. Bieberbach replied, as he put it "greatly against my inclination," since his "much-honored teacher" had "sharply attacked him." He intimated that Hilbert's national impulses were inadequate.[36]

[35] Gillispie, ed., *Dictionary of Scientific Biography*, s.v. "Painlevé, Paul" by Lucien Félix; and Lehto, pp. 32-40.

[36] A copy of these exchanges is in the Universitätsarchiv Greifswald. The intentionality of the wide distribution of Bieberbach's first letter is somewhat disputed. Compare Reid, *Hilbert*, p. 188; and Herbert Mehrtens, "Ludwig Bieberbach and Deutsche Mathematik," in *Studies in the History of Mathematics*, ed. Esther Phillips, MAA Studies in Mathematics, vol. 26 (Washington, D.C.: Mathematical Association of America, 1987), pp. 195-241 on p. 214.

Here again, German internationalism was in conflict. Erhard Schmidt hoped the Bologna Congress would be a disaster owing to the absence of the Germans.[37] In his view, neither the war nor the French-led exclusion from formerly international activity had altered the centrality of German mathematics. Schmidt looked forward to German dominated "internationalism," while Hilbert seems to have been a genuine internationalist much before that was popular. In the end, Hilbert led a delegation of seventy-six German mathematicians to Bologna, the largest non-Italian delegation; they played a prominent part in the scientific program.

As for the French, while Picard and Koenigs refused the invitation to Bologna, fifty-six French mathematicians were there, making it the largest and most international of the congresses to that time.[38] Hilbert, aged sixty and still suffering the effects of the pernicious anemia that had almost killed him three years earlier, was greeted by a standing ovation. Ever the internationalist, he responded as an elder statesman:

> Let us consider that we as mathematicians stand on the highest pinnacle of the cultivation of the exact sciences. We have no other choice than to assume this highest place, because all limits, especially national ones, are contrary to the nature of mathematics. It is a complete misunderstanding of our science to construct differences according to peoples and races, and the reasons for which this has been done are very shabby ones.
>
> Mathematics knows no races .... For mathematics, the whole cultural world is a single country.[39]

In Bologna, internationalism in the mathematical community seemed triumphant. The contretemps over the congress and over German exclusion, however, showed that this internationalism, while genuine, masked considerable problems relative to the international organization of mathematicians. Nevertheless, internationalism in the mathematical community continued, if only briefly, after 1928.

William Henry Young, an Englishman residing in Switzerland, succeeded Salvatore Pincherle as President of the International Mathematical Union. In a sense, his open-mindedness made him an ideal candidate to attempt a genuine *rapprochement* over the issue of true internationalism and German participation in both the Union and the congresses. Moreover, Young had strong ties to Germany. Despite the fact that he and his wife, the former Grace Chisholm, had lost a son in the war, they had lived in Germany before the war and had revered Felix Klein. They had also greatly valued the pre-war, German way of life. Young's efforts for internationalism ultimately came to naught, however. The political questions involved seemed to form a Gordian Knot, impossible to unravel.[40] While all agreed that international congresses should continue, many wished to end the IMU as an organization for official international mathematical collaboration.

---

[37] Erhard Schmidt to Hellmuth Kneser, undated, Nachlaß Hellmuth Kneser, in the possession of Martin Kneser.

[38] *Atti del Congresso internazionale dei matematici, Bologna, 3–10 settembre 1928*, 6 vols. (Bologna: Nicola Zanichelli, 1929), 1:63.

[39] Reid, *Hilbert,* p. 188.

[40] Lehto, pp. 50-56. See also Ivor Grattan-Guinness, "A Mathematical Union: William Henry and Grace Chisholm Young," *Annals of Science* **29** (1972):105-186, especially p. 154 and pp. 174-177.

The Americans were in the forefront of the effort to abolish the IMU. Delegates like Oswald Veblen and Roland G. D. Richardson declared that such an organization served no real purpose. Others thought that an organization of this type would inevitably become (as it had) too politicized. When the IMU as an official structure for internationalism died in 1932,[41] the congresses were the only official international activity of note. An International Congress of Mathematicians was held in Zürich in 1932, but official international mathematical collaboration died there.

In a sense, this should come as no surprise. As Paul Forman has shown with respect primarily to the Weimar physicists, but also touching other academic areas including mathematics, scientific superiority had been a *Machtersatz*—a substitute for political power—for the Germans after their defeat in World War I.[42] Brigitte Schröder-Gudehus actually traces these feelings back to 1870 and comments on the multiplication of "transnational" scientific societies at the end of the nineteenth century, which, nevertheless, did not have truly international aims. Before World War I, the primary international spirit in science, was, as the famous philologist Hermann Diels remarked, to obtain an international foundation for national interests.[43] By 1933, "if finally we ask what part of the specific vitality of Weimar scientific culture derived from a genuinely internationalist impulse," Forman concluded, "I think the answer must be: none at all."[44] A cartoon published in 1919 in the German satirical magazine, *Simplicissimus*, supports this position. It shows a German professor setting a globe spinning with a slap. The caption reads: "Demonstrate at home! As proof of his unshakable protest every morning in Berlin Professor Patrius Rumbler gives the globe a couple of resounding smacks on the face."[45] (See Figure 1.)

Thus, while "science" or "mathematics" might be "international," its investigation by humans was decidedly national. It is, therefore, not surprising that an effort at internationalization like the IMU foundered in the 1930s, while the congresses, where each nation could display its wares, succeeded.[46]

[41]For the truly Byzantine interactions of the various organizations involved in the demise of the IMU, see Lehto, and compare his chapter below.

[42]Forman, "Scientific Internationalism and the Weimar Physicists." Compare Brigitte Schröder-Gudehus, *Deutsche Wissenschaft und internationale Zusammenarbeit 1914-1928* (Geneva: Dumaret and Golay, 1966).

[43]Brigitte Schröder, "Caractéristiques des relations scientifiques internationales, 1870-1914." It may seem curious to cite Diels, a philologist, in a book about mathematicians, but so far as I know, no author has ever found a distinction between the general attitudes of scientists and mathematicians in Germany and those of their academic colleagues in other specialties. David Rowe in "'Jewish Mathematics' at Göttingen in the Era of Felix Klein" (*Isis* **77** (1986):422-429) does claim (p. 426) that Göttingen mathematicians showed that mathematicians and natural scientists did not possess the same "mandarin" attitudes as their colleagues, but what seems nearer the truth is that the Göttingen mathematicians were the exception, under the leadership of Klein and David Hilbert.

[44]Forman, p. 180.

[45]*Ibid.*, p. 150; Forman's translation.

[46]There was some success relative to international efforts directed to mathematics education. Felix Klein had spearheaded these efforts, serving as President of the International Commission on Mathematical Education until 1920. Although politics set in there, too, Henri Fehr, the Swiss founder of the journal *L'Enseignement mathématique*, was able to remark in 1936 that there was "un excellent esprit de compréhension et de collaboration [an excellent spirit of understanding and collaboration]" on issues of mathematics education internationally. Lehto, pp. 65-66.

Demonstriere zu Hause! (Zeichnung von O. Gulbransson)

Professor Thuisko Groller in Berlin versetzt zum Zeichen seines unentwegten Protestes jeden Morgen dem Globus ein paar schallende Ohrfeigen.

FIGURE 1

Hitler's Germany and Mathematical Refugees

On 30 January, 1933, Hitler became Chancellor of Germany, and on 7 April, the first decree expelling German professors (who were by law civil servants) from their jobs was issued. Some who fell under the exceptions clauses in the original law, like Richard Courant, were forcibly furloughed instead. Whatever internationalist impulses had been fostered were supplanted by aggressive nationalism; refugees followed. If one were not Jewish—in the Nazi definition of having at least one Jewish grandparent—and if one had not been active in an anti-Nazi group like the Socialist Party, then generally one could remain employed. Not all of those affected were Jewish by Nazi definition. One of the paragraphs of the 7 April law (officially for the reform of the civil service) allowed the dismissal of those who could not unreservedly support the national state; another paragraph provided for the forcible retirement of faculty on the grounds of redundancy and the need to simplify university organization. By 6 May, the decree was enlarged to cover all university faculty, but, in fact, 90% of the mathematicians emigrating from Germany were Jewish.[47]

The "non-Jewish" mathematicians, who were either dismissed or who "voluntarily" resigned and then emigrated, included, among others, Peter Thullen, Max Zorn, Hans Rademacher, Otto Neugebauer, Hans Schwerdtfeger, Hermann Weyl, and Hanna von Caemmerer. Their reasons for leaving varied. Thullen was a devout Roman Catholic who felt he could not stay in Hitler's Germany.[48] Zorn, a youthful communist, consequently could not habilitate.[49] Rademacher belonged to a minority of German academics dismissed because of a left-liberal persuasion.[50] Neugebauer was (falsely) supposed to be a communist.[51] Schwerdtfeger early felt that Hitler would be a German disaster that he could not abide.[52] Weyl had a Jewish wife.[53] Hanna von Caemmerer was not only suspected of being friendly to Jews but was actually planning to marry one, Bernhard Neumann.[54]

In all, there were three waves of dismissals. On 21 January, 1935, a new edict led to the forcible retirement or dismissal of professors in whom there was not supervening university interest in their retention. Max Dehn and Neugebauer, who had actually been in Copenhagen since 1933, were both dismissed at this time. In 1937, those who were either married or "related" to Jews were dismissed, unless they chose to divorce.[55] This new circumstance prompted Emil Artin, whose wife was a "half-Jew," to emigrate in that same year.[56]

Although these emigrations dealt a serious blow to German mathematics, a substantial and active mathematical community remained. Some non-Jewish mathematicians who were affected negatively by the Nazi regime did not emigrate. Two

---

[47]Reinhard Siegmund-Schultze, *Mathematiker auf der Flucht vor Hitler* (Braunschweig: Vieweg Verlag, 1998), p. 24.

[48]Peter Thullen, interview with the author, in Fribourg, Switzerland, 22 March, 1988.

[49]Max Zorn, interview with the author, in Bloomington, Indiana, 18 March, 1981.

[50]Siegmund-Schultze, p. 77; Ivan Niven, "The Threadbare Thirties," in *A Century of Mathematics in America, Part I*, ed. Peter L. Duren *et al.* (Providence: American Mathematical Society, 1988), pp. 209-228 on p. 222.

[51]Norbert Schappacher and Martin Kneser, "Fachverband—Institut—Staat," in *Ein Jahrhundert Mathematik, 1890–1990: Festschrift zum Jubiläum der DMV*, ed. Gerd Fischer *et al.* (Braunschweig: Vieweg Verlag, 1990), pp. 1-82 on p. 41.

[52]Siegmund-Schultze, p. 71.

[53]Norbert Schappacher, "Das Mathematische Institut der Universität Göttingen," in *Die Universität Göttingen unter dem Nationalsozialismus*, ed. Heinrich Becker *et al.* (Munich: K. G. Saur, 1987), pp. 345-373. See p. 352 for the relevant passage from Weyl's letter of resignation in 1933. Weyl was not being pressured to resign at this time, although he presumably would have been in 1937.

[54]M. F. Newman and G. E. Wall, "Hanna Neumann," *Journal of the Australian Mathematical Society* **17** (1974):1-28. Von Caemmerer's father was from a family of Prussian officers; her mother was a German Huguenot. She and Neumann met in January 1933, became engaged secretly in England (where he was then living) at Easter 1934, and emigrated. They secretly married in 1938. As in Weyl's case, von Caemmerer's departure was, technically speaking, "voluntary."

[55]Divorce, however, left the former spouse without even the minimal protection of an "Aryan" partner.

[56]See Schappacher and Kneser on these "waves of dismissal." The concentration here on the emigrés, especially those who came to the United States, owes to their putative influence in creating an "international mathematical community."

(quite different) examples are Erich Kamke and Helmut Ulm. Kamke, who had a "half-Jewish" wife, resigned his post in 1937, but stayed in Germany.[57] Ulm had developed friendly relationships with many of his professors, all of whom were Jewish, and could not habilitate until the war's end.[58] There were even some non-Jews who returned to Hitler's Germany, having somewhat naïvely misjudged the situation. Two examples among famous mathematicians are Carl Ludwig Siegel and Eberhard Hopf. Siegel managed to reemigrate on the last boat from Norway; Hopf remained trapped, to judge by his postwar views, in Germany.[59] Those who comprised the German mathematical community during the Nazi regime largely continued to carry on their researches and their broader mathematical activities.

As a case in point, the first ICM to be held after Hitler's rise to power took place in Oslo in 1936, and while international politics definitely affected the audience, the show went on. At the last minute, Stalin refused to allow Russian mathematicians to attend, perhaps owing to the fact that Leon Trotsky was living in Oslo then.[60] Italy also boycotted the congress because Mussolini was offended by the—in fact, largely ineffective—sanctions imposed by the League of Nations after his invasion of Ethiopia.[61] As for the Germans, some came from Nazi Germany, but numerous German emigrés also attended.[62]

The German participation in the Oslo congress suggests that the idea of "internationalism" persisted under the Nazis, even if German hegemony was a given. In 1939, for example, Francesco Severi had apparently laid plans for an "international mathematical congress" to convene in Rome from 22-28 October, 1939 with the participation of a German delegation headed by Helmut Hasse. This, however, was not a universally recognized international congress; that was to be the congress then set to meet in Cambridge, Massachusetts in 1940 in response to an invitation extended by the American Mathematical Society in Oslo in 1936.[63] The onset of the war ultimately meant that neither congress took place as planned, although the Rome congress did finally meet from 8-15 November, 1942, and Hasse did lead the Germans. Other national delegations included ones from Romania and Bulgaria, but communication apparently proved difficult unless one spoke Italian. Wilhelm Blaschke, a member of the German contingent, spoke Italian fluently and noted that German was poorly understood by the Italians. Although French was probably more accessible, Blaschke had already officially remarked in the autumn of 1941 that it would be unfortunate if German professors attempted to lecture in

---

[57] *Ibid.*, p. 49.

[58] See Helmut Ulm, Personalakten Ulm, Universität Münster.

[59] On Siegel, see, for example, Siegmund-Schultze, p. 211. On Hopf, see Eberhard Hopf to Richard Courant, 23 June, 1945 (five-page letter written in English), Courant Papers, New York University.

[60] Lehto, p. 69. Some Soviets may have attended, including the well-known Fourier analyst, Nina Bary. See *Comptes rendus du Congrès international des mathématiciens, Oslo*, 2 vols. (Oslo: A. W. Brøggers, 1937), pp. 29 and 39. (The latter reference indicates the attendance of eleven from the U.S.S.R.) A. Gelfond and A. Khinchine, both Soviet, were prevented from coming (*ibid.*, p. 10), despite the fact that Khinchine is listed as attending (*ibid.*, p. 33).

[61] Lehto, pp. 67-68. Although Vito Volterra and four other Italians are listed as attending, Volterra clearly did not. He had refused to take an oath of loyalty to Mussolini and consequently had been dismissed from any academic position. Compare *Comptes rendus du Congrès international des mathématiciens, Oslo*, pp. 37 and 39, and Lehto, pp. 70-71, especially note 2.

[62] Lehto, p. 69. See also the chapter by Siegmund-Schultze above.

[63] *Ibid.*, p. 71. Compare also Table 4 in the Appendix to Siegmund-Schultze's chapter above.

French (presumably referring to the war).[64] The pseudo-internationalism of this 1942 "congress" attests not only to the on-going participation of German mathematicians in events outside Nazi Germany but also to how commonplace the notion of multinational mathematical gatherings had become.

As this "business as usual" suggests, German mathematicians and, on the whole, the German professoriat seem to have been willing, at least at first, to adapt to the Nazi regime. Although there were many variations of behavior, from pro-Nazi ideologues like Bieberbach or Oswald Teichmüller, to cynical opportunists like Blaschke, to Prussian bureaucrats who actively supported the state like Hasse, to anti-Nazis like Erich Hecke, Kurt Reidemeister, or Kamke, all, except the first type, tried to keep politics out of mathematics. Most German mathematicians were not interested in being ideologues.

Indeed, this essentially apolitical stance stemmed from the way in which the notion of *Lern-und Lehrfreiheit* was actually implemented in the German context. The scholar was independent as a scholar, but he was a state servant as a civil servant. In return for having classroom and scholarly independence, the professor would not become politically involved. Thus, the "apolitical German professor" was born.[65] This peculiarly German view of academic freedom thus provides a partial explanation for the initial silence of much of the German professoriat in the face of Hitler's government and the dismissal of their colleagues. While some individual mathematicians were concerned and made an effort to help, the response of the German Mathematical Society in 1933 was one of total passivity.[66]

The United States: Country of Refuge for Mathematicians

The principal country of refuge for those leaving Germany was the United States, although Great Britain also played a substantial role. A few refugees went elsewhere with varying success, for example, Stefan Cohn-Vossen, Fritz Noether, and Stefan Bergmann went to the Soviet Union, Friedrich Levi fled to Calcutta, and Peter Thullen emigrated to Ecuador. These tertiary loci, however, played no role early on in creating an international community of mathematicians.

In the United States, the situation was not an easy one for those who immigrated.[67] First, there was the worldwide depression. Second, most of the emigrés

---

[64]See Blaschke's report dated 5 December, 1941 of lecture trips to western France and Sicily (where he had lectured both to the *Luftwaffe* and at universities in October 1941) in Personalakten Blaschke, Universitätsarchiv Hamburg. See also his report on the Rome Congress dated December 1942 in *ibid.*

[65]Fritz Stern, "The Political Consequences of the Unpolitical German," *History* **3** (1969):104-134, reprinted in Fritz Stern, *The Failure of Illiberalism* (New York: Alfred Knopf, 1972), pp. 3-35; and Eric Voegelin, "Die deutsche Universität und die Ordnung der deutschen Gesellschaft," in *Die deutsche Universität im Dritten Reich* (Munich: Piper Verlag, 1966), pp. 241-282.

It is true that many professors opposed, more or less vehemently, the Weimar Republic, but that is at least partly because, operating in an unpolitical ideal world, they viewed the Weimar government, the heir of the Versailles *Diktat*, as inappropriately imposing French parliamentary forms on Germany. In fact, one group of prominent professors supported the Weimar government because it was the "sensible [*vernunftig*]" thing to do, not because they believed in true republican parliamentary forms. They were known as *Vernunftrepublikaner*.

[66]For example, see "Geschäftssitzung der DMV an 20.9.1933 in Würzburg," *Jahresbericht der Deutschen Mathematiker Vereinigung* **43** (2) (1934):80.

[67]Much has been written on this issue. In addition to the book by Siegmund-Schultze cited above, see Nathan Reingold, "Refugee Mathematicians in the United States of America, 1933-1934: Reception and Reaction," *Annals of Science* **38** (1981):313-338; reprinted in *A Century of*

were Jewish, at least in the sense of having some Jewish antecedents, and there was more than a touch of antisemitism in the U.S. Only after World War II did American antisemitism begin to decline in society as a whole and in academe in particular.[68] Third, there was a worry about foreigners in general. How many foreigners did one want in a department? How did one deal with the potential language problems? All of these factors combined resulted in a lack of American academic jobs for mathematicians, an undercurrent of antisemitism, and a natural nationalism. Foreigners seeking refuge in the United States, most of whom were Jews, thus had to prove their ability to participate usefully in American academic life regardless of how distinguished they had been abroad. Providing such proof often proved arduous; good jobs were not abundant.

Despite these complicating factors, the United States and Great Britain ultimately absorbed more refugee mathematicians than had been thought possible in 1933. In March 1939, Oswald Veblen believed that the "saturation point" for displaced scholars had almost been reached in the "more prominent" universities but that there was still room in "less well known academic institutions." Placing emigrés in the latter category of schools, he felt, would both benefit the emigré and improve the host institution.[69] Six months later, with the onset of the war in Europe and with the increasing possibility of U.S. involvement, the emigrés were actively welcomed. This was especially true after American involvement began.[70]

Once in the United States, what kind of mathematical environment did these emigrés encounter? Lipman Bers noted that although jobs were difficult to come by in the U.S. in the thirties, the scientific level of American mathematics was high.[71] In particular, there were strong trends in both algebra and topology. At the University of Chicago, E. H. Moore's first Ph.D. student had been the redoubtable L. E. Dickson. In his 1896 dissertation, Dickson had laid the groundwork for what became his classic 1901 book, *Linear Groups with an Exposition of the Galois Field Theory*; he went on to do seminal work in the theory of algebras and their arithmetics.[72] One of Dickson's most prominent students, A. Adrian Albert, earned his Chicago Ph.D. in 1928 and similarly proceeded to establish himself as an international leader in theory of algebras. In fact, Dickson and his progeny as well

---

Mathematics in America, Part I, pp. 175-200; and Robin Rider, "Alarm and Opportunity: Emigration of Mathematicians and Physicists to Britain and the United States, 1933–1945," *Historical Studies in the Physical Sciences* **15** (1984):107-176.

[68]Since antisemitism has essentially disappeared among mathematicians in the U.S., it is perhaps difficult to realize that it was not only alive and well in the 1930s but also supported at many levels of society from the State Department on down. On this phenomenon, see, for example, David S. Wyman, *Abandonment of the Jews: America and the Holocaust, 1941–1945* (New York: Pantheon Books, 1984).

[69]Siegmund-Schultze, p. 176.

[70]Still, while the fear that emigrés would take jobs away from Americans vanished, those who had not become U.S. citizens aroused increased suspicion as "citizens" of an enemy state. See *Ibid.*, p. 198.

[71]Lipman Bers, "The European Mathematicians' Migration to America," in *A Century of Mathematics in America-Part I*, ed. Peter L. Duren *et al.* (Providence: American Mathematical Society, 1988), pp. 231-243 on p. 238.

[72]Leonard E. Dickson, *Linear Groups with an Exposition of the Galois Field Theory* (Leipzig: B. G. Teubner Verlag, 1901); *Algebras and Their Arithmetics* (Chicago: University of Chicago Press, 1923); and *Algebren und ihre Zahlentheorie* (Zürich: Orell Füssli, 1927). See also Della Dumbaugh Fenster, "Leonard Eugene Dickson and His Work in the Arithmetics of Algebras," *Archive for History of Exact Sciences* **52** (1998):119-159, and compare Fenster's chapter in the present volume.

as George A. Miller and his students dominated algebraic research in the United States during the fifty years prior to 1938.[73]

One mathematical immigrant who found this algebraic climate conducive, Emmy Noether, arrived in the United States in 1933. Reinhard Siegmund-Schultze has argued that her influence was responsible for the direction in which American algebra ultimately moved. [74] Noether's influence actually began well before her arrival on American shores and extended long after her death in 1935. In fact, she was influential not so much through her personality as through her ideas. As E. T. Bell noted in his study of "Fifty Years of Algebra in America, 1888-1938," the rise in America of "recent abstract algebra" had begun in the 1920s and had become dominant by the 1930s; it was also around 1920 that Emmy Noether's interests had turned in an increasingly abstract direction.[75] Thus, her interests meshed with the developing researches of American algebraists, which, as Emil Artin observed, were far more abstract before the 1930s than those of their European counterparts. American algebraists were thus poised to weave together the strands of the "Schur school," Dickson's approach, and Noether's point of view.[76]

Like Dickson in algebra, two other early students of E. H. Moore, Oswald Veblen and Robert L. Moore, established American research traditions, in geometry and topology in the case of Veblen, and in topology in that of Moore. Topology (or "analysis situs" as it was still sometimes called as late as 1931) was distinctly European in origin. Arguably begun by Henri Poincaré, some of its key ideas can be traced to earlier in the nineteenth century, such as the notion of homeomorphism in the work of August Ferdinand Möbius. Veblen, who became an important statesman of mathematics from his positions first at Princeton University and then at the Institute for Advanced Study, wrote what was perhaps the first significant treatise on topology, his *Analysis Situs* of 1922.[77] This, however, was soon superseded by the *Lehrbuch der Topologie* (1934) of Herbert Seifert and William Threlfall and by the *Topologie* (1935) of Paul Alexandrov and Heinz Hopf.[78] While Americans like Veblen and James W. Alexander made significant contributions to algebraic topology in the first four decades of the twentieth century, it was far from being an "American subject." Veblen's fellow student, R. L. Moore, on the other hand, animated a very American school of general and point-set topology first at the University of Pennsylvania and later at the University of Texas at Austin. This school, based on the highly idiosyncratic "Moore method" of teaching, thrived in the U.S. but had rather little influence elsewhere. In particular, Moore's topological study of real three-space and kindred issues seems to have been relegated

---

[73]Miller was Felix Klein's mathematical "grandson" via Frank Nelson Cole. On algebra in America, see Eric Temple Bell, "Fifty Years of Algebra in America, 1888-1938," in *Semicentennial Addresses of the American Mathematical Society* (New York: American Mathematical Society, 1938; reprint ed., New York: Arno Press, 1980), pp. 1-34 on pp. 7-8. See also Parshall and Rowe, pp. 381 and 445.

[74]Siegmund-Schultze, pp. 240-246.

[75]Bell, p. 7 (tables) and pp. 32-33.

[76]Emil Artin, "The Influence of J. H. M. Wedderburn on the Development of Modern Algebra," *Bulletin of American Mathematical Society* **56** (1950):65-72. The principal member of the school of Isaai Schur to emigrate to the United States was Richard Brauer.

[77]Oswald Veblen, *Analysis Situs* (New York: American Mathematical Society, 1922).

[78]Herbert Seifert and William Threlfall, *Lehrbuch der Topologie* (Leipzig and Berlin: B. G. Teubner Verlag, 1934); and Paul Alexandrov and Heinz Hopf, *Topologie* (Berlin: Springer-Verlag, 1935).

to the sidelines as interesting but not central. This seems at least partly because Moore's axiomatic system faded in importance compared to (or lacked the universal acceptance of) those developed in Europe.[79]

As these cases illustrate, there was an established American mathematical research community by the 1930s, but, compared to Europe, the United States was weak in both classical analysis and the applications of mathematics. Given that America was supposed to be the land of practicality and applications of science, the latter weakness seemed particularly striking. As Mina Rees judiciously remarked, despite strong groups at New York University (NYU) under Richard Courant and at Brown University under Wilhelm Prager (both Courant and Prager were German emigrés), neither statistics nor "applied mathematics [were] strongly represented at American universities" in the 1930s and 1940s.[80]

The European emigrés brought with them to the United States a concern for classical analysis (Karl Löwner, George Pólya, Gabor Szegö, Antoni Zygmund, among many others), a concern for probability and statistics (especially Richard von Mises, Willy Feller, Abraham Wald, and Jerzy Neyman) and other applied mathematics (Courant, Friedrichs, Prager, Theodor von Kármán, for example). Technically, however, neither Pólya nor Neyman were emigrés fleeing Hitler. Neyman was born in Moldavia of Polish parentage, trained in Russia, and came to the U.S. permanently from England in 1938, while Pólya left Switzerland in 1940 for a to-be-permanent stay in the U.S. because he did not know how the war would affect Switzerland.[81]

With America's involvement in World War II, the emigré mathematicians who had been adopted initially for moral and ethical reasons, were suddenly no longer a burden but were an asset with their applied orientation. They had arrived at an opportune time for the United States. Concern for applications would, however, have undoubtedly come with the war. Even in Germany, mathematicians were concerned that applications were not sufficiently promoted. As late as 1941, some German mathematicians used an article by the soon-to-be director of Bell Laboratories, Thornton Fry, to advocate greater attention to applications.[82] At the time, the U.S. was still technically a neutral nation, but Germany was in its second year of war and about to open a second front. It was only in 1940 that applied mathematics became a potential primary subject for university study by future mathematics teachers in Germany. A year later, Ernst Mohr, while giving the applied mathematics curriculum in Breslau (modern Wroclaw), complained about the lack of adequate teaching personnel occasioned by the war.[83] Thus, activity in the United States could be a stimulus for German activity; it was not always the

---

[79]On the schools that developed around Veblen and Moore, see Parshall and Rowe, pp. 448-450.

[80]Mina Rees, "The Mathematical Sciences and World War Service," *American Mathematical Monthly* **87** (1980):607-621; reprinted in *A Century of Mathematics in America-Part I*, pp. 275-290 on p. 275.

[81]On Neyman, see Constance Reid, *Neyman–from Life* (New York: Springer-Verlag, 1982), pp. 5, 13-14, and 155-157. On Pólya, consult Siegmund-Schultze, p. 10.

[82]Thornton Fry, "Industrial Mathematics," *Bell System Technical Journal* **20** (3) (1941):255-292. For German use of this article, see Ludwig Prandtl Nachlaß in the archives of the Max-Planck Institut für Strömungsforschung, Mappe #1, "Briefwechsel mit den Herren Professoren Seewald, Süss, Doetsch, Wegner, Blaschke."

[83]Ernst Mohr, "Über den gegenwartigen Universitätsunterricht in der angewandten Mathematik," *Deutsche Mathematik* **6** (1941):413-504.

other way around. Both countries—worried as they were about the possibility, even likelihood, of eventual American involvement in the war—served as a propaganda foil for the other.

In the United States, applied mathematics was revolutionized at NYU by Courant and at Stanford by Szegö, while von Kármán proved instrumental at the California Institute of Technology. Other emigrés had similar influence. For example, Hans Rademacher founded a whole school of analytic number theory in America following his establishment first at Swarthmore College and then at the University of Pennsylvania. Emigration thus had a profound effect on the American mathematical research community, creating a more homogeneously knowledgeable body of researchers, but was this tantamount to an international community?

Internationalism under American Leadership

The emigrés and the war certainly served to reduce both American suspicion of foreigners and academic antisemitism. Moreover, from an American point of view, the effects of the emigration of European mathematicians provided a base from which the international community could grow. Still, Americans had been in the lead in destroying the international mathematical structures that had existed after World War I. More accurately, men like Richardson, Veblen, and Norbert Wiener, believing that these structures had already been destroyed by politics, wished to bury them completely since attempts at reviving them seemed hopeless. After World War II, however, political considerations in mathematical internationalism, and especially later in connection with the Cold War, once again appeared, and yet attitudes toward the defeated Axis powers seemed different.

What accounted for the difference in American behavior after World War I and after World War II? It is certainly the case that during World War I academics on both sides participated in plenty of home-front exhortation; both sides massacred each other unmercifully at places like Verdun and Passchendaele; there were not many refugees, let alone enforced intellectual refugees, from one side to the other. In this light, it may be argued that World War I was viewed by its victors as the criminal action of the legitimate states of Germany, Austro-Hungary, and Turkey. After the war, dominant factions in the victorious countries thus held that the vanquished merited punishment as states in all spheres.

In contrast, World War II was viewed by its victors as the consequence of a criminal take-over of the German government. The emigrés of all sorts, by their very presence, contributed to this idea. The very need of the Allied side to absorb refugees led to a reduction of xenophobia and antisemitism. Among mathematicians, Richard Courant provides a case in point. Courant had been an enthusiastic supporter of World War I, had been mobilized, had risen to the rank of lieutenant, had attempted to invent earth telegraphy for military use, and, in 1915, had been wounded by an English bullet.[84] Twenty years later, he was a refugee in the United States. In another six years, he was working on military projects for that country, and although he had no religious practice, he was a Jew by German (and American) standards. Neither the latter fact nor his provenance inhibited his American

[84] Constance Reid, *Courant in Göttingen and New York* (New York: Springer-Verlag, 1976), pp. 49-68.

activity, however.[85] Like Courant, few of the refugees returned to Europe after the conclusion of World War II. As for the Germans and Austrians who survived, and who were not emigrés, those deemed in the U.S. to have supported the Nazi government (like Hasse) were ostracized by many after the war. Thus, in the aftermath of World War II, the victors perceived individuals, not nations, as guilty.[86] Individuals from all nations could thus make common cause mathematically.

Another factor in the growing mathematical internationalism after World War II was the economic and mathematical attractiveness (for research) of the United States. The latter was no doubt an effect of the emigrés and their students. The emigrés had almost instantaneously increased substantially the density of good research in America in many areas of mathematics. Because most did not return to Europe, they continued to enrich American mathematics, helping to make the United States a leading—if not the leading—postwar mathematical country and a magnet for foreign mathematicians after 1945. In this light, it was just as natural for the United States to take a leadership role in fostering a new internationalism as it had been for Germany in the years before 1914. It was also natural for the United States to earn a national reputation thereby.[87]

The emigrés from Nazi Germany undeniably served as a catalyst for the emergence both of the United States as a mathematical force internationally and of an international community of mathematicians, but, as should now be clear, their presence was not the only factor affecting these changes. At a deeper level, politics played a significant role in the restriction as well as in the expansion of an international mathematical community. Members of the mathematical community sometimes like to think that politics are not their concern, but politics have fundamentally shaped the mathematical world they so much take for granted.

References

Unprinted Sources

Blaschke, Wilhelm, Personalakte Blaschke, Universitätsarchiv Hamburg.
Courant, Richard. Courant Papers. New York University.
Kneser, Hellmuth. Nachlaß. In the possession of Martin Kneser.

[85]This is not to deny the difficulties many of the emigrés had in the United States for ethnic, social, and even pedagogical reasons, but they were absorbed. See Siegmund-Schultze, Chapter 9; Bers, pp. 234-237; and Niven, pp. 222-228.

[86]The behavior of the victorious Allies after the war added to this sense, with the *Fragebogen* and denazification campaign as well as with the policy of entrusting university governance to (supervised) academics who had been judged free of any Nazi taint. (Mathematicians in this category included Erich Hecke and Heinrich Behnke.) This, of course, has nothing to do with the fact that within Germany, many of those individuals who collaborated with the Nazis eventually were judged sufficiently unburdened by Nazi guilt to be reinstated in office. In the contemporary German jargon, they received "*Persilscheine*" or were "washed white" (Persil is the brand name of a popular German laundry detergent). The focus of "denazification" was on how seriously individuals were compromised by their involvement with the criminal regime just defeated.

[87] For a brief period in the postwar era, however, there was a sense of French mathematical intellectual dominance with the spread of the gospel of Bourbaki (the gospel itself having originated in the 1930s). Yet, the fact that two influential members of Bourbaki, André Weil and Claude Chevalley, both spent significant parts of their careers in the United States—Weil first at Chicago and then in Princeton from 1947 onward and Chevalley from 1938 to 1957—further supports the American claim to dominance.

Prandtl, Ludwig. Nachlaß. Max-Planck Institut für Stromungsforschung, Mappe #1.

Russell, Bertrand. Russell Correspondence. Russell Archive. McMaster University.

Schemann, L. Nachlaß. Handschriftenabteilung. Universitätsbibliothek Freiburg im Breisgau.

Thullen, Peter. Interview with the author. 22 March, 1988 in Fribourg, Switzerland.

Ulm, Helmut. Personalakte Ulm. Universitätsarchiv Munster.

Universitätsarchiv Greifswald. Copies of round-robin letters of Ludwig Bieberbach and David Hilbert, 1928.

Zorn, Max. Interview with the author. 18 March, 1991 in Bloomington, Indiana.

Published Sources

Alexandrov, Paul and Hopf, Heinz. *Topologie.* Berlin: Springer-Verlag, 1935.

Arnold, Matthew. *Higher Schools in Germany.* London: Macmillan, 1925.

Artin, Emil. "The Influence of J. H. M. Wedderburn on the Development of Modern Algebra." *Bulletin of the American Mathematical Society* **56** (1950):65-72.

Atti del Congresso internazionale dei matematici, Bologna, 3–10 settembre 1928. 6 Vols. Bologna: Nicola Zanichelli, 1929.

Bell, Eric Temple. "Fifty Years of Algebra in America: 1888–1938." In *Semicentennial Addresses of the American Mathematical Society.* New York: American Mathematical Society, 1938; Reprint Ed., New York: Arno Press, 1980, pp. 1-34.

Bers, Lipman. "The European Mathematicians' Migration in America." In *A Century of Mathematics in America—Part I.* Ed. Peter L. Duren *et al.* Providence: American Mathematical Society, 1988, pp. 231-243.

Biermann, Kurt R. *Die Mathematik und ihre Dozenten an der Berliner Universitat 1810–1933.* Berlin: Akademie Verlag, 1988.

Böhme, Klaus. *Aufrufe und Reden deutscher Professoren im Ersten Weltkrieg.* Stuttgart: Phillip Reclam, 1975.

Chapman, John Jay. *Deutschland über Alles or Germany Speaks.* New York and London: C. P. Putnam, 1914.

Clark, Ronald. *Einstein: The Life and Times.* New York: World Publishing, 1971.

Comptes rendus du Congrès international des mathématiciens. 2 Vols. Oslo: A. W. Brøggers, 1937.

Dauben, Joseph W. "Mathematicians and World War I: The International Diplomacy of G. H. Hardy and Gösta Mittag-Leffler." *Historia Mathematica* **7** (1980):261-288.

Dickson, Leonard E. *Algebras and Their Arithmetics.* Chicago: University of Chicago Press, 1923.

______. *Linear Groups with an Exposition of the Galois Field Theory.* Leipzig: B. G. Teubner Verlag, 1901.

______. *Algebren und ihre Zahlentheorie.* Zürich: Orell Füssli, 1927

Fenster, Della Dumbaugh. "Leonard Eugene Dickson and His Work in the Arithmetics of Algebras." *Archive for History of Exact Sciences* **52** (1998):119-159.

Forman, Paul. "Scientific Internationalism and the Weimar Physicists: The Ideology and Its Manipulation in Germany after World War I." *Isis* **64** (1973):151-180.

Frank, Philipp. *Einstein, His Life and Times.* New York: Alfred A. Knopf, 1947.

Fry, Thornton. "Industrial Mathematics." *Bell System Technical Journal* **20** (3) (1941):255-292.

"Geschäftssitzung der DMV an 20.9.1933 in Würzburg." *Jahresbericht der Deutschen Mathematiker-Vereinigung* **43** (2) (1934):80.

Gillispie, Charles C., Ed. *Dictionary of Scientific Biography.* 16 Vols. and 2 Supps. New York: Charles Scribner's Sons, 1970–1990.

Grattan-Guinness, Ivor. "A Mathematical Union: William Henry and Grace Chisholm Young." *Annals of Science* **29** 1972:105-186.

Hardy, Godfrey H. "The *J*-Type and the *S*-Type among Mathematicians." *Nature* **134** (1934):250.

Hardy, Godfrey H. and Riesz, Marcel. *The General Theory of Dirichlet Series.* Cambridge Mathematical Tracts. Vol. 18. Cambridge: University Press, 1915.

Knightley, Phillip. *The First Casualty.* New York: Harcourt, Brace, Jovanovich, 1975.

Koblitz, Ann Hibner. *A Convergence of Lives: Sofia Kovalevskaia, Scientist, Writer, Revolutionary.* Boston: Birkhäuser Verlag, 1983.

Lehto, Olli. *Mathematics Without Borders: A History of the International Mathematical Union* New York: Springer-Verlag, 1991.

Mehrtens, Herbert. "Ludwig Bieberbach and Deutsche Mathematik." In *Studies in the History of Modern Mathematics.* Ed. Esther Phillips. MAA Studies in Mathematics. Vol. 26. Washington, D.C.: Mathematical Association of America, 1987, pp. 195-241.

Mohr, Ernst. "Über den gegenwartigen Universitätsunterrich in der angewandten Mathematik." *Deutsche Mathematik* **6** (1941):413-504.

Newman, M. F. Wall, G. E. "Hanna Neumann." *Journal of the Australian Mathematical Society* **17** (1974):1-28.

Niven, Ivan. "The Threadbare Thirties." In *A Century of Mathematics in America–Part I.* Ed. Peter L. Duren *et al.* Providence: American Mathematical Society, 1988, pp. 209-228.

Osgood, William Fogg. *Lehrbuch der Funktionentheorie.* Leipzig: B. G. Teubner Verlag, 1906.

Parshall, Karen Hunger. "Mathematics in National Contexts (1875–1900): An International Overview." In *Proceedings of the International Congress of Mathematicians, Zürich, Switzerland, 1994.* 2 Vols. Basel: Birkhäuser Verlag, 1995, 2:1581-1591.

______. "Towards a History of Nineteenth-Century Invariant Theory." In *The History of Modern Mathematics.* Ed. David E. Rowe and John McCleary. 2 Vols. Boston: Academic Press, Inc., 1989, 1:157-206.

Parshall, Karen Hunger and Rowe, David E. *The Emergence of the American Mathematical Research Community, 1876–1900: J. J. Sylvester, Felix Klein, and E. H. Moore.* HMATH. Vol. 8. Providence: American Mathematical Society and London: London Mathematical Society, 1994.

Pyenson, Lewis. *Cultural Imperialism and Exact Science: German Expansion Overseas, 1900–1930.* New York: P. Lang, 1985.

Quinn, Susan. *Marie Curie: A Life* New York: Simon and Schuster, 1995.

Rees, Mina. "The Mathematical Sciences and World War Service." *American Mathematical Monthly* **87** (1980):607-621. Reprinted in *A Century of Mathematics in America—Part I.* Ed. Peter L. Duren *et al.* Providence: American Mathematical Society, 1988, pp. 275-289.

Reid, Constance. *Courant in Göttingen and New York.* New York: Springer-Verlag, 1976.

———. *Hilbert.* New York: Springer-Verlag, 1973.

———. *Neyman–From Life.* New York: Springer-Verlag, 1982.

Reingold, Nathan. "Refugee Mathematicians in the United States of America 1933–1941: Reception and Reaction." *Annals of Science* **38** (1981):313-338. Reprinted in *A Century of Mathematics in America—Part I.* Ed. Peter L. Duren *et al.* Providence: American Mathematical Society, 1988, pp. 175-200.

Rider, Robin. "Alarm and Opportunity: Emigration of Mathematicians and Physicists to Britain and the United States, 1933–1945." *Historical Studies in the Physical Sciences* **15** (1984):107-176.

Rowe, David E. "Die Wirkung deutscher Mathematiker auf die amerikanische Mathematik, 1815–1900." *Mitteilungen der Mathematischen Gesellschaft DDR.* **3-4** (1988):72-96.

———. "'Jewish Mathematics' at Göttingen in the Era of Felix Klein." *Isis* **77** (1986):422-429.

Schappacher, Norbert. "Das Mathematische Institut." In *Die Universität Göttingen ünter dem Nationalsozialismus.* Ed. Heinrich Becker *et al.* Munich: K. G. Saur, 1987, pp. 245-273.

Schappacher, Norbert and Kneser, Martin. "Fachverband–Institut–Staat." In *Ein Jahrhundert Mathematik, 1890–1990: Festschrift zum Jubiläum der DMV.* Ed. Gerd Fischer *et al.* Braunschweig: Vieweg Verlag, 1990, pp. 1-82.

Scharlau, Winfried. Ed. *Mathematische Institut in Deutschland 1800–1945.* Braunschweig: Vieweg Verlag, 1990.

Schröder, Brigitte. "Caractéristiques des relations scientifiques internationale: 1870-1914." *Cahiers d'histoire mondiale* **10** (1966):161-177.

Schröder-Gudehus, Brigitte. *Deutsche Wissenschaft und internationale Zusammenarbeit 1914–1928.* Geneva: Dumaret and Golay, 1966.

Seifert, Herbert and Threlfall, William. *Lehrbuch der Topologie.* Leipzig and Berlin: B. G. Teubner Verlag, 1934.

Siegmund-Schultze, Reinhard. *Mathematiker auf der Flucht vor Hitler.* Braunschweig: Vieweg Verlag, 1988.

Stern, Fritz. "The Political Consequences of the Unpolitical German." *History* **3** (1969):104-134. Reprinted in Stern, Fritz. *The Failure of Illiberalism.* New York: Alfred A. Knopf, 1972, pp. 3-25.

Tobies, Renate. *Felix Klein.* Leipzig: B. G. Teubner Verlag, 1981.

Tuchman, Barbara. *The Guns of August.* New York: Macmillan & Co., 1962.

Veblen, Oswald. *Analysis Situs.* New York: American Mathematical Society, 1922.

Voegelin, Eric. "Die deutsche Universität und die Ordnung der deutschen Gesellschaft." In *Die deutsche Universität im Dritten Reich.* Munich: Piper Verlag, 1966.

Willmore, J. Selden. *The Great Crime and its Moral.* New York: George H. Doran, 1917.

Wyman, David S. *Abandonment of the Jews: America and the Holocaust 1941–1945.* New York: Pantheon Books, 1984.

CHAPTER 18

The Formation of the International Mathematical Union

Olli Lehto*
University of Helsinki (Finland)

Introduction

The subject of the International Mathematical Union (IMU) is a strangely paradoxical one within the context of the present volume. It would seem almost axiomatic that any work concerning the evolution of an international mathematical community up to 1945 should contain some reference to the role that the IMU has played in this development. Yet, the original IMU—in existence from 1920 to 1932—was a failure; today's successful Union only dates from 1952.[1]

The history of the first International Mathematical Union is, however, unusually compelling, coming as it did in an era of major conflicts between world powers. The present chapter highlights this story, providing the background to the formation of the original IMU, the details of its troubled existence, and the causes of its eventual demise. Based on a variety of sources, both published and archival, it pieces together the story of how ideas of the global cultivation of mathematics across national borders gradually began to take shape over a century ago, and how these ideas developed, amidst political difficulties and serious setbacks, into the original IMU. The picture that emerges is not one of unfettered international collaboration for the good of mathematics, but rather one that, sadly, echoes the troubled international politics of the times.

Background

The origins of the International Mathematical Union are connected with the general scientific developments in Europe during the nineteenth century. As documented throughout the present volume, the timing of these developments varied

---

*For more on the history of the International Mathematical Union, see Olli Lehto, *Mathematics Without Borders: A History of the International Mathematical Union* (New York: Springer-Verlag, 1998). This monograph's "Notes" contain a more detailed list of references than the one given here.

[1]Even today, the majority of the international mathematical community is largely unaware of the IMU's function, and fewer still have shown interest in its history. It was not until the early 1990s that the Union decided to sort and catalog its vast, but largely unorganized, archive and expressed the wish that a history of the IMU should be written. These tasks were undertaken by the present author. It soon became apparent that the documentary history of the IMU was divided into two distinctly different periods. For the years from 1952 on, the volume of archival material was overwhelming, whereas there was not a single document in the IMU archives dating from before 1952. Two years of searching produced a substantial amount of relevant papers and, although not comprehensive, enough to trace the story with a fair degree of accuracy.

from country to country, but the general trend was clear. As a result, the number of scientists multiplied, and national scientific societies were founded in rapidly increasing numbers. Following the expansion of national science, cooperation across borders began to assume organized forms. At first, joint scientific work was often concerned with well-defined projects where the need for international collaboration was obvious.[2] In this context the word "international" was, initially, largely a synonym for "European."

Ironically, the trend toward international cooperation in science coincided with a deterioration in the generally stable political situation that had existed for over half a century. Near the end of the nineteenth century, the coalitions between the great powers were taking shape, and their conflicts of interests became more and more accentuated. Military crises became frequent, and armament programs were intensified. In an attempt to counter these developments, increasing internationalism assumed many forms aimed at improving understanding between peoples. One example was the first modern Olympic Games, held in Athens in 1896. The world's scientific community likewise increased its efforts to work in concert across national borders, as evinced by the inauguration of the truly international Nobel Prizes in 1901.

In mathematics, the first important step in organized international cooperation was in the field of publications. One of the by-products of the scientific and mathematical expansion during the nineteenth century was a rapid increase in the number of mathematical journals and, consequently, the number of papers appearing each year.[3] The *Jahrbuch über die Fortschritte der Mathematik* was established to serve the increasingly necessary purpose of reviewing papers appearing in mathematical research journals around the world. It quickly became indispensible for mathematical research.

The *Jahrbuch* was founded in Germany and its editors were German, but German reviewers were soon joined by mathematicians from other countries. There was one notable exception. The first volume of the *Jahrbuch*—which was dated 1868 but which was only printed in 1871—appeared at a time when the aftermath of the Franco-Prussian War hampered collaboration between German and French mathematicians. Although French papers were reviewed, the French were not initially among the reviewers. By the 1890s, however, Franco-German mathematical relations had improved to the extent that French mathematicians were able to take an active part in a German bibliographical project, the *Encyklopädie der mathematischen Wissenschaften* begun in 1894. This led to highly fruitful joint international work in which emphasis was placed on German-French cooperation.

The need for organized mathematical collaboration beyond the purely bibliographical was felt early on by Georg Cantor. Not only was he one of the driving forces behind the foundation of a mathematical society in Germany, but he also proposed a joint German-French mathematical conference to be held at a neutral

[2]Frank Greenaway, *Science International: A History of the International Council of Scientific Unions* (Cambridge: University Press, 1996).

[3]Compare the chapters by Jesper Lützen, Sloan Despeaux, and June Barrow-Green in the present volume. See also Hélène Gispert, *La France mathématique: La Société mathématique de France (1872-1914)* (Paris: Société française d'histoire des sciences et des techniques & Société mathématique de France, 1991). In addition to discussing the scope of mathematical production with the aid of the statistics provided by the *Jahrbuch über die Fortschritte der Mathematik*, Gispert also considered the distribution of mathematical publications by country, with emphasis on the contributions of French mathematics.

site, perhaps in Belgium, Switzerland, or the Netherlands. By 1890, this idea had evolved into the more far-reaching idea of an international congress. In August of that year, Walther von Dyck told Felix Klein that "G. Cantor wrote me recently about very high-flown plans concerning international congresses of mathematicians." Von Dyck, however, went on to express doubts about the plan: "I really do not know whether that is a real need."[4]

Between 1894 and 1896, Cantor was involved in active correspondence on the subject of a possible international congress with a number of mathematicians. These included Charles Hermite, Camille Jordan, Henri Poincaré, Charles-Ange Laisant, and Émile Lemoine in France, Aleksandr Vasil'ev in Russia, and von Dyck and Klein in Germany.[5] Klein had actually first expressed the idea publicly at the Congress of Mathematics and Astronomy held in Chicago as part of the World's Columbian Exposition in 1893. Although it could scarcely be called international (the number of European mathematicians in attendance was just four), the Chicago meeting was a significant historical event, being a mathematical conference with participants from two continents.[6]

In his opening address, Klein pointed out how the unprecedented growth in mathematical areas had resulted in an even greater need for scientific cooperation between nations. "What was formerly begun by a single mastermind, we must now seek to accomplish by united efforts and cooperation," he argued. "But our mathematicians must go further still. They must form international unions, and I trust that this present World's Congress at Chicago will be a step in this direction."[7]

Due to the efforts of Klein, Cantor, and others, the years from 1894 to 1896 saw support for an international congress gather momentum. In France, in the first volume of the journal *L'Intermédiaire des mathématiciens* in 1894, the editors, Laisant and Lemoine, echoed the sentiments expressed by Klein in Chicago, going on to urge the organization of an international meeting of mathematicians.[8] A year later, the French, German, and American mathematical societies all formally endorsed the idea of an international congress. An enthusiastic Laisant was soon able to claim that "some of the most brilliant scholars have exhibited a real passion for it."[9]

All that remained was to decide on a suitable time and location for the event. In September 1895, Cantor was still speaking about either Switzerland or Belgium, but by early 1896, by virtue of its tradition of promoting international interests, its

[4] "G. Cantor schreib mir in letzter Zeit über sehr hochfliegende Pläne betr. *internationaler* Mathematikercongresse. Ich weiß wirklich nicht, ob das ein wirkliches Bedürfnis ist." Walter Purkert and Hans Joachim Ilgauds, *Georg Cantor 1845–1918* (Basel-Boston-Stuttgart: Birkhäuser Verlag, 1987), p. 127. Here, as elsewhere, direct English quotations from non-English sources are translations by the author.

[5] Georg Cantor, *Briefe*, ed. Herbert Meschkowski and Wilfried Nilson (Berlin: Springer-Verlag, 1991), pp. 350-353, 376-378, and 384-385.

[6] See Karen Hunger Parshall and David E. Rowe, *The Emergence of the American Mathematical Research Community, 1876-1900: J. J. Sylvester, Felix Klein, and E. H. Moore*, HMATH, vol. 8 (Providence: American Mathematical Society and London: London Mathematical Society, 1994), chapter 7.

[7] E. H. Moore, Oskar Bolza, Heinrich Maschke, and Henry S. White, ed., *Mathematical Papers Read at the International Mathematical Congress Held in Connection with the World's Columbian Exposition, Chicago 1893* (New York: Macmillan and Co., 1896), p. 135.

[8] *L'Intermédiaire des mathématiciens* **1** (1894):vi and 113.

[9] Charles-Ange Laisant, "Les mathématiques au Congrès de l'Association française pour l'avancement des sciences à Bordeaux," *Revue générale des sciences* (Janvier 1896):31-34.

central European location, and the fact that it represented a neutral zone between the two mathematical superpowers, Switzerland had become the favored option. That year, Swiss mathematicians formally agreed to stage the event in Zürich, and the date was set for August 1897.

The First International Congresses of Mathematicians

The first International Congress of Mathematicians (ICM) was held 9–11 August, 1897 at the *Eidgenössisches Polytechnikum* in Zürich. Not surprisingly, the 208 mathematicians in attendance were predominantly central European, with sixty Swiss, forty-one Germans, twenty-three French, twenty Italians, and seventeen Austro-Hungarians. Although sixteen countries were represented, it would be more accurate to describe the participants as European than as international, since a mere seven were from outside Europe, and all of those were from the United States. (Only four of the mathematicians were women.) Of the thirty-four lectures, seventeen were in French, fourteen in German, three in Italian, and none in English.[10]

The structure of the mathematical program was determined by the decision to divide the subject into five sections: arithmetic and algebra, analysis and the theory of functions, geometry, mechanics and mathematical physics, and history and bibliography. The overall responsibility for the arrangements lay with the Swiss organizers, but they soon enlarged their organizing committee to include foreign members such as Klein and Poincaré.

From their inception, the congresses were meant to form a permanent institution to be held every three or four years. The highly positive contemporary reaction to the Zürich Congress ensured that there was no doubt as to whether such meetings would continue. On inviting the members to the next ICM, Émile Picard said that the success of the first meeting guaranteed the future of the institution that had just been founded.

A significant decision of the Congress was that special commissions could be set up to handle various joint affairs and to present their reports to the next Congress. In Zürich, ambitious goals were proposed for such commissions. One was to give surveys of the developments of the various parts of mathematics; another was to report on mathematical bibliography and on attempts that could be made to achieve a simpler, more rational, and uniform terminology; a third was to advise how to give the ICMs a more permanent character through archives, libraries, publications, and a secretariat. Ultimately, only one commission was set up, but in it, an embryonic form of the International Mathematical Union can be said to have come into existence in 1897. Unfortunately, its early results were meager. At the following Congress, the second, held in Paris in 1900,[11] the commission announced

---

[10] *Verhandlungen des ersten Internationalen Mathematiker-Kongresses in Zürich vom 9. bis 11. August 1897*, ed. Ferdinand Rudio (Leipzig: B. G. Teubner Verlag, 1898). The main source of information for the ICMs are the official *Proceedings* published after each Congress. See also Donald J. Albers, Gerald L. Alexanderson, and Constance Reid, *International Mathematical Congresses: An Illustrated History 1893–1986* (New York: Springer-Verlag, 1987); and June Barrow-Green, "International Congresses of Mathematicians from Zürich 1897 to Cambridge 1912," *The Mathematical Intelligencer* **16** (2) (1994):38-41.

[11] Until World War I, the ICMs were numbered: the 1897 Congress in Zürich was the first, the 1900 ICM in Paris the second, 1904 in Heidelberg the third, 1908 in Rome the fourth, and 1912 in Cambridge, England the fifth. The latter Congress marked a considerable increase in non-European participants. They numbered eighty-two out of a total of 574, with six from Asia, two

that it had nothing to report. The Paris Congress was nevertheless immortalized in the history of mathematics for David Hilbert's introduction of his legendary twenty-three open problems for the twentieth century.

At the fourth Congress in Rome, an Italian participant, Alberto Conti, crystallized the ideas of Klein and Cantor from the previous decade by formally proposing the establishment of an international union of mathematicians. The question was taken up at the 1912 ICM in Cambridge, but the idea of a permanent mathematical association did not progress. The British President of the Congress, Sir George Darwin, summed up the feelings of the participants by expressing the opinion that "our existing arrangements for periodical Congresses meet the requirements of the case better than would a permanent organization of the kind suggested."[12]

If the time was not yet right for a mathematical union, the Commission on the Teaching of Mathematics, which had been set up at the 1908 ICM in Rome, could report on a good start: "Several countries, in one way or another, have recognized officially the work, and have contributed financial support. About 150 reports have been published, and about 50 will appear later."[13] This was a resounding success, especially considering that, during this politically tense period, attempts to achieve international cooperation in other fields of education had all ended in failure. With Felix Klein as chair, the Commission amassed "an amount of information beyond belief,"[14] even managing to continue its work throughout World War I. By 1920, it had produced no fewer than 187 volumes, containing 310 reports from eighteen countries.

Even as the Commission on the Teaching of Mathematics was making significant strides, the future of the Congresses themselves seemed in jeopardy. At the 1912 ICM in Cambridge, the Congress accepted an invitation from Gösta Mittag-Leffler to hold the sixth ICM in Stockholm in 1916. While the onset of World War I made it impossible for the meeting to go ahead, Mittag-Leffler revived the issue on 11 November, 1918, the very day hostilities ended.[15] Given that peace had been achieved, he felt that the time was right for a return to the international cooperation of the recent past. But Mittag-Leffler and like-minded mathematicians were in for a rude awakening. The world of 1918 was very different from that of four years before. The war had destroyed the idealism and solidarity of the prewar internationalist spirit. This had serious consequences for the international community of mathematicians.

World War I and its Aftermath

As early as November 1916, two years before the end of the war, the Paris Academy of Sciences had started to plan postwar international science policy. The initiative was taken by the Permanent Secretary of the Academy, the mathematician, Gaston Darboux, who had contacted the Royal Society in London. After

from Africa, sixty-seven from North America, and seven from South America. See *Proceedings of the Fifth International Congress of Mathematicians, Cambridge, 22-28 August 1912*, ed. Ernest W. Hobson and Augustus E. H. Love, 2 vols. (Cambridge: University Press, 1913), 1:1-63 on p. 28.

[12] *Ibid.*, p. 40.

[13] *Ibid.*, p. 38.

[14] A. Geoffrey Howson, "Seventy-Five Years of ICMI," *Educational Studies in Mathematics* **15** (1984):75-93 on p. 78

[15] Gösta Mittag-Leffler to Niels E. Nörlund, 11 November, 1918, Archives of the Mittag-Leffler Institute, Djursholm, Sweden.

his sudden death in 1917, Darboux was succeeded at the Academy by another mathematician, Émile Picard, who became one of the chief architects of postwar international science policy. In a letter to the Secretary of the Royal Society, Sir Arthur Schuster, Picard formulated the predominant question of the time on the Allied side: "Do we, or do we not, want to resume personal relations with our enemies?"[16] Picard himself answered vigorously in the negative.[17]

The guidelines of the new science policy were fixed at an Inter-Allied Conference on International Scientific Organizations, held at the Royal Society in October 1918 and attended by the representatives of eight Allied countries.[18] It was unanimously decided that the nations at war with the Central Powers should withdraw from the existing international scientific associations and establish new ones without delay, with the eventual cooperation of neutral nations. This meant, in particular, that Germany was to be barred from all organized international scientific cooperation.[19]

In July of the following year, these ideas took shape at a large meeting in Brussels. A new organization, the International Research Council (IRC), was established for the sciences. The IRC was to coordinate efforts in the different branches of science and its applications, to initiate the formation of international unions deemed to be useful to the progress of science, and to enter into relations with the governments of the countries adhering to the IRC in order to promote investigations falling within the competence of the Council. Picard, who had taken an active part in the preparations of the new Council, was elected its first President.

The German view regarding these new arrangements was clear and definite. Learned societies in France and Britain, supported by their governments, had founded the International Research Council for the purpose of undermining the position of German science. In this way, the influence of German science on international cultural life could be eliminated.[20] Those countries that had remained neutral during the war now faced an awkward decision. While disapproving of the ostracism manifested by the IRC, they rejected the alternative of a scientific coalition with Germany and opted to join the Council.[21]

It was at the Brussels meeting that the first steps towards the formation of the International Mathematical Union were taken. Under the new Council, disciplinary unions could be formed, and at a special meeting chaired by the Belgian mathematician, Charles-Jean de la Vallée Poussin, the draft statutes of a new union for mathematics were approved.[22] This Union actually came into being officially at the

---

[16] "Veut-on, oui ou non, reprendre des relations personnelles avec nos ennemis?" See Lehto, p. 16.

[17] "Summary of Correspondence Relating to Conference on International Conventions after the War," handwritten by Sir Arthur Schuster but undated and unsigned, Library of the Royal Society of London.

[18] Preliminary Report of Inter-Allied Conference on International Scientific Organizations, held at the Royal Society on 9-11 October, 1918, Library of the Royal Society of London.

[19] *Ibid.*

[20] "Die deutsche Wissenschaft und das Ausland," Denkschrift der Reichszentrale für naturwissenschaftliche Berichterstattung vom 29. Januar, 1925, Archives of the Berlin-Brandenburgische Akademie der Wissenschaften.

[21] Brigitte Schröder-Gudehus, *Les scientifiques et la paix: La communauté scientifique internationale au cours des années 20* (Montreal: Les Presses de l'Université de Montréal, 1978), p. 130

[22] Sir Arthur Schuster, ed., *Conseil international de recherches 1919: Constitutive Assembly Held at Brussels, 18-28 July 1919*, vol. 1 (London: Harrison & Sons, 1920), p. 26. The participants were delegates who were mathematicians or close to mathematics. The "Projet de statuts pour

first postwar International Congress of Mathematicians, held a year later in 1920. Despite the previous decision to hold the meeting in Stockholm, the French insisted that the Congress should take place in Strasbourg, in celebration of the deliverance of Alsace from Germany. Thus, even before its birth, the IMU was denied any hope of political neutrality.

The Birth of the IMU

The International Mathematical Union was created on 20 September, 1920 with eleven founding member countries: France, Britain, Italy, Belgium, the United States, Greece, Portugal, Serbia, Japan, Czechoslovakia, and Poland. All were victorious Allies, except for Czechoslovakia and Poland. On the day following the foundation of the IMU, the neutral countries whose mathematicians the French had allowed to participate in the Congress were invited to join the Union. The former Central Powers—Germany, Austria, Hungary, and Bulgaria—were barred from membership. Russia, which had withdrawn from the war in 1917, was not asked to join.

Given the circumstances surrounding it, the Strasbourg ICM was, not surprisingly, controversial. Because the defeated Central Powers were explicitly barred from participating, Mittag-Leffler refused to recognize the 1920 Congress as an international event. In his words, "This Congress is a French affair which can by no means cancel the International Congress in Stockholm."[23] It was also later alleged that the decision to hold the ICM in Strasbourg was made without consulting either the United States or the United Kingdom.[24] Thus, resentment against the IMU and the circumstances of its foundation was present from the very beginning.

According to the statutes adopted in Strasbourg, the purpose of the Union was to initiate and promote international cooperation in mathematics and to provide for: 1) the encouragement of the pure science, 2) the correlation of pure mathematics with other branches of science, 3) the direction and progress of teaching, 4) the coordination of the preparation and publication of abstracts of papers, tables,

---

une Union internationale de mathématiciens" is on pp. 185-189; its English translation is on pp. 247-250.

[23]"Ce congrès est une affaire française qui ne peut nullement annuler le congrès international à Stockholm." See Gösta Mittag-Leffler to Eliakim Hastings Moore, 8 March, 1921, Archives of the Mittag-Leffler Institute, Djursholm, Sweden. A few years later, Mittag-Leffler discussed international congresses in his address at the 1925 Scandinavian Congress of Mathematicians in Copenhagen. In his account, the fifth ICM, in 1912, was followed by the sixth in Toronto in 1924 (of which he had approved). He made no mention whatsoever of the Strasbourg Congress. His address is published in Swedish in the proceedings of the Copenhagen congress, and in German as "Entstehung und Entwicklung der internationalen und skandinavischen Mathematikerkongresse," in *Commentationes Physico-Mathematicae*, vol. 3 (Helsinki: Societas Scientiarum Fennicae, 1926), pp. 1-20.

The invitation to Strasbourg was sent under the imprimatur of the *Congrès International des Mathématiciens.* Mittag-Leffler (and independently, William Henry Young) felt that this was a *lapsus linguae.* Since the Strasbourg Congress was not open to all mathematicians, the word "*des* [of the]" should be replaced by "*de* [of]." The Congress accepted the proposed change, and in the first note in the *Comptes rendus* the form *de* was used. However, in the *Proceedings* the word *des* was restored. In 1924, when the former enemies were still excluded, the wording "International Mathematical Congress" was used. The linguistic conundrum disappeared in 1928, when the Congress was again open to all mathematicians, irrespective of nationality.

[24]Raymond Clare Archibald, *A Semicentennial History of the American Mathematical Society, 1888-1938* (New York: American Mathematical Society, 1938), pp. 18-20.

graphs, and the construction of appliances and models, etc., and 5) the organization of international conferences or congresses. After the list of objectives, the statutes exhibited the only visible connection with the International Research Council: "The admission of countries to the Union shall be subject to the Regulations of the IRC."[25] The framers in Strasbourg also made the provision that the statutes would be valid until the end of 1931, after which time they would, with the consent of the adhering countries, be extendable for another twelve-year period.

With these rules in place, the General Assembly elected the following to serve as the inaugural Executive Committee of the IMU: President, Charles-Jean de la Vallée Poussin (Belgium); Honorary Presidents, Camille Jordan (France), Horace Lamb (UK), Émile Picard (France), and Vito Volterra (Italy); Vice-Presidents, Paul Appell (France), Luigi Bianchi (Italy), Leonard Eugene Dickson (USA), Joseph Larmor (UK), and William Henry Young (UK); Secretary General, Gabriel Koenigs (France); and Treasurer, A. Demoulin (Belgium).[26] The composition of this committee, four from France, three from Britain, two from Belgium, two from Italy, and one from the United States, gives some indication of which countries were effectively in charge of the Union.

The IMU was authorized to decide about the date and the location of future ICMs. It was also decided that the Congresses could be attended only by mathematicians from countries that were members of the parent organization, the International Research Council. This rule was applied flexibly so that participants of the 1924 ICM were allowed from countries which were *eligible* to join the IRC. Mittag-Leffler continued to insist that an earlier decision to hold an ICM in Stockholm be honored in the near future. He was not present in Strasbourg, however, and his policy of *rapprochement* was not popular in 1920.

A profound ideological change had thus taken place since the war. Picard, who was elected President of the Strasbourg Congress, reiterated in his opening address the reasons why cooperation with scientists of the Central Powers was not permissible. "[T]o pardon certain crimes," he stated, "is to become an accomplice in them."[27] His only hint at a compromise was the remark that "our successors will see whether a sufficiently long time and a sincere repentance could permit mending the relations that the tragedy of the past years has broken, and whether those now excluded from the accord of civilized nations are worthy of reentering it."[28]

The Commission on the Teaching of Mathematics, which the International Congress had set up in Rome in 1908 and whose mandate had been extended at the 1912 ICM to the next Congress, had made good progress and had worked to some extent even during the war. Now, however, the new ideology as articulated by Picard put an end to its activities. Secretary General Koenigs stated that the dissolution of the Commission was inevitable; no initiative was taken in Strasbourg to form a new commission.[29]

---

[25]See Lehto, p. 25.

[26]*Comptes rendus du Congrès international des mathématiciens, Strasbourg 22-30 septembre 1920*, ed. Henri Villat (Toulouse: n. p., 1921). A report of the Congress is given on pp. i-xlvii.

[27]" ... pardonner à certains crimes, c'est s'en faire le complice." Émile Picard, "Allocution de M. Émile Picard," *op. cit.*, pp.xxx–xxxiii on p. xxxiii.

[28]" ... nos successeurs verrons si un temps suffisament long et un repentir sincère pourront permettre de les reprendre un jour, et si ceux qui se sont exclus du concert des nations civilisées sont dignes d'y rentrer." *Ibid.*, p. xxxiii.

[29]Compare Henri Fehr, "La Commission internationale de l'enseignement mathématique de 1908 à 1920," *L'Enseignement mathématique* **16** (1920):305-318.

Opposition to the Policy of Exclusion

Of the war's victors who dictated postwar international science policy, France held the leading place by right of its efforts and losses. Two particular mathematicians in key administrative positions, Picard, the President of the International Research Council, and Koenigs, the Secretary General of the IMU, were forceful advocates of the policy of excluding Germany from international cooperation. But France was not alone in wanting to exclude defeated nations.

In Britain, the Royal Society had played an active role. The declaration of the 1918 London meeting, which had introduced the new policy, came from a British pen. The United States was also well represented at the important meetings in London and Brussels, and the American delegations approved the policy of exclusion without objections. George Ellery Hale, the influential Foreign Secretary of the U.S. National Academy of Sciences, offered the pragmatic explanation that, with the Germans at international meetings, the possibility of a return to old cordial relations would be postponed rather than hastened because it would be impossible to avoid acrimonious discussions related to the war.[30]

From the beginning, however, there was also loud opposition. In England, the distinguished British mathematician, G. H. Hardy, argued forcefully for the restoration of prewar scientific arrangements: "This seems to me worth saying on account of the many imbecilities printed during the last year [1918] by preeminent men of science in England and France."[31] In his opinion, the British policy offered an example of how a small, determined minority could prevail over an indifferent, disinterested majority.

As could be expected, critical voices were also heard from countries that had been neutral during the war. As mentioned above, a particularly active promotor of reconciliation was the Swedish mathematician, Gösta Mittag-Leffler. He believed not only that Britain and Italy were prepared to reestablish scientific contacts but also that there was a significant minority in France, led by Paul Appell and Paul Painlevé,[32] opposed to the policy of discrimination. Picard, however, remained adamantly opposed, and Mittag-Leffler spared no words in criticizing him.[33] Scientific contacts, he said, "should be exempted from all political folly."[34] However, the views of neutral countries did not carry much weight at this time.

Despite the fact that it had been decided in Strasbourg that the 1924 ICM would be held in New York, in 1922 the United States withdrew its offer to organize it. Conditions were felt to have changed so much that financial backing would have been unobtainable in the United States for a gathering with the restrictions on participation imposed by the IMU.[35] This potentially embarrassing situation was

---

[30]George Ellery Hale to Arthur Schuster, 17 June, 1918, Library of the Royal Society of London.

[31]Joseph W. Dauben, "Mathematicians and World War I: The International Diplomacy of G. H. Hardy and Gösta Mittag-Leffler as Reflected in Their Personal Correspondence," *Historia Mathematica* **7** (1980):261-288 on p. 264. Compare also the chapter by Sanford Segal in the present volume on Hardy's reconciliatory views.

[32]Paul Painlevé, a noted mathematician, held ministerial posts in French governments several times between 1915 and 1933, serving as Prime Minister in 1917 and again in 1925.

[33]See, for example, Gösta Mittag-Leffler to Ernst Lindelöf, 27 November, 1919 and Gösta Mittag-Leffler to G. H. Hardy, 3 October, 1921, Archives of the Mittag-Leffler Institute, Djursholm, Sweden.

[34]Lehto, p. 31.

[35]Archibald, p. 19.

resolved by Canada, which, under the initiative of John C. Fields, offered to stage the event in Toronto. Mathematicians from the former Central Powers were again absent, but the IMU Secretary General Koenigs allowed participation from Spain, India, Russia, Ukraine, and Georgia, despite the fact that these countries were not members of the International Research Council.

Many Americans who attended discovered for the first time when they arrived in Toronto that Germans had been excluded, and much indignation was apparently expressed. At the meeting of the General Assembly of the IMU, the American delegation offered a resolution—endorsed by Italy, the Netherlands, Sweden, Denmark, Norway and the United Kingdom—requesting the International Research Council to consider the removal of the restrictions on membership imposed by the rules of the Council. The following year, at the General Assembly of the IRC, the United Kingdom explicitly announced that it was time to delete all political clauses from the statutes of the Council and its Unions. At the vote, this view received a majority, but not the required "at least two-thirds."[36]

After the Toronto Congress, which he deliberately boycotted, Hardy proved prophetic. He believed that the 1924 ICM would be the last boycotted congress of mathematicians. In other words, if the International Research Council did not remove the ban on German science, the IMU would collapse or degenerate into a purely Franco-Belgian affair.[37]

The march of events proved rapid, indeed. The Locarno Treaty, signed in October 1925, eased tensions between France and Germany. In accordance with the change of political atmosphere, it was felt that the IRC's very existence was jeopardized unless the discriminatory membership clauses were removed. The Council was summoned to an Extraordinary General Assembly in Brussels in the summer of 1926. Intense activity preceded this meeting. On 1 January, 1926, the Council of the American Mathematical Society resolved "that the Society desires to have no official representation on the American Section of the International Mathematical Union after 1 July 1926, unless the International Research Council at its meeting in June amends its rules so that membership in the Union may be entirely international."[38] In France, pressure was exerted on Picard and other recalcitrant ultras by the government and many colleagues. Painlevé exploited his position as mathematician and politician by suggesting a joint meeting of four French and four German scientific representatives in some neutral location. As he acknowledged, "[t]he impossibility of scientific boycott has been realized, and there is a sincere wish in France for cooperation."[39]

The Extraordinary General Assembly of the IRC lasted exactly one hour. The Assembly decided unanimously not merely to admit, but to *invite* Germany, Austria, Hungary and Bulgaria to become members of the Council and its Unions.[40]

---

[36] "Conseil international de recherches: Troisième assemblée générale tenue au Palais des Académies, Bruxelles, du 7 au 9 juillet 1925, Procès-verbal des séances," Library of the Royal Society of London.

[37] [Godfrey H. Hardy], "Quotations—International Congresses (Professor G. H. Hardy in *The Scientific Worker*)," *Science* **60** (26 December, 1924):591-592.

[38] Archibald, p. 20.

[39] Schröder-Gudehus, p. 269.

[40] "Conseil international de recherches: Assemblée générale extraordinaire tenue au Palais des Académies, Bruxelles, le 29 juin 1926, Procès-verbal de la séance," Library of the Royal Society of London.

Unfortunately, the invitation did not bring the results that the IRC desired. German scientists had not forgotten the boycott to which they had been subjected. They maintained that from its inception, the purpose of the French-dominated IRC, still under the leadership of the unrelentingly anti-German Picard, had been to undermine German science. In spite of the German Foreign Ministry's insistence that membership was necessary for political reasons, German scientists declined to join.[41] Despite its best efforts, the IRC, and therefore the IMU, were still not truly international.

The Dissolution of the IMU

The Italian Salvatore Pincherle had been elected President of the IMU for the eight-year period from 1924 to 1932, with Gabriel Koenigs continuing as Secretary General.[42] Before long, however, the two men were on a collision course over the policy for the 1928 International Congress to be held in Bologna, Italy. Since Germany had been invited to join the IMU, there was a strong feeling among mathematicians worldwide that the Italian organizers should return to prewar tradition and make the Congress open to all mathematicians, irrespective of nationality. As Chair of the Italian organizing committee and President of the IMU, Pincherle was in the key position. He was firmly of the opinion that no restrictions should be imposed.

Formally, an open Congress violated the rules of the IMU. As noted, Koenigs had adopted a flexible interpretation in 1924 of the clause that participants should be from member countries of the International Research Council. With the question of Germany's admission at issue, however, the blatantly anti-German Koenigs protested vehemently. He declared that since the Bologna Congress admitted Germans, it was not an IMU Congress. He did not issue invitations to the IMU General Assembly in Bologna, and in circular letters, he advised member countries of the IMU to boycott the Congress. Pincherle tried to get support for an open Congress from Picard, who was an Honorary President of the IMU and still the President of the International Research Council, but Picard sided with Koenigs.[43] Yet Pincherle decided to ignore Koenigs's protests. In doing so, he was supported by the opinion of the great majority of mathematicians worldwide.

In Germany, the invitation to the Congress was received with mixed feelings. The influential Prussian Academy ruled that German mathematicians could attend at their own discretion. It would be in the German interest, it held, for purely scientific reasons, since German participation would weaken the prestige of the International Research Council, detach it from important international events, and render its function superfluous.[44] As it turned out, this is exactly what happened to the IMU.

---

[41]Conrad Grau, "Die Wissenschaftsakademien in der deutschen Gesellschaft: Das Kartell von 1893 bis 1940," *Acta Historica Leopoldina* **22** (1995):31-56.

[42]*Proceedings of the International Mathematical Congress Held in Toronto, 11-16 August 1924*, ed. John C. Fields, 2 vols. (Toronto: The University of Toronto Press, 1928), 1:11-81.

[43]*Atti del Congresso internazionale dei matematici, Bologna, 3-10 Settembre 1928*, 6 vols. (Bologna: Nicola Zanichelli Editore, 1929), 1:1-131 on pp. 8-9.

[44]Letter of 16 May, 1928 from the Gesellschaft der Wissenschaften zu Göttingen and the answer of 22 May, 1928 from the Preussische Akademie der Wissenschaften, Archives of the Berlin-Brandenburgische Akademie der Wissenschaften.

Among German mathematicians, a relatively small but authoritative group, led by Ludwig Bieberbach, launched a campaign against the Bologna Congress based on the thesis that the Congress was connected with the IMU and the IRC, both still hostile to Germany. Hilbert represented the opposite view, arguing that Bieberbach's approach would damage German science. Hilbert's view prevailed. In Bologna, the Germans formed the largest national contingent after the Italians and were warmly welcomed. Mathematically, the ICM was a great success. The number of participants was a record high, including a good number of mathematicians from France despite Koenigs's opposition.[45]

In contrast, the IMU suffered a great loss of prestige. Doubts about the necessity and usefulness of the Union were gathering strength. Pincherle, who had been elected its President for eight years, resigned in Bologna after serving only half of his term. Furthermore, the Congress explicitly cut off all connections with the now discredited Union.

During the following four-year period from 1928 to 1932, the IMU was literally adrift. It should have revised its statutes, which had expired at the end of 1931, but it failed to do so. Koenigs held the office of the Secretary General until his death in 1931, but in the new climate, the anti-German policy he pursued had become untenable. After the Bologna Congress, Koenigs ceased working for the Union and even stopped answering letters. The new President, William Henry Young, was asked to appoint a new Secretary General but felt that he was not entitled to do so without the consent of the incumbent. As a result, due preparations were not undertaken for the IMU General Assembly which convened in 1932 in Zürich.

Even before the Zürich Congress, the American delegates had concluded that the IMU had no problems important enough to warrant its existence,[46] and at the meeting, they aggressively maintained that the Union was useless. As a temporary measure, it was finally proposed that a commission be set up to investigate the question of permanent international collaboration in mathematics and to present its conclusions to the 1936 International Congress. During this four-year interval, the IMU would be suspended. This proposal was accepted by twenty-three votes in favor, sixteen opposed, and five abstentions.[47] However, by 1936, the political situation in the world had worsened so much that the commission (which included several German members) could not recommend the founding of a new Union to the International Congress, a decision that inevitably meant the end of the original IMU.[48]

Conclusion

To understand the reasons for the failure of the IMU, the whole history prior to 1932 must be taken into consideration. The first important step towards organized

---

[45]On the attitudes, German and otherwise, regarding the Bologna Congress, compare Sanford Segal's chapter in the present volume.

[46]Roland G. D. Richardson, "International Congress of Mathematicians, Zürich 1932," *Bulletin of the American Mathematical Society* **38** (1932):769-774.

[47]Henri Fehr, "Union internationale mathématique: Troisième assemblée générale tenue à Zurich le 11 septembre 1932: Résumé du compte rendu rédigé par le secrétaire de l'Assemblée M. Valiron," *L'Enseignement mathématique* **31** (1933):276-278.

[48]*Comptes rendus du Congrès international des mathématiciens, Oslo 1936*, 2 vols. (Oslo: A. W. Brøggers Boktrykkeri A/S, 1937), 1:1-57.

international cooperation in mathematics was the formation of the permanent institution of International Congresses of Mathematicians. This idea came from within the mathematical community and was widely supported in continental Europe and endorsed by Britain and the United States. Until World War I, it was felt that an international mathematical association was not needed because "the Congresses met the requirements of the case better."[49]

After the war, the IMU was founded as a by-product of the new arrangements concerning international scientific cooperation. Mathematicians of the Allied countries did not oppose the IMU, but its formation did not take place as a result of their initiative. The guiding principle to isolate the defeated nations from the world scientific community was derived from the Versailles Peace Treaty, and the essential structure of the statutes was composed by the academies of the three great victors of the war, France, Britain, and the United States.

The introduction of politics into the IMU caused friction from the very beginning. Political discussions occupied a large part of the Union's energies, creating disagreements and tensions that increased with the years. In such a situation, the Union should have offered some tangible mathematical attraction, but it failed to do so.

All of the scientific unions that were founded in the 1920s followed the policy of the parent organization, the International Research Council. Of these, it is interesting that the IMU was the only one that did not survive. The reason lies in the fact that, since mathematicians had the option of using their International Congresses as the prime forum for international collaboration, the politically contaminated Union could be scrapped. Thus, when the ICMs regained their apolitical status in 1928, the IMU was simply pushed aside. The Union was thus deprived of all important mathematical activities; these were transferred to the Congresses, effectively rendering the IMU obsolete.

Epilogue: The New IMU

After 1932, there was no International Mathematical Union, but the Congresses continued as before. At the 1936 ICM in Oslo, it was decided to hold the 1940 Congress in the United States in Cambridge, Massachusetts. Preparations were well advanced when World War II broke out in September 1939. The organization of the Congress was immediately suspended, and an Emergency Executive Committee was set up "to take the initiative for resumption of activity."[50] This the Committee did in 1946, an important year because decisions about the international policy to be followed in science, in general, and in mathematics, in particular, were then made. Cooperation in science was to be based on unrestricted internationalism. In this respect, the situation after World War II stands in marked contrast to that after the First World War.

In April 1946, the Emergency Executive Committee for the ICM, headed by Marston Morse, reported to the American Mathematical Society (AMS) that it was interested in the revival of plans for the Congress only if it could be an open

---

[49] *Proceedings of the Fifth International Congress of Mathematicians, Cambridge, 22-28 August 1912*, 1:40.

[50] Everett Pitcher, *A History of the Second Fifty Years 1939-1988* (Providence: American Mathematical Society, 1988), pp. 147-148.

meeting to which all mathematicians would be invited irrespective of national allegiance. The Council of the AMS agreed with this principle, which set the tone for international mathematical cooperation after 1945.[51]

In 1931, the International Research Council had been disbanded and replaced by the International Council of Scientific Unions. With the formation of the ICSU, each of the unions had autonomy in the management of its own affairs. A union could admit any country to membership irrespective of whether that country had belonged to the ICSU. The ICSU declared itself open to scientists throughout the world. This basic principle was reconfirmed at its first postwar General Assembly in July 1946.

The mathematical community of the United States felt that, as an organizer of the ICM, it could study the possibilities of re-forming an International Mathematical Union. In the summer of 1948, the responsibility for all preparations concerning the planned Union was delegated to a three-man committee, consisting of Marshall Stone (Chair), John Robert Kline, and Marston Morse. After careful preparations by worldwide correspondence under Stone's direction, draft statutes were presented to the "Union Conference" in New York in August 1950. This conference was attended by the delegates of the National Committees for Mathematics of twenty-two countries. Consensus about the statutes and by-laws was reached, and it was decided that the new IMU would come into existence as soon as it had ten member countries. This quota was reached in September 1951. Among the first ten to join were Germany and Japan.[52]

The activities of the IMU began after the First General Assembly held in Rome in March 1952. By the end of the 1950s, several important targets had been reached: the IMU was readmitted to the ICSU; a subcommission, the International Commission on Mathematical Instruction, was established to continue the work of the old Commission on the Teaching of Mathematics; the Soviet Union and other socialist countries of Europe became members; the first *World Directory of Mathematicians* appeared; and the development was initiated that gave the IMU sole responsibility for the mathematical program of the International Congresses.

Indeed, the story of the new International Mathematical Union since 1952 is no less eventful than that of the old one. The Cold War, a divided Germany, conflict between The People's Republic of China and Taiwan, and problems in the emerging African nations, all at various times posed a threat to the ideal of international cooperation. The difference is that throughout the period since 1952, the Union managed to foster and maintain itself as a forum for promoting relationships between research mathematicians and educators regardless of national or political affiliation. It recognized mathematical achievement on a global scale and provided, in the form of the International Congresses, a venue where, if only once every four years, mathematicians could feel part of a truly international community.

---

[51] *Ibid.*, p. 148.

[52] Marshall Stone Papers Pertaining to the International Mathematical Union, Brown University Library, Providence, Rhode Island.

References

Archival Sources

Conseil International de Recherches. Assemblée générale extraordinaire tenue au Palais des Académies, Bruxelles, le 29 juin 1926. Procès-verbal de la séance. Library of the Royal Society of London.

Conseil International de Recherches. Troisième assemblée générale tenue au Palais des Académies, Bruxelles, du 7 au 9 juillet 1925. Procès-verbal des séances. Library of the Royal Society of London.

"Die Deutsche Wissenschaft und das Ausland." Denkschrift der Reichszentrale für naturwissenschaftliche Berichterstattung vom 29. Januar 1925. Archives of the Berlin-Brandenburgische Akademie der Wissenschaften.

Gesellschaft der Wissenschaften zu Göttingen. Letter of 16 May 1928; and the reply from the Preussische Akademie der Wissenschaften. 22 May, 1928. Archives of the Berlin-Brandenburgische Akademie der Wissenschaften.

George Ellery Hale to Arthur Schuster. 17 June, 1918. Archives of the Royal Society of London.

Gösta Mittag-Leffler to Eliakim Hastings Moore. 8 March, 1921. Archives of the Mittag-Leffler Institute. Djursholm, Sweden.

Gösta Mittag-Leffler to Niels E. Nörlund. 11 November, 1918. Archives of the Mittag-Leffler Institute. Djursholm, Sweden.

Preliminary Report of Inter-Allied Conference on International Scientific Organizations Held at the Royal Society on 9-11 October 1918. Library of the Royal Society of London.

Marshall Stone Papers Pertaining to the International Mathematical Union. Brown University Library. Providence, Rhode Island.

"Summary of Correspondence Relating to Conference on International Conventions after the War." Handwritten, but undated and unsigned, by Sir Arthur Schuster. Library of the Royal Society of London.

Printed Sources

Albers, Donald J., Alexanderson, Gerald L., and Reid, Constance. *International Mathematical Congresses: An Illustrated History 1893–1986.* New York: Springer-Verlag, 1987.

Archibald, Raymond Clare. *A Semicentennial History of the American Mathematical Society 1888–1938.* New York: American Mathematical Society, 1938.

Atti del Congresso internazionale dei matematici, Bologna, 3-10 Settembre 1928. 6 Vols. Bologna: Nicola Zanichelli Editore, 1929.

Barrow-Green, June. "International Congresses of Mathematicians from Zürich 1897 to Cambridge 1912." *The Mathematical Intelligencer* **16** (2) (1994):38-41.

Cantor, Georg. *Briefe.* Ed. Herbert Meschkowski and Wilfried Nilson. Berlin: Springer-Verlag, 1991.

Comptes rendus du Congrès international des mathématiciens, Oslo 1936. 2 Vols. Oslo: A. W. Brøggers Boktrykkeri A/S, 1937.

Dauben, Joseph W. "Mathematicians and World War I: The International Diplomacy of G. H. Hardy and Gösta Mittag-Leffler as Reflected in Their Personal Correspondence." *Historia Mathematica* **7** (1980):261-288.

Fehr, Henri. "La Commission internationale de l'enseignement mathématique de 1908 à 1920." *L'Enseignement mathématique* **16** (1920):305-318.

______. "Union internationale mathématique: Troisième assemblée générale tenue à Zurich le 11 septembre 1932. Résumé du compte rendu rédigé par le secrétaire de l'Assemblée M. Valiron." *L'Enseignement mathématique* **31** (1933):276-278.

Fields, John C., Ed. *Proceedings of the International Mathematical Congress Held in Toronto, 11-16 August 1924*. 2 Vols. Toronto: The University of Toronto Press, 1928.

Gispert, Hélène. *La France mathématique: La Société mathématique de France (1872-1914)*. Paris: Société française d'histoire des sciences et des techniques & Société mathématique de France, 1991.

Grau, Conrad. "Die Wissenschaftsakademien in der deutschen Gesellschaft: Das Kartell von 1893 bis 1940." *Acta Historica Leopoldina* **22** (1995):31-56.

Greenaway, Frank. *Science International: A History of the International Council of Scientific Unions*. Cambridge: University Press, 1996.

[Hardy, Godfrey H.]. "Quotations: International Congresses (G. H. Hardy in *The Scientific Worker*." *Science* **60** (26 December, 1924):591-592.

Hobson, Ernest W. and Love, Augustus E. H., Ed. *Proceedings of the Fifth International Congress of Mathematicians, Cambridge, 22-28 August 1912*. Cambridge: University Press, 1913.

Howson, A. Geoffrey. "Seventy-Five Years of ICMI." *Educational Studies in Mathematics* **15** (1984):75-93.

Laisant, Charles-Ange. "Les mathématiques au Congrès de l'Association française pour l'avancement des sciences à Bordeaux." *Revue générale des sciences* (Janvier 1896):31-34.

Lehto, Olli. *Mathematics Without Borders: A History of the International Mathematical Union*. New York: Springer-Verlag, 1998.

L'Intermédiaire des mathématiciens **1** (1894).

Moore, Eliakim H.; Bolza, Oskar; Maschke, Heinrich; and White, Henry S., Ed. *Mathematical Papers Read at the International Mathematical Congress Held in Connection with the World's Columbian Exposition, Chicago 1893*. New York: Macmillan and Co., 1896.

Parshall, Karen Hunger and Rowe, David E. *The Emergence of the American Mathematical Research Community, 1876-1900: J. J. Sylvester, Felix Klein, and E. H. Moore*. HMATH. Vol. 8. Providence: American Mathematical Society and London: London Mathematical Society, 1994.

Picard, Émile. "Allocution to m. Émile Picard." In *Comptes rendus du Congrès international des mathématiciens, Oslo 1936*. 2 Vols. Oslo: A. W. Brøggers Boktrykkeri A/S, 1937, pp. xxxi-xxxiii.

Pitcher, Everett. *A History of the Second Fifty Years 1939-1988*. Providence: American Mathematical Society, 1988.

Purkert, Walter and Ilgauds, Hans Joachim. *Georg Cantor 1845–1918*. Basel-Boston-Stuttgart: Birkhäuser Verlag, 1987.

Richardson, Roland G. D. "International Congress of Mathematicians, Zürich 1932." *Bulletin of the American Mathematical Society* **38** (November 1932):769-774.

Rudio, Ferdinand, Ed. *Verhandlungen des ersten Internationalen Mathematiker-Kongresses in Zürich vom 9. bis 11. August 1897.* Leipzig: B. G. Teubner Verlag, 1898.

Schröder-Gudehus, Brigitte. *Les scientifiques et la paix: La communauté scientifique internationale au cours des années 20.* Montreal: Les Presses de l'Université de Montréal, 1978.

Schuster, Arthur, Ed. *Conseil international de recherches 1919.* Vol. I. *Constitutive Assembly held at Brussels, 18-28 July 1919: Report and Proceedings.* London: Harrison & Sons, 1920.

Villat, Henri, Ed. *Comptes rendus du Congrès international des mathématiciens, Strasbourg 22-30 septembre 1920.* Toulouse: N.p., 1921.

Index